Irrigation Show 2016

AA005356

Las Vegas, Nevada, USA
6 - 7 December 2016

ISBN: 978-1-5108-4188-8

Printed from e-media with permission by:

Curran Associates, Inc.
57 Morehouse Lane
Red Hook, NY 12571

Some format issues inherent in the e-media version may also appear in this print version.

Printed by Curran Associates, Inc. (2017)

Irrigation Association (IA)
8280 Willow Oaks Corporate Drive
Suite 400
Fairfax, Virginia 22031
USA

Phone: 703.536.7080
Fax: 703.536.7019

info@irrigationshow.org

Additional copies of this publication are available from:

Curran Associates, Inc.
57 Morehouse Lane
Red Hook, NY 12571 USA
Phone: 845-758-0400
Fax: 845-758-2633
Email: curran@proceedings.com
Web: www.proceedings.com

2016 Irrigation Show Technical Program, Las Vegas, Nevada

Agricultural Sessions December 6-7, 2016

Landscape Sessions

Managing Drip Filter Backflush Water

Charles M. Burt, Ph.D., P.E.
Chairman, Irrigation Training & Research Center (ITRC), California Polytechnic State University (Cal Poly), San Luis Obispo, CA 93407-0730, cburt@calpoly.edu

Abstract. *The paper discusses the principles of backflushing different types of filters, as well as pressure and flow requirements for proper backflushing of various filters. Additionally, it discusses options for what to do with the backflush water. A prototype ITRC design for cleaning and recycling backflush water is presented.*

Primary Keywords. Drip, filter, backflush, microirrigation, trickle, recycle

Secondary Keywords. Conservation, design, water treatment

Product Keywords. Drip/micro systems – agricultural; water filters, strainers & sand separators

Introduction

This paper will focus on three specific aspects of backflushing:

1. Principles of backflushing different types of filters
2. Backflush flow rate requirements versus pressure requirements
3. Disposal of backflush water

Principles of Backflushing

Filters are designed to catch solid particles and retain them, allowing cleaner water to pass through the filter and into the irrigation system. In partnership with filtration, backflushing is a mechanism in which a flow rate moves through the filter surface in reverse, removing the accumulated particles from the filter. That backflush water is isolated from the irrigation water, and is discharged into the air at some point.

Filters with No Backflush

Some filters do not have "backflushing". One example of such a filter can be seen in Figure 1. The photo shows a perforated horizontal cylindrical screen, inside a housing, in which dirty water enters from the center and then passes through the screen, then between the screen and the housing, and then out of the housing. Many of these filters are cleaned by sending a high flow rate along the complete length of the filter and out the downstream end. There is no reverse flow in this case.

Figure 1. Typical horizontal cylindrical filter with a coarse, perforated stainless steel screen, typically used for sprinkler applications. Algae that passed through the screen can be seen on the downstream side of the screen. The complete screen has been removed from the filter housing. A 4" flush-through (not backflush) valve and hose manifold can be seen for each cylinder.

Automatic Self-Cleaning Filters

Automatic self-cleaning "vacuum-scanning" screen filters use a special type of backflushing. As with most screens, dirty water enters the center of a tubular screen, turns at right angles, and passes through the screen material, leaving the contaminants on the inside surface. These self-cleaning screens are often equipped with relatively fine mesh (100-150 mesh) fabric that is reinforced to withstand a large pressure difference when dirty.

Automatic screen filters have a rotating mechanism inside them that can "vacuum" the contaminants off the surface of the screen when it gets dirty. There is no actual vacuum, but rather a hollow rotating shaft with nozzles positioned very close to the dirty inside of the tubular screen. When the flush cycle is activated, water flows into the nozzles, then into the shaft, and is discharged into the atmosphere. The close tolerance between the wand inlet and the screen surface causes water to flow in reverse through the screen when the nozzle rotates past a point. The majority of the water continues to pass through the screen and into the irrigation system. These rotary cleaning tubular screens have been available for many years, but the new designs are far superior to previous ones.

There is a tight clearance between the nozzles and the screen, plus the "suction" holes in the spinning wands have small diameters. Therefore, excellent pre-filtration is absolutely required. A typical backflush flow rate is about 15-40% of the dirty water inflow rate.

Figure 2. Example of a rotating wand design for a screen. Courtesy of Amiad.

Media Tanks and Disc Filters

Media tanks and disc filters require a substantial reverse flow rate during backflush. A typical backflush flow rate per tank is about the same as the dirty water tank filtering flow rate. Figure 3 illustrates the concept of backflushing of media tanks.

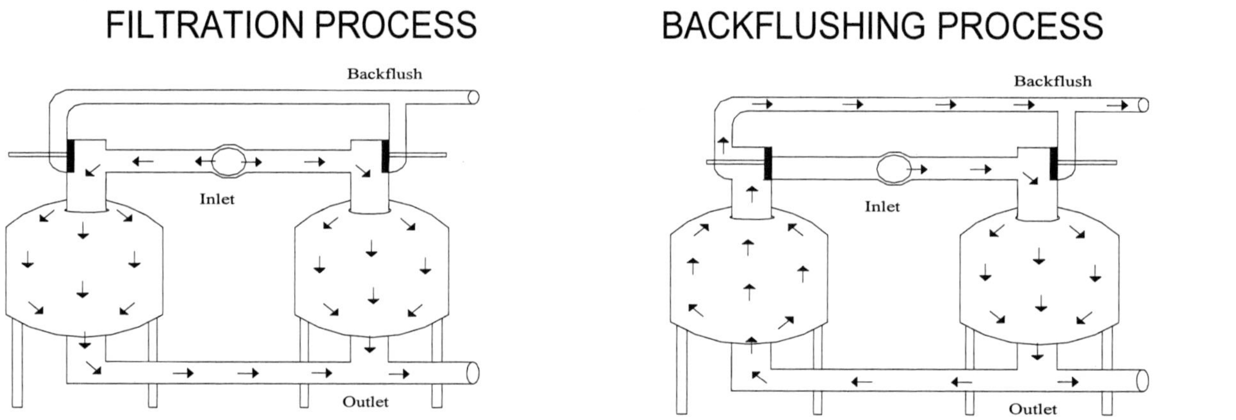

Figure 3. Filtration and backflushing processes with media tanks. Sketch courtesy of Yardney.

.

Backflush Flow Rate Requirements versus Pressure Requirements

In many drip/micro systems, the pressure requirements for backflushing will define the pressure requirement from a pump.

For example, the suction-scanning screens require 30-38 psi in the filter to overcome the friction loss that will occur through the screen, nozzle, shaft, and valve pathway with the required backflush flow. The concept is relatively simple: the water that backflushes through the screen must have sufficient velocity to clean the screen. Given the pathway sizes, lengths, and configurations, a certain amount of friction will occur at that flow rate. Some vendors have supplied booster pumps that are attached to the discharge of the backflush valve. The idea is that if the booster pump can exert a suction of 15 psi, as an example, the filter backflush will function well with only (38 psi – 15 psi = 23 psi) inside the filter body.

Disc filters are somewhat similar to the suction-scanning screens in that a certain internal pressure is needed to expand the filter discs and move the water through the manufactured pathways. Many disc filters require 30-35 psi inside the filter to achieve effective backflushing flow rates.

Media tanks have been sold for years with an understanding that they also require 30-35 psi for backflushing. However, that is completely erroneous if the media tanks backflush settings are correctly adjusted, and if the backflushing piping is reasonably sized. As with all backflushing, the key is to obtain a sufficient flow rate. In the case of media tanks, the flow rate must be sufficient to expand the media (sand) and move the contaminants upward and out of the tank backflush valves. Extensive media tank testing by the Cal Poly ITRC has conclusively shown that backflushing a hydraulic pathway through the tank, underdrain, media, and backflush filter may require as much as 13 psi with a very restrictive valve; most designs require about half of that.

The confusion about a high backflush pressure requirement for media tanks evidently arises from two things:

1. Some systems have very small backflush pipelines that may travel long distances and even uphill, without air vents. Therefore, a large amount of pressure is required to overcome friction, air locks, and elevation change. This is just a bad design.
2. When media tanks are improperly backflushed, the media can become almost cemented. People know that they can break the cemented media apart by using large pressures at the bottom of the underdrains.

Backflush Water – Reducing the Volume

Backflush water disposal can cause headaches for designers and operators. Dirty water combined with the need for fine drip tape filtration can produce situations that require very frequent backflushing – resulting in large volumes of water that needs to be disposed of.

The first step in backflush water management is to minimize the volume of backflush water needed, yet still attain the degree of filtration required. This is done in two ways:

1. Select a brand/model of media tank with a very efficient backflush operation. Some brands/models require up to three times as much backflush volume per day as others. Interestingly, the models that are least efficient will also discharge the most media along with the backflush water.
2. Use pre-treatment to reduce the dirt load entering the filters. For example, a typical 48" media tank will have a backflush flow rate of about 220 GPM, which only provides a velocity of about 0.04 feet/second above the media during backflushing. This certainly will not remove sand and many other contaminants. Sand is seen in backflush water with filters that have very non-uniform backflush flow patterns, in which much higher-than-average velocities occur in some "hot spots". Note: There are numerous ways to pre-filter water before it enters the media/suction-scanning/disc filters that are mean to be "polishing" filters. Those pre-filtration techniques will not be covered here.

A third option has evidently not been used in agricultural drip filtration. That option uses air to create additional turbulence during backflushing, which might result in less volume of water needed.

Backflush Water Disposal

One concern about backflush water disposal is the ultimate destination of chemicals that may be injected upstream of the filters. For most organic fertilizers and other compounds such as gypsum, a sure path to emitter plugging is to inject downstream of the filters. Therefore, one should consider two options:

1. Stop injecting the chemicals sufficiently early prior to backflushing. This can be accomplished with some commercial backflush controllers. Of course, if the filters are almost constantly in a backflushing mode this is a poor solution. See the comments on pre-filtration requirements.
2. Recirculate the backflush water after cleaning it.

There appear to be four common ways to dispose of backflush water:

1. Dump the water into a field or into a drainage ditch. For many areas, there simply is not enough water available for this to be a viable option.
2. Dump the filter backflush water into the supply canal. Let your downstream neighbor deal with it. This is a common practice in California.
3. Recirculate the backflush water into the reservoir from which the original water came. Hope that some miracle will cause the backflush water to clean itself in the reservoir even though that didn't happen the first time.
4. Clean the backflush water with a specially designed automatic overflow screen, and pump the relatively clean water back into the system, upstream of the drip system filters. An example of this is seen in Figure 4.

Figure 4. A media tank filtration system that uses a small commercial (FV&C) overflow screen to clean the backflush water before recycling it

For the cases in which the backflush water is returned to a reservoir or a canal, ITRC has experimented with a "mushroom" or "turbulent fountain" screen through which the backflush water must pass. This design is common for cleaning canal water prior to entering center pivot systems in the Pacific Northwest, albeit with a much coarser screen. The screen seen in Figure 5, which was an early prototype by ITRC, used a 150 mesh stainless steel screen.

Figure 5. ITRC mushroom screen for cleaning and recirculation of filter backflush water.

Unfortunately, the 150 mesh screen did not work well in the field. Over time, it plugged up and much of the backflush water simply flowed over the outer edge of the screen unless it was cleaned before every backflush cycle. ITRC plans to experiment with some revised designs. It appears the commercial designs with bottom-side spray wands – minus the pump that is used to re-pressurize the water – may be the best solution.

REVERSING YIELD LOSS BY MANAGING DISTRIBUTION UNIFORMITY

Chris Binfield, BS Biochemistry
Meras Engineering Inc, Modesto, CA, cbinfield@meras.com

Abstract: Poor System Distribution Uniformity (DU) has been shown to negatively impact crop yields in numerous studies conducted over the past several decades. By managing DU and utilizing other tools for identifying uneven application of water across a field, these yield losses can be reversed. Performing DU analysis along with physical and visual inspections of the system components is the first step. Taking corrective action to restore the DU to at, or near design performance is the next step. Research shows that correcting pressure variations and/or remediating systems to eliminate plugging have the greatest impact on improving DU.

Managing your DU through ongoing maintenance and periodic inspections is the key to reversing yield losses permanently. Don't let poor DU be a silent killer of your yields and profits, take control of your system by managing your DU and maintaining your micro irrigation system with straightforward process described here; measure, remediate, maintain, and measure again.

I have never met a grower who does not agree with the statement, "Water and inputs directly affect yield." No water equals no crop, and not enough water equals not enough yield. Truth is that yield improvement is the biggest reason why growers install micro irrigation in the first place. However, it is surprising how few users of micro irrigation are regularly measuring the effectiveness of their irrigation systems. Most users are turning systems on for a set number of hours and assuming that the water and inputs are reaching the crop. Some use moisture monitoring to verify that water is getting where it is needed. More and more growers now also have flow meters to measure the volume of water applied. But when was the last time you had a Distribution Uniformity (DU) test performed to ensure even application of water across your field?

Data from numerous studies over the last few decades establish a direct link between yield loss and low DU. Furthermore, several studies document the decline of DU in an average system over time. Lack of maintenance, poor water quality, inadequate filtration, system design, or system operation can all lead to DU decreases. Poor DU is a silent killer of yields and therefore profits. There is a clear opportunity here to reverse yield losses by measuring DU and also other qualitative factors to determine overall irrigation system performance.

So what is DU (distribution uniformity)? It is the measure of how uniformly water is applied to the area being watered. There are several methods for measuring DU, but the most widely accepted method today is published by the ITRC of Cal Poly in San Luis Obispo (Burt 2004: Burt et al. 1992). In this method the Global or System DU is calculated by mathematically combining several Component DU values together and is expressed as a fraction between 0 and 1. (We frequently see people convert this number to a percentage and discuss a DU of say 85%, but officially it's a fraction of 1, ie 0.85.)

The four DU Components are pressure differences, unequal drainage, unequal application rates or unequal spacing, and "other" causes which are any factor that would cause flow rate differences at identical pressures like plugging, wear, or manufacturing variation. Unfortunately simply measuring the flow rates of emitters across a field does not tell you whether the differences are due to pressure variation

Title: Reversing Yield Loss by Managing Distribution Uniformity or other causes like plugging or wear. In an ITRC study performed on 329 fields researchers found that 45% of the non-uniformity was due to pressure differences, 52% was due to "other" causes such as plugging. Only 2% of the variation was due unequal application rates (emitter spacing) and 1% due to unequal drainage. Therefore the ITRC Rapid Technique (Burt 2004) requires measurements of pressure differences in the field and the "other" causes of flow rate differences not related to pressure such as plugging, wear, and manufacturing variation. Each component is calculated using the lowest 25% of readings divided by the average of the total. This results in a designation of "lower quartile" or DU_{lq}. The global or system DU of the lower Quarter is calculated as:

$$DU_{lq\Delta q\ global} = DU_{lq\Delta p} \times DU_{lq\ Other}$$

Management of your DU is important to determine whether every portion of your crop is receiving the same amount of water and nutrients. Using the ITRC rapid procedure will enable you to determine whether or not the DU deficiency is due to uneven pressures in your system or another factor like plugging. It is one of the few quantitative measurement tools available that allows a grower to understand the uniformity of water and input applications; and it has been tested extensively and proven to be accurate when performed correctly.

DU testing needs to be performed frequently enough to ensure that any changes in system performance can be identified and corrected. Once a year is probably a minimum, but there are growers who perform DU analysis weekly or monthly. Most commonly we see a test at the beginning of the season, a few weeks after the hottest portion of the season begins, and at the end of the season.

If you get a lower than desired DU number at any time, take corrective actions to improve your system performance. Approximately half the time a low DU is due to pressure differences ($DU_{lq\Delta p}$) throughout the field, and half the time it is due to "other" ($DU_{lq\ Other}$) problems, primarily plugging.

Pressure differences may be an original system design problem and may be costly to repair. However, these pressure differences can also be due to simpler problems such as:

1. Poorly functioning pressure regulating valves
2. Plugging at the riser screens resulting in uneven line pressures
3. Poor pumping performance out of the well due to plugging or lack of maintenance.
4. Inadequate discharge pressure out of the supply pump.

These types of pressure problems are all correctable and many of them are not particularly costly to remedy, especially when compared to the yield losses they can cause.

If the DU deficiencies in your system are caused by DU "other" such as plugging, there is corrective action available as well. Please note that on rare occasions, "other" sources of non-uniformity may be due to wear on the emitters or sprinklers from chemigation with abrasive products like gypsum, or due to manufacturing variation. However, most commonly we see these DU losses caused by plugging in two key categories, microbiological plugging, and mineral or chemical precipitation. Additionally on buried tape (SDI), root intrusion can also be a source of emitter plugging.

Once plugging has occurred and it has decreased your DU, the system must be remediated in order to correct the issue. The type of remediation performed depends on the nature of the foulant that is plugging your system. If the foulant is microbiological (see Figure 1), you will need to remediate with a biocide, typically, bleach, chlorine dioxide, Peracetic acid, hydrogen peroxide, copper, or ozone. If the foulant is mineral in nature most likely an acid will be needed to dissolve the mineral back into solution and flush it from your system. We see sulfuric acid, nitric acid, N-phuric, and citric acid as well as some designer safe acids being used commonly. If you have root intrusion, both bleach and sulfuric acid have proven to be effective if applied timely and properly. Again, the nature of the foulant (stuff plugging up your system) must first be identified. Here are some tools we use to identify foulants:

1. Inspect riser screens
2. Inspect filtration devices
3. Flush out the hoses and capture the water in a plastic bottle to look for foulants
4. Cut out sections of tubing and dissect for inspection
5. Shave down emitters to inspect emitter pathways.
6. Destruct and inspect sprinklers and button emitters for inspection
7. Dig up sections of SDI (buried drip tape) and inspect for foulants or root intrusion

The remediation procedure, including choice of chemistry that will have the greatest efficacy and be most cost effective, can be somewhat complex. Also the concentration used, the length of application time, and the number of applications necessary can vary from location to location. There are some basic process steps however:

1. When performing a remediation, flush the system completely ahead of time to remove any loose foulant that will increase the chemical demand and decrease the efficacy.
2. Startup the system and get the selected chemistry flowing immediately at the correct dosage range (do not exceed lablel application rates if using a biocide as they are considered pesticides and must be applied at or below label rates.) If using an acid follow the micro irrigation system manufacturer's specifications on the desired pH range and exposure time; typically not less than 2 pH, but not more than 3 pH. Recommended exposure times vary. (see Netafim Drip Irrigation System Maintenance Handbook, 2014)
3. Next open up the individual lines to speed the delivery of the full concentration of chemistry to the end of the system and measure for proper residual at the furthest point.
4. Close the lines again to bring the system up to full pressure, now the remediation has actually begun and all emitters should be seeing the same concentration of chemistry.
5. Run for the desired length of time, typically two hours, shut down the system and allow it to soak overnight.
6. Flush the system and repeat the entire remediation procedure one more time.

Using the good old 80/20 rule, this procedure is effective most of the time if using the right chemistry at the right dosage. However, we have seen systems that needed continuous biocide dosing for 72 hours in order to restore the DU to desired ranges. Some systems need to be remediated multiple times before the DU is restored. Unfortunately every system is different, every foulant is different, and there is no such thing as a silver bullet that works every time. If you find that you have plugging issues in need of remediation, we highly recommend working with a reputable company that you trust to assist in your efforts, or to provide a turn-key solution.

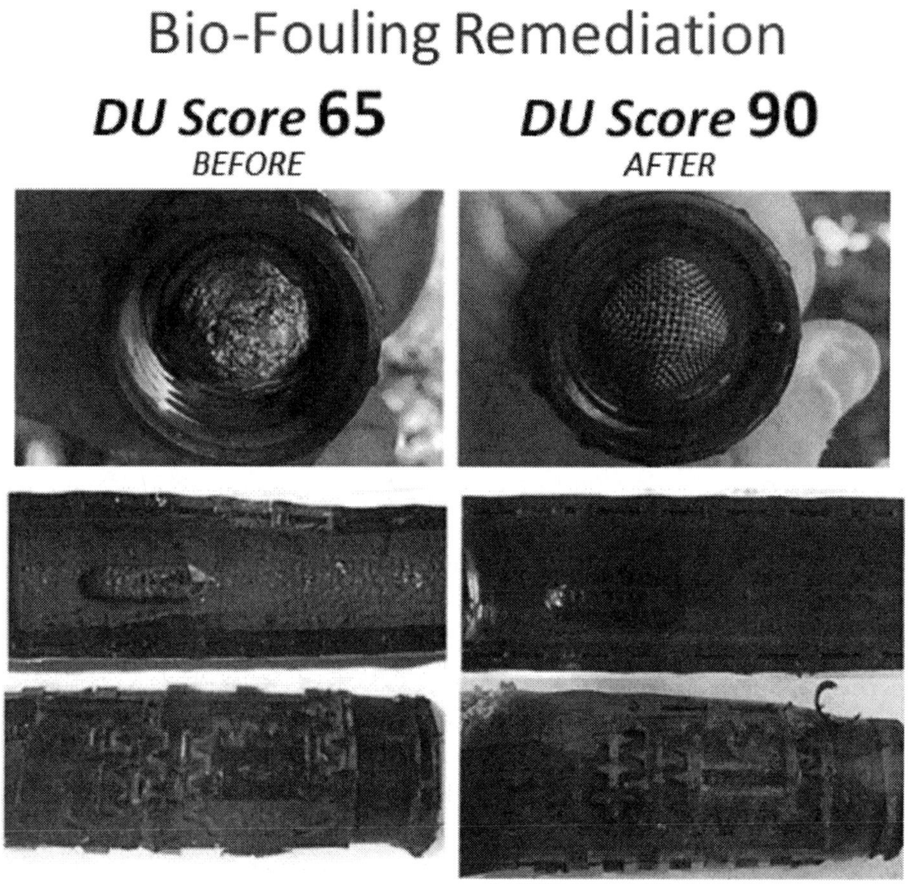

Figure 1

Once you have invested in restoring your DU to acceptable ranges, you will want to maintain that system performance over time in order to maximize your yields. Therefore system maintenance is incredibly important ongoing. Regular periodic flushing is one of the simplest and most underutilized tools to ensure good performance. Here are a few key items to check routinely:

1. Flush the entire system, mains, sub-mains, laterals, and individual lines in that order.
2. Check and clean riser screens and ensure that they are performing correctly and free from plugging.
3. Keep your filters clean and ensure proper back flushing and performance.
4. Check pressure regulator valves for proper performance.
5. Spot check system pressures to ensure that they are consistent and gauges are functioning correctly.
6. Maintenance: sprinkler, emitter types, and / or nozzle sizes must match.
7. Maintenance: fix breaks and leaks immediately.
8. Perform moisture monitoring to ensure water is being delivered as you expect
9. Use vegetative density analysis like NDVI to identify areas with poor plant vigor
10. Walk the fields and look for signs of plant stress

System maintenance using chemigation to prevent plugging before it occurs can be highly effective if properly managed. Again, the nature of the foulant causing the plugging must be known; is it microbiological, or chemical / mineral? Also note that sometimes it is both! Once the foulant is identified and the proper chemistry and dosage rate has been selected, a maintenance program can be set up. Typically this involves a simple chemical metering pump, a flow switch, pressure switch, or flowmeter to activate the chemical pump automatically every time you irrigate.

Unfortunately like any piece of equipment, these systems need regular maintenance and calibration in order to be effective in preventing plugging and maintaining high DU and yield. Also required dosage and choice of chemistry can change as water quality, ambient temperature, or other factors change on your property. A good service company can be essential to help ensure the success of your maintenance program and maintaining adequate DU to ensure yield losses are permanently reversed.

It is an established fact that water and inputs drive yield, and therefore lack of water and inputs, or uneven application of water and inputs result in yield losses. In order to identify the uneven application of water, we use a tool called Distribution Uniformity (DU) analysis. This tool allows a grower to determine whether water is being uniformly applied across a given area, and identify whether the cause of uneven distribution is due to pressure differences or "other" causes like plugging. Growers can also use qualitative inspections like inspection of filters, riser screens and emitter or hose destruction and inspection to identify the foulants that can cause plugging. Walking the fields and inspecting plant vigor, use of NDVI, inspection of wetting patterns, and soil moisture sampling can all lead to the identification of non-uniform water application.

Once a low DU has been discovered and the cause identified, corrective action can be taken. If the cause is uneven pressures an Irrigation specialist may be able to help identify the mechanical problems and offer a solution for corrective action.

If the cause of the low DU is plugging and the foulant has been identified, remediation through chemigation is usually effective at restoring the DU to acceptable levels. One you got it up, keep it up! Regular physical maintenacnce of your filter systems, flushing of your system and lines, and pressure checks make a huge difference. A chemical maintenance program can be a very effective tool at maintaining high DU numbers if properly applied.

You must inspect what you expect, regular DU checks, physical inspections of qualitative factors and moisture monitoring are the keys to ensuring that we all reverse yield losses by maintaining high DU.

Title: Reversing Yield Loss by Managing Distribution Uniformity

Effects of root-zone micro-irrigation on Cabernet Sauvignon

Pete W. Jacoby[1], Xiaochi Ma[2], and Jeremy R. Thompson[2]

[1]Professor and [2]Graduate Research Assistants

Department of Crop and Soil Science, Washington State University

[1]IA member and corresponding author, (509) 335-3495, jacoby@wsu.edu

Abstract. *During 2015, vines receiving season-long drip irrigation delivered into the lower root zone via hard plastic tubes yielded 70 % of commercial production while receiving only 15% of the water as the vines receiving full commercial rates of surface drip irrigation. Vines receiving direct root-zone (DRZ) irrigation at rates reduced to 60, 30, and 15% of commercial drip irrigation (DI) produced individual clusters with higher numbers of berries, yet smaller in size, than did clusters from vines receiving full rates of surface drip irrigation. Preliminary findings suggest that a new form of subsurface micro-irrigation may have potential to not only conserve more water than surface drip irrigation, but may produce grapes of a higher and more desirable quality for producing premium red wines. Relative water use efficiency during 2016 was 1.5, 2.5, and 5.0 times greater for grapes produced by sub-surface irrigation at rates of 60, 30, and 15%, respectively, than for surface drip irrigation. Interrupted irrigation delivery did not provide any advantage over uninterrupted delivery. Treatments were repeated during 2016 and fruit quality at harvest was analyzed for several key factors attributed to high quality wine grapes.*

Keywords: *Subsurface drip irrigation, root-zone, grapevine, water use efficiency, direct root zone irrigation, micro-irrigation, fruit quality, plant stress, deficit irrigation, pulse irrigation, wine quality*

Introduction

Production of wine grapes in the state of Washington has experienced more than 8 percent growth each year for the past decade, accounting for more than 50,000 acres (20,250 ha) currently and placing Washington in second place behind California in U.S. wine grape production. The primary limiting factor to this phenomenal growth is available water for irrigation which is essential in the desert environment in which most grapes are produced. Expansion of wine grape acreage will result either through conversion of other irrigated crops to wine grapes or through the development of more efficient water management and irrigation techniques.

The hot days and cool nights of this desert region have been credited for the production of high quality premium red wines such as Cabernet Sauvignon that have created an increasingly high demand for Washington wines. Irrigation is typically applied to wine grapes as surface drip which is considered an efficient and effective method. However, in the hot, dry region of Washington State, water is lost to both soil surface evaporation and weeds. Additionally, vine root systems become mostly concentrated in the upper 18-20 inches (0.5 m) of the soil profile under current irrigation strategies (Stevens and Douglas, 1994; Bauerle *et al.*, 2008; Davenport *et al.*, 2008). This condition is compounded by the fact that wine grapes in this region are mostly own-rooted from varieties that typically develop shallow to medium depth root systems (Keller *et al.*, 2012). These growth patterns may be more susceptible to both cold intolerance and drought during infrequent, short-term climatic perturbations such as experienced in 2010 and 2015.

Deficit irrigation is known to enhance a number of compounds related to production of premium red wines and substantial literature has been published on this topic that is summarized in detail by Chaves *et al.* (2010). Likewise, subsurface drip irrigation has emerged as an effective water saving strategy in a number of row crops (Lamm *et al.*, 2010; 2015).

To investigate opportunities to increase water use efficiency, we initiated a coordinated, trans-disciplinary research effort in 2014 with three broad objectives: 1) determine the potential and feasibility for use of a new form of sub-surface drip irrigation which might improve water conservation while improving the use of deficit irrigation in the production of high quality premium red wine grapes; 2) gain a better understanding of the potential advantages of developing deeper root systems of vines to obtain moisture from the available soil profile in the rhizosphere; and, 3) evaluate the potential of remote sensing to monitor water stress on the whole vineyard scale when applying direct root-zone micro-irrigation at rates much lower than used to meet commercial wine grape production goals.

Methods and Materials

Industry Collaboration. A 4-person stakeholder advisory group was established involving active growers and managers representing leading red wine grape producers in Washington. This group was convened quarterly to confer and advise the project Principal Investigator (PI) and team members on critical industry needs and opportunities. This group also assisted in identifying collaborators to host research experiments in commercial vineyards. These stakeholders also provided letters of support for grant proposals submitted to sponsoring agencies. Additionally, the project PI also became an active member of the National Grape and Wine Initiative and participated in board meetings and education meetings addressing issues of the wine grape industry to both gain knowledge and potential research collaborators.

Research Site Description. Treatments were installed in a commercial block of Cabernet Sauvignon wine grapes located on Kiona Vineyards, Block 2 (46°16'59" N, 119°26'33" W) in the Red Mountain American Viticulture Area (AVA) near Benton City, WA in early 2015. Soil on the experimental site is of the Aridisol order and classified as a Hezel loamy fine sand (Xeric Torriorthents) on a terrace landform with parent materials being eolian sands over silty glacio-fluvial sediments deposited at the end of the most recent ice age. These soils are well-drained, subject to wind erosion, and relatively infertile, containing very low amounts of organic matter. Depth to nearest water table is more than 80 inches (>2 m). Normal annual precipitation is 8.83 inches (224 mm) and occurs mainly as rainfall during the dormant growing season. Summer temperatures average 70°F (21° C) with mid-day temperatures reaching 90°F (32°C) or more and cooling during the night, typical of a desert climate. These conditions favor the development of high quality red wine grapes under irrigation. The research site was planted to Cabernet Sauvignon (Clone 2) of own-rooted vines on a spacing of 8' (2.5 m) between rows and 6' (1.8 m) between vines. The vineyard was 8 years old at the beginning of the 2015 growing season.

Experimental Design and Treatments. The experimental design is a randomized complete block with two main effect treatments, irrigation rate and depth of delivery sub-surface. A split plot design was super-imposed to compare pulse irrigation and constant delivery. Each treatment plot involved 15 vines (5 vines x 3 rows) with the center-most 3 vines designated for physiological measurements while being buffered by other vines receiving the same treatment. Each of 18 treatment plots was replicated 3 times (810 vines) and compared with 12 plots (180 vines) receiving full commercial rate of irrigation via surface drip and designate as the control treatment when comparing production and water stress with that of the direct root-zone treatments. Irrigation scheduling was determined by the vineyard manager according to long-standing guidelines used to meet commercial production goals. Irrigations were applied more frequently in 2015 which was the hottest and driest growing season on record. During most of the 2015 growing season, irrigation was applied in a 20 hour set every four days. By contrast, the 2016 growing season was characterized by near normal temperature and precipitation and irrigation sets were more varied in timing and duration, according to the temperature and growing conditions of the vines. Direct root-zone irrigation applications occurred on the same dates as commercial plots during both years, but water amounts were reduced to approximately 60, 30, or 15 percent the amount of the commercial rate through the use of battery powered controllers (Galcon Kfar Blum, type 11000L). Actual water amounts applied to treatment and control plots were quantified by small mechanical water meters (D.L. Jerman Co., Hackensack, NJ) read after each irrigation event. Controllers were programmed to irrigation during the evening hours during the same time schedule every day throughout the growing season. Commercial irrigation sets were scheduled to run throughout the entire evening during each event.

Direct root-zone delivery device. Growers in Washington have attempted to use subsurface micro-irrigation applied through buried lines, but have found this technique generally unacceptable owing to clogging of emitters and damage by burrowing rodents. To achieve direct root-zone irrigation without the use of buried driplines, a 1 inch (25.4 mm) diameter hole was bored vertically to a depth of 1, 2, or 3 foot (0.3, 0.6, and 0.9 m, respectively) about 1.5 foot (ca. 0.5 m) either side from the base of each vine and beneath the trellis wire and suspended irrigation dripline. A length of PVC tube was inserted into each hole. Each section of PVC pipe was previously cut to length to reach the desired depth while extending above ground for a given distance and then split about 6 inches (15 cm) from the lower end with a band saw to allow sufficient water passage to move into the soil to reduce the backing of water up the tube. A PVC cap, previously drilled to allow passage of a one foot (30.5 cm) length of ¼ inch (6.35 mm) diameter micro-tubing through the hole prior to attaching one end to a barbed connector and inserted into the main horizontal water line, and attaching a pressure compensating drip emitter at the other end before placing the emitter snuggly into the top of the PVC tube. The cap was then secured over the end of the tube to prevent dirt or debris to reach the emitter. Emitters were selected to deliver 0.5 gallon (0.6 liter) per hour, thus delivering 1.0 gallon per vine per hour.

Installation of all treatments involving 990 vines was completed prior to beginning of the 2015 growing season. Additionally, soil moisture access tubes were installed in designated plots to monitor soil water dynamics both temporally and spatially across depths within the top 6.5 feet (2.0 meters). Soil moisture content was determined by electronic capacitance probes [SynTek Diviner (spot reading) and SynTek EnviroSCAN (continuous reading), Stepney, S.A., Australia.]. Irrigation events were monitored by data transmitted from the EnviroSCAN probes via cellphone transmitters to a base server (Tuctronics, Walla Walla, WA).

Monitoring of plant water stress. Vine water stress was determined by periodically sampling the center three vines of direct root-zone treatments and control vines by measuring mid-day leaf stem water potential using the pressure bomb method (Scholander *et al.*, 1965). Leaves were selected on the east side of vines (most sun drenched prior to sampling), then covered with a plastic bag inside an aluminum foil exterior envelope and allowed to equilibrate for about an hour prior to sampling. The encased leaf was detached from the vine by severing the petiole with a razor blade, then quickly removed from the bag and inserted into the pressure chamber, while shielding the leaf from direct sunlight, then pressurized with nitrogen gas to the point of water movement from the cut petiole. Pressures were measured in bars then converted to mega-pascals. These data were correlated with multi-spectral digital images obtained through aerial and ground-based platforms during both growing seasons (data not shown). Visual observations of plant condition were documented with digital photos to document phenological impacts on vines from treatments as referenced by Keller (2005)

Determination of grape production and quality. In 2015, grape clusters from designated vines were collected to determine estimates of production, as well as number and size of berries in each treatment and control block. Sampling was executed during two stages of maturity, early-*veraison* and mid-*veraison*. At the 2015 harvest date (September 26), all grapes were harvested from replicated rows of each treatment and weighed by individual vine. In 2016, cluster sampling was not repeated, but all vines in the treatment plots and half of the control plots were harvested on September 24 and weighed. Samples from specific treatments were collected two weeks prior to harvest and submitted to a private commercial laboratory for analytical determination of a number of characteristics associated with quality red wines.

Results and Discussion

Treatments were successfully implemented during both 2015 and 2016 growing seasons. Data presented has not yet been statistically analyzed. Direct root-zone treatments (DRZ) were effective in delivering water to the designated depths. On occasion, we experienced some water backing up in the tube, but this problem was attributed to soil plugging at the time of tube installation, and once dislodged, was not a lingering problem. Water leakage was often observed at the dripline point of insertion by the barbed connector. Sealant was used and provided temporary relief, however this problem persisted for the duration of the study. It is felt that the tool used to punch the initial hole into the dripline was a major source of this problem, as we did not experience it at two other research locations. Operating pressure of the primary pressure pump could also be a source of the problem, but we have not yet addressed this possibility. The electronic controllers worked flawlessly, as did the small mechanical meters. We discovered that the meters performed with most accurately when placed face up rather than to the side, owing to an internal design factor.

Comparison of pulse and constant irrigation delivery revealed no differences in either plant physiological stress or fruit production; therefore, treatment data from this split plot design was pooled. Likewise, depth of water delivery did not produce any consistent advantages across application rates or time (Table 1).

Table 1. Seasonal irrigation delivery and water use efficiency based on grape production during 2015 and 2016 comparing commercial surface drip irrigation with season-long deficit irrigation imposed by direct root-zone micro-irrigation delivered subsurface from 1-3' depths at rates of 60, 30, or 15% the rate of surface drip irrigation.

	Irrigation Treatments			
	Surface Drip (DI)	DRZ		
	(100 %)	(60 %)	(30%)	(15%)
2015 Water Use (acre ft.)	1.35	0.81	0.40	0.20
Water/vine each event	16.25	9.75	4.88	2.44
Grape production (tons/ac)	4.54	4.08	3.40	3.18
Production Efficiency (lbs./acre inch applied)	560	840	1400	8271
Relative Efficiency	1.0	1.5	2.5	4.7
2016 Water Use (acre ft.)	1.37	0.84	0.43	0.23
Water/vine each event	17.59	10.27	5.13	2.57
Grape production (tons/ac)	6.73	3.79	2.96	2.20
Production Efficiency (lbs./acre inch applied)	818	752	1147	1598
Relative Efficiency	1.0	0.9	1.4	2.0

Our original objective was to ascertain the greatest degree of water conservation that could be achieved while maintaining health and productivity of the vine. For that reason, season-long deficit irrigation was used for all DRZ treatments, although such a strategy is not typically used in commercial grape production. The 2015 growing season was later determined to be the hottest and driest on record for the area. Fruit production at the commercial irrigation rate and applied by surface drip averaged 10 pounds (4.5 kg) per vine, while DRZ irrigation applied 1-3 feet subsurface at reduced rates of ca. 60, 30, and 15% of full commercial rate produced an average of 9.0, 7.6, and 7.2 pounds per vine, respectively (Table 1). Concern was expressed by the grower and members of the stakeholder advisory group that some stress effects from 2015 might be reflected in both fruit production and plant vigor during the following year. These concerns proved valid during 2016 and were reflected in lower overall fruit production in the DRZ treatments than during the previous year, despite the fact that growing conditions were more favorable in 2016 than in 2015, and there was wider disparity in fruit production between the commercial treatment plots and the DRZ plots than occurred during the previous growing season (Table 1). While pruning weights from the 2016 treatment vines have not been obtained at this time, there were visual differences in shoot lengths and condition among the treatments throughout the 2016 growing season.

Plant water stress, as measured by obtaining xylem pressure potentials among the treatment vines, showed obvious differences among the irrigation delivery rates when measured at 3 dates during the growing season (Table 2). Plant water stress increased proportionately with decreasing irrigation rate and progression of the growing season. Similar measurements were made at only one date during 2015 and also showed progressively more stress with lowering rates of irrigation (data not shown).

Table 2. Plant water stress as determined by leaf stem xylem potential during 2016 growing season contrasting commercial surface drip irrigation with season-long deficit irrigation imposed by direct root-zone micro-irrigation delivered subsurface from 1-3' depths at rates of 60, 30, or 15% the rate of surface drip irrigation.

	Irrigation Treatments			
	Surface Drip (DI)	-DRZ-		
	(100 %)	(60 %)	(30%)	(15%)
Date	Xylem Pressure Potential (-kPa)			
June 3	-528.62	-592.95	-640.66	-781.17
July 7	-635.01	-825.40	-924.59	-1187.96
August 10	-868.74	-1176.93	-1521.67	-1592.69

Water use efficiency, determined as amount of fruit produced per unit of water applied, increased progressively with reduced rates of irrigation in 2015 (Table 3). This trend was repeated in 2016, but was not as pronounced, largely owing to a much higher rate of grape production that occurred from the commercial plots in 2016 than in 2015. Some of this difference may also be attributed to carry-forward effect from the lower water applications during 2015 in the DRZ treatments.

Table 3. Grape production from plots receiving full commercial irrigation applied as surface drip (SD) and applied as direct root-zone micro-irrigation (DRZ) at season-long reduced rates of ca. 60, 30, and 15 % of full commercial rate during 2015 and 2016.

	Irrigation Treatments							
	Surface Drip (DI)		-DRZ-					
	(100 %)		(60 %)		(30%)		(15%)	
2015	**Wt. per Vine**							
	Lbs.	kg.	Lbs.	kg.	Lbs.	kg.	Lbs.	kg.
Surface Drip	10.0	4.55						
DRZ at -1'			8.6	3.92	6.6	2.98	6.8	3.09
DRZ at -2'			9.1	4.11	7.4	3.36	7.8	3.55
DRZ at -3'			9.3	4.21	8.8	3.99	7.1	3.21
Mean	**10.0**	**4.55**	**9.0**	**4.08**	**7.6**	**3.44**	**7.2**	**3.28**
2016								
	Lbs.	kg.	Lbs.	kg.	Lbs.	kg.	Lbs.	kg.
Surface Drip	14.8	6.73						
DRZ at -1'			8.6	3.90	6.9	3.11	5.4	2.45
DRZ at -2'			8.0	3.62	6.3	2.85	4.6	2.09
DRZ at -3'			8.5	3.84	6.4	2.92	4.5	2.08
Mean	**14.8**	**6.73**	**8.4**	**3.79**	**6.5**	**2.96**	**4.9**	**2.20**

In 2015, cluster samples from the DRZ treatments showed that cluster weights were slightly lower, but grapes were more numerous, yet smaller in size than in the clusters from vines receiving the higher irrigation rate. These findings suggested that the grapes from the lower irrigation rates might have greater potential to produce higher quality red wine, owing to higher concentration of anthocyanins, tannins and sugars. In 2016, similar effects were noted, but not documented, for grapes receiving the DRZ treatments. Replicated cluster samples were obtained from the commercial and DRZ treatment plots and submitted to a private, commercial analytical lab for determination of a dozen components and ratios. Data summarized in Table 4 illustrates four of these components. Acidity became progressively reduced below the 60% irrigation rate, while sugars (Brix), tannins, and anthocyanins all trended higher with decreasing rate of irrigation. These results are in line with the findings of Casassa *et al.* (2015) who noted that efforts to derive benefits in grape quality and water savings through greatly reduced irrigation levels should recognize the potential for yield reductions and/or physiological impacts on vines. Results from our study provides evidence that use of efficient irrigation application such as DRZ could both sustain vines and produce grapes during drought conditions while yielding grapes with potential to produce premium quality red wines in the hands of skilled viticulturists and enologists.

Table 4. Comparison of selected chemical components influencing red wine quality. Analyses of Cabernet Sauvignon grapes grown under full and reduced rates of irrigation season-long during 2016. Reduced irrigation rates were applied via direct root-zone micro-irrigation (DRZ) delivered 2 feet (61 cm) subsurface.

Component	Surface drip (DI)	DRZ		
	Control (100 %)	High (60 %)	Moderate (30%)	Low (15%)
pH	3.41	3.36	3.48	3.55
Brix	25.5	27.1	27.6	28.6
Tannins	403	594	600	741
Anthocyanins	1015	1242	1298	1480

Conclusions

A new form of subsurface micro-irrigation was developed to achieve direct root-zone subsurface delivery or drip irrigation to wine grapes in Washington State. Vines were maintained for production, albeit less than achieved under commercial surface irrigation at full irrigation delivery needed to meet production goals. Rates of 60, 30, and 15% of commercial irrigation produced ca. 90, 75, and 70%, respectively, of the commercial grape production weight. Second year production rates dropped to 57, 44, and 33% that of commercial production at the same rate of irrigation applied the previous year (60, 30, and 15% of commercial irrigation rate, respectively). Carry-forward effects from previous year water stress, improved weather, and longer commercial irrigation sets during 2016 may have contributed to lower second year production ratios. No obvious advantages were found for using pulse irrigation delivery over continuous application sets. No consistent patterns were observed to favor a specific depth among 1, 2, or 3 ft. subsurface delivery points.

Acknowledgements

This research was supported in part by the following: 1) Northwest Center for Small Fruit Research, Agreement No. 59-5358-4-010 and grant # 2072-21000-047-16. 2) Washington State Department of Agriculture – Specialty Crop Block Grant Program – Agreement K1768. The authors also express sincere thanks to Scott Williams, general manager – Kiona Winery and Vineyards, Benton City, WA for providing access to land and resources for this project. We also thank Rick Hamman of Hogue Ranches, Bill Riley of Ste. Michael Wine Estates, Julia Kock of Klipsun Vineyards, and Jim Holmes of Ciel du Cheval Vineyard for serving as advisors to this research project and others. Appreciation is also extended to Drs. Markus Keller and R. Troy Peters for subject matter consultation related to various aspects of this research.

References

Bauerle, T.L., J.H. Richards, D.R. Smart and D.M. Eissenstat. 2008. Importance of internal hydraulic redistribution for prolonging the lifespan of roots in dry soil. *Plant, Cell & Environ.* 31: 177-186.

Casassa, L.F., M. Keller and J.F. Harbertson. 2015. Regulated deficit irrigation alters anthocyanins, tannins and sensory properties of Cabernet Sauvignon grapes and wine. *Molecules* 20: 7820-7844.

Chaves, M.M., O. Zarrouk, R. Franciso, J.M. Costa, T. Santos, A.P. Regalado, M.L. Rodriques and C.M. Lopes. 2010. Grapevine under deficit irrigation: hints from physiological and molecular data. *Annals Bot.* 105: 661-676.

Davenport, J.R., R.G. Stevens and K.M. Whitley. 2008. Spatial and temporal distribution of soil moisture in drip-irrigated vineyards. *HortSci.* 43: 229-235.

Keller, M. 2005. Deficit irrigation and vine mineral nutrition. *Am. J. Enol. Vitic.* 56: 267-283.

Keller, M., L.J. Mills and J.F. Harbertson. 2012. Rootstock effects on deficit-irrigated winegrapes in a dry climate: vigor, yield formation, and fruit ripening. *Am. J. Enol. Vitic.* 63: 29-39.

Lamm, F.R., J.P. Bordovsky, L.J. Schwanki, G.L. Grabow, J. Enciso-Medina, R.T. Peters, P.D. Colaissi, T.P. Trooien and D.O. Porter. 2012. Subsurface drip irrigation: status of the technology in 2010. *Trans. ASABE* 55: 483-491.

Lamm, F.R., K.C. Stone, M.D. Dukes, T.A. Howell, J.W. Robbins and B.Q. Mecham. 2015. Emerging technologies for sustainable irrigation. ASABE/IA Symposium. *Trans. ASABE* 59: 155-161.

Scholander, P., E. Bradstreet, E. Hemminsen, and H. Hammel. 1965. Sap pressure in vascular plants: negative hydrostatic pressure can be measured in plants. *Science* 148: 339-346.

Stevens, R.M. and T. Douglas. 1994. Distribution of grapevine roots and salt under drip and full-ground cover micro-jet irrigation systems. *Irrig.Sci.* 12:181-186.

This page intentionally left blank.

Soil Nitrate Levels for Surface-Drip Irrigated Cauliflower

Josue Samano-Monroy, MS
California State University, Fresno. 5370 N. Chestnut M/S OF 18, Fresno, CA. 93740.
(559)-579-8362

Dave Goorahoo, PhD
California State University, Fresno. 2415 E. San Ramon Ave MS AS 72. Fresno, CA. 93740-8033.
(559)-278-8448.

Florence Cassel Sharma, PhD
California State University, Fresno. 2415 E. San Ramon Ave MS AS 72. Fresno, CA. 93740.
(559)-278-7955.

Touyee Thao, BS
California State University, Fresno. 5370 N. Chestnut M/S OF 18, Fresno, CA. 93740.
(559)-390-1096.

Abstract. *Nitrate contamination in water is an unresolved environmental issue, with high levels having been detected in California's drinking water. Cauliflower is a shallow rooted crop with high demand for Nitrogen (N), thereby providing a challenge to optimizing yield while minimizing nitrate leaching. Nitrate levels were measured within the top four feet of a sandy loam used for surface drip irrigated cauliflower fertilized at three N rates with organic soybean meal (ORG) and conventional UAN. Soil nitrate contents in response to fertilizer showed a higher NO3-N content as compared to Control plots. There was no significant difference between ORG and UAN treated plots, suggesting that nitrate leaching can occur in either case with the use of nitrogen fertilizers. There was an interaction effect of fertilizer type x rate with the greatest soil nitrate content occurring within the 12-24 inches of soil for the plots fertilized with 225 lbs N/acre of UAN32.*

Keywords: Nitrate leaching, surface drip irrigation, soil nitrate, organic fertilizer, urea-ammonium-nitrate, soybean meal.

Introduction

Cauliflower production in California accounts for 86% of the total US production (Geisseler & Horwath, 2015). The harvested area in California for 2010 was 32,900 acres, and the production accounted for 210 million dollars (NASS, 2011). In 2014, the production was 313,6000 tons with an income of 309 million dollars (CDFA, 2015). The Central Coast, South Coast, South Eastern Desert and the San Joaquin Valley (SJV) are the most important producing regions. Around 85% of the total production in the state is located in the coastal regions. Arizona is the second largest producer state, with the regions of the Yuma Valley accounting for only 9% of the US cauliflower production (NASS, 2011).

Cauliflower is practically transplanted and harvested year-round in California, and fields are subjected to high N fertilizer applications to ensure profitable yield. This factor added to the sandy texture of the soils can contribute to ground water nitrate (NO_3) loading. Nitrate- nitrogen (NO_3-N) is a byproduct of the N fertilizers, and excess amounts in water can be harmful to the environment and the human health. The California Department of Public Health has set the safety threshold value for the NO_3-N concentration in drinking water to be under 10 mg/l, and total NO_3 to be at a maximum of 45 mg/l (CDPH, 2014).

Surface drip irrigation has proven to be an efficient tool to manage water and nutrient application. (Thompson et, al. 2000). The adoption of drip irrigation offers a powerful crop management tool and it can also increase water and N fertilizer use efficiency. According to the 4Rs for nutrient stewardship the goal is to minimize groundwater pollution by applying the right source of nutrients, at the right rate, in the right place and at the right time (Bruulsema, 2009). In order to mitigate nitrate leaching from cropland is very important to understand plant-soil-water relationships and to apply the 4Rs rules to irrigation practices in what is commonly refer to best fertilizer management practices (BFMP) (Rigby & Cáceres, 2001).

Objective

Based on the identified priorities for BFMP, the objective of this research was to quantify pre-plant and post-harvest soil NO_3 levels within the top 4 feet of soil for cauliflower grown with an organic soybean meal (ORG) and a conventional urea ammonium nitrate (UAN) fertilizers.

Materials and Methods

The study was located at the California State University, Fresno Farm, on a Hanford fine sandy loam soil for two cauliflower crops with cultivar "incline", planted in Fall 2014 and Fall 2015. The nutrient sources comprised of soybean based organic fertilizer 7-1-2 (ORG) and the conventional urea-ammonium-nitrate (UAN-32) applied at three N fertilizer rates; 75, 150 and 225 lbs/N acre and a Control with no fertilizer addition. Hence, there were seven treatments with the following codes; Control (no fertilizer application), ORG1 (organic fertilizer at 75 lbs/N acre), ORG2 (organic fertilizer at 150 lbs/N acre), ORG3 (organic fertilizer at 225 lbs/N acre), UAN1 (UAN-32 at 75 lbs/N acre), UAN2 (UAN-32 at 150 lbs/N acre) and UAN3 (UAN-32 at 225 lbs/N acre) replicated five times, resulting in a total of 35 plots.

The field was irrigated with a surface drip irrigation system consisting of two lines per bed located in the inner part of the bed with 12 inches of separation. The drip tape was a Eurodrip™ 5/8 "seamless classic, 10 mil, 12" inches emitters spacing, 0.4 gph at 10psi or 0.58 gpm/100ft at 10psi. An Orbit™ 4 station Easy-Dial Electrical Timer was installed to control the irrigation. A manifold with two manual valves, one automatic valve, filter, pressure gauge and flow Meter was also installed as a part of the irrigation system. Irrigation scheduling was based on meeting 100% of crop evapotranspiration (ETc).

Soils were sampled to determine the existing amount of NO_3-N present at the moment of planting using the approach described by Carter (1993). Soil samples were taken pre-planting and post-harvest at four depths; 12, 24, 36 and 48 inches in each of the 35 plots. At each sampling event 140 soil samples were collected for a total of 560 samples over the two years of study.

Soil NO_3-N levels were determined in extracts using the SEAL AQ2 Discrete Analyzer designed for environmental samples including water, soil and plant extracts. The AQ2 uses a 100% optical quality glass cuvette used for precise absorbance measurements, 10mm optimum path length, reagent wedges with on-board cooling, use only 20ul-400ul reagent per test, disposable reaction wells, cadmium coil for reduction of nitrate/nitrite determination, and a flexible software to manage the analyzer and indicate the desired test.

Results

Overall, fertilizer types had a significant effect ($P< 0.001$) on soil NO_3-N concentrations, In 2014, the average NO_3-N concentrations were 1.76 ± 1.35, 11.92 ± 0.78 and 11.71 ± 0.78 respectively for Control, Organic and Conventional treated plots (Table 1). However, the Organic and Conventional plots were not significantly different from each other (Figure 1). For the 2015 study, the mean soil NO_3-N concentrations were: 0.27 ± 0.75, 5.20 ± 0.43 and 5.52 ± 0.43 respectively for Control, Organic and Conventional plots (Table 1), with no significant differences between the Organic and Conventional plots (Figure 2). Generally, the fertilized plots showed significantly higher soil NO_3-N content as compared to Control (Table 1). This difference in the NO_3-N concentrations is due to the fact that plant uptake consumes part of the nitrogen available in the soil, while other portions might be lost either by leaching, denitrification and volatilization (Hartz, 2007).

Table 1: Average (± S.E.) soil NO_3-N content in response to fertilizer type in 2014 and 2015.

	2014	2015
Fertilizer type	NO_3-N mg/l	NO_3-N mg/l
Control	1.76 ± 1.35	0.27 ± 0.75
Organic	11.92 ± 0.78	5.20 ± 0.43
Conventional	11.71 ± 0.78	5.52 ± 0.43

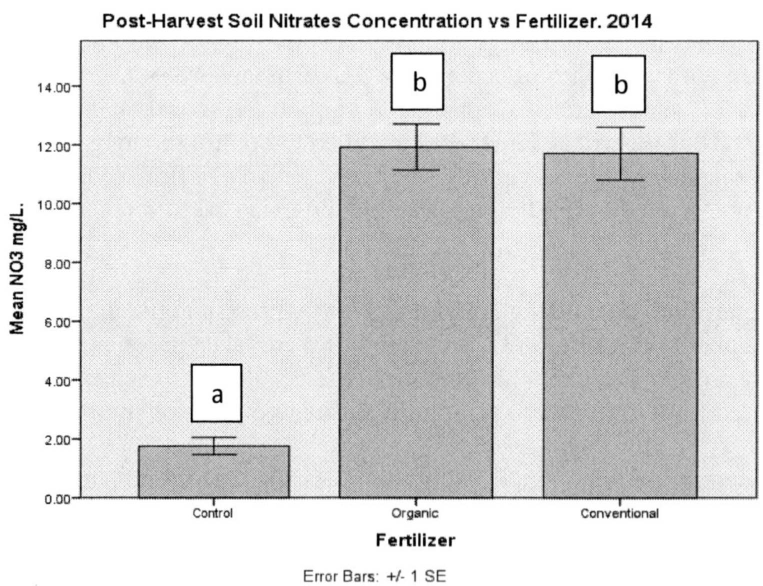

Figure 1: Soil NO_3-N concentrations (mg/l) in response to fertilizer type for 2014.

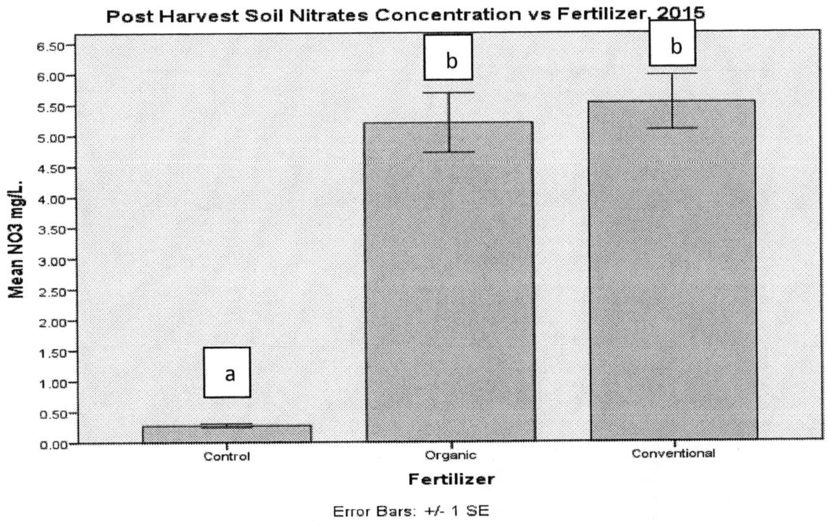

Figure 2: Soil NO_3-N concentrations (mg/l) in response to fertilizer type for 2015.

Soil samples taken at four depths determined that for the higher N rates treatments- ORG2, ORG3, UAN2 and UAN3 there was a trend towards a higher NO_3-N concentration in the 36 and 48 inches as compared to the concentration in the top 12 and 24 inches (Table 2). The elevated concentrations in some plots within the top 12-in of soil were probably the result of mineralization, whereas the higher concentrations within the 36-48-in could be as a result of NO_3-N leaching. Similar results for the NO_3-N concentrations at different depths were reported by Jaynes et al. (2001) on a rotation cropping system in which high NO_3-N concentrations at the top layers of the soil for some years were attributed to nitrogen mineralization. And in years with higher precipitation rates, soil NO_3-N concentrations were higher deeper in the soil horizon, attributed to NO_3-N leaching.

Table 2: Average NO_3-N (±S.E.) concentrations (mg/l) for each treatment at four depths; 12, 24, 36 and 48 inches in 2014.

Treatments	**Control**	**ORG1**	**ORG2**	**ORG3**	**UAN1**	**UAN2**	**UAN3**
Depth (in)							
12	2.7	6.4	12.5	18.7	8.4	5.2	18.5
	(±0.74)	(±3.13)	(±2.94)	(±0.92)	(±1.93)	(±2.67)	(±0.92)
24	1.4	10.6	11.1	10.3	6.7	8.5	17.1
	(±0.63)	(±3.74)	(±2.60)	(±0.73)	(±1.78)	(±2.20)	(±0.83)
36	1.9	12.0	12.3	15.6	11.2	8.5	17.7
	(±0.39)	(±3.00)	(±2.49)	(±2.72)	(±2.85)	(±2.54)	(±0.69)
48	1.0	7.3	7.6	11.2	7.3	8.6	23.2
	(±0.28)	(±2.02)	(±3.51)	(±1.48)	(±1.67)	(±3.06)	(±0.79)

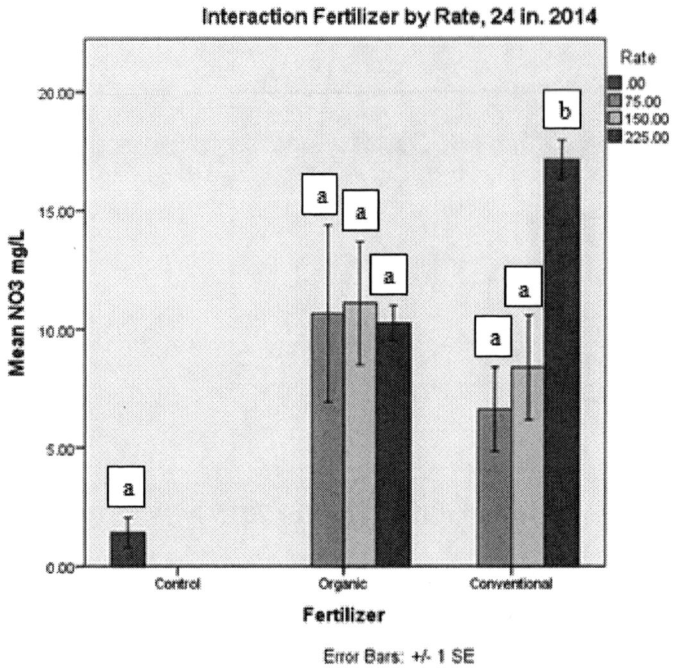

Figure 3: Mean soil NO_3-N concentrations (mg/l) within the 12-24 inches as a function of fertilizer x rate interaction in 2014.

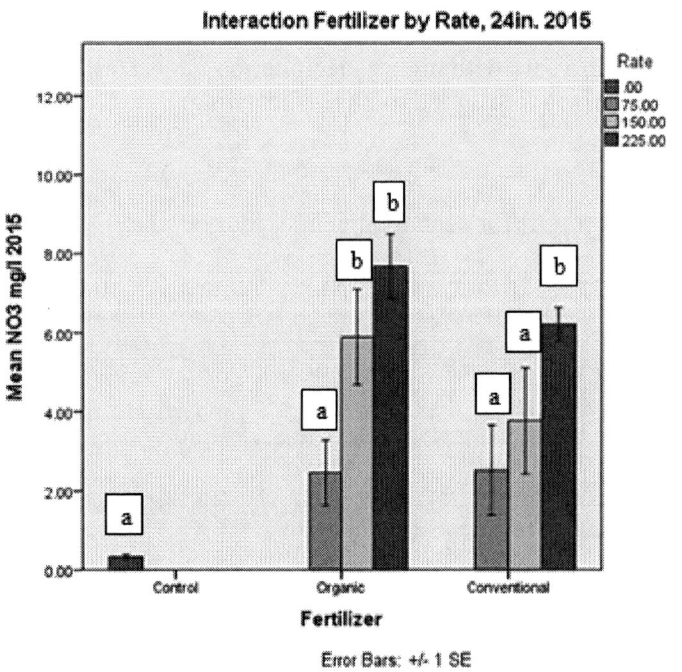

Figure 4: Mean soil NO_3-N concentrations (mg/l) within the 12-24 inches as a function of fertilizer x rate interaction in 2015.

In the top 24-inch of soil for the field study in 2014, there was a significant interaction (P= 0.043) between Conventional fertilizer type by the 225 lbs/N acre N fertilization rate (Figure 3). For the 2015 field study there was no significant interaction (P> 0.05) among the fertilizer treatments (Figure 4) within the top 24 inch of soil.

Conclusions

Soil nitrate contents in response to fertilizer type showed a higher NO_3-N content as compared to Control plots with no fertilizer addition. However, there was no significant difference between organic (ORG) and conventional (UAN) fertilizers, which suggest that the nitrate leaching might occur in either case with the use of nitrogen fertilizers.

Nitrate content in the soil as a function of depth did not show a significant different among the treatments. Generally, there was an interaction between UAN-32 fertilizer and the highest fertilizer rate of 225 lbs/N acre for the 0-12, 12-24, 24-36 and 36-48 inches of soil.

When combined with the appropriate fertilizer types and application rates, surface drip irrigation is a potentially useful tool to help mitigate the nitrate leaching in a sandy loam soil used to grow shallow rooted cauliflower.

References

Carter, M. R. (Ed.). (1993). Soil sampling and methods of analysis. CRC Press.

CDFA. (2015). California Agricultural Statistics Review. 2014-2015. CDFA. http://www.cdfa.ca.gov/Statistics/ (accessed 15 Feb. 2016)

CDPH. (2014). Nitrate Fact Sheet. http://www.waterboards.ca.gov/drinking_water/certlic/drinkingwater/documents/nitrate/Fact%20Sheet%20-%20Nitrate_May2014%20update.pdf (accessed 15 March. 2016)

Bruulsema, T. (2009). Recommendation development under 4R nutrient stewardship. In *Proceedings North Central Extension-Industry Soil Fertility Conference* (Vol. 25).

Geisseler, D., & Horwath, W. R. (2015) Cauliflower Production in California. On line: https://apps1.cdfa.ca.gov/FertilizerResearch/docs/Cauliflower_Production_CA.pdf (accessed on 30th March 2016).

Hartz, T. K. (2007). Efficient nitrogen management for cool-season vegetables. University of California, Department of plant Science. UC Davis http://vric. ucdavis.edu/pdf/fertilization_EfficientNitrogenManagementforCoolSeasonveg etable2007. pdf.

Rigby, D., & Cáceres, D. (2001). Organic farming and the sustainability of agricultural systems. Agricultural systems, 68(1), 21-40.

Thompson, T. L., Doerge, T. A., & Godin, R. E. (2000). Nitrogen and Water Interactions in Subsurface Drip-Irrigated Cauliflower II. Agronomic, Economic, and Environmental Outcomes. *Soil Science Society of America Journal*, *64*(1), 412-418.

USDA-NASS. (2011). Cauliflower production in California. USDA-NASS. http://www.nass.usda.gov/StatisticsbyState/California/Publications/Vegetables/201101vegrv.pdf (accessed 05 Feb. 2016)

AirJection Irrigation Mitigates Denitrification and Leaching

Dave Goorahoo, PhD
California State University, Fresno. 2415 E. San Ramon Ave MS AS 72. Fresno, CA. 93740-8033.

Adrian Unc, PhD
Boreal Ecosystems Research Initiative, Memorial University of Newfoundland, Grenfell Campus.
20 University Drive, Corner Brook NL, A2H 5G4, Canada.

Crystal McCall, MS
Boreal Ecosystems Research Initiative, Memorial University of Newfoundland, Grenfell Campus
20 University Drive, Corner Brook NL, A2H 5G4, Canada.

Josue Samano-Monroy, MS
California State University, Fresno. 5370 N. Chestnut M/S OF 18, Fresno, CA. 93740.

Florence Cassel Sharma, PhD
California State University, Fresno. 2415 E. San Ramon Ave MS AS 72. Fresno, CA. 93740.

Touyee Thao, BS
California State University, Fresno. 5370 N. Chestnut M/S OF 18, Fresno, CA. 93740.

Govind Seepersad, PhD
The University of the West Indies, St. Augustine, Trinidad & Tobago, West Indies.

Abstract. *Using a high efficiency venturi to inject air into water delivered through subsurface drip irrigation, commonly referred to as AirJection® Irrigation, has been shown to result in increased yields for a variety of crops. Studies have also indicated that the technology can positively affect photosynthetic activity, soil respiration rates, and stomatal conductance. In the current study, we compared the relative quantity of a series of genes known to be involved in the nitrogen cycle for soils subjected to AirJection® Irrigation for at least five years, with those that were not aerated. DNA was extracted using a PowerSoil™ kit and gene quantification was obtained via polymerase chain reaction. Distribution of the tested genes within the microbial populations was very distinct among the aerated and non-aerated soils. AirJection Irrigation had a clear selective impact on the distribution of the tested genes among the soil microbial population. While AirJection did not impact N fixation or ammonia oxidation, it did significantly change the denitrification genes population in manner that can positively affect nitrogen use efficiency. Furthermore, with judicious water management within the root zone, AirJection Irrigation can favor the dominance of bacteria that enhance plant nitrate uptake with a potential reduction in nitrate leaching.*

Keywords: Airjection Irrigation, Nitrate leaching, N fixation, Denitrification, Oxygation

Introduction

Injection of air into the root zone environment has shown to enhanced crop productivity. However, the cost of an air-only injection system separate from the irrigation system, had previously remained cost-prohibitive. More than 75 years ago, Durell (1941) wrote, "a study of suitable oxygen carriers, which could be applied as fertilizer, and which would release oxygen slowly to the soil during the growing season, may be worthwhile." With the acceptance of subsurface drip irrigation (SDI) by commercial growers, implementation of an air injection system has become economically feasible. Nonetheless, the design of an air-injection system through a SDI tape requires thorough analysis and understanding of air movement within the soil profile and at the soil surface. When air alone is supplied to the SDI system it emits as a vertical "stream," moving above the emitter outlet directly to the soil surface. As a consequence, the air affected soil volume is probably limited to a chimney column directly above the emitter outlet. Balancing the air/water relationships as well as changing soil temperature could affect growing conditions, yield, and time of harvest, particularly in locations with limited growing seasons. The concept of aerating the irrigation water increases the potential for the air to travel within the root zone, thereby positively affecting plant growth.

Through work in other areas, the Mazzei Corporation has developed high efficiency venturi injectors capable of aerating water with fine air bubbles. By combining the Mazzei injectors with SDI, it is possible to deliver "aerated water" close to the root zone. The technology has now been patented and is referred to as ***Air-jection® Irrigation***. In summary, the system allows for a fluid mixture to be delivered to the root zone of the plant, via the irrigation systems, in what can best be characterized as an air/water slurry. In previous work with growers on a commercial test plot basis, Air-jection has demonstrated bell pepper yield increases of 13 percent and 8 percent for premium and processed bell peppers, respectively. Findings from the initial CSU-Fresno study by Goorahoo et al (2001) justified follow-up fieldwork on larger commercial plots. On average, AirJection® Irrigation has resulted in a13-18% yield increase in fresh market tomatoes, cantaloupes, honeydews, broccoli, strawberries and sweet corn (Goorahoo et al., 2008). Similar results have been obtained by a research group at Queensland University in Australia (Bhattarai, et al., 2004, 2005 & 2006), where the technology has been called "Oxygation". Our work on organic farming systems indicated that AirJection® Irrigation also positively affected photosynthetic and soil respiration rates, stomatal conductance, leaf scale water use efficiency, plant tissue nitrate concentrations, and shoot and root biomass (Reddy, 2008).

Objective

In our ongoing research, we are evaluating the impact of AirJection® Irrigation on yield and soil salinity for tomatoes grown on salt affected heavy clay soils. The specific objectives are to: determine the impact of AirJection® Irrigation on yield and Brix level of fresh-market tomatoes grown on a salt affected heavy clay soils; and, evaluate the impact of Air-jection® Irrigation on the spatial and temporal variability of salinity levels as measured by the apparent electrical conductivity of the soil. Concurrently, we are also attempting to evaluate the long term impacts of Air-jection® Irrigation on the component so the nitrogen (N) cycle for soils used for vegetable crops. Hence, the objective of the current study was to quantify the relative proportion of a series of genes known to be involved in the N cycle for soils collected from non-aerated fields and from those subjected to Air-jection® Irrigation for at least five years.

Materials and Methods

The study was located at a commercial vegetable grower in Mendota, California USA, a Panoche clay soil. A replicated (four times) soil sampling protocol was implemented in 2015 in which soils were collected within the 0-6 and 6-12 inches depths in adjacent fields that were non aerated (water only) and those subjected to Aijection Irrigation (Table 1). Samples were collected at distances of approximately 1/4 (Head), 1/2 (Middle) and 3/4 (Tail) from the irrigation inlet along the distance of the drip tape run length.

Table 1. Summary of treatments used in evaluation of soil DNA series

Irrigation type	Distance	Depth	Treatment
Water	H	6in	WH6in
		12in	WH12in
	M	6in	WM6in
		12in	WM12in
	T	6in	WT6in
		12in	WT12in
AirJection	H	6in	AH6in
		12in	AH12in
	M	6in	AM6in
		12in	AM12in
	T	6in	AT6in
		12in	AT12in

Nitrogen cycling genes were selected to describe the entire N cycle from nitrogen fixation (nifH) to nitrification (ammonia oxidation), and denitrification (Nitrate, nitrite and nitrous oxide reduction) (Table 2). Bacterial quantification was carried out using protocols employing the primers described by Hilty et al. (2010). DNA was extracted using the PowerSoil™ extraction kit (MoBio Laboratories, Carlsbad, CA) according to the manufacturers protocol.

Table 2. Tested genes

Role	Bacteria	Archaea
Nitrogen	*nifH* (nitrogenase reductase)	-
Ammonia oxidation	*amoA* (ammonia monooxygenase)	-
	-	Arch *amoA* (archaea
	-	Cren *amoA* (Crenarchota
Nitrate reduction	*narG* (Proteobacterial Membrane-	-
	napA (Proteobacterial Periplasmic	-
Nitrite reduction	*nirK* - Denitrifying nitrite reductase	-
	nirS - Denitrifying nitrite reductase	-
Nitrous oxide	*nosZ* - nitrous oxide reductase 1	-
Total	*16S rDNA region*	

A pooled sample of DNA extracted from all 48 soil samples were made for primer pair thermal gradient optimization tests. Dilutions (undiluted, 1/10, 1/20, 1/40, 1/100) of the pooled sample were used as template to assess the optimal annealing temperature of each of the primer pairs. The assay recipe and protocol for use with BioRad QX200™ ddPCR™ EvaGreen Supermix was in accordance to manufacturers instruction and was modified by inserting a thermal gradient between 52 °C and 64 °C for 1 min extension time in place of the annealing step. This insured that the correct dilution (which turned out to be 1/10 for the 16S primers and no dilution for all others) as well as the correct annealing temperature were used for each primer pair to achieve optimum results.

Data were tested individually for each gene, employing a standard t-test to assess the differences of the means; equal or non-equal variance was considered as appropriate. Principal components analysis (PCoA) was employed to visualize the trends within the entire dataset.

Results

Abundance of the tested genes was normalized to the abundance of the bacterial or archaeal indicator ribosomal genes (16S rDNA). Thus results represent the possible intensity of the respective function among the bacterial or archaeal populations but not necessarily the absolute counts of each gene per unit soil mass or volume. A principal component analysis was carried out (Figure 1) in which the proportional gene quantities were considered as independent variables with the treatment as dependent variables. This analysis allowed for an evaluation of the relationship between the independent variables and also their relationship with the treatments. There was a clear separation between the AirJection and the control treatments. This was obviously associated with a decrease in the proportional counts for three denitrification genes describing the entire denitrification pathway. Generally, AirJection minimizes the count of genes known to effect denitrification along the entire denitrification sequence (i.e. inhibits reduction of nitrate, nitrite and nitrous oxide = likely less NOx gaseous losses = likely increased nitrate-N availability to plants)

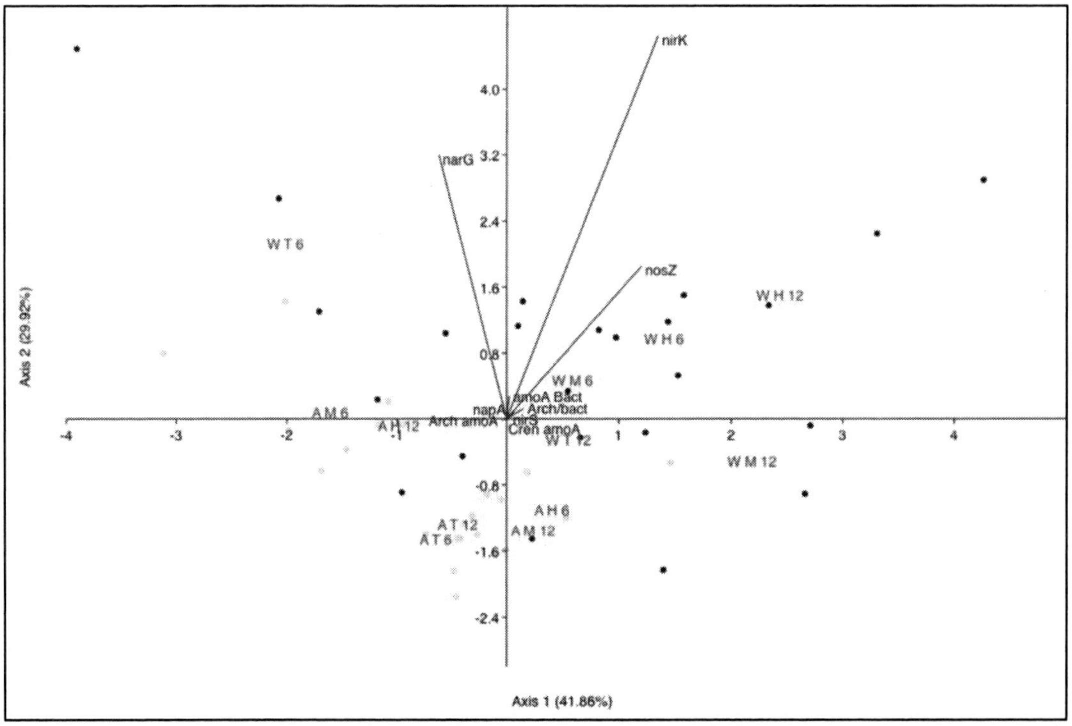

Figure 1. Principal component analysis. Blue dots = AirJection (A) treatment samples (A); Black dots = water (W) only treated samples (control); H, M and T = location along the delivery irrigation line; 6 and 12 = sample depth in inches; Red labels located at the centroid of the samples describing the respective treatment. The graph describes about 72% of the total variability. Lines indicate the direction and the discriminant capacity of the tested genes (all genes were normalized in units of gene per bacterial count).

N fixation: The *t-test* analyses indicated that the means for the AirJection and control data were statistically significantly different (pH_0=0.07), this was mainly due to a couple of extreme data points in treatment W-H 12in (Figure 2). Thus it may be stated that there is not sufficient evidence to indicate a decline in nitrogen fixation due to AirJection, however, the trend justifies future testing.

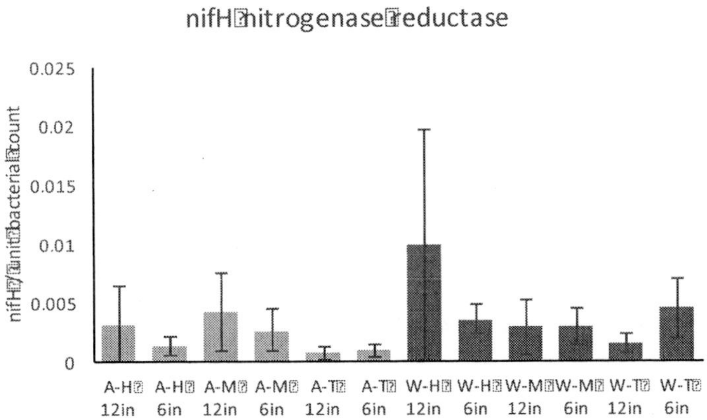

Figure 2. Distribution of nifH proportional gene counts across treatments. Error bars describe the statistical 95% Confidence Interval

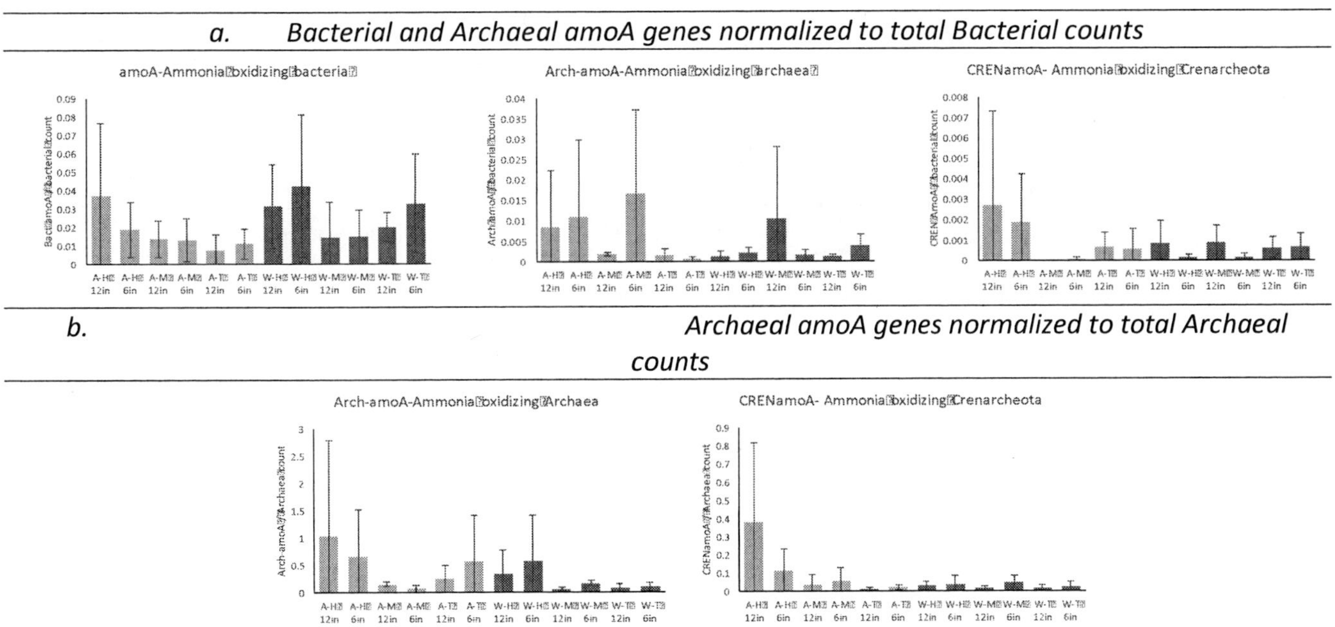

Figure 3. Ammonia monooxygenase distribution; top row describes its intensity (*amoA*) within or against (*Arch-amoA* and *CRENamoA*) bacterial population; the second row describes its intensity within (*Arch-amoA*) or against (*CRENamoA*) Archaeal population. Error bars describe the statistical 95% Confidence Interval.

Nitrification: Ammonia oxidation potential was tested for both Bacteria and Archaea (Figure 3). Bacterial nitrification activity was similar between AirJection and control (pH_0=0.18). Archaeal denitrification per unit bacteria was also not distinct across treatments (pH_0=0.27 for *Arch-amoA*;

pH_0=0.36 for *CRENamoA*). It should be noted that this comparison integrates both changes in archaeal counts and changes in archaeal associated nitrification and therefore reflects the contribution of archaeal nitrification to the entire microbial population (Figure 3a). A verification of the changes in nitrification among the archaeal population only (Figure 3b) while might show a trend for more activity in the AirJection tests, such trend is not statistically significant (pH_0=0.22 for *Arch-amoA*; pH_0=0.1 for *CRENamoA*). Our results indicate that aeration of the irrigation water was not sufficient to minimize the density of nitrification genes in the population. Nevertheless, these tests did not verify the actual gene expression which would more accurately describe the role of bacteria and archaea in nitrification under the treatment.

<u>Denitrification</u>: The genes density for this N gaseous loss, whose expression is favored under anaerobic conditions, were the most clearly affected by AirJection treatment (Figure 4).

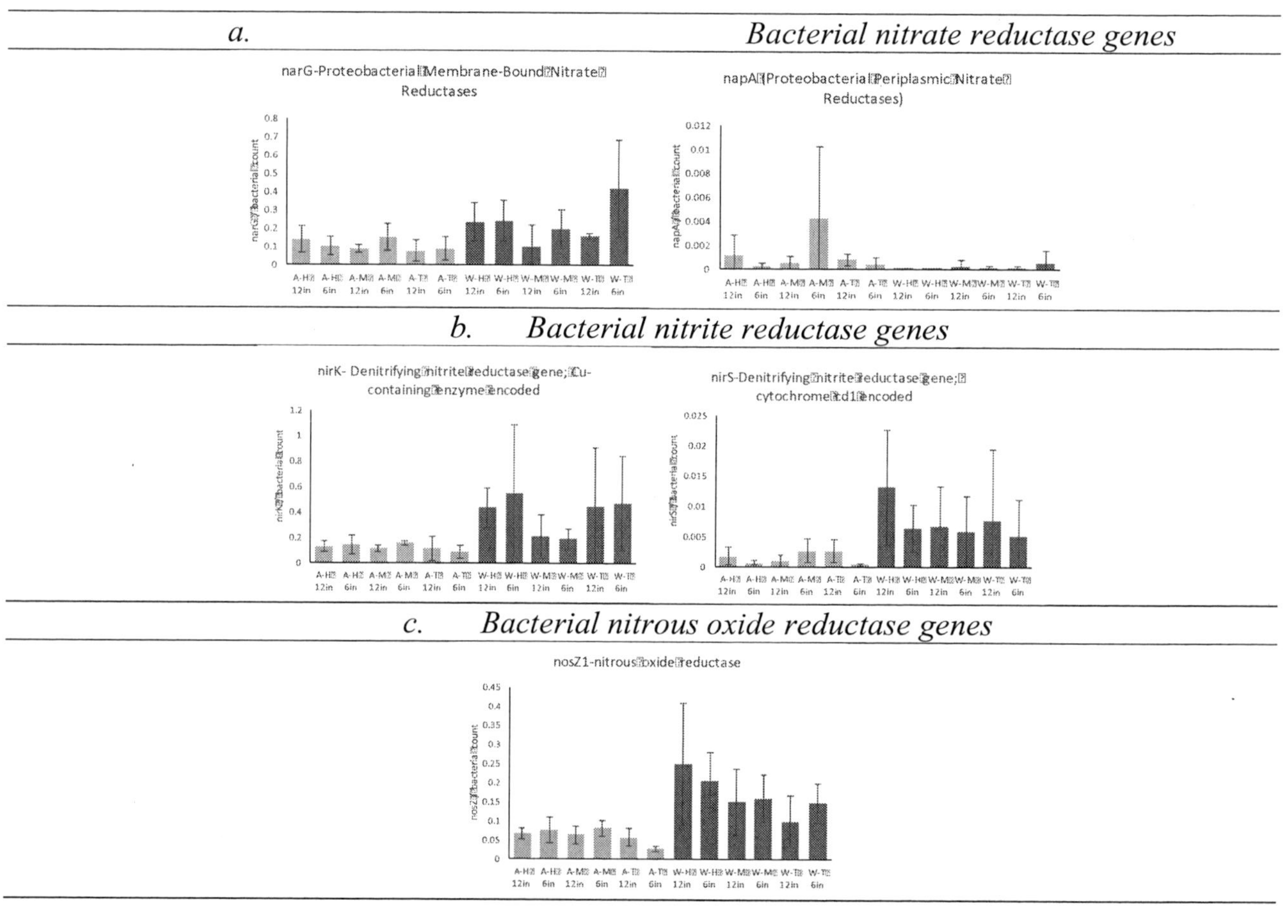

Figure 4. Distribution of genes involved in the denitrification pathway; nitrate reductases (*narG* and *napA*), nitrite reductases (*nirk* and *nirS*), and nitrous oxide reductase (n*osZ*). Error bars describe the statistical 95% Confidence Interval.

Nitrate reductase participate in the reduction of nitrate (NO_3) to nitrite (NO_2). Of the two tested genes one, narG, was depressed in the AirJection treatment a trend statistically significantly different from control (pH_0=0.002). The second relevant gene, napA, was enhanced by the treatment but not this was statistically not significant (pH_0=0.08) (Figure 4a). It is considered that narG is active at high nitrate concentration while napA may more actively reduce nitrate when the nitrate concentration is reduced (Stewart et al., 2002). Thus in our control experiment the larger narG suggest active denitrification linked to likely large nitrate availability and more anaerobic conditions. The depressed napA activity is likely linked to the large nitrate concentration and high narG activity. A reduction of narG activity under aerobiosis (i.e.AirJection) might occasionally enhance napA activity to the detriment of narG. Nevertheless the consistent decrease of narG gene copies with AirJection suggest continuous selective pressure that limits nitrate reduction potential within the microbial population.

Nitrate reductase participate in the reduction of nitrite (NO_3) to nitrous oxide (N_2O). For both genes tested here there was a significant decrease in copy number in the bacterial population ($pH_0=0.001$ for nirK and $pH_0<0.001$ for nirS). The results are self-evident; aerobic conditions associated with AirJection depleted these genes (Figure 4b) from the microbial population clearly decreasing the capacity of these microbes to reduce nitrite. This is likely a cascade effect whereby lower nitrate reduction potential (see narG above) produces less nitrate that may be available for further oxygen loss.

Concluding Remarks

- The relative quantity for a series of genes known to be involved in nitrogen cycle was estimated for soils collected from non-aerated fields and those subjected to Airjection Irrigation for at least five years.
- The ratio between total archaea and total bacteria were estimated. Archaea have been shown to be active in matter cycling in soils and to be more resilient than bacteria. On the other hand, bacteria dominance indicates a likely shift to more luxurious growth conditions.
- Nitrogen fixation potential was evaluated using the most commonly known relevant gene, *nifH*. Ammonification genes (ammonia monooxygenases), related to the rates of mineralization of organic matter, were tested for both Bacteria (*amoA*) and Archaea (Arch-*amoA,* CREN-*amoA*).
- Denitrification potential, common in anaerobic conditions and a process directly linked to gaseous losses of nitrogen (as NOx's), was verified through the quantification of a number of genes related to various metabolic pathways known to be of relevance. Thus nitrate reductase genes (narG, napA), nitrite reductase genes (nirS, nirK), and nitrous oxide reductase (nosZ) were evaluated.
- The distribution of the tested genes within the microbial populations was very distinct among the two treatments, aerated and non-aerated. This indicates that AirJection had a clear selective impact on the distribution of the tested genes among the population. It may thus be hypothesized that total diversity also changed.
- Generally, AirJection Irrigation led to a proportional increase of Bacteria versus Archaea. While the AirJection Irrigation did not have a significant impact on nitrogen fixation or ammonia oxidation. the practice of adding aerated water via the buried drip line did have a significant impact on denitrification genes suggesting lower NOx production potential and thus likely increased availability of nitrate in the root zone. This might be hypothesized to enhance nitrogen use efficiency potential with AirJection, and with the judicious water management within the root zone, plant nitrate uptake can be enhanced with a potential reduction in nitrate leaching.

References

Bhattarai, S.P., S. Huber, and D.J. Midmore. 2004. Aerated subsurface irrigation water gives growth and yield benefits to zucchini, vegetable soybean and cotton in heavy clay soils. Ann. Appl. Biol. 144: 285-298.

Bhattarai, S.P., L. Pendergast, D. J. Midmore. 2006 Root aeration improves yield and water use efficiency of tomato in heavy clay and saline soils. Scientia Horticulturae. 108: 278-288.

Bhattarai, S.P., N. Su, and D.J. Midmore. 2005. Oxygation unlocks yield potentials of crops in oxygen limited soil environments. Advances in Agronomy. 88: 313-377.

Durell, W. D. (1941). The effect of aeration on growth of the tomato in nutrient solution. *Plant physiology*, *16*(2), 327.

Goorahoo D., G. Carstensen and A. Mazzei. 2001. A Pilot Study on the Impact of Air Injected into Water Delivered through Subsurface Drip Irrigation Tape on the Growth and Yield of Bell Peppers. California Agricultural Technology Institute (CATI) Publication # 010201.

Goorahoo D., D. Adhikari, D. Zoldoske, F. Cassel S., A. Mazzei, and R. Fanucchi. 2008. Potential for AirJection® Irrigation in Strawberry Production. pp 152-155 In : Takeda, F., D.T. Handley, and E.B. Poling (ed.). Proc. 2007 N. American Strawberry Symposium. North American Strawberry Growers Association, Kemptville, ON Canada.

Reddy, N. 2008. Optimizing Nitrogen Rates for Organic Vegetables Subjected to Airjection® Irrigation. Thesis submitted in partial fulfillment of Master's degree in the Plant Science Department at California State University-Fresno. May 2008.

This page intentionally left blank.

A sowing method for SDI to increase emergence rate in corn

Yan Mo, PhD in reading, College of Water Resources and Civil Engineering, China Agricultural University, Beijing 100083, China, moyansdi@cau.edu.cn

Guangyong Li, Professor, College of Water Resources and Civil Engineering, China Agricultural University, Beijing 100083, China, lgycau@163.com

Dan Wang, PhD in reading, College of Water Resources and Civil Engineering, China Agricultural University, Beijing 100083, China, wangdan_9090@cau.edu.cn

Mingkun Cai, Master in reading, College of Water Resources and Civil Engineering, China Agricultural University, Beijing 100083, China, cmk1948@163.com

As a result of driplines being buried below the plow layer, it is hard for crop germination by using subsurface drip irrigation. This study were conducted from 2015 to 2016, including two sowing methods, namely alternate row/ bed planting (AP) with a 10 cm deep trapezoidal furrow; seeds were then sown in 5 cm deep soil below the furrow bottom and flat planting (FP), at two dripline burial depths (30 and 35 cm) with pre-emergence irrigation amounts from 15 mm to 75 mm. The following results were obtained: AP significantly increased the 5 cm soil depth moisture content below the seeds. The emergence rate, yield, water use efficiency and nitrogen partial factor productivity under AP increased by an average of 24.0%, 10.0%, 8.1% and 9.6%, respectively, compared with FP. Overall, AP for subsurface drip irrigation can considerably promote spring corn germination.
alternate row/ bed planting, flat planting, soil moisture content, emergence rate, yield, water use efficiency

1. Introduction

The corn belt in Northeast China is one of the country's corn commodity production zones, with annual corn output of more than 42 million tons and corn-planted acreage covering more than 5.1 million ha (accounting for 70% of the total production of grain crops) (Ma *et al.*, 2008). Under global warming conditions, the local limited rainfall from late April to mid-June cannot support the germination and seedling growth of spring corn, thus adversely affecting the germination and yield of this crop (Li *et al.*, 2010).

Subsurface drip irrigation is currently the most advanced water-saving irrigation method. Compared with other irrigation methods, subsurface drip irrigation can maintain and even increase the yield of more than 30 types of crops, including corn, alfalfa, cotton, tomato, sweet corn, etc., by requiring less water in most cases (Adamsen, 1992; Alam *et al.*, 2002; Bar-Yosef *et al.*, 1989; Camp *et al.*, 1989; Phene *et al.*, 1987; Plaut *et al.*, 1985; Wood and Finger, 2006). Considering the long-term use of the subsurface drip irrigation system, the dripline must be buried below the plow layer (Camp and Lamm, 2003). In a silt-loam experimental cornfield at Kansas State University, most of the driplines were buried at a depth of 40-45 cm, thereby remaining constantly in dry soil surface and avoiding moisture evaporation and weed growth (Lamm *et al.*, 1997; Lamm and Trooien, 2003). However, the low soil moisture content of topsoil in the 0-10 cm layer results in germination difficulty because of gravity (Lamm and Trooien, 2005), particularly in the severe spring droughts of arid and semiarid regions. Germination using subsurface drip irrigation is primarily affected by the distance between the seed and the dripline, which is closely related to the depth at which the dripline is buried (Charlesworth and Muirhead, 2003; Pablo *et al.*, 2007; Patel and Rajput, 2007). The relationship between the dripline depth and germination has been a matter of great concern among scholars around the world in recent years.

To ensure the uniformity of the emergence rate for different dripline depths, a large amount of water could be used to wet the soil around the seed (Bordovsky and Porter, 2003; Henggeler, 1995; Howell *et al.*, 1997). During irrigation, a low limit of the soil matrix potential in the 20 cm soil layer was maintained at the same level. The emergence rates of potato with dripline depths of 10-50 cm reached 100%. The deeper the dripline is buried, the larger the quantity of irrigation that is needed, which will cause a slower increase in ground temperature. Lower temperatures result in a delay in germination (Liu *et al.*, 2015). Excessive irrigation may also cause deep percolation, which affects the groundwater environment and results in soil compaction, thus affecting ventilation and leading to crop yield reductions (Colaizzi *et al.*, 2004). A number of scholars have agreed that uniformity of the emergence rate can be maintained through not allowing irrigation during seed germination (Lamm *et al.*, 2010; Lamm and Trooien, 2005) or transplanting during seedling stage (Leskovar *et al.*, 2001; Machado *et al.*, 2003). No significant effects were observed in the yield or water use efficiency of sunflower, soybean, sorghum (Lamm *et al.*, 2010), corn (Lamm and Trooien, 2005), tomato or melon (Leskovar *et al.*, 2001; Machado *et al.*, 2003).

The low emergence rate caused by inadequate irrigation may appreciably affect the yield and WUE_{ETc}. The emergence rate, yield and WUE_{ETc} of corn was greater under a dripline depth of 15 cm than under burial depths of 20-30 cm, with only the surface of the 15 cm treatment wetted during the pre-emergence irrigation (Pablo *et al.*, 2007). The dripline depth should similarly be no greater than 20 cm for tomato or seed germination, yield and WUE_{ETc} might be affected (Marouelli and Silva, 2002; Schwankl *et al.*, 1990).

In California, less than 10% of farmers adopted subsurface drip irrigation for crop establishment, with dripline depths of no more than 10 cm (Burt and Styles, 1999). Other farmers used sprinkler irrigation systems to guarantee germination. To ensure the emergence rates of the Hami melon and broccoli, the pre-emergence water amount using subsurface drip irrigation was increased by 185 and 230 mm compared with the sprinkler irrigation (Roberts *et al.*, 2008). When the pre-emergence irrigation amount was same, the emergence rate of turf grass with sprinkler irrigation increased by 25.6% compared with subsurface drip irrigation (Schiavon *et al.*, 2015). Guaranteeing the germination rate with sprinkler irrigation costs an additional US$ 400-800 ha^{-1} $crop^{-1}$, and the return on field crops, such as corn and cotton, is extremely low (Lamm *et al.*, 2012). Several scholars recommend installing and recording subsurface driplines using Real-Time Kinematic-Global Positioning System-guided tractors, thus achieving shallow burial of drip irrigation pipes without damage from farm machinery (Bordovsky, 2006; Heidman *et al.*, 2003; Lamm *et al.*, 2012). However, for most Chinese farmers, equipment costs are exceedingly high (Ji and Zhou, 2014; Li and Lin, 2006). In addition, some researchers propose placing subsurface driplines above a V-shaped impermeable material to improve the wetted width, to decrease the deep percolation and, finally, to solve the problem of germination with subsurface drip irrigation (Barth, 1999; Welsh *et al.*, 1995). While the effect was not obvious, the corresponding cost was higher, and the process involved more difficult construction (Brown *et al.*, 1996; Charlesworth and Muirhead, 2003).

When soil tillage is used, the dripline must be put below the plow layer. The deeper the dripline is buried, the harder the emergence. There is no cheap and convenient method that can guarantee the crop emergence rate with little water. The objective of this article was to propose a new subsurface drip irrigation sowing method called alternate row/ bed planting and by comparing the emergence rate, yield, water use efficiency and nitrogen partial factor productivity of spring corn under the same irrigation and fertilizer amount, to develop a proper sowing method for subsurface drip irrigation.

2. Materials and methods

2.1 Experimental site

The experimental plots are located in Chifeng City, in Eastern Inner Mongolia (42°57′ N, 119°19′ E, altitude 625 m), China. This location has a semiarid continental monsoon climate with a mean annual temperature of 11 °C and a mean annual precipitation of 343 mm (primarily from June to August). The effective precipitation for spring corn during the growth stage in 2015 was 180 mm, and no effective precipitation was observed during the beginning of May to the middle of June while the effective precipitation in 2016 was 23.9 mm. The soil texture of the 0-40 cm layer of the experimental plots was classified as a sandy loam, whereas the 40-60 cm section was classified as loam. The mean dry bulk density was 1.51 g/ cm^3, and the mean field volume capacity was 32.06%. The mean soil organic matter was 6.74 g/ kg, and the contents of total nitrogen, nitrate nitrogen, ammonium nitrogen, available potassium, and available phosphorus were 0.41 g/ kg, 94.15 mg/ kg, 24.92 mg/ kg, 289.7 mg/ kg and 18.3 mg/ kg, respectively.

2.2 Experimental design

The tested plants——"Xianyu 335" spring corn, which were sown on May 8, 2015 and May 5, 2016, germinated on May 30, 2015 and May 25, 2016, and were harvested on October 5, 2015. Wide/ narrow planting rows were 67 cm × 53 cm, with row spacing of 20 cm and a planting density of 82,500 plants/ ha.

Considering the long-term use of subsurface drip irrigation system, the locally popular rotary tillage depths (20-25 cm) and subsurface dripline should be buried as shallow as possible (Camp, 1998). The experimental plot had two different dripline depths (30 (D30) and 35 cm (D35)) and two sowing methods, namely alternate row/ bed planting (AP) and flat planting (FP). This experiment had four treatments: (1) alternate row/ bed planting with a dripline depth of 30 cm (APD30) (see Fig. 1a); (2) alternate row/ bed planting with a dripline depth of 35 cm (APD35) (see Fig. 1a); (3) flat planting with a dripline depth of 30 cm (FPD30) (see Fig. 1b); and (4) flat planting with a dripline depth of 35 cm (FPD35) (see Fig. 1b). A total of 12 plots (3 replicates for each treatment) were randomly arranged.

As shown in Fig. 1a, AP refers to using a plough ahead of the seeding nozzle to make a trapezoidal furrow (east-west) (as shown in the red line) before sowing, then sowing the seeds in 5 cm deep soil below the furrow bottom. Four rows were sown once. The top edge of the trapezoidal trench was set at 30 cm; the bottom edge was set at 6 cm; the vertical distance between the top and the bottom edge was set at 21cm (see Fig. 1a). As a result of a furrow bottom depth of 10 cm relative to the original soil surface (i.e., the soil surface of the FP), the distance between the seed and the dripline was small: D_{APD30} (31 cm) <

D_{APD35} (33 cm) < D_{FPD30} (36 cm) < D_{FPD35} (40 cm). The image of AP is shown in Fig. 2. FP referred to the conventional method of sowing a seed in 5 cm deep soil on a flat field.

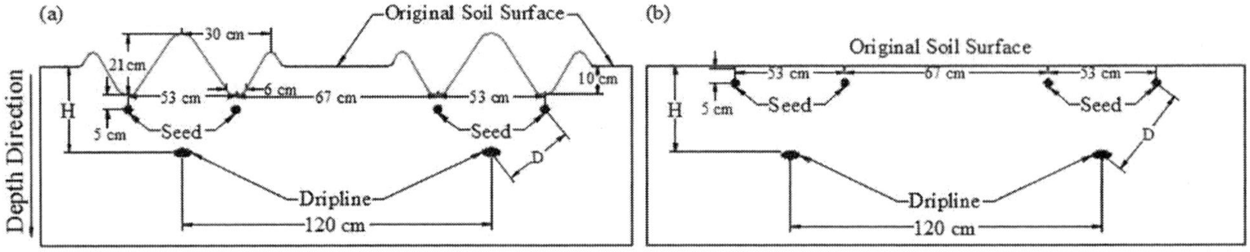

Fig. 1 Sectional drawing of alternate row/ bed planting (AP) (a) and flat planting (FP) (b) with a dripline depth of 30 and 35 cm H, dripline depth: 30 and 35 cm; D, distance between seed and dripline.

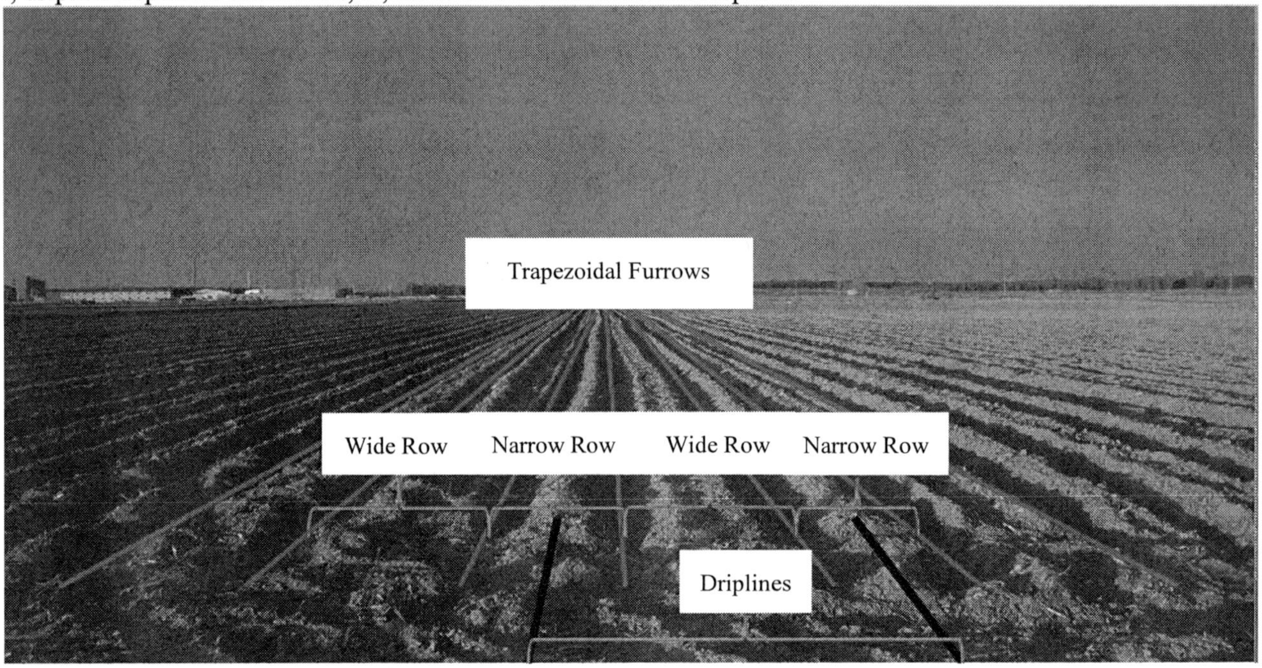

Fig. 2 Image of alternate row/ bed planting

As shown in Table 1, the pre-emergence irrigation amount in 2015 for all treatments was 25 mm. In 2016, the irrigation amount was increased by three levels for D30 and D35. This effect on corn emergence by alternate row/ bed planting will be further investigated to determine the appropriate pre-emergence irrigation amount. For all experiments, to prevent the influence of rainfall, the area was covered with tarpaulins during precipitation until the rain stopped.

Table 1. Pre-emergence irrigation amount of Year 2015 and 2016 for Spring Corn

<table>
<tr><th rowspan="2">Treatments</th><th colspan="2">Pre-emergence irrigation amount (mm)</th></tr>
<tr><th>Year 2015</th><th>Year 2016</th></tr>
<tr><td rowspan="2">APD30</td><td rowspan="8">25</td><td>15</td></tr>
<tr><td>25</td></tr>
<tr><td rowspan="2">FPD30</td><td>45</td></tr>
<tr><td>60</td></tr>
<tr><td rowspan="2">APD35</td><td>25</td></tr>
<tr><td>45</td></tr>
<tr><td rowspan="2">FPD35</td><td>60</td></tr>
<tr><td>75</td></tr>
</table>

Each plot of 8.0 m × 50 m included six driplines with spacing of 1.2 m. The driplines were parallel to the corn planting ridges and in the middle of the narrow row. The driplines were provided by NATEFIM, with a wall thickness of 0.38 mm, a diameter of 16 mm, emitter spacing of 30 cm, and emitter rated flow of 1.05 L/ h. A complete set of pressure gauges, water meters, and valves were separately installed along the edge of a field to monitor the amount of water applied to the field and the flow rate of the subsurface drip irrigation system.

2.3 Irrigation and fertilization

After emergence, when the measured volumetric water content was between 70% and 75% of the field capacity, irrigation was necessary. The amount of water applied to the field was calculated by formula (1):

$$I' = 85\% \mathrm{ET}' - P' \qquad (1)$$

where I' is irrigation amount (mm); ET' is evapotranspiration (mm), which can be calculated by formula (2); P' is effective precipitation (mm).

$$\mathrm{ET}' = K_c \times ET_0 \qquad (2)$$

where ET_0 is reference evapotranspiration (mm/ d) calculated by the Penman-Monteith formula; and K_c is crop coefficient. Based on the results from several researchers in the same experimental plots, the corn planting process usually has a seedling stage $K_c = 0.7$, a jointing stage $K_c = 1.0$, a tasseling stage $K_c = 1.2$, a filling stage $K_c = 0.9$, and a milk stage $K_c = 0.5$ (Mi, 2013; Xu, 2014; Yuan, 2015).

The amount of nitrogen applied in the four treatments was identical (290 kg/ha; 20% applied as the base fertilizer and 80% applied with irrigation water). Self-priming pumps ($H = 38$ m, $Q = 3$ m^3/ h) and the pattern of "1/4 W-1/2 N-1/4 W" were used to ensure uniform fertilization, which consisted of clean water irrigation for 1/4 of the duration, fertilizer application for 1/2 of the duration, and flushing the pipework with clean water for the remaining 1/4 duration (Li *et al.*, 2003). The details regarding the irrigation and fertilization systems in 2015 are provided in Table 2. The water applied on July 24 and

August 29 was minimal as it was used to fertilize. In 2016, the fertilization schedule was identical to that in 2015.

Table 2. Irrigation and Fertilization Schedule for Spring Corn of Year 2015

Sequence of Irrigation and Fertilization	Date	Irrigation Amount (mm)	Amount of Nitrogen Applied (kg ha^{-1})
Seed fertilizer			58
Pre-emergence irrigation	5/12	25	
1	7/9	25	58
2	7/24	7	72.5
3	8/7	45	43.5
4	8/18	10	
5	8/29	5	43.5
6	9/4	15	
7	9/11	30	14.5
Total		162	290

2.4 Monitoring indicators

2.4.1 Weather and soil moisture content

There was an automatic weather station (ET107; produced by Campbell Scientific, America) which was 100 meters from the experiment plot that could monitor and acquire the temperature, wind speed, wind direction, relative humidity, radiation, and other meteorological data within a time interval of one hour, continuously.

Three sets of EnviroSMART moisture sensors (produced by Sentek, Australia) were set between the wide row, narrow row and two corn plants, called A, B and C, respectively, and recorded once every one hour. As shown in Fig. 3, each of the sensors had four probes, numbered from 1# to 4#. Each probe recorded the volumetric water content (VWC) from a soil volume outside the access tube, which has a sphere of influence of 10 cm vertical height and 10 cm radial distance from the outer wall of the access tube and the precision of the sensor was 0.01cm^3/ cm^3. As shown in Fig. 3(a), the sensor B for AP was installed vertically downward at the bottom of the trapezoidal furrow, such that probes 1-4# could monitor the VWC at depths of 10, 20, 40, and 60 cm vertically below the furrow bottom. As shown in Fig. 3(b), the sensors for FP were installed with the flat field as a benchmark, such that probes 1-4# could monitor the VWC at depths of 10, 20, 40, and 60 cm vertically below the flat field (i.e., the original soil surface). Both sensors B for AP and FP could monitor the VWC at depths of 5, 15, 35, and 55 cm vertically below the seed. When calculating the seasonal change in soil water storage at 0-60 cm deep soil and then calculating the seasonal corn evapotranspiration, the average values from sensors A, B and C were used. When analyzing the soil moisture variation below the seed or corn, the values from sensor B were used. Moreover, augers were also used to measure the soil moisture content to correct the data gathered by the EnviroSMART moisture sensors.

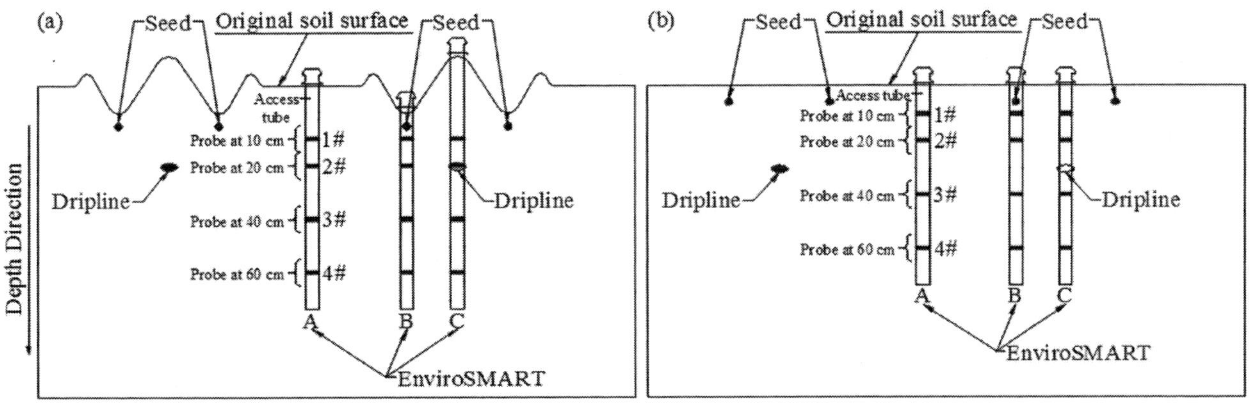

Fig. 3 EnviroSMART moisture sensor installations of alternate row/ bed planting (AP) (a) and flat planting (FP) (b)

2.4.2 Corn growth indexes at the seeding stage

The germination rate is the main indicator of the soil moisture content after sowing, and it is defined as the ratio between the emergence number and the sown seed number. The emergence number is determined under the condition in which corn plants have 2 cm of topsoil. Ten days after sowing (May 18, 2015 and May 15, 2016), the observation of the germination rate begins, and it ends after 2 weeks.

Five typical corn plants at the seedling stage were sampled on June 7, 2015 and June 5, 2016 to measure the plant height, stem diameter, and the width and length of all expanding leaves. Next, the leaf area index (LAI) was ascertained by using a correction coefficient of 0.75.

2.4.3 Seed and yield test

In year 2015, for the corn test, we eliminated the two rows outside of the plot and picked all of the plants in the mid-10 m of the four rows at the middle of every plot to measure their ear length, ear diameter, bare top length, kernels per ear, hundred-grain weight (after air-drying), ear weight, and yield (converted into the moisture content using 14% of the standard mass).

2.4.4 The index calculation and statistical methods

The seasonal crop evapotranspiration (ET_c) (mm) is calculated using the water balance equation (3), and the water use efficiency of corn (WUE_{ETc}) (kg ha^{-1} mm^{-1}) is calculated as follows:

$$\mathrm{ET_c} = P + I + U - R - \Delta S \tag{3}$$

$$\mathrm{WUE_{ETc}} = Y / \mathrm{ET_c} \tag{4}$$

where P is the seasonal effective precipitation (mm); I is the seasonal irrigation (mm); U is the seasonal upward capillary flow into the root zone (mm) (The capillary rise was negligible since the groundwater table was 20-30 m below the soil surface); R is the seasonal runoff (mm) (Runoff was never observed in the field); ΔS is the seasonal change in soil water storage at 0-60 cm (mm); and Y is the corn yield (kg ha^{-1}).

The nitrogen partial factor productivity (NPFP) (kg kg^{-1}) is calculated as follows:

$$\mathrm{NPFP} = Y / N \tag{5}$$

where N is the seasonal nitrogen applied amount (kg ha^{-1}).

All of the experimental data were statistically analyzed using SPSS17.0 and Microsoft Excel.

3. Results

3.1 Soil volumetric moisture content

3.1.1 Soil moisture change after sowing and after pre-emergence irrigation

In both year, after sowing, the AP soil moisture at a depth of 5 cm vertically below seed increased by 15.0%, on average, compared to FP. In 2015, forty-eight hours after a pre-emergence irrigation of 25 mm, only the soil moisture content of APD30 at 5 cm below seed increased by 10.9% while the soil moisture contents of the remaining three treatments decreased. Similarly, in 2016, only the soil moisture content of APD30 changed dramatically after pre-emergence irrigation of 15 mm and 25 mm. For the other three pre-emergence irrigation treatments, the AP soil moisture was larger than FP. (Table 3 and Table 4).

Table 3. Soil moisture content (Vol. %) 5 cm vertically below the seeds after sowing and after pre-emergence irrigation (year 2015)

Treatments	pre-emergence irrigation amount (mm)	After sowing	48 hours after pre-emergence irrigation
APD30	25	16.24	18.01
FPD30		14.47	14.39
APD35		16.82	15.98
FPD35		13.61	13.38

Table 4. Soil moisture content (Vol. %) 5 cm vertically below the seeds after sowing and after pre-emergence irrigation (year 2016)

Treatments	pre-emergence irrigation amount (mm)	After sowing	After pre-emergence irrigation	24 hours after pre-emergence irrigation
APD30	15	19.67	22.78	22.85
FPD30		17.80	14.44	20.70
APD30	25	19.67	29.96	25.01
FPD30		17.80	17.07	21.33
APD35		21.14	18.30	25.08
FPD35		18.62	15.25	18.47
APD30	45	19.67	34.22	30.12
FPD30		17.80	24.40	19.02
APD35		21.14	28.39	25.39
FPD35		18.62	26.86	22.48
APD30	60	19.67	33.33	32.94
FPD30		17.80	23.85	19.42
APD35		21.14	30.68	25.58
FPD35		18.62	27.27	22.89

APD35	75	21.14	32.05	23.22
FPD35		18.62	26.00	23.28

3.1.2 Other growth stages

The EnviroSMART moisture sensors were completely installed in early July and continuous recording of all soil volumetric moisture contents from July 5 to October 5 are shown in Fig. 4. The figures show that the moisture contents of soil at 10, 20, and 40 cm changed dramatically. The soil moisture content at 20 cm changed the most significantly, whereas the soil moisture content at 60 cm was relatively stable. Compared with topsoil (10 and 20 cm), the moisture content of deep soil (40 and 60 cm) increased by 54.2% on average.

In top soils of 10 and 20 cm, the mean soil moisture content of the two dripline depths under AP increased by 23.8% and 21.2%, respectively, compared with FP. Similarly, the mean soil moisture content of the two sowing methods under D30 increased by 13.8% and 13.9% compared with D35. For deep soils of 40 and 60 cm, the mean soil moisture content of the two dripline buried depths under FP increased by 9.1% and 0.6%, respectively, compared with AP. Similarly, the mean soil moisture content of the two sowing methods under D35 increased by 18.0% and 6.9% compared with D30.

After forty-eight hours of 30 mm irrigation on September 11, the soil moisture content in 10 cm soil under APD30 and APD35 increased by 8.9% and 6.0%, respectively, whereas FPD30 and FPD35 decreased by 3.5% and 0.2%, respectively. Forty-eight hours after 45 mm irrigation on August 7, the soil moisture content of 10 cm soil under APD30, APD35, and FPD30 increased by 4.1%, 5.5% and 1.4%, respectively, whereas FPD35 decreased by 3.7%.

Forty-eight hours after 29 mm precipitation on July 21, under a soil moisture content of 10 cm, AP and FP increased by 23.7% and 18.1%, respectively. Under a soil moisture content of 20 cm, AP and FP increased by 26.6% and 11.6%, respectively. Under a soil moisture content of 40 cm, AP and FP increased by 0.2% and 3.7%, respectively. Under a soil moisture content of 60 cm, AP decreased by 0.1% and FP increased by 1.5%, respectively. Forty-eight hours after 38 mm precipitation on July 29, for a soil moisture content of 10 cm, AP and FP increased by 40.3% and 24.0%, respectively. Under a soil moisture content of 20 cm, AP and FP increased by 28.1% and 19.5%, respectively. Under a soil moisture content of 40 cm, AP and FP increased by 3.7% and 10.8%, respectively. And under a soil moisture content of 60 cm, AP and FP increased by 0.9% and 3.5%, respectively.

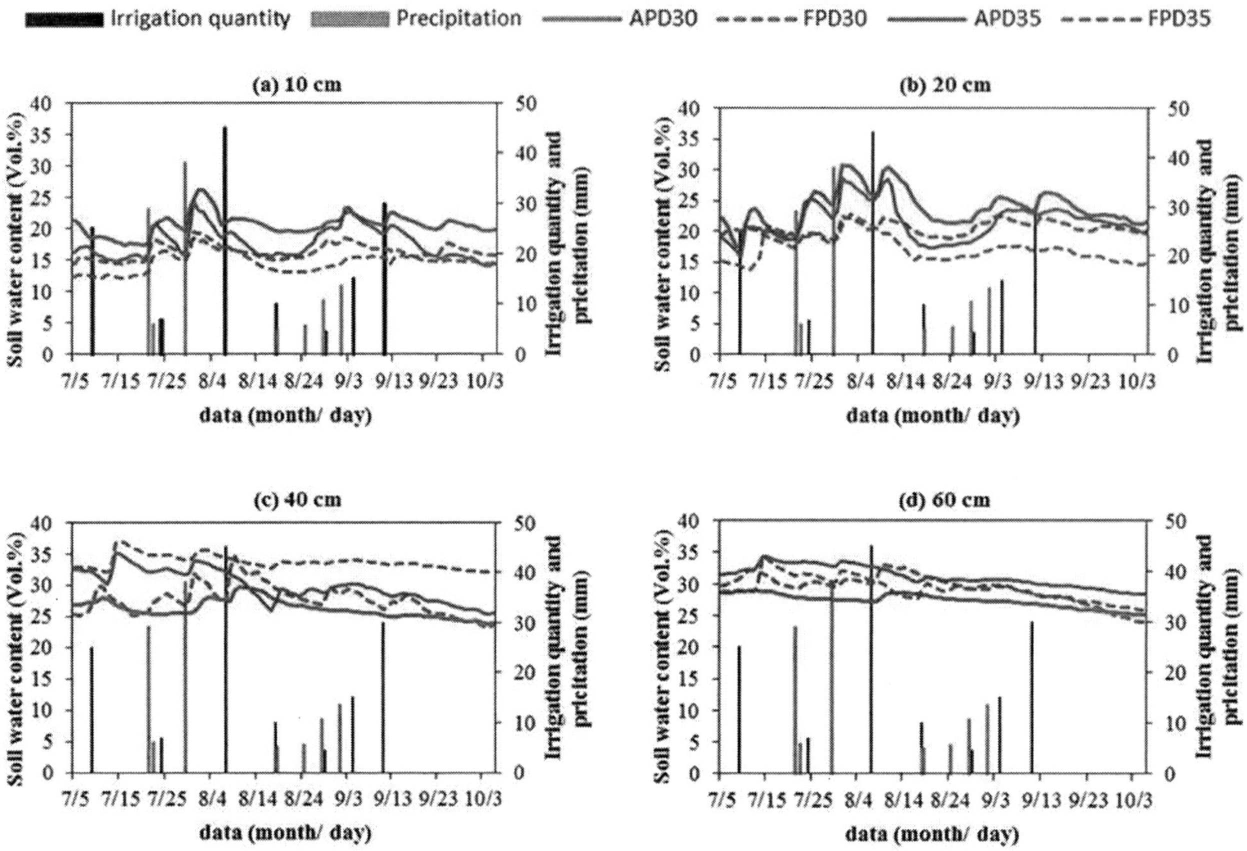

Fig. 4 Changes in the soil moisture content (Vol. %) of the four treatments at 10 (a), 20 (b), 40 (c), and 60 cm (d) from July 5 to October 5

APD30, alternative row/ bed planting with dripline depth of 30 cm; APD35, alternative row/ bed planting with dripline depth of 35 cm; FPD30, flat planting with dripline depth of 30 cm; FPD30, flat planting with dripline depth of 35 cm.

3.2 Emergence rate, plant height, stem diameter and LAI at the seedling stage

As the pre-emergence irrigation amount increased, the plant height, stem diameter, and LAI at the seedling stage of AP were initially increased and decreased later, whereas FP continued to increase (Table 5). For APD30, when the irrigation amount was 25 mm, the emergence rate was greater than 90% and later reached 100% with the irrigation amount approaching 45 mm. For APD35, when the irrigation amount was 25 mm, the emergence rate was greater than 90% and reached a maximum of 93%. Furthermore, for FPD30 and FPD35, the emergence rate was always less than 80%. Compared with FP, the mean emergence rate of AP increased by 23.3%. The larger the irrigation amount, the smaller emergence rate difference between APD30 and APD35. Under the same irrigation amount, there was a small emergence rate difference between FPD30 and FPD35. For D30, an irrigation rate of 15 mm had a substantial effect on the emergence rate, compared with irrigation amounts of 25, 45, and 60 mm. However, there was no significant difference for these three irrigation amounts. For D35, on the other hand, the irrigation amount had no significant impact on the emergence rate. For other physiological indicators, the pre-emergence irrigation amount mostly reached significant levels.

Table 5. Influences of pre-emergence irrigation amount on the emergence rate, plant height, stem diameter and LAI at the seedling stage

Pre-emergence irrigation amount (mm)	Emergence rate (%)	Plant height (cm)	Stem diameter (cm)	LAI
APD30				
15	81a	36.3a	1.10a	0.19a
25-Year 2016 (Year 2015)	93b (91)	39.3ab (41.8)	1.21ab (1.20)	0.26b (0.28)
45	100b	42.6b	1.29bc	0.38c
60	95b	42.1b	1.39c	0.33c
Mean	92	40.1	1.25	0.29
LSD ($p < 0.05$)	**	**	**	**
FPD30				
15	64a	26.0a	0.80	0.08a
25-Year 2016 (Year 2015)	74b (79)	31.9b (30.8)	0.94 (0.93)	0.13b (0.15)
45	75b	33.6b	0.91	0.15b
60	76b	31.1b	0.97	0.14b
Mean	72	30.6	0.90	0.13
LSD ($p < 0.05$)	**	**	NS	**
APD35				
25-Year 2016 (Year 2015)	82 (81)	36.6a (38.5)	1.13 (1.10)	0.21a (0.22)
45	92	39.8ab	1.17	0.27ab
60	93	41.7b	1.24	0.32ab
75	90	38.1b	1.17	0.28b
Mean	89	39.0	1.18	0.27
LSD ($p < 0.05$)	NS	*	NS	*
FPD35				
25-Year 2016 (Year 2015)	73 (74)	32.3a (30.1)	0.92a (0.97)	0.14a (0.12)
45	75	33.0ab	0.99ab	0.15a
60	74	35.8bc	1.12b	0.20b
75	77	34.5c	1.16b	0.22b
Mean	75	33.9	1.05	0.18
LSD ($p < 0.05$)	NS	**	*	**

Table 6 shows a significance analysis of the sowing methods, dripline depths, and their interaction on corn seedling physiology indicators under the same pre-emergence irrigation amount in 2015 and 2016. Sowing methods had a significant effect on the corn seedling physiological indicators under different irrigation amounts. With an increase in the irrigation amount, the dripline depth had a substantial impact on the emergence rate and plant height. Furthermore, for an irrigation amount of 60 mm, interaction of the planting method and dripline depth had a considerable effect on all indicators.

Table 6. Influences of sowing methods and dripline depths on the emergence rate, plant height, stem diameter, and LAI at the seedling stage for different pre-emergence irrigation amount

Treatment	Emergence rate (%)	Plant height (cm)	Stem diameter (cm)	LAI
Year 2015 25mm				
F value for dripline depth	7.683	3.858	0.364	5.878
LSD ($p < 0.05$)	*	NS	NS	*
F value for sowing method	12.444	86.684	13.091	40.209
LSD ($p < 0.05$)	**	**	**	**
F value dripline depth × sowing method	1.016	1.592	1.455	0.557
LSD ($p < 0.05$)	NS	NS	NS	NS
Year 2016 25mm				
F value for dripline depth	3.138	1.246	1.460	2.881
LSD ($p < 0.05$)	NS	NS	NS	NS
F value for sowing method	17.085	29.888	37.549	82.881
LSD ($p < 0.05$)	**	**	**	**
F value dripline depth × sowing method	1.898	2.165	0.365	6.881
LSD ($p < 0.05$)	NS	NS	NS	*
Year 2016 45mm				
F value for dripline depth	2.921	2.413	0.222	10.116
LSD ($p < 0.05$)	NS	NS	NS	*
F value for sowing method	73.014	53.593	62.720	116.053
LSD ($p < 0.05$)	**	**	**	**
F value dripline depth × sowing method	2.472	1.087	9.102	11.463
LSD ($p < 0.05$)	NS	NS	*	**
Year 2016 60mm				
F value for dripline depth	12.676	18.184	0.011	3.267
LSD ($p < 0.05$)	**	**	NS	NS
F value for sowing method	69.014	271.233	74.136	129.067
LSD ($p < 0.05$)	**	**	**	**
F value dripline depth × sowing method	24.845	25.021	24.960	8.067
LSD ($p < 0.05$)	**	**	**	*

APD30, alternative row/ bed planting with dripline depth of 30 cm; APD35, alternative row/ bed planting with dripline depth of 35 cm; FPD30, flat planting with dripline depth of 30 cm; FPD30, flat planting with dripline depth of 35 cm; LAI, leaf area index.

* indicates a significant differences ($P<0.05$), ** indicates extremely significant differences ($p <0.01$), and NS indicates insignificant differences.

3.3 Yield and its components

Table 7 shows the significance results for corn yield and its components. The yield of APD30 was the highest, at 14.7 t/ ha, and the yield of FPD35 was the lowest, at 11.6 t/ ha. For D30, the effective ears per ha and yield under AP increased by 12.6% and 14.8%, respectively, compared with those under FP. For D35, the effective ears per ha and yield under AP increased by 10.3% and 5.2%, respectively, compared with those under FP. For D30, the kernels per ear under AP increased by 6.9% compared with that under FP. However, insignificant differences were observed for the kernels per ear at a burial depth of 35 cm. For D30 and D35, the hundred-grain weights under FP increased by 4.5% and 8.5%, respectively, compared with those under AP, indicating significant differences. For D30, the ear diameter, effective ears per ha, kernels per ear, and yield increased by 2.4%, 11.0%, 3.5%, and 16.0%, respectively, compared with D35, indicating significant differences.

Table 7. Statistical analysis of spring corn yield and its components

Treatment	Ear Length (cm)	Ear diameter (cm)	Bare top length (cm)	Effective ears per ha	Kernels per ear	Hundred-grain weight (g)	Yield ($t\ ha^{-1}$)
APD30	19.93	5.07	1.97	77617	635	30.16	14.7
FPD30	19.47	5.17	2.13	68930	594	31.52	12.8
Mean	19.70	5.12	2.05	73274	615	30.84	13.8
F value	0.925	4.5	0.205	17.026	9.145	1.857	53.14
APD35	19.00	5.00	1.87	69268	593	29.71	12.2
FPD35	19.30	5.00	1.83	62801	596	32.25	11.6
Mean	19.15	5.00	1.85	66035	594	30.98	11.9
F value	5.4	0	0.034	8.687	0.13	17.052	0.774
F value for dripline depth	4.797	9.8	0.954	22.668	6.314	0.055	29.515
LSD ($p < 0.05$)	NS	*	NS	**	*	NS	**
F value for sowing method	0.11	1.8	0.106	24.835	5.677	11.039	12.101
LSD ($p < 0.05$)	NS	NS	NS	**	*	*	**
F value for dripline depth × sowing method	2.33	1.8	0.238	0.533	7.629	1	3.386
LSD ($p < 0.05$)	NS	NS	NS	NS	*	NS	NS

APD30, alternative row/ bed planting with dripline buried depth of 30 cm; APD35, alternative row/ bed planting with dripline buried depth of 35 cm; FPD30, flat planting with dripline buried depth of 30 cm; FPD30, flat planting with dripline buried depth of 35 cm;

* indicates a significant differences ($P<0.05$), ** indicates extremely significant differences ($p <0.01$), and NS indicates insignificant differences.

3.4 Water use efficiency and nitrogen partial factor productivity

Table 8 shows that for D30 and D35, the ET_c under AP increased by 2.4% and 0.7%, respectively, compared with those under FP that had no significant difference. However, the dripline depth had a significant influence on the ET_c. For D30, the mean ET_c of the two sowing methods increased by 3.7%, compared with D35. Water use efficiency (WUE_{ETc}) is consisted by yield and ET_c and nitrogen partial factor productivity (NPFP) is consisted by yield and seasonal nitrogen applied amount (i.e., 290kg/ ha). The sowing methods and dripline depths significantly affected the WUE_{ETc} and NPFP. The WUE_{ETc} under AP increased by 8.2% on average compared with that under FP, and D30 increased by 11.8% on average compared with D35. The NPFP under AP increased by 9.8% on average compared with that under FP, and D30 increased by 15.6% on average compared with D35.

Table 8. Statistical analysis of ET_c, WUE_{ETc} and NPFP

Treatment	ET_c (mm)	WUE_{ETc} (kg ha^{-1} mm^{-1})	NPFP/ (kg kg^{-1})
APD30	409.4	35.9	50.6
FPD30	400.0	32.1	44.3
Mean	404.7	34.0	47.4
F value	1.270	10.758	52.933
APD35	391.6	31.1	42.0
FPD35	389.0	29.8	40.0
Mean	390.3	30.4	41.0
F value	0.098	0.407	0.759
F value for dripline depth	5.914	9.355	29.438
LSD ($p < 0.05$)	*	*	**
F value for sowing method	1.043	4.773	12.027
LSD ($p < 0.05$)	NS	**	**
F value for dripline depth × sowing method	0.338	1.162	3.407
LSD ($p < 0.05$)	NS	NS	NS

APD30, alternative row/ bed planting with dripline buried depth of 30 cm; APD35, alternative row/ bed planting with dripline buried depth of 35 cm; FPD30, flat planting with dripline buried depth of 30 cm; FPD30, flat planting with dripline buried depth of 35 cm; ET_c, seasonal crop evapotranspiration; WUE_{ETc}, water use efficiency. NPFP, nitrogen partial factor productivity.

* indicates a significant differences (P<0.05), ** indicates extremely significant differences (p <0.01), and NS indicates insignificant differences.

4. Discussion

When the subsurface dripline was buried below the plow layer, the moisture content of the soil under the dripline was higher than that of the soil above the dripline because of gravity (Cote et al., 2003; Schiavon *et al.*, 2015), thereby causing the following problems: (1) neither the water nor fertilizer requirements at the seed germination and early growth stages of corn are met because of water deficiency in the moisture content of the soil around the seed; (2) deep percolation of water and fertilizer may result. The new sowing method of AP made a trapezoidal furrow with 10 cm depth, then seeds were sown in 5 cm deep soil below the furrow bottom which was similar to the method Lamm *et al.* (2012) recommended to move the dry and loose soil near the surface to the traffic furrow, and to sow into the wetter and firmer soil, it could provide a wetter soil environment for germination. Besides, AP could also shortened the distance between the seed and the dripline. Therefore, the soil water under AP would be easier to reach for seeds under the same irrigation amount. In both 2015 and 2016, only the APD30 soil water could reach up to 5 cm vertically below the seed after pre-emergence irrigation of 25 mm. The closer distance between the seeds and the dripline, the higher the soil moisture content around the seeds, the higher the rate of emergence rate and the better the growth for early corns. So the corn emergence rate, plant height, stem diameter and LAI at the seedling stage under AP were greater than FP.

From July 5 to October 5, the moisture contents of soil at 0-40 cm was affected by irrigation, rainfall, evaporation, and corn consumption changed dramatically. These results Liu and Li (2009) said that in the topsoil (10-20 cm), the moisture content significantly decreases with the increased dripline depth. However, in deep soil (20-70 cm), the moisture content increased with the increasing dripline depth. Due to the topsoil (10 and 20 cm) of AP was closer to dripline, while the deep soil (40 and 60 cm) was far away from the dripline, the average topsoil moisture content of AP was greater than that of FP, yet the deep soil moisture content of AP was smaller than that of FP.

The ridge and furrow rainfall harvesting system allows for collection and storage of rainwater and considerably increases the water storage capacity (Bu *et al.*, 2013; Dahiya *et al.*, 2007; Hu *et al.*, 2014; Li *et al.*, 2009). Similarly, the sowing method of AP enlarged the contact area between the soil surface and the rainwater, then storing the water in the soil below the furrow bottom where corn grows. This result was obtained by comparing the changes in soil moisture content after two precipitation events of 29 and 38 mm. The comparison showed that the mean moisture of soil at 0-60 cm of AP were 1.46 and 1.26 times bigger than FP, respectively.

Several scholars believe that if the burial depth of the dripline is too shallow, the topsoil will get readily wet, and the existing moisture will be prone to evaporation loss (Lamm *et al.*, 2006). Lamm *et al.* (2010) noted that the evaporation under dripline depths of 40 cm was less than at buried depths of 20 and 30 cm. The result of the mean ET_c of two sowing methods for D30 significantly increased, by 3.7%, compared with D35, which agreed with the above conclusions. Due to the shortened distance between the seed and the dripline under AP, the water could more easily reach the soil surface and could accelerate evaporation. Besides, the corn under AP exhibited better outcomes than FP, the total actual water consumption of corn under AP would be higher than FP.

However, the proportion of the furrow bottom in the total area of the soil surface was small and the furrow in the present experiment was situated east-west and perpendicular to the direction of the locally common northerly wind. Therefore, a high evaporation resistance could reduce evaporation losses (Li *et al.*, 2009). Furthermore, the poor corn cover of FP would increase evaporation. Although the ET_c under AP increased by 2.4% (D30) and 0.7% (D35), compared with FP, respectively, no statistically significant difference was observed.

When crop establishment is non-limiting (i.e., the emergence rate is uniform), the dripline depth, in most cases, has no remarkable direct influence on the yield or WUE_{ETc} (Lamm *et al.*, 2010; Lamm and Trooien, 2005; Liu and Li, 2009; Machado *et al.*, 2003). With respect to seed germination, a closer distance between the seed and the dripline implies a higher emergence rate, higher yield and improved water use efficiency (Charlesworth and Muirhead, 2003; Lamm *et al.*, 2012; Pablo *et al.*, 2007). Therefore, in the present experiment, AP was determined to be superior to FP in terms of the emergence rate and effective ears per ha. The yield under AP was significantly larger than FP for both dripline depths. The number of kernels per ear under D30 was considerably larger than that under D35, which is consistent with the findings of Lamm and Trooien (2005), who reported that, for the number of kernels per ear, the burial depth of the drip irrigation pipe at 20-30 cm was considerably higher than that at 41-61 cm. As a result, a low emergence rate can provide better air permeability and light transmittance, which can promote single corn growth and a change in the law of hundred-grain weights that was contrary to the emergence rate: FPD35 > FPD30 > APD30 > APD35.

Due to the no significant difference of ET_c between AP and FP, The WUE_{ETc} was principally determined by the yield and increased with yield and the WUE_{ETc} of AP was significant higher than FP. In addition, when the nitrogen applied amount in all treatments were the same, it could be concluded from formula (5) that NPFP also increased with the increase of Yield, so the NPFP of AP was higher than FP and the NPFP for D30 was higher than D35. The sowing methods and dripline depths had a significant effect on WUE_{ETc} and NPFP.

5. Conclusions

This study proposed a new sowing method for subsurface drip irrigation called alternate row/ bed planting (AP), which involved removal of the dry topsoil and then sowing seeds below the trapezoidal furrow bottom, which could also shorten the distance between the seed and the dripline. After sowing and irrigation, the soil moisture content vertically below the seed of alternate row/ bed planting (AP) was higher than flat planting (FP). Thus, the emergence rate of AP was significantly higher than FP under both dripline depths. As corn roots are shallow during the early growth stage, the growth of corn can be considerably fostered by AP. Furthermore, AP changed the geometry of the topsoil, allowing the collection and storage of rainwater after rain and increasing the topsoil moisture content. The mean yield, WUE_{ETc} and NPFP of two dripline depths under AP increased by 10.2%, 8.2% and 9.8%, respectively, compared with FP. Therefore, this sowing method is suitable for regions affected by spring droughts.

Acknowledgements

We are grateful for financial support from the National Science and Technology Project in the 12th Five-Year Plan of Chinese Ministry of Science and Technology (2014BAD12B00) and the Foundation for Innovative Research Groups of the National Natural Science Foundation of China (51321001).

References

Adamsen, F.J., 1992. Irrigation method and water quality effects on corn yield in the mid-Atlantic coastal plain. Agron J, 84, 837-843.

Alam, M., Trooien, T.P., Dumler, T.J., Rogers, D.H., 2002. Using Subsurface Drip Irrigation for Alfalfa. Journal of the American Water Resources Association, 38, 1715-1721.

Bar-Yosef, B., Sagiv, B., Markovitch, T., 1989. Sweet corn response to surface and subsurface trickle phosphorus fertigation. Agron J, 81, 443-447.

Barth, H.K., 1999. Sustainable and effective irrigation through a new subsoil irrigation system (SIS). Agricultural Water Management, 40, 283-290.

Bordovsky, J.P., 2006. Variability of subsurface drip irrigation (SDI) tape placement using RTK-GPS guided equipment, Proc. 2006 Beltwide Cotton Conf., San Antonio, TX.

Bordovsky, J.P., Porter, D., 2003. Cotton Response to Pre-plant Irrigation Level and Irrigation Capacity Using Spray, LEPA, and Subsurface Drip Irrigation, ASAE Paper No. 032008. ASAE, St. Joseph, Mich, pp. 27-30.

Brown, K.W., Thomas, J.C., Friedman, S., Meiri, A., 1996. Wetting patterns associated with directed subsurface irrigation, in: Joseph, S. (Ed.), Irrigation scheduling: proceedings of the international conference. ASAE, San Antonio, Texas, pp. 3-6.

Bu, L.D., Liu, J.L., Zhu, L., Luo, S.S., Chen, X.P., Li, S.Q., Hill, R.L., Zhao, Y., 2013. The effects of mulching on maize growth, yield and water use in a semi-arid region. Agricultural Water Management, 123, 71-78.

Burt, C., Styles, S.W., 1999. Drip and Micro Irrigation for Trees, Vines, and Row Crops: Design and Management (with Special Sections on SDI). Irrigation Training and Research Center, Bioresource and Agricultural Engineering Department, California Polytechnic State University.

Camp, C.R., 1998. Subsurface drip irrigation: a review. Transactions of the ASAE, 41, 1353-1367.

Camp, C.R., Lamm, F.R., 2003. Irrigation systems, subsurface drip. Encyclopedia of water science, 560-564.

Camp, C.R., Sadler, E.J., Busscher, W.J., 1989. Subsurface and Alternate-Middle Micro Irrigation for the Southeastern Coastal-Plain. Transactions of the ASAE, 32, 451-456.

Charlesworth, P.B., Muirhead, W.A., 2003. Crop establishment using subsurface drip irrigation: a comparison of point and area sources. Irrigation Science, 22, 171-176.

Colaizzi, P.D., Schneider, A.D., Evett, S.R., Howell, T.A., 2004. Comparison of SDI, LEPA, and spray irrigation performance for grain sorghum. Transactions of the ASAE, 47, 1477-1492.

Cote, C.M., Bristow, K.L., Charlesworth, P.B., Cook, F.J., Thorburn, P.J., 2003. Analysis of soil wetting and solute transport in subsurface trickle irrigation. Irrigation Science, 22, 143-156.

Dahiya, R., Ingwersen, J., Streck, T., 2007. The effect of mulching and tillage on the water and temperature regimes of a loess soil: Experimental findings and modeling. Soil Till Res, 96, 52-63.

Heidman, B.C., Abidine, A.Z., Upadhyaya, S.K., Hills, D.J., Robert, P.C., 2003. Application of RTK GPS based auto-guidance system in agricultural production, Proceedings of the 6th International

Conference on Precision Agriculture and Other Precision Resources Management, Minneapolis, MN, USA, 14-17 July, 2002., pp. 1205-1214.

Henggeler, J., 1995. A history of drip-irrigated cotton in Texas, Microirrigation for a Changing World: Conserving Resources/Preserving the Environment. Proc. Fifth International Microirrigation Congress, pp. 669-674.

Howell, T.A., Schneider, A., Evett, S., 1997. Subsurface and surface microirrigation of corn—Southern High Plains. Transactions of the ASAE, 40, 635-641.

Hu, Q., Pan, F., Pan, X., Zhang, D., Yang, N., Pan, Z., Zhao, P., Tuo, D., 2014. Effects of a ridge-furrow micro-field rainwater-harvesting system on potato yield in a semi-arid region. Field Crops Research, 166, 92-101.

Ji, C., Zhou, J., 2014. Current Situation of Navigation Technologies for Agricultural Machinery. Transactions of the Chinese Society for Agricultural Machinery, 45, 44-54.

Lamm, F.R., Ayars, J.E., Nakayama, F.S., 2006. Microirrigation for crop production: design, operation, and management. Elsevier, Amsterdam, Holland.

Lamm, F.R., Bordovsky, J.P., Schwankl, L.J., Grabow, G.L., Enciso-Medina, J., Peters, R.T., Colaizzi, P.D., Trooien, T.P., Porter, D.O., 2012. Subsurface Drip Irrigation: Status of the Technology in 2010. Transactions of the ASAE, 55, 483-491.

Lamm, F.R., Kheira, A.A.A., Trooien, T.P., 2010. Sunflower, soybean, and grain sorghum crop production as affected by dripline depth. Appl Eng Agric, 26, 873-882.

Lamm, F.R., Stone, L.R., Manges, H.L., O'Brien, D.M., 1997. Optimum lateral spacing for subsurface drip-irrigated corn. Transactions of the ASAE, 40, 1021-1027.

Lamm, F.R., Trooien, T.P., 2003. Subsurface drip irrigation for corn production: a review of 10 years of research in Kansas. Irrigation Science, 22, 195-200.

Lamm, F.R., Trooien, T.P., 2005. Dripline depth effects on corn production when crop establishment is nonlimiting. Appl Eng Agric, 21, 835-840.

Leskovar, D.I., Ward, J.C., Sprague, R.W., Meiri, A., 2001. Yield, quality, and water use efficiency of muskmelon are affected by irrigation and transplanting versus direct seeding. Hortscience, 36, 286-291.

Li, C., Sun, J., Zhou, X., Qiang, X., Guo, D., 2009. Research progress of water and heat transfer and change under furrow irrigation. Transactions of the Chinese Society of Agricultural Engineering, 25, 1-5.

Li, J., Lin, M., 2006. Research progress of automatic guidance technologies applied in agricultural engineering. Transactions of the Chinese Society of Agricultural Engineering, 22, 232-236.

Li, J., Wang, Y.Q., Qu, Z.Y., Ma, L.Z., 2010. Temporal and spatial distribution characteristics of drought disaster in the Inner Mongolia Autonomous Region. Agric. Res. Arid Areas, 266-272.

Li, J.S., Zhang, J.J., Xue, K.Z., 2003. Principles and applications of fertigation through drip irrigation systems. Beijing: China Agricultural Science and Technology Press.

Liu, X.F., Wan, S.Q., Feng, L., Jiang, S.F., Kang, Y.H., Liu, S.P., 2015. Response of potato yield and irrigation water use efficiency under subsurface drip irrigation at various lateral depths. Journal of Irrigation and Drainage, 34, 63-66.

Liu, Y.C., Li, J.S., 2009. Effects of lateral depth and layered-textural soil on water and nitrate dynamics and root distribution for drip fertigated tomato. Journal of Hydraulic Engineering, 40, 782-790.

Ma, S.Q., Wang, Q., Luo, X.L., 2008. Effect of climate change on maize (Zea mays) growth and yield based on stage sowing. Acta Ecologica Sinica, 28, 2131-2139.

Machado, R.M., Do RosArio, M., Oliveira, G., Portas, C.A., 2003. Tomato root distribution, yield and fruit quality under subsurface drip irrigation. Plant & Soil, 255, 333-341.

Marouelli, W.A., Silva, W.L.C., 2002. Drip line placement depth for processing tomatoes crop. Horticultura Brasileira, 20, 206-210.

Mi, R.G.L., 2013. The Cultivation Pattern of Maize Matching to Trickle Irrigation in the Chifeng Region, College of Water Resources and Civil Engineering. China Agriculture University.

Pablo, R.G., O'Neill, M.K., McCaslin, B.D., Remmenga, M.D., Keenan, J.G., Onken, B.M., 2007. Evaluation of corn grain yield and water use efficiency using subsurface drip irrigation. J Sustain Agr, 30, 153-172.

Patel, N., Rajput, T.B.S., 2007. Effect of drip tape placement depth and irrigation level on yield of potato. Agricultural Water Management, 88, 209-223.

Phene, C.J., Davis, K.R., Hutmacher, R.B., Mccormick, R.L., 1987. Advantages of subsurface irrigation for processing tomatoes. Acta Horticulturae, 200, 101-113.

Plaut, Z., Rom, M., Meiri, A., 1985. Cotton response to subsurface trickle irrigation, Drip/Trickle Irrigation Congress. ASAE, St. Joseph, Mich, pp. 916-920.

Roberts, T.L., White, S.A., Warrick, A.W., Thompson, T.L., 2008. Tape depth and germination method influence patterns of salt accumulation with subsurface drip irrigation. Agricultural Water Management, 95, 669-677.

Schiavon, M., Serena, M., Leinauer, B., Sallenave, R., Baird, J.H., 2015. Seeding Date and Irrigation System Effects on Establishment of Warm-Season Turfgrasses. Agron J, 107, 880-886.

Schwankl, L., Grattan, S., Miyao, E., 1990. Drip irrigation burial depth and seed planting depth effects on tomato germination, Irrigation Symp. ASAE, St. Joseph, Mich, pp. 682-687.

Welsh, D., Kreuter, U., Byles, J., 1995. Enhancing subsurface drip irrigation through vector flow, Proceedings of the 5th International Microirrigation Congress. ASAE, Orlando, FA, pp. 2-6.

Wood, M.L., Finger, L., 2006. Influence of irrigation method on water use and production of perennial pastures in northern Victoria. Animal Production Science, 46, 1605-1614.

Xu, H.C., 2014. The Cultivation Pattern of Maize Matching to Mulched Trickle Irrigation in the Region of Chifeng, Inner Mongolia, College of Water Resourse and Civil Engineering. China Agriculture University.

Yuan, J.J., 2015. The effects of different subsurface drip tape types and irrigation scheduling on maize growth, College of Water Resourse and Civil Engineering. China Agriculture University.

This page intentionally left blank.

Prospects on Drip Irrigation Development in Xinjiang, China

Shalamu Abudu[1], Zhuping Sheng[1], Jiancun He[2], Chunliang Cui[2], Bayinmengke[2]

1.Texas A&M AgriLife Research Center at El Paso, 1380 A&M Circle, El Paso, TX 79927-5020; PH (915) 859-9111; FAX (915) 859-1078; e-mail: shalamu.abudu@ag.tamu.edu
2.Xinjiang Water Resources Research Institute, No. 73 Hongyanchi North Road, Urumqi, Xinjiang China, Phone: 86-991-8565833, E-mail: shalamu3@qq.com

Abstract. *China's Xinjiang Uyghur Autonomous Region is one of the extreme arid regions in the world and has been suffering from severe water scarcity problems for decades. In the last 20 years, application of drip irrigation technique has been expanded rapidly in the region from zero hectare in early 1990s to more than three million hectares at present which accounts for approximately half of the cultivated area of the region. While the widespread adoption of drip irrigation temporarily addressed the water scarcity issues and improved the crop production, it brought issues that should be addressed and resolved in the near future to ensure the sustainable development of agriculture and food safety in the region. This paper summarized the current status, and identified problems and challenges that widespread adoption of drip irrigation has brought to the agriculture production and environment at a regional scale. A number of technological and policy solutions were also identified through the study and several integrated water management strategies were proposed for the sustainable agricultural production and environmental protection in the region.*

Key words: Drip Irrigation, Adoption, Development, Arid Region, Xinjiang

1. Introduction

The arid regions occupy a vast area in northwestern China with the total area of 2.5 million km^2 or one-quarter of Chinese territory. These regions include the western part of Inner Mongolia, the northern part of Ningxia Hui Autonomous Region, most of Qinghai and Gansu provinces and the Xinjiang Uyghur Autonomous Region (As shown yellowish area in Figure 1). In these regions, mean annual rainfall is less than 250 mm, further reduced (50-150 mm) in the western plains and reaches the lowest (less than 25 mm) in the Taklimakan Desert in Xinjiang. The annual evaporation is more than 1,400 mm in general, and about 2,000-3,000 mm in desert areas. Because of the arid climate, about 70 per cent of the total arid regions are unusable areas such as sandy deserts, gravel deserts, and other wildernesses (Chen, 2014). Compared to other region of China, the arid northwestern China is relatively less developed and the local economy depends only on irrigated agriculture and animal husbandry. Water is not only the most precious natural resource in this region but also the most important environmental factor of the ecosystem. Since ancient times, water utilization has always had a decisive impact on local socioeconomic development. But the increased intensity of human activities and overdraft of water resources caused, and quickly spread, agro-environmental degradation, including salinization, vegetation degeneration, and sandy desertification. The shortage of water resources has become a "bottleneck" restricting agricultural production and economic development. To ensure national food security, China has been developing water-saving agriculture since last decades in these regions. Particularly, the central government has been mainly promoting the adaptation of drip irrigation technology in the Xinjiang Uyghur Autonomous Region (Xinjiang) where is the country's main cotton and grain production area.

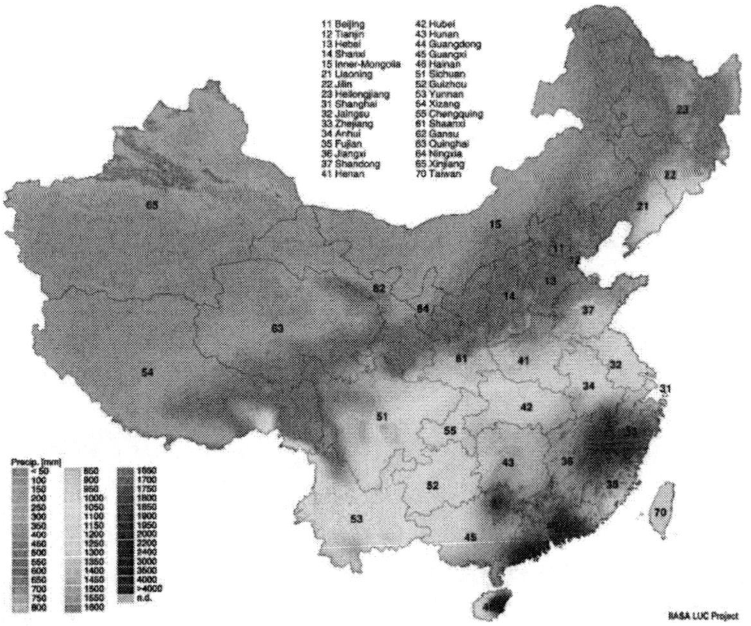

Figure 1. Average annual precipitation in China (Institute of Soil Science, 1986)

As one of the world's largest arid places, Xinjiang is situated in northwest China. Located in the hinterland of the Eurasian continent, it has world's second largest Desert-Taklimakan desert that located in the southern part of Xinjiang (as shown in Figure 2). The topography of Xinjiang features three mountain ranges and two basins: The Altai Mountains in the north, the Tianshan Mountains running through the middle of the region, the Kunlun Mountains in the south, Dzungarian Basin and Tarim Basin between the three mountain ranges. Xinjiang's total annual water resources of 83.2 billion m^3, the unit area of water production is only $5m^3$ / km^2, ranks the third from the last in the country. The total water resources utilized in social and economic sectors is 61.7 billion m^3, of which the agricultural water consumption is 59.18 billion m^3 that accounts for 95.8% of total water in use (Gao and Shi, 1992; Liu et al., 2013). In 2014, total economic and social water consumption in Xinjiang is 58.18 billion m^3, of which agricultural water consumption is 55.09 billion m^3 (94.7%). As of 2011, the total cultivated land in Xinjiang is 4.12 million hectares, accounting for only 2.5% of the total area (Xinjiang Water Resources Research Institute, 2015). The desert, barren land, and other unused areas totaled around 102 million hectares which accounts for the large proportion of unusable land in the region (Chen, 2014). Hence, Xinjiang's agriculture is typically an oasis agriculture under the dual constraints of water shortage and landscape structure that limit agricultural production. With its population growth and intensifying agricultural activity, the Xinjiang region is facing threats of water security as other parts of the world that have similar climate and environmental conditions, such as Israel. Another challenge that will only worsen the region's growing water woes: melting and shrinking glaciers. The Tarim basin, the main agricultural region in Xinjiang, is one of the driest geographical places that relies heavily upon water sourced from the melting glaciers in surrounding mountains. Climate related drought and human activities have significantly contributed to the region's dwindling water supply.

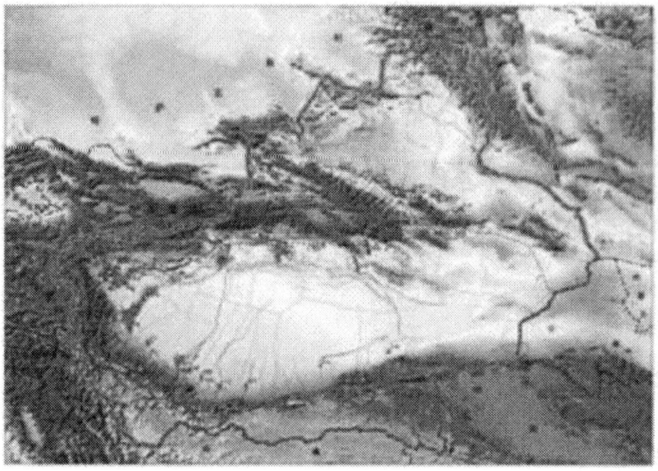

Figure 2. Topographic map of Xinjiang, China with green color shows cultivated area (Xinjiang Institute of Ecology and Geography, 2009).

In the last decades, the canal restoration, improved drainage systems and a switch to drip irrigation has helped to enhance water conservation practices within the area. However, water use efficiency is still low, agricultural irrigation water accounts for more than 90% of the total water consumption and produces only 20% of the total GDP of the region (Chen, 2014). Hence with all the odds, it seems that widespread application of efficient irrigation technologies is the only option to conserve the limited water and land resources in the region. The first drip irrigation program was introduced from Israel in 1994 to start use of the world's most efficient irrigation method in the region (Abudu, 1997). For the last 20 years, with the help of Israeli drip irrigation technology and with the implementation of the national western development strategy, the widespread utilization of drip irrigation in Xinjiang has been undergone intensively. The area irrigated with drip irrigation technique has been expanded rapidly from zero hectares in early 1990s to more than three million hectares at present. The Xinjiang has become world's largest drip-irrigated region in terms of total area and diversity of crops. Even though the widespread adoption of drip irrigation temporarily addressed the water scarcity issues and improved the crop production, it also brought issues that should be addressed and resolved to ensure the sustainable development of agriculture and food safety in the region. This paper provides an overview of the current status of drip irrigation in Xinjiang Region, identifies challenges in technological and institutional issues and proposes measures to assure the sustainable utilization of water resources and development of agriculture in the region.

2. Adaptation and Development

2.1 Periods of Development

The adaptation of drip technology in Xinjiang has been experiencing three major periods from the early 1990s to the present. They can be categorized as Demonstration and Adjustment period (1993-2000), Large-scale Extension period (2000-2010) and Progression and Upgrading period (2010-present).

The first period-the demonstration and adjustment period is mainly characterized by the introducing the drip technology from Israel, establishing demonstration projects and conducting pertinent research on the adjustment of the drip technology based on the local climatic, hydrological, environmental, agricultural and socio-economic conditions. During this period from the early 1990s to 2000, the drip irrigation technology was introduced from Israel and established demonstration projects in different areas of

Xinjiang starting from driest areas such as Turpan, Kumul gradually to southern Xinjiang, and to Tianshan North belt economic areas with large scale irrigated cash crops. During this period, about dozens drip irrigation projects were completed and the effects of drip irrigation on different crop yield was tested. Most of the extension work was focused on the testing, experiment and development of some cost-effective drip irrigation products such as basic screen filter and sand separators. During this time, the drip irrigation was mainly installed in orchards such as grapes, pearls, and field crops such as cotton due to high cost imported driplines. Considering the cost of head control system, the groundwater was used as the water source in almost all the demonstration projects due to its low filtration system cost as compared to the canal water which requires expensive filtration system for drip irrigation. The total drip irrigated area in this period is under 100,000 hectares in the whole region mainly due to the high cost of the system that can use only imported driplines and filtration systems.

The second period-the large-scale extension period is characterized by the expansion of the drip system at a rochet high speed due to the lowered cost of the system and direct provision of subsidies for the farmers from the central and regional governments during 2000-2010.The main feature of this period is local companies was able produce dripline and tapes at a very low cost. In addition, the technical assistance from Israel government and institutions moved from inner provinces to Xinjiang Region. For example, the Sino-Israeli Demonstration and Training center for Agriculture in Arid Zone was established in Xinjiang and it was the first cooperative project between China and Israel in north-west of China at the governmental level. It was the most advanced agricultural base of water-saving in dry land in China. The project has been operated over 10 years, and more than 20,000 people visited and around 6,000 people participated in the training activities that were organized there. The establishment of the farm played important role in promoting drip irrigation in the region from both technical and management perspectives. With the research and development (R & D) and industrialization of water-saving agriculture, many local irrigation manufactures start to produce better quality drip irrigation products including driplines, tapes, filtration and controlling systems. A series of water-saving products and complete sets of equipment were initially formed with Chinese characteristics and independent intellectual property rights, which promoted the rapid development of water-saving leading enterprises like Xinjiang Tianye, Fujian Yatong and so on (Wu, 2004; Wu, 2010). Dripline and drip tape producing lines were also established in huge scales, which made the cost of driplines drop from 0.20-0.40US$/m to 0.06-0.10 US$/m, with drip tapes dropping from 0.05 - 0.08 US$/m down to the 0.001-0.002 US$/m (Xinjiang Water Resources Research Institute, 2015). With the development and production of local filtration equipment, the cost of filtration system has also been reduced to one third of filtration products that manufactured abroad such as AMIAD, ARKAL filtration systems. Thus, the drip irrigation system cost was dropped drastically in this period, is about one third of the prices at end of 1990s, which in turn facilitate the wide-spread application of drip systems all over the Xinjiang. As can be seen from Figure 3 that, the total drip irrigated area increased sharply from 0.17 million hectares in 2005 to nearly 1 million hectares in the end of 2010, and this is equivalent to one fourth of total cultivated area in the whole region.

The third period-progression and upgrading period from 2010 to present is characterized by the utilization of better quality of drip irrigation products and widespread application of automatic controlled drip irrigation system. During this period, with the expansion of drip-irrigated agriculture in the Xinjiang Region, some technological and institutional concerns and issues related to large scale drip irrigation drew the attention of government and stakeholders including farmers, companies, researchers, and environmentalists. And with the growing cost of labors, a special attention is given for using better quality filtrations systems, automatic self-cleaning filtrations systems, better quality driplines and drip tapes and even semi- and fully automatic controlled systems in the drip irrigated agriculture in the region. As seen in Figure 3, the total drip irrigated area is keep expanding from 1 million hectares in 2010 to for almost 2million hectares in the end of 2015, with more than half cultivated area in Xinjiang. At present, drip irrigation is used to irrigate all the crop types successfully, including cash crops to field crops and all types of orchards. The Xinjiang region has basically formed a government subsidies-oriented and farmer's investment-voluntary financial supporting system to supplement the diversified and efficient water-saving construction system. The widespread application of the drip technology has changed the traditional agricultural practices in the region. With the expansion of drip irrigated agricultural area, the integration of new cultivation techniques has been underway with drip irrigation with help of researchers from numerous disciplines including the water conservancy sectors, agronomy, agricultural machinery, fertilizers, and even computer information technology. During this period, the drip irrigation development in Xinjiang is looking more in integration of different technologies and knowledge related to agricultural production and environment to improve crop production and quality, treating system as whole watershed instead of a single drip irrigation project. Special attention is gradually given to reintroduction of advanced agro technology from the abroad, particularly learning from Israeli experience of integrated water resources management under changing irrigation method in a regional level.

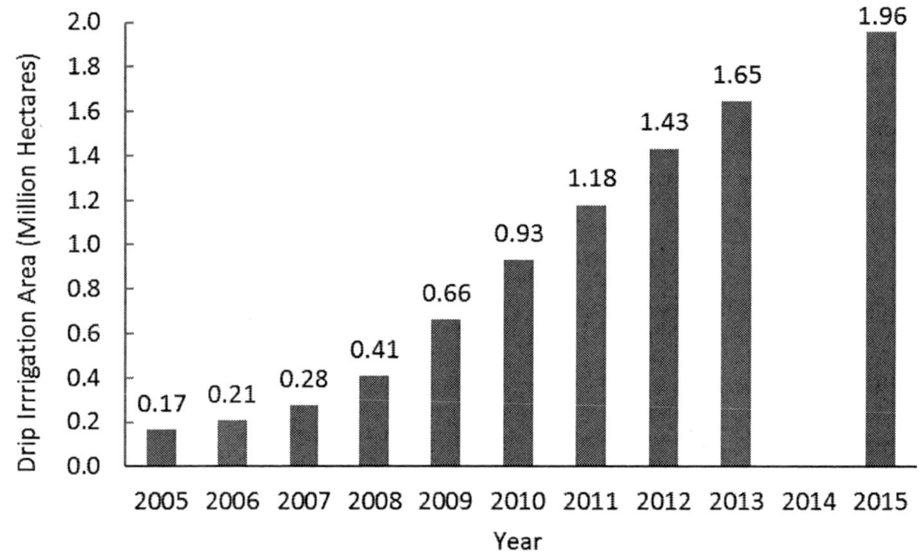

Figure 3. Development of drip irrigation area from 2005 to 2015 (Xinjiang Water Resources Research Institute, 2015)

2.2 Drivers for Development

As described earlier, the unique natural conditions of Xinjiang demand use of drip irrigation technology in the region. Although a significant cost reduction has taken place through increasing locally manufactured drip irrigation products, but the cost of investment is still far beyond the farmer's financial ability to construct drip systems by their own. Cremades et al. (2015) conducted comprehensive research on the importance of governmental support measures and economic incentives for the adoption of modern irrigation technology in China. Their results showed that the government policies and incentive mechanisms played a significant role in promoting the adoption of modern irrigation technology in China. The twenty years of drip irrigation practices in Xinjiang also indicates that the governmental support is an important factor in farmers' decisions whether or not to adopt drip irrigation technology. Governmental policies in promoting adoption of drip irrigation technology played a major role in adaptation and development of the water saving technology in Xinjiang region by providing favorable economic and technical support directly to farmers. Consistent encouragement and support has been given to local government, research and design institutes and manufacturers by the central and regional government in aiming to adapt drip irrigation technology in the whole region for the past 20 years and more. To overcome economic constraints, government direct provision of subsidies has proven to be an important policy measure in increasing the adoption level of drip irrigation technology.

In terms of technical support, providing knowledge and technical advice through extension service activities are effective ways to increase the adoption level of modern irrigation technology. In the last few decades, a rational economic incentive for farmers in the region was gradually set through various positive measures and policies such as water price reforming, marketing of drip irrigated fruits in addition to governmental support, which in turn are other important factors that influenced farmer's technology adaption behavior.

3. Major Challenges

With large-scale application of the drip technology in the region, many new issues and potential problems have surfaced that could pose great impacts on sustainable agricultural production in the region. These issues include, but are not limited to, the effects of drip irrigation on soil salinity, low irrigation uniformity due to low quality of products and poor design, lack of management of the existing systems, and the coupling drip irrigation technique with other agricultural practices such as tillage, crop structure, harvesting and soil management. Following are few important issues and challenges that should be resolved in order to assure the sustainable utilization of the techniques in the region's agricultural production.

3.1 Low Water Productivity

The challenges the Xinjiang Region faces in terms of water availability for the agricultural sector are impaired by the sector's low irrigation efficiency. Overexploitation of water resources, including excessive diversions from rivers, and overdraft of groundwater resources, causing decline of groundwater levels, is a common problem in the region (Ye et al., 2015). Average water productivity for grains is reported to be around 0.7-0.8 kg/m^3, which is much lower than the levels of 2.0-2.5 kg/m^3 recorded in the industrialized countries (Chen, 2014). Even after large-scale utilization of drip irrigation, the irrigation efficiency is still around 0.5 in the region, particularly in the southern Tarim Basin. One of the causes is that water delivery systems including main canals in Xinjiang have the long-distance delivery with low efficiency. According to the field test and evaluation in recent years, the concrete lining of the main canal can reduce water loss by 75% in Tarim Basin (Xinjiang Water Resources Research Institute, 2015), this indicates that mere on-farm application of drip systems cannot guarantee high water use efficiency. Increasing irrigation system delivery efficiencies and improving water productivity are key to better managing water resources in agriculture in the region. Measures should be taken to improve water delivery system to enhance water efficiency for the whole region.

3.2 Poor Management

With almost half of the cultivated area are irrigated with drip irrigation, the water productivity should have been higher as compared to other parts of China. However, lack of or poor management of irrigation water limited realization of the existing drip irrigation system. It attributes to: 1) the lack of practical planning and design of the drip irrigation system (Ma et al., 2010); thus, the farmers are hardly taking advantage of the constructed drip system; 2) low product quality not only increases the cost of the maintenance, but also affects the crop production and farmer's income directly, which in turn sometimes results in the abandonment of the system by the farmers and changing back to flood irrigation; and 3) weak farmer organizations. Water Users Associations (WUA) fill an organizational and institutional gap in the irrigation management system and provide significant benefits such as improving irrigation systems operation and maintenance, contributing to water savings, reducing water conflicts, and ensuring better water fee collection rates. Most of the cooperatives currently in practice are the "company + household" model, with a disproportional influence by companies (or by larger households) in the ownership, management, and decision-making. This structure also appears to be favored by local governments who tend to extend services and support to such cooperatives. WUAs coverage remains limited in the region and many existing WUAs continue to face financial, legal, and institutional challenges, threatening their sustainability.

3.3 Unbalanced Spatial Development

Development of the drip irrigation varies spatially. Until recently, most of the drip irrigated area are located in the east and central Xinjiang where is economically developed as whole. In those areas farmers have a better income and are able to invest to some portion of the drip systems, which in turn benefits the farmers with higher productivity. However, the drip irrigation area only accounts for 14% of the total irrigated area in the vast less developed area of southern Xinjiang (Xinjiang Water Resources Research Institute, 2015), where farmers are not able to invest even a small portion of the system. In these years, the government's incentives dedicated to this area with the highest subsidies that almost cover the total cost of the drip systems.

3.4 Inadequate Scientific Research and Technical Standardization

The scientific research activities have been far behind the rate of extension of drip irrigation area. Due to the lack of stable scientific research investment mechanism, and scientific and technological innovation, it had been difficult to form a sustained scientific and technological support system. As a result, many scientific and technological issues, such as environmental impacts of drip irrigation in watershed scale, better crop drip irrigation schedule and management that suitable for the diversified natural conditions of region, and development and utilization of low-energy, cost-effective drip products, remained unaddressed. These technological challenges need to be resolved in order to ensure sustainable agricultural production and food security in the region. Another issue currently affecting the quality and extension of the drip systems in the region is the standardization of basic technology including design, products and technological measures for construction, maintenance of the drip system. Many of the existing water-saving technologies have not been standardized and technical guidance has been lacking, which makes it difficult to adapt to different local conditions such as meteorology, crop, water source and type of irrigation project, which makes it difficult to popularize and extend the technology.

4. Prospective

Water is and will continue to be one of the most challenging natural resources issues in the arid regions of China, particularly in Xinjiang Uyghur Autonomous Region that account for one sixth of total area of China. For sustainable agricultural development and, hence, economic growth and society's progress, water is the key to success in this region. As an advanced irrigation technology, drip irrigation has become a modern agricultural technology platform in Xinjiang region to facilitate precision agriculture and increase agricultural productivity. With precision irrigation and fertilization, a high-water use efficiency can be achieved with water and energy savings. The 20-year development of drip irrigation in the region proved that only through large scale, sustainable utilization of drip irrigation, the Xinjiang's agriculture moves toward to modernization and high productivity.

Currently, the water saving irrigation, particularly the drip irrigation, is becoming primary irrigation method in Xinjiang's agriculture. Large-scale extension and sustainable utilization of drip irrigation systems in Xinjiang are two major tasks that will be going on in near future. To accomplish this, proper strategies and measures should be taken not only in resolving current challenges and issues, but also in addressing and finding solutions for the potential problems that we may encounter in the near future for sustainable development of advanced irrigation technologies in the region. These strategies and measures could include but not be limited to following aspects:

1. Special attention should be given to effective and integrated management of the drip systems at a basin level in addition to the government's favorable subsides policy and technical support from multidisciplinary research institutions.
2. Establish and strengthen of water user's associations for better operation and maintenance of on-farm irrigation infrastructure and improved water management. Improved irrigation infrastructure with widespread adaptation of drip irrigation will be handed over to local water user's associations after completion of construction. Water user's associations will be trained and supported to ensure that they have adequate resources and capacity to operate and manage the irrigation systems.
3. Achieve the maximum benefit of drip irrigation system by improving water use efficiency with adaption and further development of drip irrigation in a basin scale and by improving water delivery system efficiency through constructing new high-efficient water conservancy infrastructure and promoting current water systems aiming to upgrade system efficiencies.
4. Government and industrial sectors should increase the investment on the scientific research and technical innovation, strengthen R&D to promote high quality of the locally made products and improve reliability of the drip irrigation system.
5. Accelerate technical standardization both of planning & design procedure and products that commonly used in the drip systems, simplify maintenance procedures with automatic control and internet technology, support extension and education to teach end users to take full advantage of the advanced irrigation system.

References

Abudu, S. (1997). Approaches to Several Problems in the Development of Drip Irrigation in Turpan Basin, Xinjiang Water Resources, 94, 39-41.

Chen, Y. (2014). Water Resources Research in Northwest China, Springer, Dordrecht.

Cremades, R., Wang, J., & Morris, J. (2015). Policies, economic incentives and the adoption of modern irrigation technology in China. Earth System Dynamics, 6(2), 399-410.

Gao, Q. and Shi, S. (1992). Water resources in the arid zone of northwest China. Journal of Desert Research (Lanzhou) 12-4:1-12

Institute of Soil Science. (1986) The soil atlas of China. Cartographic Publishing House, Beijing: 6.

Liu, M., Yang, J., Li, X., Liu, G., Yu, M., & Wang, J. (2013). Distribution and dynamics of soil water and salt under different drip irrigation regimes in northwest China. Irrigation Science, 31(4), 675-688.

Ma, Y., Wu, H., Hong, M., & Zhao, J. (2010) Drip irrigation technology development and trend analysis of Xinjiang. Water-saving irrigation, (12), 87-89.

Wu, P. T. (2010). The modern water-saving agricultural technology: Progress and focus. African Journal of Biotechnology, 9(37), 6017-6026.

Wu, P. T. (2004). Chinese water-saving agricultural science and technology strategies and regional develop models. Essays of Water saving Agriculture Forum in China.

Xinjiang Institute of Ecology and Geography. (2009). Distribution map of landform in Xinjiang, China Academy of Science, Retrieved online on October 24, 2016 from http://english.egi.cas.cn/rh/rd/xjygs/

Xinjiang Water Resources Research Institute (2015). High-efficient water saving advanced irrigation demonstration project, Design Report, Urumqi, Xinjiang, China.

Ye, J., Liu, H., He, X., Gong, P., Zhang, J., and Amat, A. (2015). Operation and Development of the Efficient Demonstration Base for Water-Saving Irrigation Management in Xinjiang. Journal of Water Resources Research, 4, 130-135.

Evaluating Center Pivot Drag-line Drip Irrigation Systems

Charles Swanson – Extension Program Specialist
Texas A&M AgriLife Extension Service
College Station, Texas

Dr. Guy Fipps, PE – Extension Agricultural Engineer
Texas A&M AgriLife Extension Service
College Station, Texas

Dr. Charles Hillyer – Extension Specialist
Texas A&M AgriLife Extension Service
Amarillo, Texas

Abstract. *Center pivot irrigation is the most prevalent irrigation technology used in Texas, accounting for about 70% of the total acreage. Except in irrigation districts, farmers continue to switch from furrow to center pivot irrigation. In spite of the costs, pivots offer many advantages over furrow irrigation including higher efficiency and lower labor requirements. Drip irrigation is often used in production of vegetables, fruits and other high value specialty crops in Texas. However many farmers are discouraged from implementing permanent drip irrigation systems on larger fields due to its costs and high maintenance requirements. Although not a new concept, some manufacturers have started marketing center pivot drag-line drip irrigation systems. These systems combine the application efficiency of drip irrigation with the operational efficiency of center pivot systems. This paper will review the operation and performance of various center pivot operated drag line drip irrigation systems and summarize farmer perceptions of this technology.*

Keywords. Center pivot, drip irrigation, drag-line systems

Background

The use of drip irrigation technology with center pivot irrigation machines is not a new concept. However, recently manufacturers and local irrigation dealers in Texas have started marketing drip irrigation packages for center pivots, often referred to as "drag-line drip irrigation systems" or "precision mobile drip irrigation". These systems drag varying lengths of drip irrigation tubing behind the machine applying water slowly on the surface. The combination of these two technologies has potential to combine the higher application efficiency related to drip with the operational and maintenance benefits of center pivot machines. This paper reviews the design and performance of three recently installed drag line systems in Texas.

System Design

Three drag-line drip irrigation systems were installed in Texas in 2015. One grower converted all 8 spans of his pivot to drip irrigation. The other two growers installed one and two spans on their existing center pivots, converting from LESA and LEPA water application technologies. Design of the three drag-line systems varied. Factors such as crop type, planting layout (ie straight or circle rows), and designer/installer influenced the design. One type uses a secondary pvc manifold positioned below the pivot main which is held in place by guidewires attached to each pivot tower span. A series of pvc pipes or flexible pivot drop hose were used to connect drip lines to the manifolds. Systems either have a 30 or 60 inch drop/drip line spacing. Drip line length varies based upon the flow rate needed and is matched to the pivot printout. These three systems used Netafim DripNet PC Dripline. Drip line flow rates varied from 1 gallon per hour per foot to 2 gallons per hour per foot.

Figure 1. Example Design of Manifold Assembly.

Filtration and pressure regulation are typically standard practice when operating conventional drip irrigation systems. However, the use of filtration and pressure regulation varied across all the installed systems. Two of the systems evaluated had filters installed. Filter location varied from one at each drip line to only 1 from each pivot drop to the manifold. Pressure regulation also varied and 2 of the systems had pressure regulators on each drop/drop line. All three installations used pressure compensating drip tubing.

System Operation & Maintenance

None of the growers reported any major maintenance problems with their systems. One grower reported that the plugs on the end of the drip lines would pop off during operation. The original compression plugs were later switched out and replaced with "twist-locking" caps and no further problems were reported. Two growers reported some minor rodent damage to the drip tubing that required repair.

The growers expressed two operational concerns about the system. Adjusting the drip lines when changing pivot direction was required to avoid damaging the crop or having the dripline become entangled in the crop canopy. One grower noted that he had to move the move the drip lines by hand at the end of the field so he could perform tillage operations.

Figure 2. Drag-line system parked on end of field.

Advantages and Disadvantages

The biggest advantage all three growers found with the use of the drag-line systems was the decreased depth of wheel tracks compared to sprinkler irrigation. One grower noted the ability to irrigate during colder weather and avoid freezing concerns. Growers also noted that use of drip irrigation reduced runoff from the field and evaporative and/or wind losses from sprinklers.

(a)  (b) 

Figures 3a & 3b. Pivot Wheel Tracks.

Figure 4. Drip line being dragged along crop row, Field Planted in Circle.

Figure 5. Drip line being dragged across crop rows, Field Planted Square.

The biggest problem observed by the grower who did not plant his crop in a circle was that the drip lines would pull across the top of the crop canopy (cotton plants) when traveling perpendicular to the rows. The grower did note some leaf damage but could not verify if it impacted crop yield.

Figure 6. Drip line being dragged over crop canopy.

Conclusion

Many growers in Texas are interested in adopting the drag-line drip irrigation concept on their farms but have reservations regarding the performance, operation, management and costs. The information collected from the three growers is helpful in addressing potential grower concerns. Further evaluation and documented successful systems are needed before many growers will implement this system. As all three systems had a different design, further evaluation of each systems design using different types of manifolds and drops is needed to determine which design is the most practical and cost beneficial for converting existing systems.

This page intentionally left blank.

Close Spacing LEPA Applicators
Improve Irrigation Efficiency

John Maurer, Manager Triple D Ranch
Twenty-five years farming, Central Nevada Regional Water Authority (CNRWA) Member, Valley Electric Association District 4 Director and Board President, hillsidehayco@gmail.com

Dan Schueler, District Manager Senninger Irrigation
Thirty-nine years in the irrigation industry, Studied at Oklahoma Panhandle State University and Agriculture at Lamar Community College, dschueler@senninger.biz

Abstract. *Close Spacing irrigation with LEPA applicators can save water, reduce energy consumption and increase yields by providing more available water to the plants at lower pressures. Following successful results in 2015 on the Triple D Ranch, comparison testing was initiated to provide more information. In 2016, testing began comparing alternating spans of moving deflector sprinklers and LDNs with LEPA pads and Shrouds in order monitor the effects of both sprinklers. Moving deflector sprinklers were at 120-inch spacing. LEPA applicators were at 40-30-inch spacing. The LEPA applicators provided more water to the plants which could be attributed to the reduced energy of the water making it more resistant to wind drift and evaporation. Low pressure operation provided additional savings in the form of reduced energy required to pump the water. The close spacing of the LEPA sprinkler heads greatly diminished burrowing varmint damage to fields by creating an undesirable habitat.*

Keywords. Close Spacing, LEPA (Low Energy Precision Application) applicators, LDN, irrigation, save water, reduce energy consumption, improve irrigation efficiency, water to plants, reduced wind-drift loss, reduced evaporation loss, low pressure, reduced varmint damage.

Background

Triple D Ranch is located in Dyer Nevada (Longitude: 118.01, Latitude: 37.615) in Esmeralda County. Elevation is 4898 feet. The climate is windy with little to no rainfall and very low humidity. The primary soil type is heavy silt clay loam. The ranch encompasses 4,600 overall acres with 52 center pivots over alfalfa. Thirty-seven center pivots irrigate with LEPA applicators excluding two test machines. Area water table levels are declining which led to investigating ways to reduce water use and still retain successful yield levels.

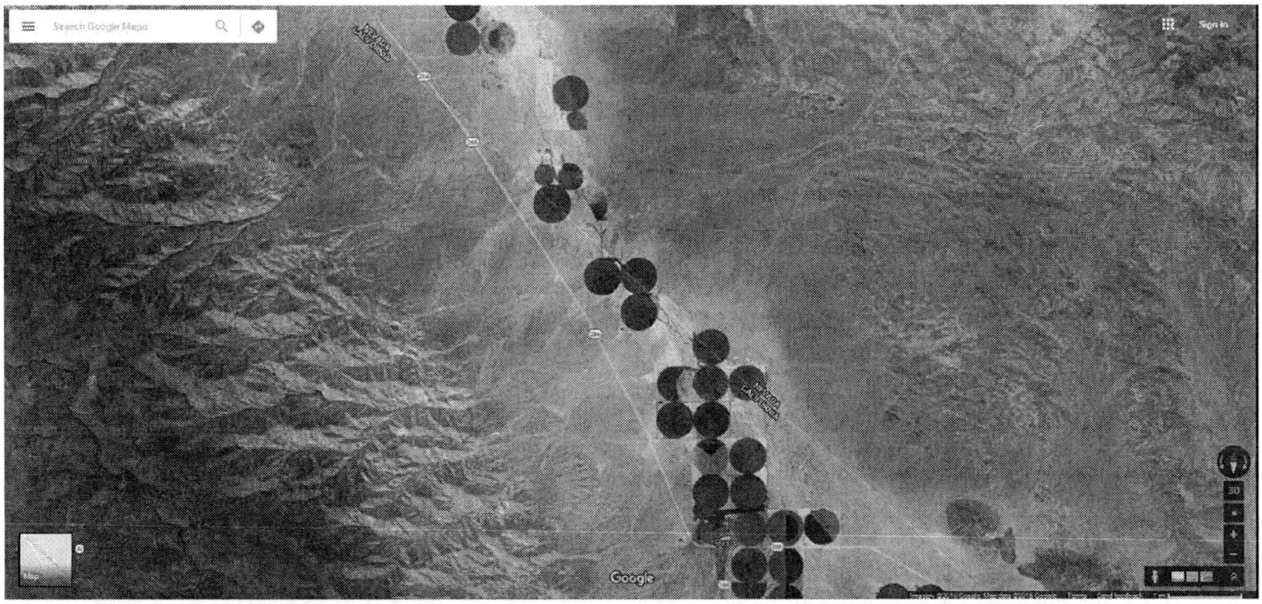

Image 1. Triple D Ranch is located in Dyer Nevada (Longitude: 118.01, Latitude: 37.615) in Esmeralda County.

Exposure to Close Spacing LEPA at World Ag Expo led to additional independent research. This included information about the success of LEPA in Texas dating back to the 80's and renewed interest and recent university studies. Investigated different product options from the Texas Senninger Irrigation District Manager. Secured product samples and tested several combinations, settling on the Senninger LDN with a single pad equipped with the CM1 insert and Shroud at 10 psi. Converted two machines in 2015 to 30" spacing between hose drops and 18" applicator height. The results were water and energy savings.

Image 2. LEPA applicators were at close spacing of 30-inches, 18-inch height, with 10-psi pressure regulators.

Image 3. LEPA applicators demonstrated increased irrigation efficiency by making more water available for plants.

Irrigation efficiency is much higher

- Less water is used – Reduction of 152.7 Acre-Feet Water Applied (control ranch) – 8.7% with LEPA; Reduction of 272.9 Acre-Feet Water Applied (all other ranches) – 17.8%
- Wind loss is reduced
- Less evaporation loss
- Electric costs are reduced as consumption is down – 170,811 kWh energy reduction = $16,620 savings/year

Additional benefits

- Higher yield production
- Savings by reducing varmint damage

In Depth Testing

In order to learn what was actually contributing to the improved results, Triple D Ranches partnered with Senninger on a new test (beginning July 2016). This would include a span to span direct comparison measuring the following: Yield, Water Penetration, Salinity, Soil temperature, Total Water applied, ET. This allowed for a more direct comparison of moving deflector sprinklers with LDN LEPA applicators within the same field reducing the variability of different soil types and plant ages.

Test Description

The test was set-up with three spans of moving deflector sprinklers and four spans of LDNs with LEPA pads (C33/CM1) and Shrouds ...

- Sprinklers were at 120-inch spacing, 36-inch height, with 10 psi pressure regulators.
- LEPA applicators were at close spacing of 40-30 inches, 18-inch height, with 10-psi pressure regulators.
- Balanced flow rates consistent with their pivot span location.
- Two soil moisture probes employed per pivot – one under span with moving deflector sprinklers and one under span with LEPA LDNs.

Table 1. Test Pivot Specifications. Pivots have similar characteristics. There were slight variations in the soil. The crop was the same seed variety and the field inputs were the same.

	Pivot 2	*Pivot 7*
Area:	129.84 acres	125.82 acres
Pivot Manufacturer:	Raincat	Raincat
Machine Flow:	900 gpm	800 gpm
Pivot Pressure:	40 psi	40 psi
Machine Length:	1342 ft	1321 ft
Distance to Last Tower:	1282 ft	1261 ft
Speed of Last Tower:	12.51 ft	12.51 ft
Precipitation/Acre:	6.93 gpm	6.36 gpm
Time for Coverage:	10.73 hrs	10.56 hrs
Soil:	Silt clay loam	Heavy silt clay loam
Elevation Change:	17 ft	6 ft
Crop:	Alfalfa	
Field Inputs:	Sulfur, Nitrogen, Zinc, Phosphorus	

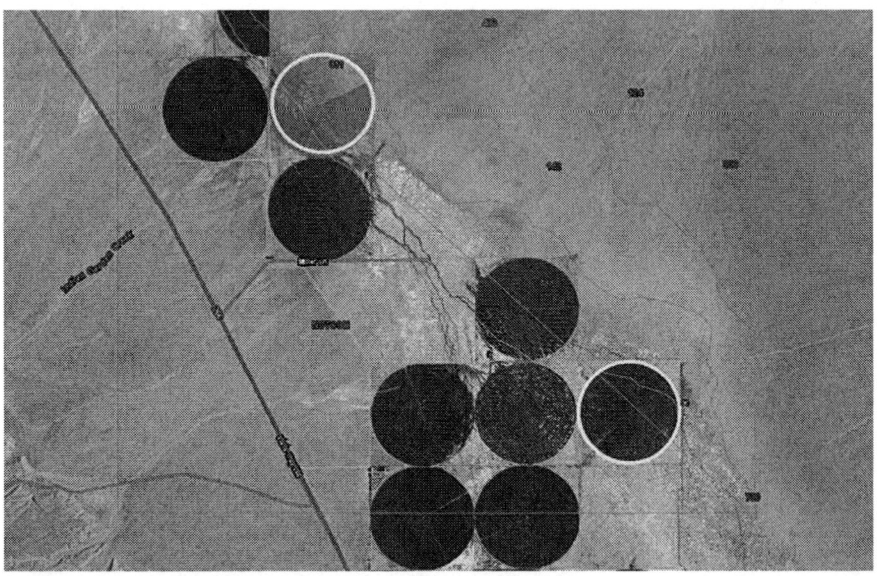

Image 4. Locations of Pivot 2 (North) and Pivot 7 (Oasis).

Climate Characteristics

- Mean annual precipitation: 4 to 9 inches
- Mean annual air temperature: 50 to 54 degrees F
- Frost-free period: 130 to 155 days

Soil Water Content (inches)

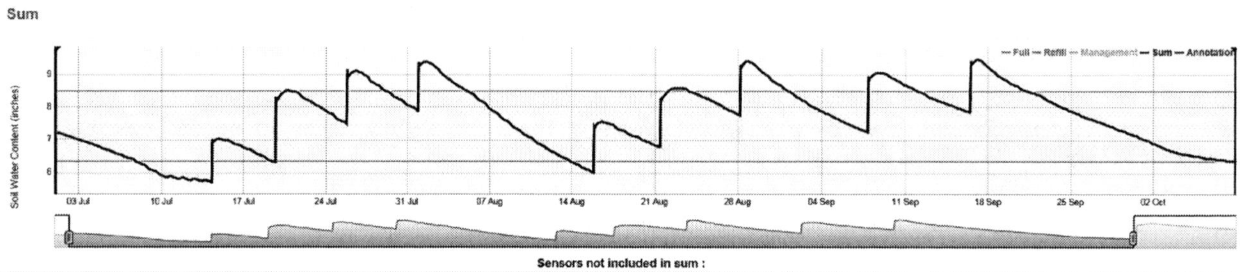

Figure 1. Test Pivot 2 span results for Close Spacing LEPA LDNs.

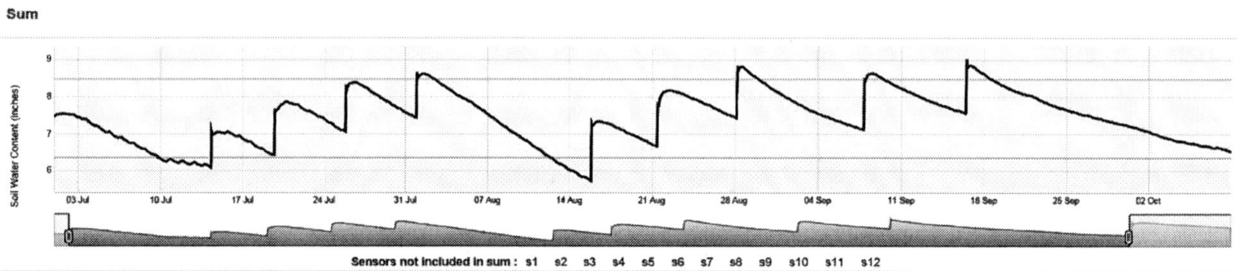

Figure 2. Test Pivot 2 span results of sprinklers.

Conclusion

Close Spacing irrigation using shrouded LEPA applicators improves irrigation efficiency in this semi-arid climate. Losses to wind drift and evaporation are minimized which means more water is available for plant use. Low pressure operation saves pumping energy. This application creates an undesirable habitat for field varmints and thereby reduces labor and repair costs. With only two cuttings, the impact on yield is yet inconclusive. Further research in the 2017 season is needed to adjust the irrigation prescription to reduce the amount of water applied with LEPA LDNs to better determine specific water and energy savings and yield variation.

References

Crop Metrics is managing the soil moisture probes. The login page shows information about the probes. http://cropmetrics.com/. October 18th, 2016.

Snyder, A. D. "Efficiency and uniformity of the LEPA and spray sprinkler methods: A review." ResearchGate. Transactions of the ASAE, pp. 937-944, Vol. 43, No. 4, July 2000. www.researchgate.net/publication/43255576_Efficiency_and_uniformity_of_the_LEPA_and_spray_sprinkler_methods_A_review. October 18th, 2016.

National Resources Conservation Service. United States Department of Agriculture. www.nrcs.usda.gov. October 18th, 2016.

Station Monthly Time Series. *Western Regional Climate Center.* October 18th, 2016. www.wrcc.dri.edu/cgi-bin/rawMAIN.pl?nvdyer. October 18th, 2016.

Lowering Average Application Rate (AAR) Expands Potential of Center Pivots

Jon R. Johnston

President, Irrigation Accessories Co. – IACO
PO Box 820118, Vancouver WA 98662
jon@boombacks.com

Abstract: *Technical advancements have eliminated the single greatest limitation of Center Pivots. According to Dr. Brad King "The main disadvantage of Center Pivot irrigation systems is the high water application rates under their outer spans to compensate for the increased rate of travel." In lowering the AAR on the last third of any Center Pivot by installing 15' Boombacks on every outlet in opposing directions, Center Pivots are now a viable option for any soil type.*

AAR is an often misunderstood concept. Even seasoned irrigators may think the speed of rotation can effect AAR. We have created an innovative learning program with custom animation that simplifies this complex concept.

We have incorporated research on AAR by Dr. Howard Neibling and on runoff by Dr. Troy Peters. We also include the experience of two of the nation's largest growers in California who now use Center Pivots where they believed it impossible, reducing water usage by 33% and increasing yields by 37%.

This presentation brings together university research, real world farm data and modern teaching techniques to explain how using Boombacks to lower AAR will expand the use of Center Pivots to what was previously thought unsuitable land - such as tight soils, hilly ground or on any soil with a low infiltration rate.

Keywords: Boomback, Boombacks, Boom, Booms, Offset, Offsets, Center Pivots, Pivots, Average Application Rate, Sprinklers, Efficiency, Reducing Waste, Reducing Pollution, Conservation, Conserving Water, Optimal Performance, Energy Savings, Wetted Footprint, Water Savings, Increased Yields, Increased Crop Quality, Increased Crop Uniformity, Runoff Eliminated, Increased Soak Time.

Introduction: In 1997 Bradley King and Dennis Kincaid with the University of Idaho published a paper entitled ***Optimal Performance from Center Pivot Sprinkler Systems***. In this paper they stated, *"The main disadvantage of Center Pivot irrigation systems is the high water application rates under their outer spans to compensate for the increased rate of travel."*

On a ¼ mile pivot, with 180 foot spans, the last tower travels eight times faster than the first tower. Sprinkler packages are designed with incrementally increasing nozzle sizes to compensate for the increasing rate of travel and the increased area of coverage.

That speed difference translates into 2.35 feet per minute at the first tower and 18.8 feet per minute on the last. In the same amount of time, the first tower is covering just 2.34 acres compared to 35.04 acres on the last tower, nearly 15 time more area covered in a single revolution.

Pivot Rotation Speed and Area Covered:

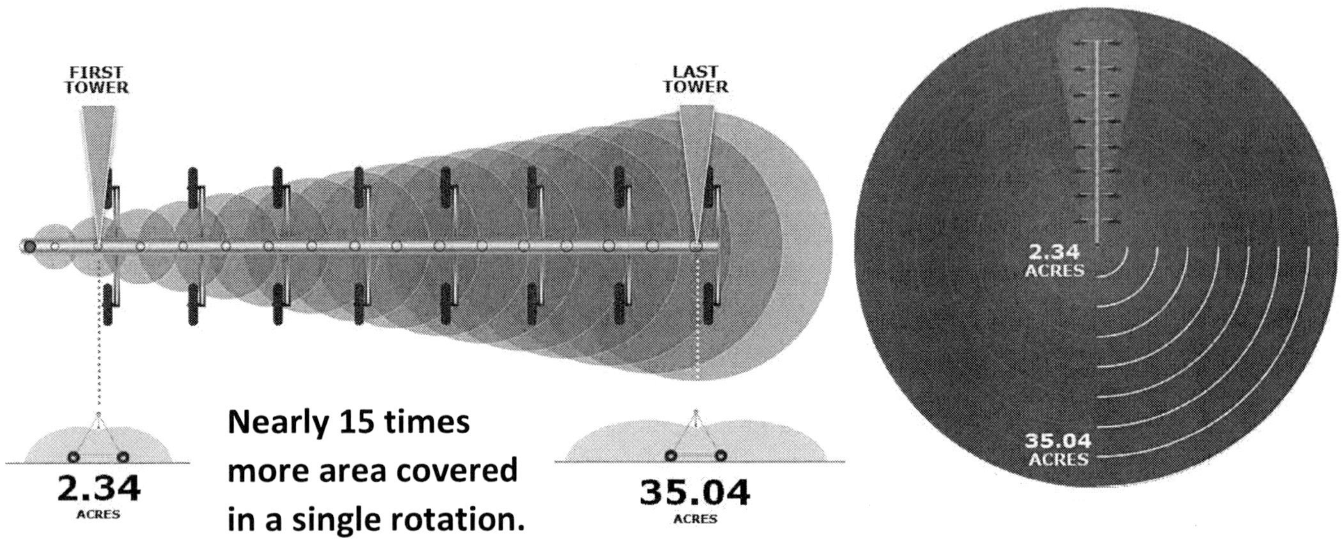

In that same paper King and Kincaid determined *"Application rates under the outer spans of the standard quarter-mile-long low pressure center pivot normally exceed infiltration rate and result in runoff."*

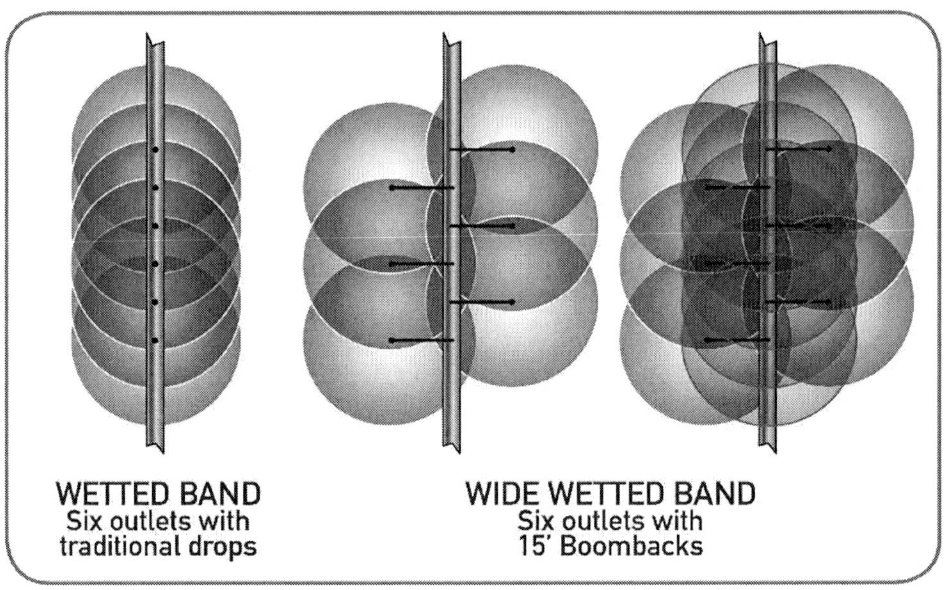

But they did not just identify a problem. They recommended a solution. *"The application rate of low pressure spray sprinklers can be reduced by using offset booms on alternate sides of the center pivot lateral."*

Boombacks, sometimes called offset booms, are add-on accessories for Pivots or Linears to re-locate or offset the sprinkler 15' away from the center line of the span.

Innovative Boom Technology, putting Boombacks on every outlet – alternating in opposing directions - on roughly the last third of a pivot, can eliminate runoff and greatly reduce the occurrence of soil sealing - also known as crusting.

Spreading out the sprinklers using Boombacks, *even with the larger nozzle sizes*, can lower the *Average Application Rate* and decrease the *Application Intensity*.

Changing from drops to Boombacks gains a 66% increase in wetted footprint. You do not change the *amount of water* you are putting on, but you do change the *amount of time* it takes to put that water on.

Boombacks increase the amount of time it takes to put on the same amount of water. It takes longer for the Boomback spray pattern to move past the same point as the traditional drop spray pattern. By using Boombacks to create a larger wetted footprint, soak time is increased. Dr. Howard Neibling created a graph to depict the advantages of Boombacks. His graph depicts infiltration rates. The dotted line is the infiltration rate of the soil. It will vary by the type of soil, but all soil types have the same basic curve. The line may be higher up on the graph for tight clay soils or lower down on the graph for looser sandy soils. The bottom of the graph represents time. As indicated, in the beginning the soil can absorb more water.

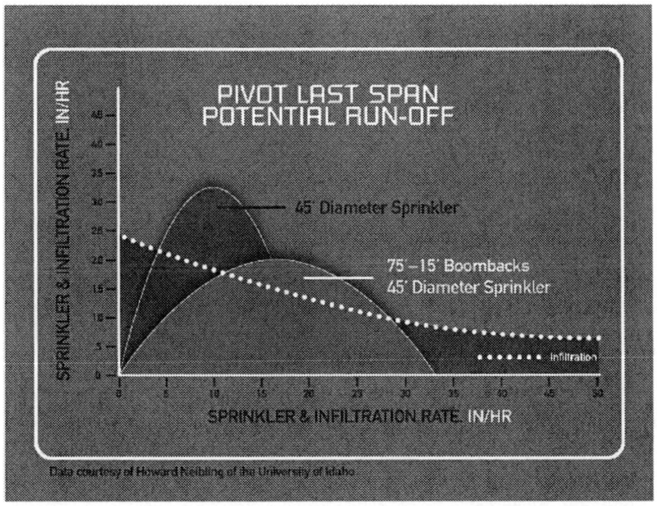

Over time the ground becomes saturated and is no longer able to absorb water at the same rate. The curve of the infiltration rate almost flattens out over time. The first arc is from a 45 foot diameter sprinkler. Anything above the dotted infiltration rate line is potential runoff. The objective is to flatten out the application curve to match the soil infiltration rate.

Typically, the first thing irrigators do when they see runoff is to speed up the pivot. What happens when you speed up the pivot? You *reduce* the amount of water being applied. You cannot control the Average Application rate by the speed of the pivot. The Average Application Rate is a constant determined by contributing factors other than the speed of the pivot. The average application rate is the same if the pivot is sitting still or moving at maximum speed. The only thing you control with the speed of the pivot is the depth of water being applied.

Speeding up the Pivot may have eliminated any runoff, but now you are not getting the root penetration or the appropriate amount of water necessary to maximize yields. On tight soils, such as clay, or hilly ground, or on any other land experiencing runoff, speeding up the Pivot is not a solution.

Using Boombacks to increase soak time also lowers application intensity, reducing the likelihood of soil sealing. A point under any traditional drop spray pattern will be watered by five different sprinklers. All sprinkler manufactures recommend a 200% overlap for optimal uniformity. The impact to the ground by five sprinklers increases application intensity and can lead to soil sealing or crusting. Using Boombacks can reduce the application intensity significantly by spreading out the sprinklers.

A Theoretical Example: Consider a ¼ mile pivot, with traditional drops moving at average speed, putting on 1.25 inches of water. Presumably, the calculations show this is the required amount of water the crop needs. If everything works correctly there is no runoff - the application rate has matched the infiltration rate of the soil.

But what if there is runoff? What if the water being applied is not being absorbed into the ground? Then the crop is distressed, not getting the root penetration required for maximum yields. In addition, runoff erodes fertile soil. It also carries fertilizer and other chemicals into the surrounding soil, polluting groundwater and streams.

By adding Boombacks to the above example, you are still applying the same amount of water, 1.25 inches, but over a longer period of time. We have increased the wetted diameter, lowered the Average Application Rate, and more closely matched the soil's infiltration rate. We increased root penetration, eliminated runoff and are poised for maximum yields. We also lowered the Instantaneous Application Rate, greatly reducing the chances of soil sealing. We are using water more effectively.

Average Application Rate is the rate at which the depth of water increases if applied uniformly throughout the wetted area.

The formula is:

$$\textbf{AAR (in./hr.)} = \frac{\textbf{(Flowrate x dfp)}}{\textbf{72 x Cov}}$$

There is one constant, 72, and three variables. 1. **Flowrate:** the system flow, gallons per minute per acre. 2. **Dfp:** distant from the pivot point in feet. 3. **cov:** the sprinkler throw diameter (coverage) in feet. It is the *Flowrate* ***times*** the *distance from the pivot* ***divided*** by 72 ***times*** the *throw diameter* of the sprinkler.

Here is a specific example. On this particular pivot we have a Flowrate of 6 gallons per minute per acre. The pivot is 1300 feet long. Using Nelson Rotators with an Orange plate and a 20 PSI regulator at 9 feet height on traditional drops, gives a 72 foot throw diameter. Doing the math, we have a 1.5 inches per hour Average Application Rate.

$$\textbf{AAR (in./hr.)} = \frac{(6 \times 1300)}{72 \times 72} = \textbf{1.5 Inches per Hour}$$

Here is that same example with 15' Boombacks. We still have the 6 gpm/acre and the 1300 feet from pivot point, but now we have a wider wetted band. Because of the 15' Boombacks, we have increased the throw diameter by 30 feet. So, instead of the 72 feet, the throw diameter is now 102 feet.

$$\textbf{AAR (in./hr.)} = \frac{(6 \times 1300)}{72 \times 102} = \textbf{1 Inch per Hour}$$

We now have an Average Application Rate of 1 inch per hour, a 33% decrease.

We've lowered the Average Application Rate and the Instantaneous Application Rate. We've increased the soak time, eliminated runoff, and reduced the chances of soil sealing.

A 2014 study by Dr. Troy Peters of Washington State University (WSU) at Prosser, Washington: **Efficacy of Boom Systems in controlling runoff under center pivots and linear move irrigation systems,** looked at how much Boombacks reduce runoff compared to traditional drops. The study does not look at yields or crop quality. For this particular study, Dr. Peters quantified how much runoff is actually reduced by using Boombacks.

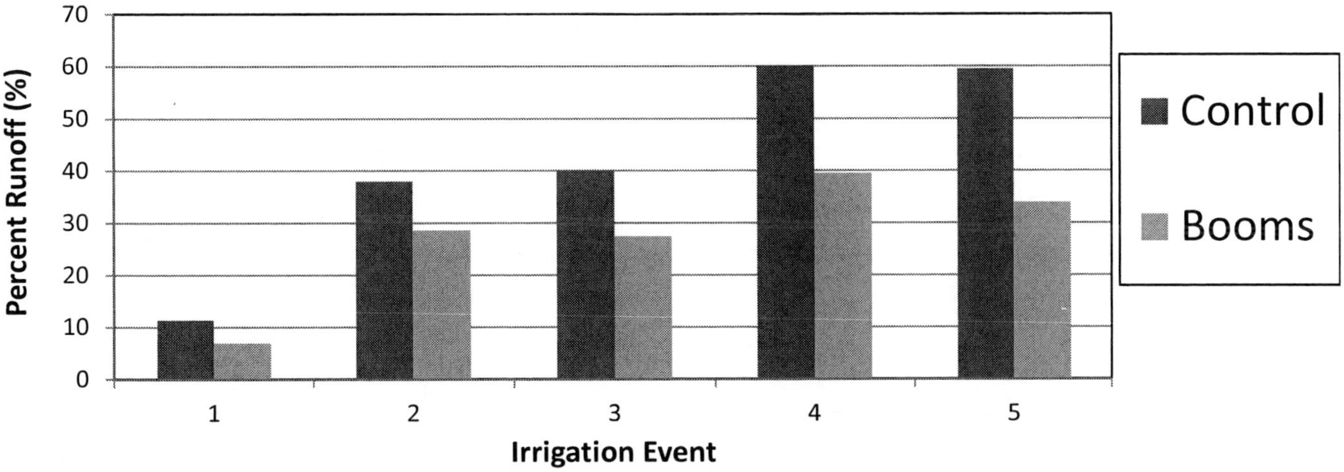

The numbers on the bottom of the chart depict the irrigation event. The dark blue bar represents runoff from traditional drops. The light gray represents runoff from Boombacks. For this study, WSU overwatered until they had some runoff to measure from both systems so that they could compare the difference. The first event yielded a 4% reduction using Boombacks. Referring back to the graph representing soil absorption rates, you will recall that when the ground is dry it can absorb more water. Once they reached the 5th event, the reduction in runoff by using Boombacks had already jumped to a significant 24%.

Dr. Peters' study was published and is available online at the ASABE Technical Library, Volume 30, Issue 5 of *Applied Engineering in Agriculture*.

Real World Results: A large grower in California wanted to germinate carrots seeds with a pivot. The application rate required was too high for the ground to absorb - especially at the end of the pivot. They installed Boombacks on every outlet in opposing directions on the last third of a Pivot. Over a three year period, the solution was so effective they installed several more pivots. After the first three years, an interview with the grower revealed valuable results. Here is what they had to say.

"We have compiled data on all the different ways of irrigating, Pivots, Pivots with Boombacks, Handlines, Wheelines, Drip and Flood. We determined using Pivots with Boombacks is more than a solution for carrot germination. We are now putting the water into the ground more efficiently and reaping the benefits."

- 33% Water Savings
- 37% Increase In Yields
- Increase Crop Quality
- Increase Crop Uniformity
- Runoff Practically Eliminated

"The 33% water savings is coming from the eliminated runoff. We are now achieving deeper root penetration. All the listed benefits are generated by a more efficient use of water. When compared head to head, Pivots with Boombacks are by far the most economical way to irrigate."

Conclusion: There is now the possibility to utilize the advantages of Center Pivots in areas and on soils previously thought unsuitable for this type of irrigation.

The primary need for the expanded use of Center Pivots is the dissemination of the results of these studies and the real world experience of successful and innovative growers.

A lower Average Application Rate produces an increase in yields, improved crop quality and uniformity - all while conserving water.

Using Boombacks to lower the Average Application Rate (AAR) increases the potential for Center Pivots, including areas with a low infiltration rate or anywhere runoff is a problem.

REFERENCES:

King, B.A. and D.C. Kincaid, 1997. Optimal Performance from Center Pivots Sprinkler Systems. University of Idaho, http://www.cals.uidaho.edu/edComm/pdf/BUL/BUL0797.pdf

Nakawuka, P., R. O. Okwany, T. R. Peters, K. Desta and S. H. Sadeghi, 2014. Efficacy of Boom Systems in controlling runoff under center pivots and linear move irrigation systems. Published online @ ASABE Technical Library--Volume 30, Issue 5 of *Applied Engineering in Agriculture*. http://www.boombacks.com/wp-content/uploads/Efficacy_Of_Boombacks.pdf

Hines, S. and H. Neibling, 2013. Center Pivot for Corn: Water Management and System Design Considerations in Southern Idaho. University of Idaho Extension Bulletin 881. 10 pp. http://digital.lib.uidaho.edu/cdm/ref/collection/ui_ep/id/31492

This page intentionally left blank.

A Dynamic Variable Rate Irrigation Control System

Mr. Calvin Perry

C.M. Stripling Irrigation Research Park, University of Georgia, Camilla, GA 31730.

Dr. Vasileios Liakos

Crop and Soil Sciences Department, University of Georgia, Tifton, GA 31793.

Dr. Wesley Porter

Crop and Soil Sciences Department, University of Georgia, Tifton, GA 31793.

Mr. Michael Tucker

Crop and Soil Sciences Department, University of Georgia, Tifton, GA 31793.

Dr. George Vellidis

Crop and Soil Sciences Department, University of Georgia, Tifton, GA 31793.

Abstract

Currently variable rate irrigation (VRI) prescription maps used to apply water differentially to irrigation management zones (IMZs) are static. They are developed once and used thereafter and thus do not respond to environmental variables which affect soil moisture conditions. Our approach for creating dynamic prescription maps is to use soil moisture sensors to estimate the amount of irrigation water needed to return each IMZ to an ideal soil moisture condition. The UGA Smart Sensor Array (UGA SSA) is an inexpensive wireless soil moisture sensing system which allows for a high density of sensor probes. Each probe includes three Watermark sensors. We use a modified van Genuchten model and soil matric potential data from each probe to estimate the volume of irrigation water needed to bring the soil profile of each IMZ back to 75% of field capacity. These estimates are converted into daily prescription maps which we downloaded remotely to a VRI controller thus creating a dynamic VRI control system. During 2015, we conducted an on-farm experiment to assess our system. We worked with a producer in a 230 ac (93 ha) field in southwestern Georgia. The field was divided into alternating conventional irrigation and dynamic VRI strips with each strip 120 rows wide. The conventional strips were irrigated uniformly based on the producer's recommendations. We divided the VRI strips into IMZs and after planting we installed UGA SSA probes in each of the IMZs. The data from the probes were used to develop daily irrigation scheduling recommendations for each IMZ. The recommendations were converted into a daily prescription map and downloaded remotely to the pivot VRI controller. When an irrigation event was initiated, the VRI-enabled pivot responded dynamically to soil moisture conditions. We will present the design of our dynamic VRI control system and the results from the 2015 study.

Keywords. VRI, wireless, soil moisture sensors, management zones, peanuts, center pivot

Introduction

Irrigation is becoming an essential component of farming in many areas of the world because it is a tool for ensuring food security. Irrigation not only serves to reduce risk of crop loss but also to build resiliency to climate variability and yield stability in food production systems. Irrigated agriculture provides 40% of the world's food while being used on only 18% of the cultivated land (FAO, 2015). The United Nations Food and Agricultural Organization estimates that the world currently consumes about 70% of available fresh water for irrigation (FAO, 2015). This results in growing competition for available fresh water supplies between agriculture, industry and residential uses. An indicator of this competition is that during the last few decades, ground water is depleting at an alarming rate in many agricultural areas. In addition, agriculture will need to produce more food to address the needs of a growing population. If irrigated agriculture is to expand in order to meet growing demands for food, then new irrigation practices and tools must be developed for more efficient water use. Precision irrigation is one possible approach (Vellidis et al., 2013).

Precision irrigation, like many other aspects of precision agriculture, has the goal of applying inputs, which in this case is irrigation water, where needed and when needed. The when needed is a particularly important aspect of precision irrigation because timing of irrigation applications is equally, if not more important, than the amount of irrigation water applied during a growing season (Vellidis et al., 2016). Vories et al. (2006) found that improper timing of irrigation on cotton can result in yield losses of between USD 150/ac (370/ha) to USD 750/ac (1850/ha).

Variable Rate Irrigation

Precision irrigation has its roots in variable rate irrigation (VRI) technology developed for center pivot irrigation systems by the University of Georgia (UGA) Precision Agriculture team in 2001 (Perry et al., 2002; Perry and Pocknee, 2003). The UGA Precision Agriculture team recognized that variable rate application of irrigation water was a key enabling technology for adoption of precision agriculture in the Southeast. This was because fields in this region are highly variable in soil type and texture, moisture holding capacity, and slope. Ignoring site-specific water needs while attempting to vary other inputs like fertilizers would not result in the desired efficiency gains theoretically possible by using precision agriculture. In the Southeast, irrigation of agronomic crops is now done mostly by center pivots. Conventional center pivots apply the same rate of water along the entire length of the pivot and cannot account for within-field variability or non-farmed areas. Because of this, the UGA Precision Ag team focused on development of VRI for pivots.

Several pivot irrigation manufacturers now offer their own VRI systems. VRI allows center pivots to vary water application rates along the length of the pivot by using electronic controls to cycle sprinklers and control pivot speed. Sprinklers are controlled individually or together typically in groups of 2 to 10 depending on the level of resolution desired by the farmer. Each group or bank of sprinklers represents a grid with a 1 to 10 degree arc in which the irrigation water application rate can be set as percentage of the normal application rate – for example from 0% to 200% of normal (Figures 1 and 2). The number of degrees in the arc is determined by the level of resolution desired.

Figure 1. VRI-enabled pivot at UGA's Stripling Irrigation Research Park being used to vary irrigation application rates over research plots.

A 50% application rate is half the normal rate and is achieved by cycling the sprinklers on and off every 30 seconds. A 150% application rate is achieved by leaving the sprinklers on continuously while decreasing the travel speed of the pivot by 50%. If other grids along the length of the pivot require lower application rates, the VRI controller adjusts the sprinkler cycling pattern within those grids accordingly. VRI can be installed retroactively on most existing pivots. Installation costs vary widely by brand and are also a function of the length of the pivot and the level of resolution desired by the farmer to address the variability of the field. Application rates are determined from an application or prescription map.

The prescription map for each field is typically developed jointly by the farmer and VRI dealer on desktop software (Figure 2) and then downloaded to the VRI controller on the pivot. The field is divided into irrigation management zones (IMZs) and application rates assigned to each of the IMZs using whatever information is available. At the moment, the prescription maps are static. In other words, they are typically developed once and used thereafter. Static prescription maps do not respond to environmental variables such as weather patterns and other factors which affect soil moisture condition and crop growth rates. So although VRI is a great leap forward in improving water use efficiency, the system could be greatly enhanced by having real-time information on crop water needs to drive the application rates. One approach for creating *dynamic* prescription maps is to use soil moisture sensors to estimate the amount of irrigation water needed to return each IMZ to an ideal soil moisture condition (Figure 2). The goal of this work was to develop a dynamic variable rate irrigation control system by coupling real-time soil moisture sensing networks with an irrigation scheduling decision support tool and VRI.

Methods

The operational paradigm for our dynamic VRI control system is that the field is divided into IMZs and a soil moisture sensing network with a high density of sensor nodes is installed to monitor soil condition within the zones and provide hourly soil moisture measurements to a web-based user interface. At the interface, the soil moisture data are used by an irrigation scheduling model running in the background to develop irrigation scheduling recommendations by IMZ. The recommendations are then approved by the user (farmer) and downloaded wirelessly the VRI controller on the center pivot as a precision irrigation prescription. When the center pivot irrigation system is engaged by the farmer, the pivot applies the recommended rates.

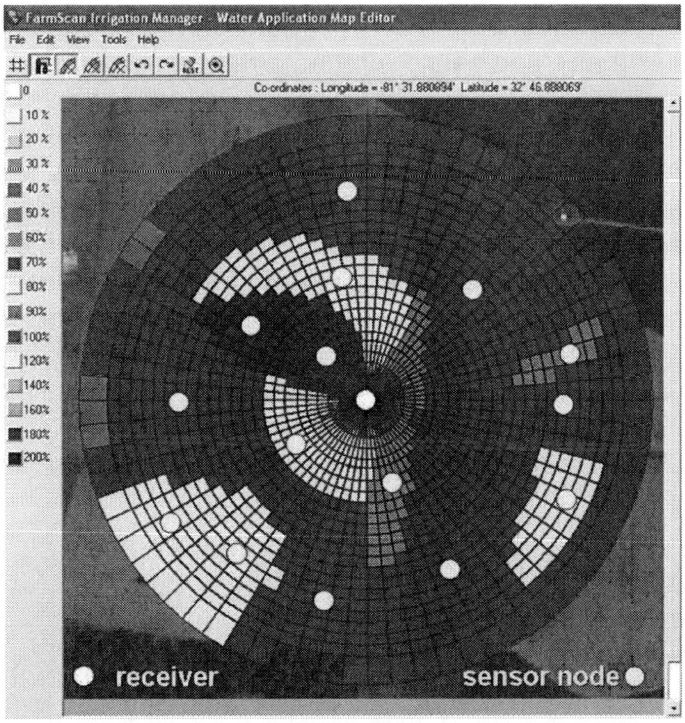

Figure 2. VRI prescription map for a 126 ac (51ha) field in Georgia. Grids represent discreet areas which can receive unique application rates. The yellow circles represent potential locations of soil moisture sensor nodes.

The UGA SSA is an inexpensive wireless soil moisture sensing system which allows for a high density of sensor nodes – a feature needed to account for soil variability and enable dynamic prescription maps. The UGA SSA was developed by the UGA Precision Ag Team and licensed to Advanced Ag Systems (Dothan, Alabama) during 2014. It became commercially available on a limited scale during 2015.

Figure 3. A UGA SSA sensor node has a low profile when installed in the field. The flexible whip antenna allows field vehicles to pass directly over the node.

The UGA SSA consists of smart sensor nodes and a base station. The term sensor node refers to the combination of electronics and sensor probes installed within a field (Figure 3). The electronics include a circuit board for data acquisition and processing and a radio frequency transmitter. In the current design, the UGA SSA supports Watermark® soil moisture sensors. Each soil moisture probe integrates up to three Watermark sensors as shown in Figure 3. In addition, each node supports two thermocouples for measuring soil and/or canopy temperature. For field crops like cotton or maize, the sensors on the probe are arranged so that when installed they are at 8, 16 and 24 in (20, 40, and 60 cm) below the soil surface although any combination of depths is possible. Soil moisture is measured in terms of soil matric potential and reported in units of kPa. A Synapse brand radio frequency (RF) transmitter is responsible for transmitting sensor data. The transmitter is an intelligent, cheap, and low-power 2.4 GHz radio module. At the center of each field, a base station receives the data from all nodes at hourly intervals. The base station stores the data on a solar-powered netbook computer and transmits the data via cellular modem to a FTP server hourly.

A wireless mesh network is used for communication between the nodes. Data are passed from one node to the other through the RF transmitter which also plays the role of a repeater. If any of the nodes stop transmitting or receiving, or if signal pathways become blocked, the operating software reconfigures signal routes in order to maintain data acquisition from the network. The published range of the RF transmitter is 1640 ft (500 m) although we have observed its range to exceed 2460 ft (750 m) under field conditions.

To overcome the attenuating effect of the plant canopy, the RF transmitter antenna is mounted on spring-loaded, hollow flexible 0.24 in (6 mm) diameter fiberglass rod (Figure 3). Variable antenna heights are used to ensure that the antenna is always above the crop canopy. Rods which are 8.2 ft (2.5 m) long are used for low-growing crops like cotton, soybeans, and peanuts and rods which are 14.8 ft (4.5 m) long are used for tall crops like corn. This design allows field equipment such as sprayers and tractors to pass directly over the sensors without damaging them. This is a feature that is typically not found on other wireless soil moisture sensors as most of those require a solar panel to power the sensor and telemetry. The UGA SSA nodes are powered by two 1.5 V alkaline batteries which in our system have a life of more than 150 days. This typically spans an entire growing season. To optimize battery life, the nodes are programmed to be in a low-current sleep mode when not transmitting. The UGA SSA is described in detail by Vellidis et al. (2013) and Liakos et al. (2015).

To date the UGA SSA has been used primarily in farm fields irrigated by center pivots. The fields have been delineated into IMZs and one to three sensor nodes installed in each IMZ to characterize soil moisture during the growing season. Ten to 12 sensor nodes are typically installed in each field. The base station is usually located at the pivot point for easy access. The base station sends the node data to an FTP server hourly using a cellular modem. The data are also stored on commercial server space which can manage geographic data with different formats including the GeoJSON (Geographic JavaScript Object Notation) format. GeoJSON is used for visual representation of the data. The FTP server stores the raw soil moisture data while the commercial server manipulates and processes the raw data, stores them after applying a classification process, and serves as the interface with users through a dedicated website (www.ugassa.org).

Web-Based User Interface and Decision Support Tool

The purpose of the web-based interface is to allow users to visualize their soil moisture data and to make irrigation recommendations. The PHP (Personal Home Page) and Javascript programming languages were utilized to create different visualizations of the soil moisture data (Figure 4). The different visualizations provide users and especially farmers with the opportunity to better understand the soil condition and IMZ delineation within their fields. The website is smartphone compliant. To avoid the confusion of using negative numbers to report matric potential, data are reported in terms of soil water tension on the website.

In addition to data visualization, the web-based user interface incorporates a decision support tool which offers irrigation recommendations for each IMZ. We use a modified Van Genuchten model to convert the soil matric potential data to volumetric water content (Liang et al., 2016). The strength of the method is that it can use data readily available from USDA-NRCS soil

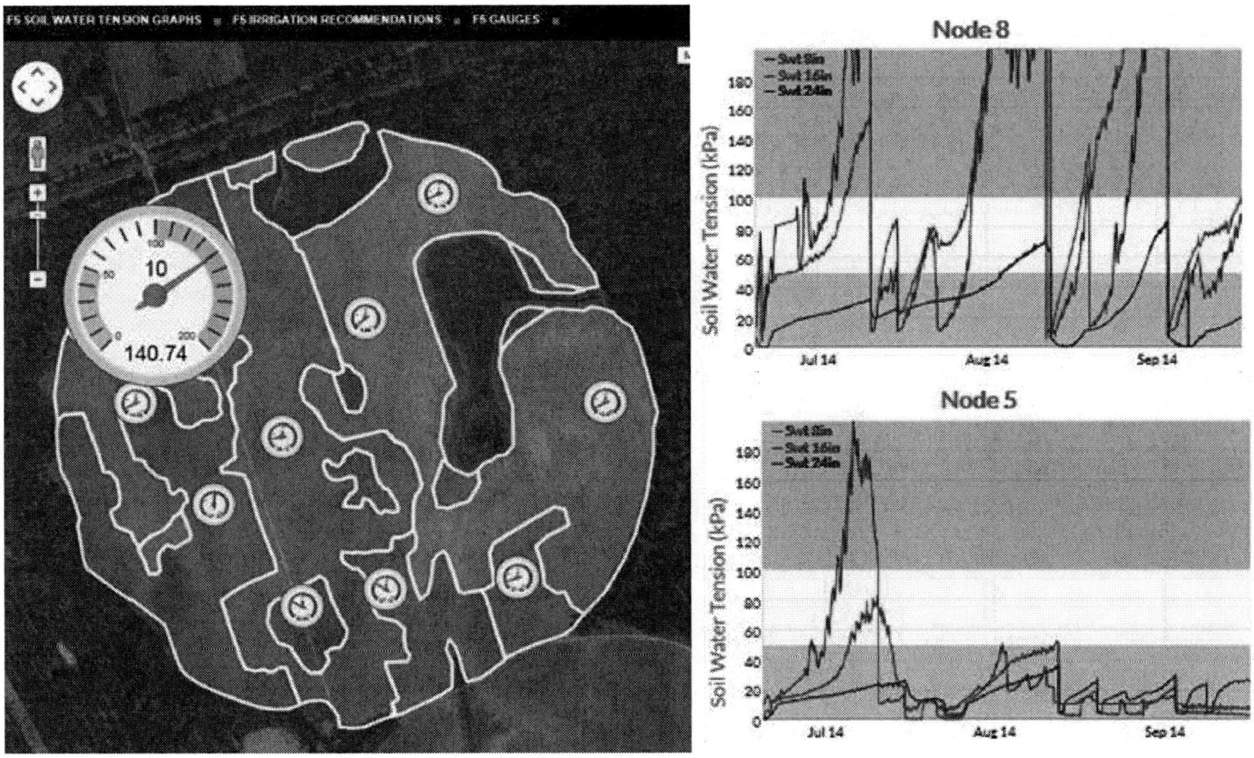

Figure 4. Two different visualizations of UGA SSA soil moisture data. On the left is current soil water tension displayed through color-coded gages. On the right are soil water tension curves for the entire growing season.

surveys to predict soil water retention curves and calculate the volumetric water content and soil water tension of a soil at field capacity. Those parameters are then used to translate measured soil water tension into irrigation recommendations which are specific to the soil moisture status of the soil. Soil properties for each IMZ are extracted from the NRCS web soil survey. Our application of the Van Genuchten model uses mean hourly soil matric potential data measured between 07:00 and 09:00 by all nodes within an IMZ to calculate the volume of irrigation water needed to bring the soil profile back to the desired soil moisture condition which could be field capacity or a percentage of field capacity (for example 75% of field capacity) (Figure 5). Each node's soil water tension value is a weighted average of the soil water tension values of the three Watermark sensors of the node. At this point, our irrigation recommendations use the same soil water tension threshold across all of the crop's phenological stages although that will be adjusted as more information becomes available from crop physiologists who are researching different irrigation thresholds (Meeks et al., 2016).

Field Testing of the Dynamic VRI Control System

During 2015, we initiated a dynamic VRI "proof-of-concept" study. We identified a producer who has fields equipped with VRI in southwestern Georgia. We used the 93 ha field shown in Figure 6 to conduct our study. The field was planted to peanuts (Arachis hypogaea). We divided the field into alternating conventional irrigation and precision irrigation strips with each strip 120 rows wide (Figure 6). We used aerial photographs, soil maps, soil electrical conductivity, topography, yield history, producers' knowledge of the fields and geostatistical software to

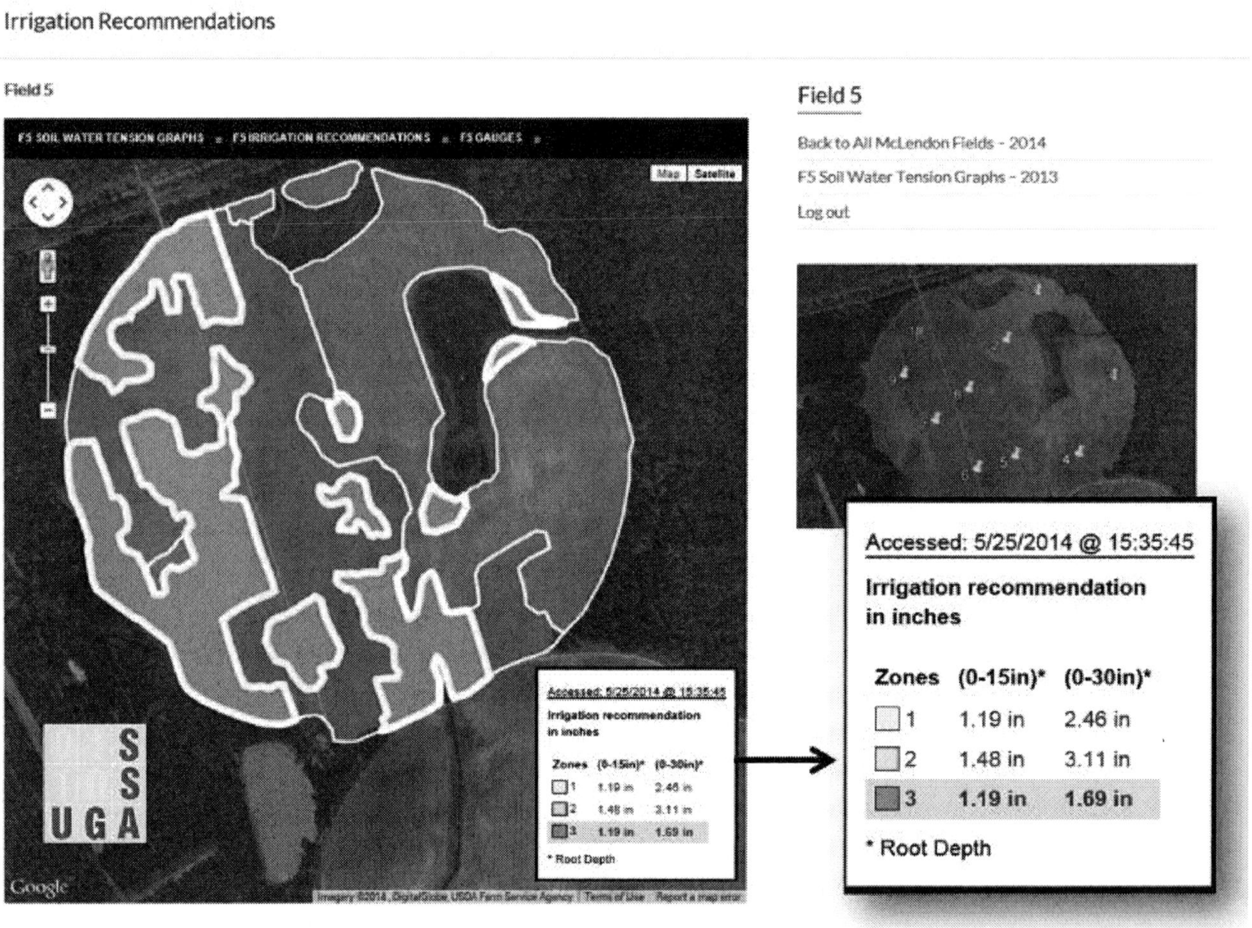

Figure 5. Irrigation recommendations are available daily for each IMZ through the UGA SSA web-based user interface.

develop irrigation management zones (IMZs) in the precision irrigation strips. After planting and establishment we installed UGA SSA sensor probes in each of the IMZs. Each probe contained three Watermark sensors. When the probes were installed the sensors were located at 4, 8, and 16 in (10, 20, and 40 cm) below the soil surface.

The data from the sensors was used to dynamically develop irrigation scheduling recommendations for each IMZ. A 50 kPa weighted mean soil water tension (SWT) was used to trigger irrigation in the VRI strips. The weighting function was (0.5×SWT at 10 cm) + (0.3×SWT at 20 cm) + (0.2×SWT at 40 cm). At each irrigation event, the mean SWT sensor data from each IMZ were automatically converted into irrigation recommendations using the decision support tool (Figure 7). The tool calculated the volume of irrigation water needed to bring the soil profile of each IMZ back to 75% of field capacity. The irrigation recommendations for each IMZ were then manually coded to the prescription map which was wirelessly downloaded to the pivot VRI controller prior to an irrigation event. In this field, approximately 72 hours were required for the center pivot irrigation system to circle the field. Because of this, a new prescription map was downloaded to the VRI controller every morning during an irrigation event. However, it was possible to download new prescription maps more frequently at hourly intervals.

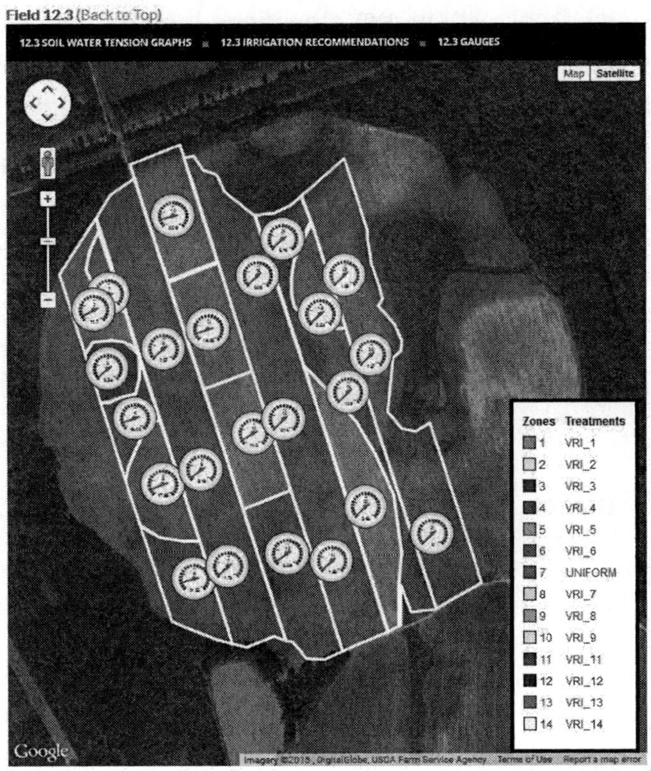

Figure 6. VRI Zones and field used for the 2015 on-farm VRI evaluation of dynamic VRI. The gages indicate the location of UGA SSA sensor nodes.

UGA SSA sensor probes were also installed in the conventional irrigation strips to monitor soil moisture conditions. The conventional strips were irrigated uniformly by the producer using Irrigator Pro (Davidson et al., 2000) for irrigation decisions. Irrigator Pro is a public domain irrigation scheduling tool developed by USDA which utilizes soil temperature, ambient temperature, and precipitation to provide yes/no irrigation decisions for peanuts. Total yield from each strip were measured by aggregating the weights of the truckloads of peanuts harvested from the strips.

Results

Precipitation during the 2015 growing season was 22 in (559 mm) which is slightly below the long-term mean precipitation for the period. As a result, irrigation during 2015 was truly supplementary to precipitation. Over the entire growing season, the dynamic VRI system (sensors + van Genuchten model + VRI) recommended an average irrigation amount of 3 in (76 mm) compared to 4.3 in (109 mm) by Irrigator Pro with approximately the same overall yields for both methods. The average yield for the dynamic VRI system strips was 4945 lb/ac (5543 kg/ha) while the average yield for Irrigator Pro strips was 4953 lb/ac (5552 kg/ha). However, there were yield differences between strips. The parallel strip design allowed us to directly compare yields between precision-irrigated and uniformly irrigated areas with similar soil and topographic properties and assess the benefits of dynamic VRI.

Because during the 2015 growing season the field received near mean precipitation, the dynamic VRI system outperformed Irrigator Pro in yield by 8.4% in the wetter areas of the field which were mostly areas of lower topographical relief. In contrast, Irrigator Pro outperformed dynamic VRI yields in sandy areas with higher elevations by 9.6% indicating that the 50 kPa irrigation trigger may have been too dry for these areas. Because the amount of plant available soil water is very small above 50 kPa in sandy soils, any delay in irrigation results in the SWT increasing rapidly and the crop experiencing water stress. In retrospect, it appears that the threshold for these areas should have been lower to account for time to irrigation. Figure 8 shows SWT graphs from two nodes in the field. The top graph is from a node in the northwestern area of the westernmost VRI strip. The SWT data line at 16 in (40 cm) (black line in Figure 8) clearly shows that for large periods of time, SWT at this depth was around 100 kPa and the plateaus on the graph indicate that the peanut roots were no longer able to extract water from the soil. In contrast, the lower graph which is from the easternmost uniform strip shows that the soil profile in this area was mostly saturated for the entire growing season.

Conclusions and Future Work

During 2015 we demonstrated that the technology and knowhow to implement dynamic VRI is available and feasible. The system performed well but our results indicate that we have more to learn about triggering irrigation in sandier soils. The harvest season was plagued by excessive rain which resulted in this field being harvested over a period of several weeks instead of the usual 3 to 4 days. Consequently, the yield difference observed could also be an artifact of harvest conditions. The experiment will be repeated in 2016 to incorporate lessons learned and to collect more data about the performance of the dynamic VRI control system. Our research goal for the next two years is to fully automate the process so that each morning, a farmer is able to view a dashboard similar to the one shown in Figure 9 and with two clicks enable dynamic VRI. By clicking the green "Download" button, the user would send the prescription map wirelessly to the VRI controller. A short video describing VRI and showing the VRI-enabled pivot used in this study is available at https://www.youtube.com/watch?v=DgexX_ITol0.

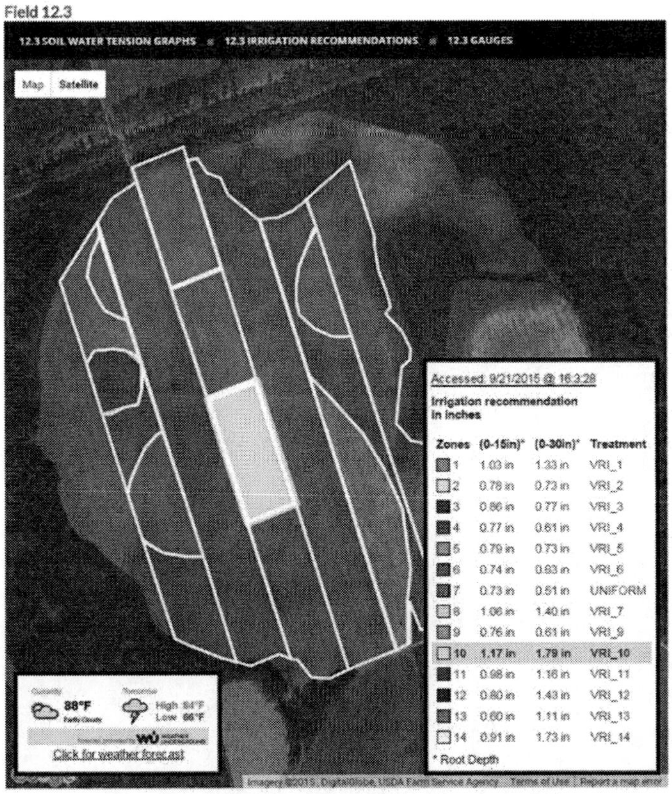

Figure 7. Dynamically developed irrigation scheduling recommendations for each IMZ. Clicking on either the zone or the recommendation will highlight both. In the figure, zone 10 is highlighted. The recommendations are to bring the soil profile to within 75% of field capacity.

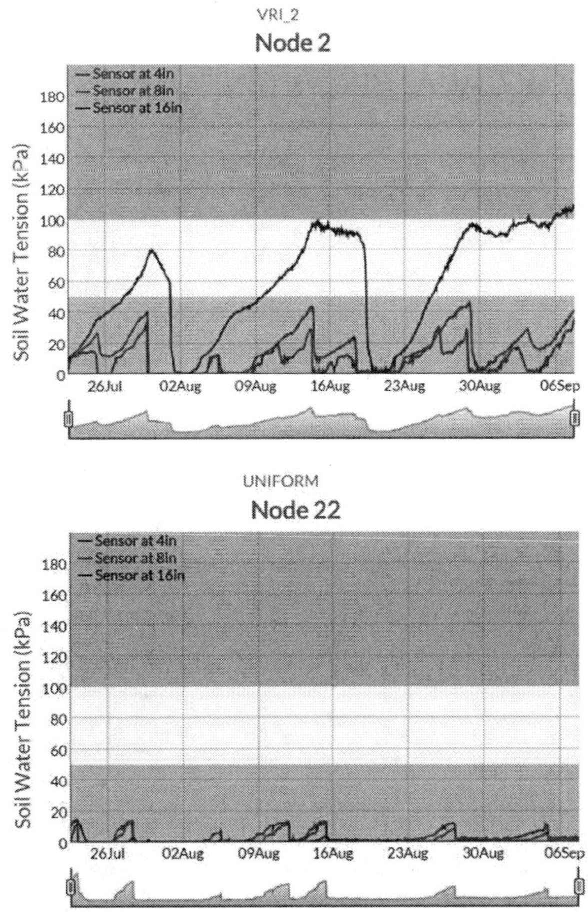

Figure 8. Season-long soil moisture data graphs from the VRI strip (top) and Uniform strip (bottom). The soil in the uniform strips is being maintained much wetter than in the VRI strips.

Acknowledgements

We thank Mr. Adam McLendon of McLendon Acres for allowing us to work in his field and engaging with us during this study. Funding for this study was provided by the National Peanut Board through the Southern Peanut Research Initiative and the Georgia Peanut Commission. Support for the development of the UGA SSA was provided by the Georgia Cotton Commission, Cotton Incorporated, the USDA-NRCS Conservation Innovation Grant (CIG) Program through the Flint River Soil and Water Conservation District, and by Hatch and State funds allocated to the Georgia Agricultural Experiment Stations. The authors would like to acknowledge Marcus, Cofield, Herman Henry, Rodney Hill, Anna Orfanou, and Dimitris Pavlou for their contributions to the project.

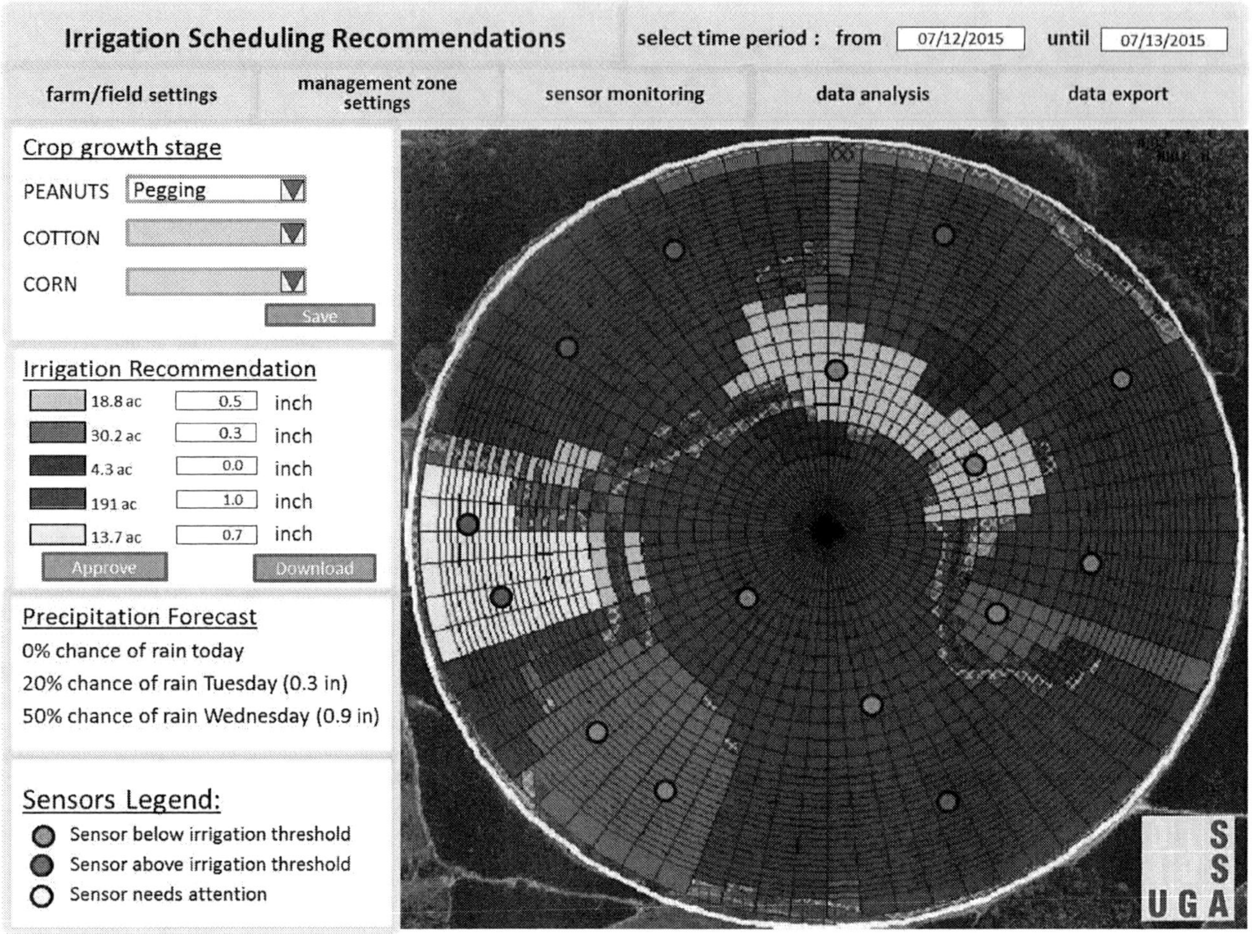

Figure 9. Mock-up of a dynamic VRI control system dashboard showing a prescription map of the field, location and status of soil moisture sensor nodes, irrigation recommendations for each IMZ, and approval and download buttons.

References

Davidson Jr., J.I., Lamb, M.C., Sternitzke, D.A., 2000. Farm suite-Irrigator Pro (Peanut irrigation software, and user's guide), The Peanut Foundation, Alexandria, VA.

FAO. 2015. Food and Agriculture Organization of the United Nations, Aquastat website. Rome, Italy: United Nations. www.fao.org/nr/water/aquastat/water_use/index.stm.

Liakos, V., G. Vellidis, M. Tucker, C. Lowrance, X. Liang. 2015. A decision support tool for managing precision irrigation with center pivots. In: J.V. Stafford (Ed.), Precision Agriculture '15 - Papers Presented the 10th European Conference on Precision Agriculture (10ECPA), Tel Aviv, Israel, p677-683, doi:10.3920/978-90-8686-814-8.

Liang, X., V. Liakos, O. Wendroth, and G. Vellidis. 2016. Scheduling irrigation using an approach based on the van Genuchten model. Agricultural Water Management (in press – accepted 23 May 2016).

Meeks, C., J. Snider, W. Porter, G. Vellidis, G. Hawkins, D. Rowland. 2016. Primed acclimation: An assessment to improve water use efficiency in a sensor-based irrigation system. In S. Boyd,

M. Huffman and B. Robertson (eds) Proceedings of the 2016 Beltwide Cotton Conference, New Orleans, LA, National Cotton Council, Memphis, TN, paper 17055, p.68.

Perry, C., Pocknee, S., Hansen, O., Kvien, C., Vellidis, G. and Hart, E. 2002. Development and Testing of a Variable Rate Irrigation Control System. ASAE Paper No. 02-2290. St. Joseph, MI: ASAE.

Perry, C. and Pocknee, S. 2003. Development of a Variable-Rate Pivot Irrigation Control System. Proceedings of the 2003 Georgia Water Resources Conference at the University of Georgia, April 23 - 24.

Vellidis, G., Tucker, M., Perry, C., Reckford, D, Butts, C., Henry, H., Liakos, V., Hill, R.W., Edwards, W., 2013. A soil moisture sensor-based variable rate irrigation scheduling system. In: J.V. Stafford (Ed.), Precision Agriculture 2013 - Proceedings of the 9th European Conference on Precision Agriculture (9ECPA), Lleida, Spain, p.713-720. doi: 10.3920/978-90-8686-778-3.

Vellidis, G., Liakos, V., C. Perry, W. Porter, and M. Tucker. 2016. Irrigation scheduling for cotton using soil moisture sensors, smartphone apps, and traditional methods. In S. Boyd, M. Huffman and B. Robertson (eds) Proceedings of the 2016 Beltwide Cotton Conference, New Orleans, LA, National Cotton Council, Memphis, TN, paper 16779, p.772-780.

Vories, E.D., Teague, T., Greene, J., Stewart, J., Clawson, E., Pringle, L., Phipps, B. 2006. Determining the optimum timing for the final irrigation on mid-south cotton. In. Proceedings National Cotton Council Beltwide Cotton Conference, January 3-6, 2006, San Antonio, Texas. 516-521, 2006 CDROM.

Using Aerial Imagery for Irrigation Management

Jake LaRue
Valmont Irrigation, Valley, NE
jlarue@valmont.com

Abstract
Water, energy and labor resources are often limited and the need for improvements in irrigation management continues. Often information for irrigation may come from field sensors, crop modeling, field scouting, watching what the neighbors do or a combination of all of these. These provide value but in most commercial cases are limited in the area of the field they adequately address and/or provide any level of current detail. Another information source gaining attention is the use of aerial imagery. Aerial imagery has the advantage of providing a complete view of the field but has many challenges. This paper will review the experiences of using aerial imagery over the last three and part of a fourth growing seasons (2013 through 2015, 2016) for irrigation management. The discussion will focus on using satellite, manned planes and UAV (drones) to collect images, methods of analysis and the challenges of each. The potential application of each for management of center pivots and sub-surface irrigated fields will be presented. Included will be some examples on how each performed and their value for irrigation management.

Keywords: Aerial image, center pivot, drones, irrigation management, satellites,

Background
Irrigation management is often called irrigation scheduling which is the process of evaluating factors and determining when to irrigate and how much water to apply (Evans 1996). Farmers' approaches to irrigation management vary greatly. Commonly used methods as identified by the USDA 2013 Irrigation Survey include but are not limited to condition of the crop, feel of the soil, soil moisture sensing device, commercial scheduling service, reports on daily crop water evapotranspiration, personal calendar schedule, computer simulation models and of course when neighbors begin to irrigate (USDA Farm and Ranch Irrigation Survey 2013). The overall driving force for adopting irrigation scheduling is economics – scheduling is used because it increases profits or decreases expenses. Nonetheless, even irrigators who find scheduling profitable will discontinue its use if it becomes too burdensome (Hennegler, 2013). Methods of irrigation scheduling can be broken down into three main categories – soil, plant and climate. One tool that can help maximize a farmer's limited crop management time while improving his decision-making ability is aerial photography (Reising, 2016). In most discussions of irrigation scheduling and management no mention is made of using aerial images.

Aerial images have been used in agriculture for many years primarily for providing general information about the field and/or crop. Use of aerial images in the irrigation industry has been confined primarily to providing information on the crop condition and performance of irrigation equipment. The primary challenges with aerial images, depending on their source, have traditionally been cost, the time lag from ordering the image until delivered to the end user and resolution. A key advantage of an aerial image is it provides a 'snap shot' of the entire field at one instant in time.

Sources of aerial images include but are not limited to satellites such as the Landsat8 which provides a variety of spectral bands, manned planes which typically provide infrared and RGB (red, green blue) color images and most recently UAV (unmanned aerial vehicle) commonly called drones which can provide a variety of image types depending on the cameras being used.

Valmont Industries Inc. has been using aerial imagery regularly for many years to help evaluate performance of different types of irrigation equipment such as but not limited to, center pivots, center pivots with corner arms and linears. In the past the primary focus has been on evaluating the sprinkler package.

With the development of VRI (variable rate irrigation) in 2009 Valmont began to use NDVI (normalized difference vegetation index) to assist with the evaluation of the crop's performance and response to VRI. In addition in most cases also collected were infrared and RGB images. All aerial images were collected using a manned plane. In 2010 to help better understand the performance and use of VRI Valmont added soil moisture sensing into key areas of the fields besides using aerial images. Until about 2011 all the images were collected using manned planes. In 2011 Valmont Industries tried a satellite service offering images with 5.0m resolution. Also in 2011 Valmont had aerial images collected and data provided on the chlorophyll and ground cover of a particular field. In each case the turnaround time was seven to twenty one days.

With all of the use of aerial images in the irrigation industry there has been little consideration of using the images for actual irrigation management primarily due to the lag time between when the image is scheduled for delivery and when it is delivered.

Methods

In 2013 Valmont expanded to try to make more use of aerial images for irrigation management. Due to the cost and time lag between images they were still ended up being only used to confirm what had already happened and not for making timely management decisions. One particular field called BF had two manned plane flights August 19th and then again on September 18th looking not only at the center pivot with a corner but also include a SDI (sub surface irrigation) areas of the field.

2014 saw the addition of UAV (unmanned aerial vehicle), drones for the collection of some aerial images. Valmont expanded the work to compare the information from the soil moisture sensors, the crop and aerial images in the BF field started in 2013 again both for the center pivot with the corner and the SDI. For the BF, three images were acquired the first at full canopy, second early reproduction and the third at early maturity. Again a manned plane was used to collect the images and the time lag precluded using for irrigation management.

In 2015 Valmont had the opportunity and tried satellite imagery again with the anticipation the images would be delivered timely with better resolution than had previously been experienced. This was in an attempt to truly move toward something approximating near real time irrigation management. The satellite imagery was to be delivered every seven to ten days with a resolution of 2.0m. Colorization to develop the NDVI map was done utilizing QGIS software by Valmont personnel. The newest colorized image was compared to the previous looking for change. Irrigation decisions were based on the change. Validation of the performance was based on field scouting, images taken from a plane (one image) and soil moisture sensors again for the BF field previously used. In addition crop yield based on the combine yield monitor was used to compare with the other information. The HP field was managed by the farmer using traditional methods and the BF field using the satellite imagery process. Unfortunately it was not possible to have replicated treatments within a single field due to the non-acceptance by the farmer.

For the 2016 crop season the same satellite company is being used as in 2015 with the expectation of improved delivery of images and faster turnaround of the colorization process. Valmont is continuing to use QGIS to create the NDVI information for the fields. In addition two other geographic areas were added with a variety of crops. Total included in the satellite project are three different areas of the United States with a total of thirteen fields and five crops including the BF field utilized in 2013 through 2015. Validation using field scouting, soil moisture sensors and crop yield has been expanded. An UAV company was contracted to do some work as well as a manned plane company for comparison.

Discussion

2013 Results - for the BF field the August 2013 flight NDVI indicated better crop health in the southern half of the field and poorer crop health in the northeast area of the center pivot. The NDVI also indicated areas of some uneven water distribution in the SDI fields on the east side, northwest corner and southwest corner. Adjustments were made to the SDI areas to compensate for what the August flight indicated. The September NDVI image was of no value from an irrigation management standpoint but did indicated the need to review why the patterns had developed seemly to indicate non-uniformity of the crop that could be associated with the center pivot corner. However the same non-uniformity did not show up in the yield data. The conclusion was the aerial images were helpful but the information was received too late to be of significant benefit in the short run. The information was valuable for planning for the next crop season. The NDVI images are shown in figures 1 and 2.

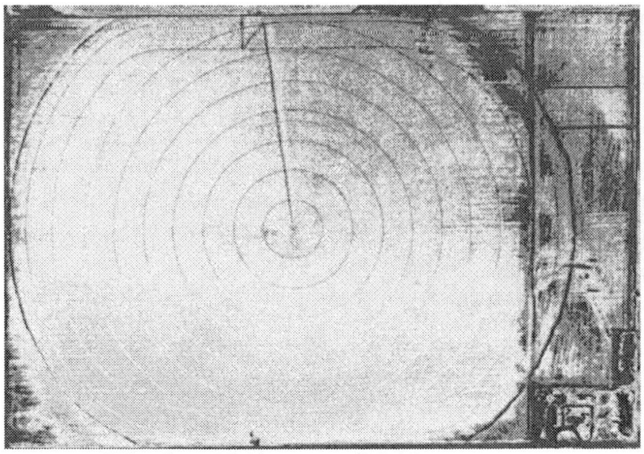

Fig 1 19/August/2013

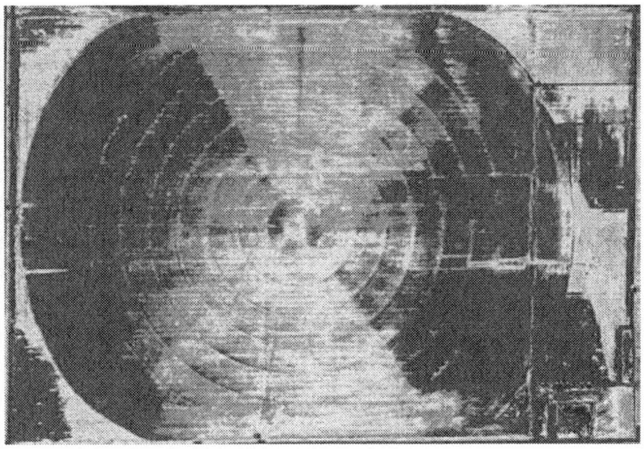

Fig 2 NDVI 18/Sept/2013

2014 information led to the following conclusions again for the BF field. Adjustments were made based on the image information but not specifically addressing weekly irrigation management. See figures 3 and 4 below. The information from the aerial images were considered valuable to the farmer but again for the longer term and not the short term. Again no non-uniformity of yield was seen that could be attributed to the center pivot corner. Scheduling of the UAV flights and the turn around to receive the NDVI images was slow. The UAV company chose to use a different colorization scheme than what was requested confusing the farmers. Also resolutions of 5cm had some interest but the file size was too large to manage easily and did not show the information most needed. The drawback of aerial images continued to be the time to receive images and the cost. Through all of this there was sufficient interest to explore if images could be used for irrigation management but needed to deliver aerial images in a more regular and routine fashion.

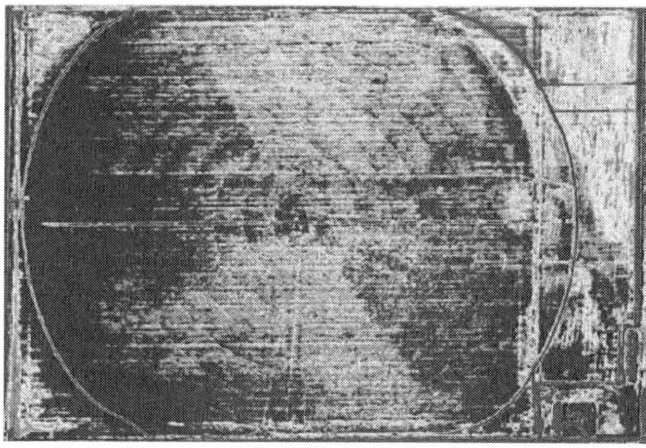

Fig. 3 NDVI 18/June/2014

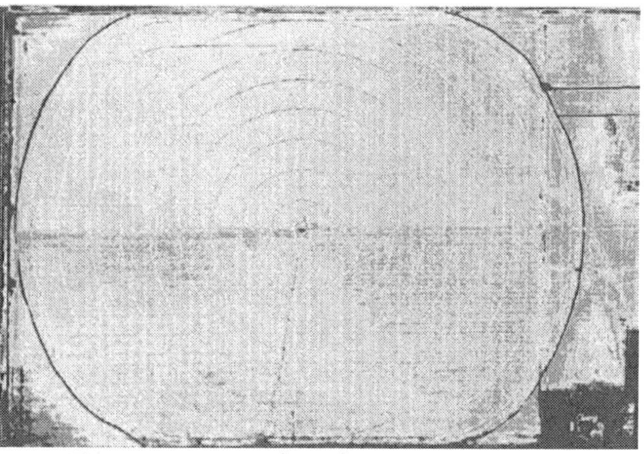

Fig 4 NDVI 22/July/2014

2015 was a difficult year for a variety of reasons. First there was a slow start using the satellite delivery of images and analysis of the aerial imagery. It was not until mid-July the image delivery began to be on a regular basis and the conversion to NDVI became a smooth process using QGIS. Second was the unanticipated amount of rainfall early in the crop season as shown in figure 5. Until there is full crop canopy it appeared aerial images providing NDVI are of limited value. A water balance was used to help manage irrigation. On at least one occasion irrigation was recommended and then the field received significant unexpected rainfall.

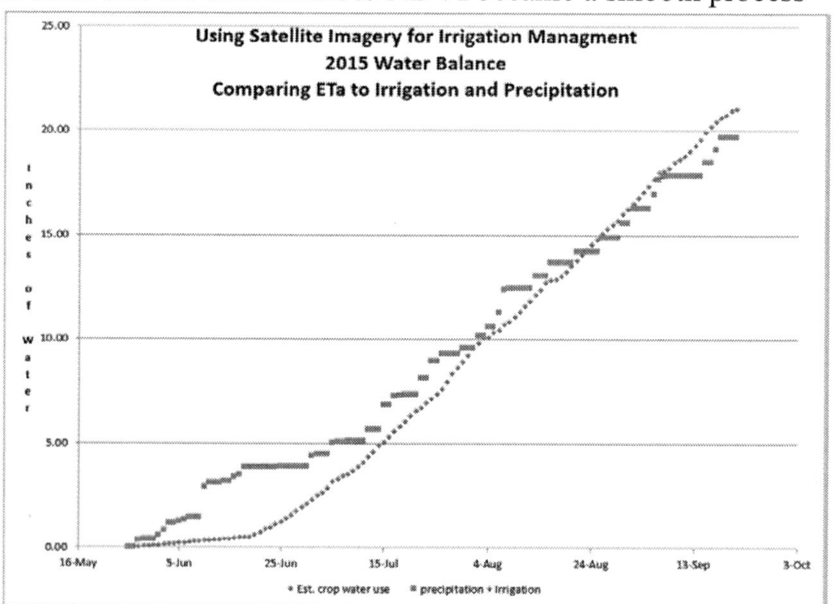

Fig. 5 Information on irrigation and weather

In review significant irrigation was not needed until later in the crop season. Figures 6, 7 and 8 are NDVI images for the BF field. In addition cloud cover was a problem during two weeks to the point no irrigation management decisions could be made based solely on aerial images and continued to operate using the water balance. Overall summation of the crop season for the BF field is shown below in table 1. In the table the BF field is called "satellite managed" and the HP field is called "farmer managed". All of the information displayed is on a per acre basis. While some less water was applied to the BF field, the better yield of the HP field provided more income to offset the cost of the irrigation. In a situation where irrigation water was limited the management using the aerial images could have proved to be more valuable to the overall outcome. The BF field received more irrigation later in the season than did the HP field

	satellite managed	farmer managed
irrigation:	8.2	9.2
rainfall:	10.2	10.2
overall average yield:	76.4	76.7
yield per unit irrigation:	9.3	8.3
yield per unit total water:	4.2	4.0
value of additional yield:	0	$2.58
cost to pump additional water:	0	$2.14

Table 1

Until the crop was well into the reproductive cycle neither the NDVI aerial images nor the soil moisture sensors indicated a significant need for irrigation. The NDVI generated from the manned plane aerial image matched well with the satellite aerial images. The yield maps indicated little variability across either field. After the crop season ended it was learned that historically the HP field has tended to out yield the BF field.

Fig. 6 NDVI 21/July/2015

Fig. 7 NDVI 28/July/2015

Fig. 8 NDVI 3/Aug/2015

In April of 2016 the aerial images began to be received and have been available generally in five to seven days. Valmont began to use MSAVI2 to provide better information on the crop prior to full canopy development and to avoid some of the saturation issues associated with NDVI. Cloud cover has been a problem at each of the three sites. Also a challenge is identifying if there is atmospheric water vapor in the upper deck resulting in false values. Figures 9 and 10 are examples of what the NDVI looked like at two different atmospheric situations. Initially it was not recognized that sufficient water vapor was in the upper deck 'sapping' too much of the reflected energy needed to produce an accurate measure of the plant health as shown in figure 9.

Fig 9, August 22nd

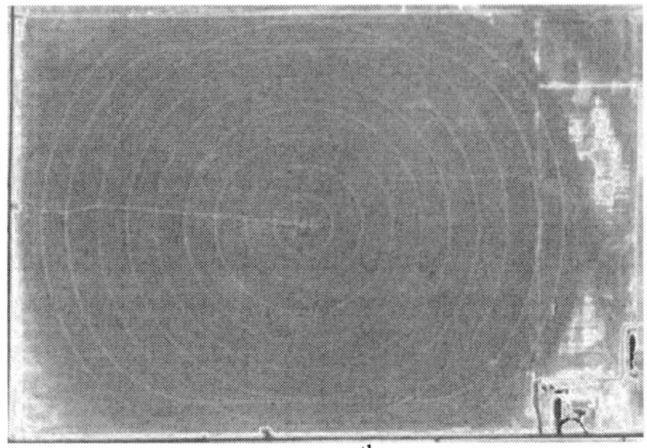

Fig. 10 September 5th

One interesting note while it was believed aerial images were of limited value prior to full crop canopy, use was made by three of the farmers to evaluate the performance of the burn down of their cover crop and status of the crop in the early season. In addition each farmer used the aerial images to evaluate early season weed pressures. The soil moisture sensor data is confirming the indications of the aerial images so far. Again the fields will be evaluated also using yield data from the combine. A 2.0m resolution can provide an indication of the stand but does not allow for individual counting of plants. The drone company contracted with went out of business before any images were delivered. Manned plane images were taken in early June and late July. Figures 11, 12 and 13 are from field HY #1 and give an indication of the information received.

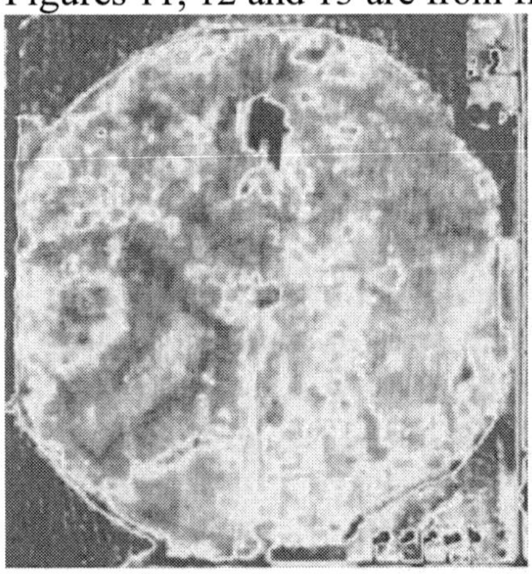

Fig. 11 28/June/2016

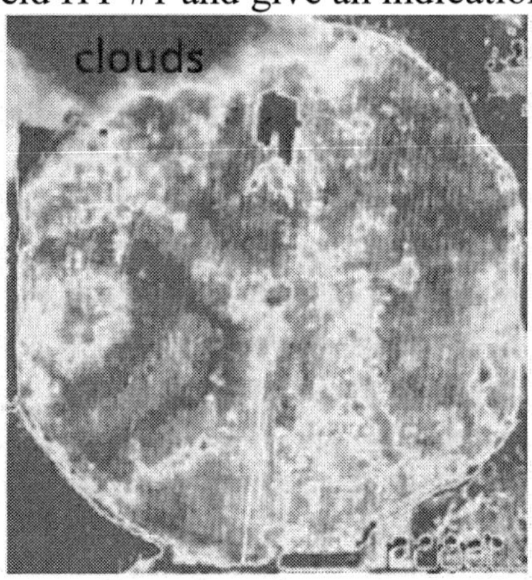

Fig. 12 03/July/2016

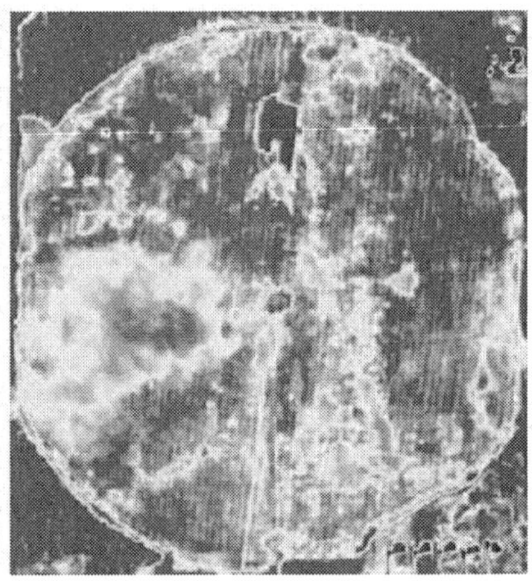

Fig. 13 05/July/2016

Summary

Many tools exist to manage irrigation. Work was done to explore the use of aerial images from satellites, manned planes and drones for irrigation management.

The common challenges of using aerial images in the past have been:

- Cost of images
- Timing of collection and turnaround time from collection to available to the farmer and/or consultant
- Resolution
- Interpretation challenges

The advantages are:

- Snap shot of the complete field
- Can see crop changes over time if collected at sufficiently close intervals during the crop season
- Automation of image management and analysis
- Minimize obstructions in the field

Work has been done with a satellite company which overcomes many of the traditional challenges of working with aerial images. In 2013 and 2014 it was obvious that receiving images every three to four weeks using manned planes or drones was insufficient to adequately manage irrigation. In 2015 due to weather conditions and slow start to image delivery and conversion the results for using satellite images was inconclusive though showed promise for dynamic irrigation management.

2016 has started well and anticipate a good test since working over a wider geographic area and with different crops. Economics of the individual fields involved are being more closely tracked. The need for solutions to automate the process of image download and analysis has become apparent as are providing irrigation recommendations.

A review of costs per aerial image seen in 2015 and 2016 indicate the following:

	$/acre/image	Minimums
Manned plane	$ 1.80	2,000 acres
UAV or drone	$ 3.80 to $ 5.40	200 to 500 acres
Satellite	Free to $ 2.48*	varies

*The cost for satellite data varies greatly due to the possibility of using public domain images such as from Landsat8, Modis, Sentinel-2 and others that are for profit companies.

The use of satellite images shows promise not only for irrigation management but can be used to identify early season field characteristics and also how the crop is maturing.

Return per acre when using aerial images also varies greatly and has been hard to determine for commercial fields. One study on remote sensing (aerial images) states “when budget assumptions are standardized the reviewed studies show that RS has the potential to improve average on-farm profit by about $12.95/acre” (Tenkorang, 2008). More work needs to be done to determine the economic value to a grower.

References

2012 Census of Agriculture, Volume 3 Special Studies Part 1, AC-12-SS-1, United States Department of Agriculture National Agricultural Statistics Service, issued November 2014

Evans, R, Cassel, D.K., Sneed, R.E., 1996, Soil, Water, and Crop Characteristics Important to Irrigation Scheduling, Publication Number AG 452-1, North Carolina Cooperative Extension

Henggeler, J., 2013, Irrigation Scheduling, University of Missouri Extension, #13

Reising, P., Steinhardt, G.C., Daughtry, C.S., 2016, Aerial Photography as a Crop Management Aid, Purdue University Cooperative Extension Service, AY-252

Tenkorang, F.; Lowenberg-DeBoer, J., 2004 On-Farm Profitability of Remote Sensing in Agriculture, The Journal of Terrestrial Observation, Volume 1, Number 1, Spring 2008

This page intentionally left blank.

Comparing Mobile Drip Irrigation to Low Elevation Spray Application in Corn

I. Kisekka, T. Oker, G. Nguyen, J. Aguilar, and D. Rogers

Abstract. *Diminishing well capacities coupled with the desire to extend the usable life of the Ogallala aquifer have stimulated the quest for efficient irrigation application technologies. Mobile Drip Irrigation (MDI) which integrates drip line onto a center pivot or lateral move system has attracted attention lately. By applying water along crop rows, it hypothesized that MDI could eliminate water losses due to spray droplet evaporation, wind drift, and reduce soil evaporation due to reduced surface wetting. A study was conducted to compare grain yield, above ground biomass, and water productivity of MDI and LESA at two irrigation capacities 2.3 and 4.6 gpm/ac. The experimental design was arranged in a randomized complete block design with four replications. Preliminary results indicate grain yield was not significantly different between MDI and LESA due to the above normal rainfall received during the 2015 growing season.*

Introduction

Diminishing well capacities coupled with the desire to extend the usable life of the Ogallala aquifer have stimulated the quest for efficient irrigation application technologies. Mobile Drip Irrigation (MDI), which integrates driplines onto a mechanical irrigation system such as a center pivot, has attracted attention lately. By applying water along crop rows, it is hypothesized that MDI could eliminate water losses due to spray droplet evaporation, water evaporation from wetted canopy, and wind drift. MDI also may reduce soil evaporation due to limited surface wetting especially before canopy closure.

The idea of replacing center pivot sprinkler nozzles with driplines is not new (Olson and Rogers, 2007; Rawlins et al., 1974 and Phene et al., 1981). However, what is new is the advancement in precision positioning of the drip line and pressure compensated emitter technology. Such emitters eliminate the need for pressure regulators since they maintain constant flow over wide pressure range as long as the minimum threshold pressure is exceeded. Another advantage of MDI is that in areas where this technology could prove very useful, such as western Kansas, many producers already own center pivots; therefore the transition from sprinklers to MDI would be relatively easy.

To quantify the benefits of MDI, a study was conducted to compare grain yield, above ground biomass, and water productivity of MDI and LESA at two irrigation capacities 2.3 and 4.6 gpm/ac.

Procedures

Experimental Site

The study was conducted at the Kansas State University Southwest Research-Extension Center (38°01'20.87" N, 100°49'26.95" W, elevation of 2,910 feet above mean sea level) near Garden City, Kansas. The soil at the study site is a deep, well-drained Ulysses silt loam. The climate of the study area is semi-arid, and average annual rainfall is 18 inches. Two independent studies were conducted to compare MDI and in-canopy spray nozzles (LESA). Study 1 compared the two application technologies at high well capacity (600 gpm) and Study 2 compared the technologies at low well capacity (300 gpm). The two well capacities were intended to mimic a range of pumping capacities experienced by producers in southwest Kansas. The experimental design in each study was a randomized complete block with four replications (each span 135 feet long was a replication having MDI and in-canopy spray nozzles) as shown in Figure. 1.

Figure 1. Drip irrigation and spray nozzles in each span of four span 560 feet center pivot at the Kansas State University, Southwest Research and Extension Center near Garden City Kansas.

Agronomic Management

The experiment was conducted in a field that was previously under fallow. The corn hybrid planted in 2015 was DKC 61-89 GENVT2P, with a relative maturity of 111 days. Planting was done on May 18, 2015 at a seeding rate of 32,000 seeds per acre using a no-till planter, planting depth was 2 inches. Nitrogen fertilizer was applied preplant at a rate of 300 pounds of N per acre as urea 46-0-0. Weed control involved application of 3 qt/a of Lumax EZ (S-metolachlor, Atrazine, Mesotrione) and 2 oz/a of Sharpen (Saflufenacial) as pre-emergence herbicide and 32 oz/a of Mad Dog Plus (Glyphosate) and Prowl H2O (Pendimethalin) as post emergence herbicides. Harvesting was done by hand by taking two 40 feet corn rows in the center of each plot at physiological maturity. A detailed description of agronomic management is reported in Kisekka et al. (2016).

Irrigation Management

Irrigation was applied using a center pivot sprinkler system (Model: Valley 8000 Polyline, 4 Tower 560 feet, Valmont Industries, Inc., Valley, Nebraska). A 130 micron disc filter with a flow rating of 200 gpm was installed at the pump station also equipped with a Variable Frequency Drive (VFD). Irrigation treatments for the two studies are listed below:

Study 1: 600 gpm well capacity

1. MDI 4.6 gal/a irrigation capacity (1 inch every 4 days)
2. In-canopy spray nozzles and 4.6 gal/a irrigation capacity (1 inch every 4 days)

Study 2: 300 gpm well capacity

1. MDI and 2.3 gal/a irrigation capacity (1 inch every 8 days)
2. In-canopy spray nozzles and 2.3 gal/acre irrigation capacity (1 inch every 8 days)

Irrigation was triggered whenever available soil water reached 60% in the top 4.0 feet of the soil profile, but irrigation frequency was limited by irrigation capacity. Soil water measurements were taken weekly using a neutron probe (CPN 503DR, CPN International, Concord, California) at 1-foot depth increments from 1 to 8 feet deep. Each irrigation event applied 1.0 inch for all treatments scheduled to be irrigated on a given day. Nozzle flow rate was confirmed using the Spot-on device.

Results and Discussion

Rainfall

Rainfall during the 2015 growing season from May 1 to October 31 exceeded the long-term average in the same period from 1950 to 2013 as shown in Figure 2. The 2015 summer growing season rainfall exceeded the long-term average by 4.2 inches. Above normal rainfall in May of 2015 ensured sufficient soil water at corn planting. Also, above normal rainfall at tasselling in July and during grain fill in August contributed substantially to crop water needs.

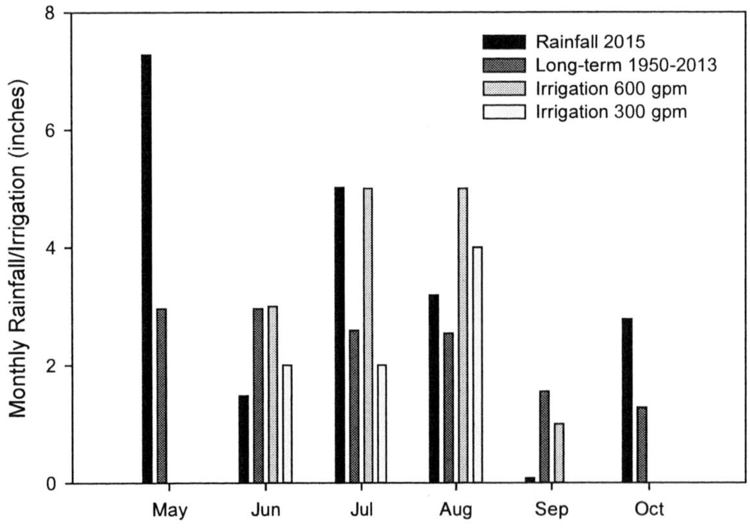

Figure 2. Growing season (May to October) rainfall for 2015 and long-term average, monthly irrigation applications for the 300 and 600 gpm studies at the Kansas State University Southwest Research-Extension Center, near Garden City, Kansas.

Yield

The effect of irrigation application method (MDI versus in-canopy spray nozzles) on yield at high (or 4.6 gpm/a) and low (2.3 gpm/a) well capacities was not statistically significant at the 5% level. The p-values were p = 0.37 and p = 0.67 for Study 1 and 2, respectively (Kisekka et al., 2016). In Study 1 (4.6 gpm/a), MDI and in-canopy spray nozzles produced yields of 247 and 255 bu/a, respectively. Under Study 2 (2.3 gpm/a) MDI and in-canopy spray nozzles produced yields of 243 and 220 bu/a, respectively. The lack of significant differences in yield could be attributed to the high rainfall received during the 2015 growing season (18 inches from May to October).

Crop Water Use

Crop water use under Study 1 was 29.8 and 29.0 inches for MDI and in-canopy spray nozzles respectively (Kisekka et al., 2016). Study 2 crop water use was 22.6 inches and 23.3 inches for MDI and in-canopy spray nozzles, respectively. The differences in seasonal crop water use (ETc) could be attributed to differences in irrigation application amounts between the two studies. Fourteen inches were applied in Study 1 while 8 inches were applied in Study 2. High irrigation amounts under Study 1 probably increased water losses in form of soil water evaporation and deep drainage. The effect of application method on water productivity and irrigation water use efficiency was also not significant at high and low well capacities (Figures 3 and 4). In Study 1, average water productivity of MDI and in-canopy spray nozzles was 8.3 and 8.9 bu/a/in, respectively. In Study 2, average water productivity of MDI and in-canopy spray nozzles was 10.7 and 9.5 bu/a/in, respectively. Irrigation water use efficiency was not significantly different in Studies 1 and 2 (Figure 4).

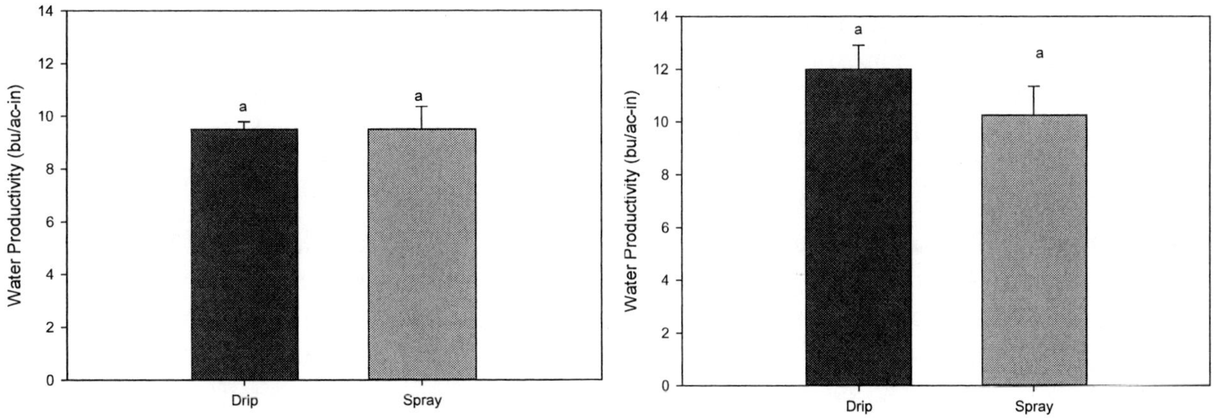

Figure 3. Water productivity of Mobile Drip Irrigation and in-canopy spray nozzles for well capacity of 600 gpm during the 2015 growing season at the Kansas State University SWREC, near Garden City, Kansas (Kisekka et al., 2016).

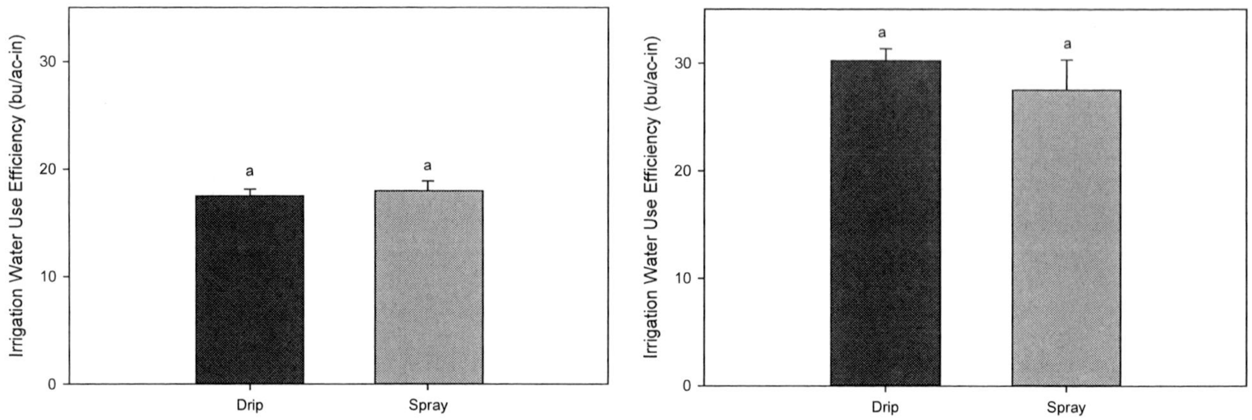

Figure 4. Irrigation water use efficiency of Mobile Drip Irrigation and spray nozzles for well capacity of 300 gpm during the 2015 growing season at the Kansas State University Southwest Research-Extension Center, near Garden City, Kansas (Kisekka et al., 2016).

Conclusion

Mobile Drip Irrigation was evaluated under high and low well capacities in corn. The effect of irrigation application method (MDI versus spray nozzles) on yield at high (600 gpm) and low (300 gpm) well capacities was not significant ($p > 0.05$) in 2015. The effect of application method on water productivity and irrigation water use efficiency was also not significant. The lack of significant differences could be attributed to the above normal rainfall received during the 2015 growing season. Water productivity and irrigation water use efficiency were higher under the 300 gpm study compared to the 600 gpm, implying that water was used more efficiently as the number of irrigation applications decreased. It is worth noting that plots under MDI did not have deep wheel tracks or rutting problems associated with sprinkler nozzles. More research is needed to confirm benefits of MDI.

Acknowledgments

The authors would like to thank Teeter Irrigation, Servi-Tech Inc., K-State Global Food Systems, U.S. Department of Agriculture Ogallala Aquifer Project, and Kansas Water Office for providing funding or material support for this project. The authors would also like to thank Mr. Dennis Tomsicek and Mr. Jaylen Koehn for their help in implementing this project and data collection and processing.

References

Kisekka, I., T. Oker, G. Nguyen, J. Aguilar and Danny Rogers. 2016. Revisiting Mobile Drip Irrigation under Limited Water. *Irrigation Science*. Under Review.

Olson, B. and D. Rogers. 2007. Center Pivot Precision Mobile Drip Irrigation. In Proceedings: Central Plains Irrigation Conference. Kearny, NE.

Phene C. J., T. A. Howell, R. D. Beck, and D. C. Sanders. 1981. A traveling trickle irrigation system for row crops. "Irrigation, the hope and the promise." Annu Techn Conf Proc Irrigation Association, Arlington, Virginia, pp 66-81.

Rawlins, S. L., G. J. Hoffman and S. D. Merrill. 1974. Traveling trickle system. In: Proc. Second Drip Irrigation Cong. pp. 184-187.

This page intentionally left blank.

Molecularly Oriented PVC Pipe (PVCO) for Water Transmission

Richard St-Aubin, P.Eng.
National Sales Manager – IPEX USA LLC
1425 North Service Road East, Unit #3
Oakville ON Canada L6H 1A7

Email: Richard.st.aubin@ipexna.com

ABSTRACT:

Molecularly Oriented PVC (PVCO) is an improved material that has many potential benefits for large diameter irrigation projects. Its unique material structure gives it higher strength, improved impact resistance and higher toughness in cold weather conditions.

Recent advances in manufacturing technology have made larger diameters of PVCO possible, with 24" diameter recently becoming available in the North American market. This paper will briefly describe the molecular orientation process, the various manufacturing processes for PVCO pipe and the resulting improvements in material properties. Finally, the benefits of using PVCO for irrigation projects will be briefly discussed.

Introduction

PVC pipe has become one of the most commonly used pipes for large diameter irrigation projects. It's light weight and ease of installation, coupled with its dependability and long life span has made it the pipe of choice for the last 20 years in irrigation. In addition, new sizes and pressure ratings of PVC pipe up to 60" diameter have been developed within the last two years, giving engineers even more flexibility when designing irrigation projects.

Molecularly Oriented PVC (PVCO) pipes are the next generation of PVC pipes. PVCO is stronger, more ductile and more impact resistant that standard PVC pipe. These improved properties make PVCO pipes even lighter and easier to handle than standard PVC pipes, while carrying the same pressure rating. In addition, PVCO's improved impact resistance and toughness makes it less sensitive to installation deficiencies and rough jobsite handling.

While PVCO is a relatively new pipe material, its manufacture and testing is governed by well-established ASTM, AWWA and CSA standards.

This paper will briefly describe the molecular orientation process, the various manufacturing processes for PVCO pipe and the resulting improvements in material properties. Finally, the benefits of using PVCO for irrigation projects will be briefly discussed.

Molecular Orientation in Polymers

Thermoplastics are made up of long chains of molecules as shown in Figure 1. These molecules are made up of carbon, hydrogen and other elements. In some materials the chains are more organized and form what is called a semi-crystalline structure, while in others the chains resemble a plate of spaghetti, as shown in Figure 2[1] . PVC has this "spaghetti like" arrangement and is classified as an

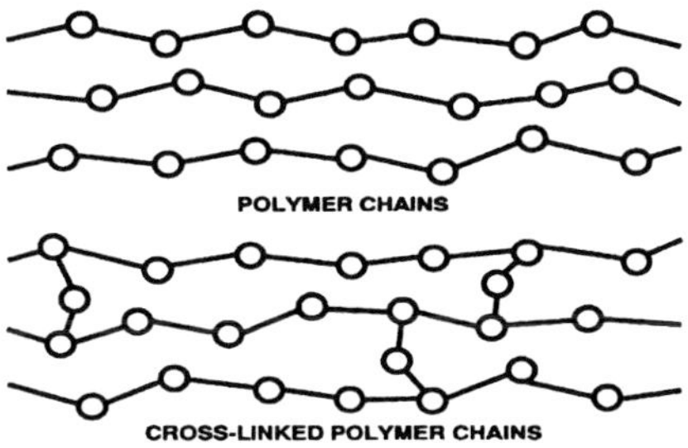

Figure 1 - Polymer Chains

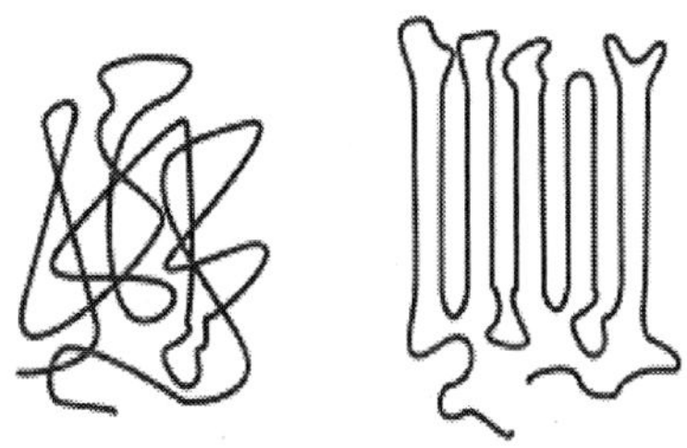
Figure 1 - Amorphous and Semi-Crystalline structures

amorphous polymer. When PVC is heated to its glass transition temperature T_g, and then stretched, those spaghetti-like chains tend to orient in the direction of the applied strain. The structure of the material itself changes, as the polymer chains "orient" in the direction of the stretching and form a lattice-like structure as shown in Figure 3[2]:

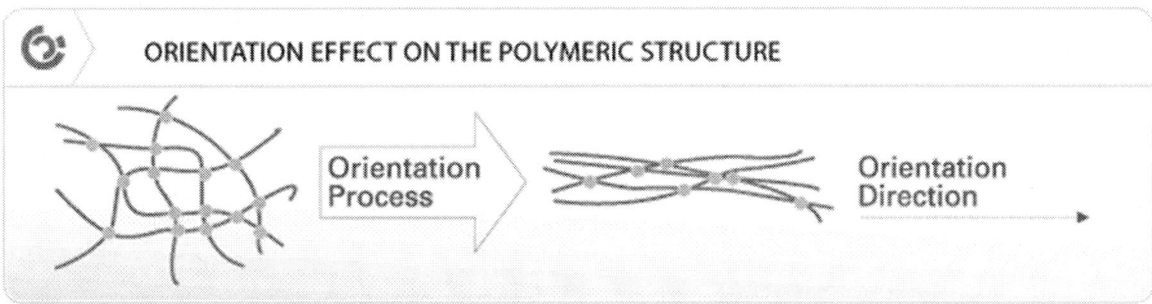

Figure 3 – Orientation Process

While all of this occurs at a molecular level, and is invisible to the naked eye, the effect on the material's structure and properties are profound. The strength of the material increases in the direction of orientation, and the material develops a "layered" structure as shown in Figure 4.

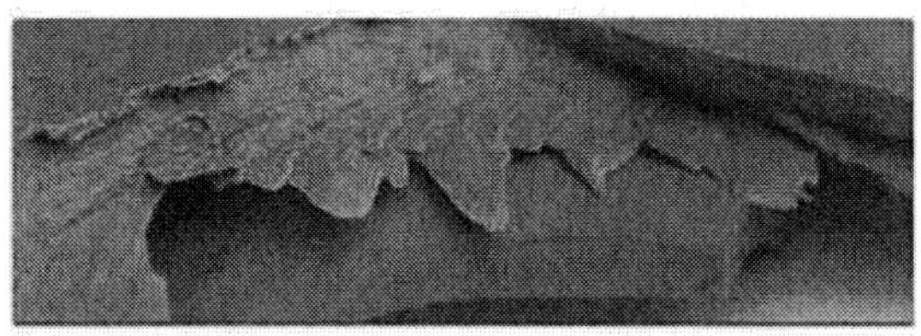
Figure 2 - Layered Structure of PVCO

[1] https://en.wikipedia.org/wiki/Crystallization_of_polymers
[2] http://www.molecor.com/en/technology/molecular-orientation

Manufacturing of PVCO Pipe

While the beneficial effects of molecular orientation for PVC pipes have been known since the early 1970's, it was not until the 1990's that the first commercial processes were developed. These early processes were only marginally successful, and while they produced an excellent product, they tended to be prone to breakdown and production delays. As a result, PVCO remained a small "niche" product across North America.

However, with the advent of more advanced manufacturing processes in the early 2000's, manufacturing PVCO became more reliable and it became possible for larger diameters of pipe to be manufactured. While the industry was limited to 12" diameter pressure pipe 10 years ago, the pace of innovation continues to accelerate, and 24" diameter PVCO pipes are now commercially available, with 30" diameters on horizon.

The process for making PVCO is simple in concept but complicated in practice. PVC pipe is taken to a precise temperature and then stretched. Once stretched the pipe must be immediately cooled to "lock in" the orientation.

There are two distinct methods for accomplishing this. On-line processes stretch the pipe as a second step in the extrusion process. A pre-form pipe is extruded and farther down the extrusion line it is continuously re-heated and stretched over a mandrel. Once it is over the mandrel it is immediately cooled. Off-line or batch processes take discrete pre-extruded preform pipes and heat the entire pipe up to the required temperature and then expand them, using either air or hot water, in a large mold. Both processes can produce excellent results, and there are advantages and disadvantages to either approach.

Figure 3 - On-line Orientation Process (Photo courtesy of Wavin Overseas B.V.)

Advanced Properties of PVCO

Strength – The stretching process dramatically increases the strength of the material. PVC and other plastic pipes are pressure rated based on their hydrostatic design basis (HDB), which is essentially the long term stress that the material can withstand for a minimum of 100,000 hours. PVC pipe meeting the AWWA C900/ 905standards has a long-term HDB of 4000 psi. The orientation process improves the long term HDB of PVCO to 7100 psi. This increased strength allows for increased inside diameters as less material can be used to achieve an equivalent pressure rating.

Toughness – PVCO has been proven to have much higher impact resistance than conventional PVC (Michel and Akkerman (2013)). While it exhibits improved impact strength across a wide range of temperatures, the difference becomes more pronounced as the temperature drops. This is illustrated graphically in Figure 6[3]. What is perhaps even more interesting is that the impact performance of PVCO appears to be unaffected by the presence of notches in the sample.

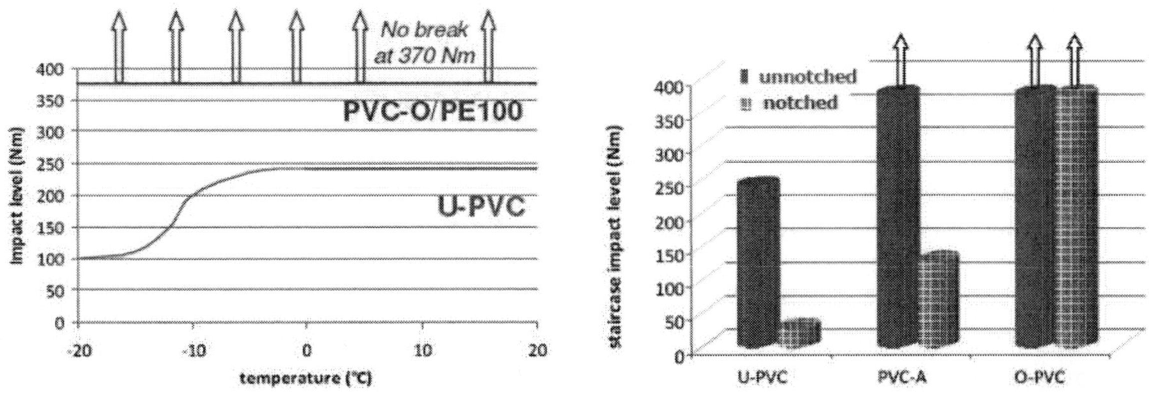

Figure 5 - Impact Resistance of PVCO

Failure Mode – The new structure created when PVC is oriented also gives the material a different failure made when compared to standard PVC Pipe. PVCO will exhibit a localized failure mode rather than a split. Figure 7 shows a 12" diameter in-service PVCO pipe that was struck by a crossbore. This type of localized failure is typical of PVCO as its layered structure tends to arrest any cracks and attenuate any propagation.

Environmental Footprint – Less material, lower weight and increased inside diameters equate to a lower life cycle cost for owners of PVCO systems. In addition, two separate studies have concluded that PVCO has the lowest embodied energy (a measure of environmental impact) of any commonly used piping material[4],[5].

Figure 4 - Crossbore Hit on PVCO

[3] Graphic taken from Catherine Michel, Johannes Akkerman; "*A Study Assessing the Performance of O-PVC in Pressure Pipes*" (2013)

[4] M. Ambrose, S. Burn; "*Embodied Energy of Pipe Networks*" (2005) CSIRO Manufacturing and Infrastructure Technology

[5] Baldasano Recio, Guererro et al; "*Estimate of the Energy Consumption and CO_2 Emission associated with the Production, Use and Final Disposal of PVC, HDPE, PP, Ductile Iron and Concrete Pipes.*" Universitat Politecnica de Catalunya, Environmental Modelling Laboratory.

Standards

The first North American standard for PVCO was ASTM F1483, which covered both CIOD and IPS outside diameters.

The majority of North American PVCO pipe is manufactured under the AWWA C909 standard (revised 2016) , which covers 4" and larger sizes.

In Canada, PVCO pipes are third-party certified to CSA B137.3.1.

Benefits for Irrigation Projects

Plastic pipes have tremendous benefits for water transmission projects: corrosion resistance, ease of installation, and excellent hydraulic properties. However, PVCO has a number of clear advantages for installers and operators of large irrigation systems:

1. Improved Cold Weather Impact Resistance – Many irrigation projects are completed during the cold winter months. As temperatures drop PVCO's impact resistance remains extremely high, dramatically reducing the possibility of damage to the pipe during rough installation.
2. Improved Notch Resistance – Rocks and other debris in pipe bedding are far less likely to damage PVCO than any other material. In many cases this means that native backfill may be suitable many projects.
3. Improved Hydraulics – the larger inside diameters and glass like inside surface associated with PVCO pipes allowing for energy savings in pumped systems.
4. Failure Mode – in the event that a PVCO system is impacted by a crossbore or a heavy equipment strike, any damage is localized and can be easily repaired using readily available fittings.

With advances in manufacturing technology, the available diameters and pressure ratings of PVCO pipes will continue to expand. 24" pipe is currently available up to a 235 psi pressure rating, while 30" pipe rated at 165 psi will enter the market within six months.

This page intentionally left blank.

Irrigation and ESA-Success in Partnerships

Jason Scott, FP-C
jscott@geoengineers.com
523 East Second Avenue
Spokane, WA 99202

Abstract

As water demand increases and availability decreases, the potential for conflict between water user groups escalates. Compounding the potential conflicts are regulatory issues associated with the Endangered Species Act (ESA). Nationwide; fish, wildlife and plants are being added to the ESA list at a rate much higher than they are being removed from it. Irrigators and agriculture producers are often assigned blame for species decline end up bearing a substantial amount of the burden for their recovery.

In many cases the blame for declines and burden of recovery are scientifically unjustified. Still, irrigators and producers are easy targets for three main reasons: 1) the scale of infrastructure and landscape are highly visible. 2) In most cases they don't have any data to refute accusations even if the accusations are without merit. 3) Irrigators and agriculture producers have done a poor job of telling their story so their operations are poorly understood.

There are success stories where agriculture interests are being accommodated and the blow lessened through key partnerships. Through those partnerships come resources to conduct studies, establish support from regulatory agencies and develop mutually beneficial solutions. Common to the partnership success stories navigating ESA issues is an understanding that nobody wants to see species go extinct and nobody wants to go hungry.

Introduction

Throughout the United States there are approximately 163 fish species/populations listed on the Endangered Species Act (ESA), 35 amphibians and over 100 aquatic invertebrates, all of which depend on the same water resources that agriculture production relies on. The total number of ESA listed species is much greater when other water dependent ESA listed species, such as birds and plants, are included. Associated with many of the ESA listings are critical habitat designations that often span broad geographic areas and even include areas where the particular species currently does not exist or have access to. Further, in some cases, recovery plans are linked to poorly defined holistic ecosystem function and linkages to non-ESA listed species. The number of species, geographic areas represented and agencies responsible for their management make for complex regulatory environment that can be difficult and frustrating to navigate through.

ESA listed species recovery is ultimately the responsibility of the US Fish and Wildlife Service (USFWS) and the National Marine Fisheries Service (NMFS). However, those federal agencies often rely on states and tribes for support and management. Collectively, those agencies are under constant litigious fire from entities that profit from lawsuits, which forces management agencies to dedicate more resources and effort to lawsuits than actual problem solving and work toward recovery. Caught in the crossfire, with very little decision making ability or input, are landowners and agriculture interests. Ironically, habitat necessary for ESA species recovery is largely on land owned by the people excluded most and who could most influence recovery. Given the legal responsibility of federal and state agencies to manage long-term sustainability of species, the enormous amount of lawsuits filed by environmental pressure groups and the relative lack of landowner/agricultural involvement in decision making begs the question: "is ESA really about species recovery?"

Assuming lawsuits are, in fact, intended to promote species recovery, then shouldn't science provide answers that guide recovery solutions? By law, ESA decisions must be based on the best science available, but this requirement is often criticized due to lack of rigor. Still, if that science is the best available, even if it is considered imperfect or incomplete, it still may be used. This conflict is complicated by the resources and time required to acquire extensive data, particularly for lesser-known species that have little or no economic value.

In general, "sound science" is held up as desirable on all sides of ESA debates. In the most basic definition, science is a way of examining phenomena to produce explanations of the "why" and "how" of these phenomena (National Academy of Sciences, 1999). Therefore, scientific knowledge is dynamic and changes as new information becomes available. Moreover, scientific conclusions are dependent on the specific question being asked, experimental methods, assumptions made and interpretation of results, all of which could be influenced by personal values and policy positions of their employer. Exacerbating the potential conflicts is the indistinct boundary between science and policy; where science is generally based on probability and policy-makers want science to provide certainty for complex decisions (Mills, 2000).

While several bills have been introduced to address ESA reform over the last decade (Corn et al. 2013), it is clear that legislation is not going to solve ESA conflicts anytime soon. Similarly, legal actions are time consuming, costly and have provided very little relief to those subjected to ESA regulation and conflict. Further, landowners and agriculture producers typically don't have the time or financial resources necessary to fully engage in ESA processes and decisions that directly affect them. However, partnerships between diverse interest groups can provide avenues for participating in decision-making processes and, in many cases, can provide expertise and financial resources that result in mutually beneficial solutions.

Partnerships

Successful partnerships are founded on similar interests and finding mutually beneficial solutions based on those interests. Regulatory/management agencies, non-profit interest groups and non-governmental organizations are becoming increasingly aware that positive relationships with landowners and producers are much more successful in addressing species recovery than a punitive approach. Partnerships allow management actions that achieve ecological resilience where multiple objectives are balanced by a single resilience strategy (Paukert and Lynch, 2016)

As it applies to this topic, one common interest landowners, agriculture producers, water users and ESA management agencies share is the availability of water. Relative to water; stream, wetland and riparian habitats provide the habitats necessary for ESA species recovery. Landowner and producer interests are associated with livelihoods and economics. However, most landowners have a strong sense of stewardship and sense of obligation to conserve the environment around them.

Success Stories

1. Yakima Basin Integrated Plan, Washington

The Yakima River Basin spans 6,100 square miles in central Washington State. Demand for irrigating over 464,000 acres requires 2.4 million acre-feet (AF) of water. Additionally, the basin supports 48 species of fish, including two ESA listed species. Bureau of Reclamation reservoirs have the capacity to store approximately 1 million AF and the balance has historically been stored in the Cascade Mountains as snowpack. However, drought conditions have reduced snowpack storage and resulting water disputes have historically been resolved through lengthy and costly lawsuits, which prevented basin stakeholders from deciding on a comprehensive plan for water development and uses.

The Yakima Integrated Plan was developed by traditional opponents who came together to determine an alternative plan for the Basin's water needs. The solutions include fish habitat restoration, increasing the stability of stream flows and ensuring the reliability of agricultural irrigation and municipal water supply. The plan was developed through a collaborative public process where stakeholders weighed their needs versus wants and came to understand the views of their traditional opposition then negotiated to reach a consensus.

2. John Day River Watershed Restoration Strategy, Oregon

The John Day River is one of the most critical watersheds for fisheries in the Columbia River Basin. The Confederated Tribes of the Warm Springs Reservation identified the need to clearly select and prioritize restoration projects based on habitat limiting factors and targeted restoration actions through a transparent process with basin stakeholders. Stakeholders who participated in strategy development process included state and federal agencies, NGO's, private interest groups and private landowners. The strategy outlined several objectives, which included recovery of culturally significant fish species (including two ESA listed species) and incorporation of stakeholder priorities to ensure benefits were mutually beneficial for basin landowners. The key to this strategic plan's success was working with stakeholders to develop specific project prioritization criteria that included both agricultural needs as well as fish needs. Landowners and stakeholders benefit by being able to access fish restoration funding to improve their operations and infrastructure and the Tribes benefit by restoring critical habitat on private lands.

3. Birch Creek Watershed Action Plan, Oregon

Birch Creek is a tributary to the Umatilla River in northeast Oregon. Approximately 87 percent of the watershed is privately owned and largely managed for agriculture production. The watershed is also home to ESA listed summer steelhead, which are the last remaining native anadromous salmonid in the basin. Over the last several decades, landowners have been plagued with sediment deposition, streambank erosion and lacking water availability that impact their agriculture operations. Those same conditions also plague habitat conditions for the steelhead population.

The Confederated Tribes of the Umatilla Indian Reservation and partnering agencies developed a collaborative approach to incorporate ecological and fisheries recovery goals with land management and use. Collaboration began at the plan's outset and the primary goal was to build community trust and create strong and lasting partnerships to address complex natural resource issues. In a community that historically was adversarial on water related issues, this collaborative approach was so successful that it garnered support from much of the community to the extent that field study crews were granted access to over 60 miles of streams on private land throughout the watershed. Collaborators in the plan development included state and federal management agencies, NGO's, a municipal government, private non-profit partners and individual landowners. Plan solutions incorporate landowner needs as well as fish needs. Through the plan development, relationships were strengthened through development of mutually beneficial solutions.

Conclusion

Aside from the few success stories summarized above, there are numerous other examples where partnerships have led to stronger relationships, better understanding and sustainable solutions. Resource management agencies should be encouraged by the success of these partnerships in pursuit of listed species recovery and landowners should be encouraged that ESA issues can be addressed outside of court rooms. Establishing partnerships proactively will continue to generate momentum through collaboration and mutually beneficial solutions. It wasn't that long ago that healthy ecosystems and agriculture production coexisted. Healthy partnerships are an effective way to work toward making that a reality again.

Literature cited

Corn, L.M., K. Alexander and E.H. Buck. 2013. The Endangered Species Act and "Sound Science." CRS Report for Congress. Congressional Research Service 7-5700. RL32992. www.crs.gov

Mills, T.J. 2000. Position Advocacy by Scientists Risks Science Credibility and May be Unethical. Northwest Science, V. 74, no. 2. p. 165-168.

National Academy of Sciences, Science and Creationism: A View from the National Academy of Sciences, Second Edition. National Academy Press, 1999. Washington DC.

Paukert, C.P. and A.J. Lynch. 2016. Effects of Climate Change on North American Inland Fishes. Fisheries. 41:329-330.

Infrared Thermometry for Deficit Irrigation of Peach Trees

D. Wang, H. Zhang, J. Gartung
USDA-ARS Water Management Research Unit
San Joaquin Valley Agricultural Sciences Center
Parlier, California

Water shortage has been a major concern for crop production in the western states of the USA and other arid regions in the world. Deficit irrigation can be used in some cropping systems as a potential water saving strategy to alleviate water shortage, however, the margin of error in irrigation management becomes smaller. For early-maturing peach varieties, it has been demonstrated that established orchards are not sensitive to moderate water stress in the non-fruit bearing postharvest growth periods. In a multi-year field study, an early-maturing peach was irrigated using furrow, drip, and micro-sprinkler systems under both full and deficit irrigation schemes. Peach tree water status was monitored using periodic stem water potential measurements and thermal infrared temperature sensors in real-time. The data was used to derive a plant water status and canopy temperature function. The functional relationship was subsequently applied in the experiment for irrigation scheduling. The field study demonstrated the feasibility of managing postharvest deficit irrigation of early season peaches using the infrared thermometry measurement. The postharvest deficit irrigation was reasonably successful for peach production with significant water savings.

1 Introduction

Deficit irrigation is a potential means of reducing total crop water consumption by irrigating at less than the full amount required by crop evapotranspiration needs. For fruiting trees such as peaches, because fruit yield and quality at harvest may not be sensitive to water stress at some developmental stages such as during non-fruit bearing postharvest season, there is more interest in applying deficit irrigation strategies (Goldhamer et al., 1999). However, deficit irrigation has not been widely used due partially to the lack of effective and fast methods of monitoring plant water stress in near real-time and determining associated risks of applying deficit irrigation. When crops are managed under deficit irrigation, the margin of error in timing and amount of water application becomes smaller before causing yield losses. Monitoring the soil and plant water status is more critical for reducing risks of a crop failure or permanent damage to the trees. However, current established techniques of monitoring the soil and plant water status such as neutron probe readings of soil water profile and pressure chamber measurements of stem water potential are labor intensive, and lack the timeliness needed for irrigation scheduling purposes.

Infrared canopy temperature was used by Jackson et al. (1981) to estimate water stress in annual crops such as wheat. The canopy temperature method was also applied to irrigation scheduling for cotton production (Wanjura and Upchurch, 1997). Using canopy temperature measurement, the canopy to air temperature difference was correlated to the vapor pressure deficit in peach trees and used to reference stomatal responses to water stress (Glenn et al., 1989). Approximately 10,000 ha of commercially-grown peach trees in central California depend on irrigation as the primary source of water in the peak summer growing season. A potential solution for managing water shortage is to use deficit irrigation during postharvest growth stages. The purpose of the multi-year field study was to develop a framework to use infrared temperature sensing to manage deficit irrigation in peach.

2 Methods

The study was initiated in 2007 in a 1.6 ha mature peach orchard located near Parlier, California, USA (Wang and Gartung, 2010). The trees were early-ripening "Crimson Lady" (*Prunus persica (L.)* Batsch) peach on "Nemaguard" rootstock planted in April 1999. The orchard was divided into separate irrigation blocks which were subjected to furrow, drip, or micro-sprinkler irrigation methods and managed under full and deficit irrigation treatments. The full and deficit treatments were carried out for furrow and drip blocks from 2007-2015 and for micro-sprinkler blocks from 2011-2015. The experimental design was a randomized block with six replications.

During the growing season, stem water potential was measured weekly or bi-weekly from both the full and deficit irrigation blocks. Infrared temperature sensors were installed in the orchard in both the full and deficit blocks irrigated by furrow, drip, or micro-sprinkler methods. These temperature sensors were mounted on galvanized metal pipes extending above the tree canopy. The center of field of view for each sensor was aimed at the middle three trees of the center row for each measurement block. A datalogger system was used to record temperature readings at 15 min internals and readings were averaged to hourly outputs for each growing season. Thermocouples were installed in the orchard to record air temperature at the same frequency as the infrared sensors, and hourly canopy-air temperature difference was computed. A linear regression was made between the canopy-air temperature difference and stem water potential measurements using data collected in 2007, 2008, and 2010. The regression equation was subsequently used to guide irrigation scheduling in 2011-2014 (Zhang and Wang, 2013).

Peach fruit was harvested each year by a commercial contract crew following typical farming procedures. Only marketable fruits were harvested and a total of two to three picks were used during each season. The total fruit weight per tree and number of peaches per tree were measured for each treatment block. Average weight per fruit or fruit size was obtained by dividing the weight per tree with number peaches per tree. Statistical comparisons were made between different irrigation methods and irrigation amounts for each year.

3 Results and discussion

Fruit yield (weight per tree) under different irrigation treatments is shown for each year for the multi-year field study (Table 1). For furrow irrigation there is no significant difference in yield between full and deficit irrigation except in 2011 and 2012. For drip, the difference is significant in 2008 and 2015. For micro-sprinkler blocks, no significant difference was found in fruit yield.

Table 1: Fruit yield (kg/tree) under different irrigation regimes*

Treatment	2008	2009	2010	2011	2012	2013	2014	2015
Furrow, full	22^{a}	12	17^{ab}	26^{a}	19^{a}	20^{a}	14^{a}	15^{a}
Furrow, deficit	22^{a}	11	19^{a}	22^{b}	15^{b}	18^{ab}	10^{ab}	13^{ab}
Drip, full	21^{a}	11	16^{b}	24^{ab}	16^{b}	18^{ab}	10^{ab}	13^{ab}
Drip, deficit	18^{b}	10	18^{ab}	22^{b}	17^{ab}	16^{b}	8^{b}	10^{c}
Micro-sprinkler, full	na**	na	na	na	16^{b}	15^{bc}	13^{a}	12^{bc}
Micro-sprinkler, deficit	na	na	na	na	16^{b}	14^{c}	12^{ab}	11^{bc}

*Different letters indicate significance at $P < 0.05$ using the Tukey's studentized range (HSD) test.
** na = data not available.

Figure 1. Correlation between stem water potential and infrared canopy – air temperature difference.

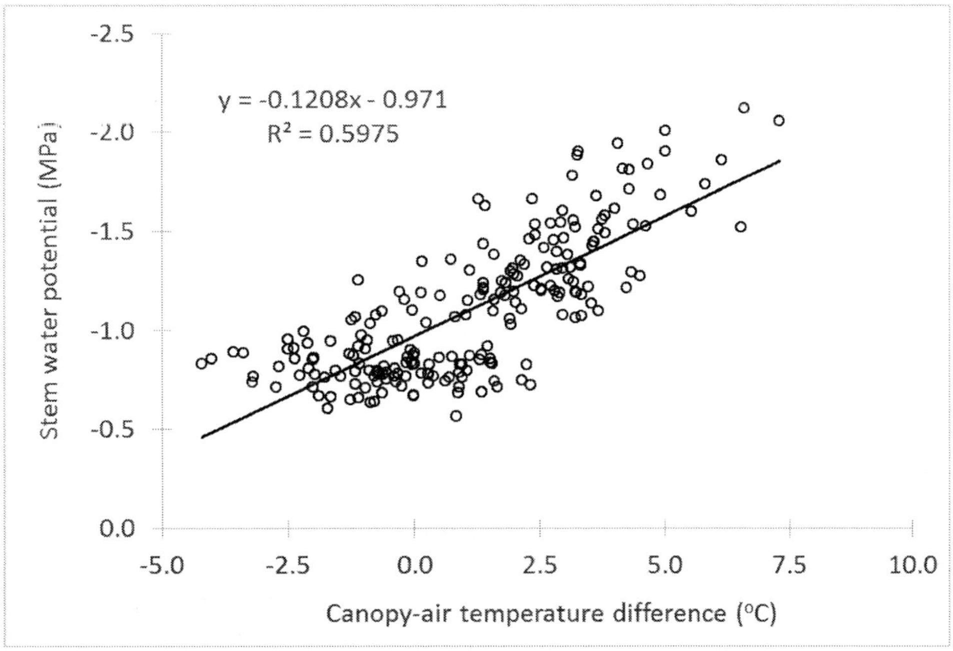

Figure 2. Fruit yield and size over previous season postharvest irrigation totals. Data from both full and deficit treatment of furrow, drip, and micro-sprinkler irrigation blocks from 2013-2015.

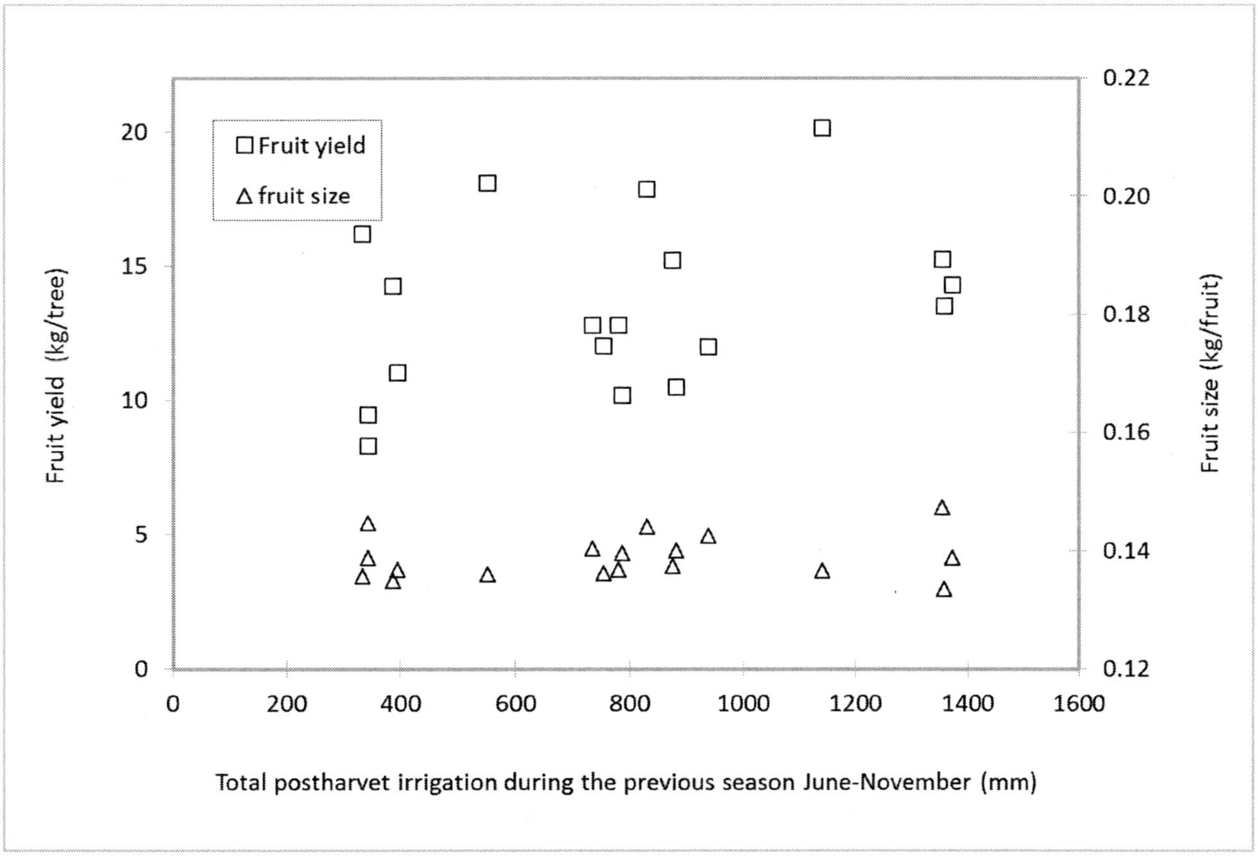

Correlation between stem water potential and canopy-air temperature difference showed a significant relationship ($R^2 = 0.6$) where more negative potential values corresponded to larger canopy-air temperature differences (Figure 1). This is expected because when plants are under water stress, stomatal resistance increases thus stem water potential is more negative. At the same time, transpiration decreases thus the canopy could be at higher temperature than the ambient air due to reduced evaporative cooling. The graph also indicates that infrared temperature measurement is not sensitive to stem water potential variations in the range of -0.5 to -1.0 MPa. This may imply that the infrared canopy temperature approach is applicable to water stressed conditions such as under deficit irrigation, but not sensitive to well-watered situations.

Average fruit yield and size of fruit from 2013 to 2015 over furrow, drip, and micro-sprinkler methods showed no significant change when cumulative irrigation increased from an average of 360 mm to 1360 mm during the postharvest season of 2012-2014 (Figure 2). The average fruit yield was 11.9, 12.9, and 14.3 kg/tree when previous year postharvest irrigation totals was 360, 823, and 1360 mm, respectively. The large variation in yield from year to year was attributed, at least partially, to orchard management practices such as annual pruning and fruit thinning (see also Table 1). The size of fruit remained nearly constant at approximately 0.14 kg/fruit.

In summary, the multi-year field study demonstrated that deficit irrigation and infrared thermal sensing are potential management strategies for reducing overall crop water use and monitoring tree water stress. The questions remain in the determination of optimum amount of water deficit without causing unacceptable yield losses or losses in product quality.

References

Glenn, D.M., Worthington, J.W., Welker, W.V., and McFarland, M.J. 1989. Estimation of peach tree water use using infrared thermometry. J. Am. Soc. Hort. Sci. 114:737-741.

Goldhamer, D.A., Fereres, E., Mata, M., Girona, J., and Cohen, M. 1999. Sensitivity of continuous and discrete plant and soil water status monitoring in peach trees subjected to deficit irrigation. J. Am. Soc. Hort. Sci. 124:437-444.

Jackson, R.D., Idso, S.B., Reginato, R.J., and Pinter, Jr., P.J. 1981. Canopy temperature as a crop water stress indicator. Water Resour. Res. 17:1133-1138.

Wang, D., and Gartung, J. 2010. Infrared canopy temperature of early-ripening peach trees under postharvest deficit irrigation. Agric. Water Manage. 97:1787-1794.

Wanjura, D.F., and Upchurch, D.R. 1997. Accounting for humidity in canopy-temperature-controlled irrigation scheduling. Agric. Water Manage. 34:217-231.

Zhang, H., and Wang, D. 2013. Management of postharvest deficit irrigation of peach trees using infrared canopy temperature. Vadose Zone J. 12(3): doi:10.2136/vzj2012.0093.

This page intentionally left blank.

Scheduling for Deficit Irrigation

Charles Hillyer, Assistant Professor - Extension Specialist
Texas A&M AgriLife
Charles.hillyer@ag.tamu.edu

John Collin English, Director of Communications
Irrigation for the Future
collin.english@gmail.com

Laureine ElKhoury, Research Associate
Dept. of Biological and Ecological Engineering, Oregon State University
elkhourl@oregonstate.edu

Marshall English, Professor emeritus
Dept. of Biological and Ecological Engineering, Oregon State University
marshall.english@oregonstate.edu

ABSTRACT

This paper deals with budgeting of irrigation water for deficit irrigation – deciding how much water to use for a season and how much to allocate to each stage of crop development. When irrigation water is applied as needed to meet crop water demands (i.e. full irrigation) the amount of water to be used is determined by the crop itself, and irrigation timing is commonly based on real time observations of field conditions. But when a crop is deliberately under-irrigated the amounts and timing of water use need to be decided in advance.

The central theme of this paper is that success in deficit irrigation management will depend more on advanced and sophisticated modeling, and less on real time monitoring technologies.

Making the best use of limited water requires sophisticated modeling to assess alternative water use strategies, plan the allocations of water through the season to implement a preferred strategy, and adapt the implementation plan to accommodate the specific circumstances of individual fields. Modeling also provides a scientific basis for interpreting feedback data from the field as the season evolves. All of this involves substantial variability and uncertainty. It is a modeling challenge.

An advanced decision support system developed explicitly for deficit irrigation management addresses these challenges. Distinguishing features include:

(1) sophisticated modeling of the disposition of applied water enables derivation of field-specific crop production functions and long-range projection of crop water availability
(2) anticipated irrigation schedules can be routinely and continuously updated to accommodate unexpected circumstances or changing constraints
(3) adaptive feedback can be used to increase analytical precision, minimize uncertainty and provide insight into field-specific relationships between water use and crop production

The characteristics and performance of this decision support system will be outlined and the utility of the information it provides will be illustrated by two case studies, one involving optimum irrigation with high pumping energy costs, the other concerning deficit irrigation of an almond orchard under severe drought conditions.

INTRODUCTION

Systematic, science based procedures for irrigation management appeared about five decades ago with the advent of *scientific irrigation scheduling* (SIS). The prevailing management paradigm of the time was *full irrigation*; the objective was to maximize crop yields while minimizing water losses. But in recent years there has been increasing interest in partial irrigation, or *deficit irrigation*, an altogether different and more challenging management paradigm (English, et als, 2002). That increasing interest is reflected in the appearance of technical bulletins on the subject from diverse institutions worldwide (e.g. FAO, 2002; UCANR, 2016).

The objective of deficit irrigation is to maximize net economic returns rather than maximizing yields *per se*. The focus on economic returns is increasingly motivated by competition for water. Farm water supplies are often simply insufficient for full irrigation, as forcefully demonstrated recently in the devastating California drought. But even when a farm has access to ample water, partial irrigation can be more profitable, especially when competing demands for water create opportunity costs for water. The forces driving this competition -- food shortages, energy costs and global water shortages -- will only grow stronger in the next few decades (English, 2010).

Deficit irrigation, the natural response to those forces, requires a fundamental change in the way irrigation is managed. Conventional, full-irrigation management has commonly relied on continuous monitoring of soil water depletion or crop stress to determine when to irrigate and how much to apply. We would characterize that approach as *'real time scheduling'*. And, significantly, with conventional SIS the total amount of water to be used for the season *is not a management decision*, it is determined by crop water demand.

Deficit irrigation management is altogether different and more complicated. The manager must decide *in advance* how much water to use for the season, when to use that water and *when to withhold it*. That requires analyzing how a sequence of irrigations will play out months into the future. For purposes of this discussion we will refer to such long range projections of irrigation schedules as 'forward scheduling'.

The central theme of this paper is that success in deficit irrigation management depends less on real time scheduling technologies and more on advanced and sophisticated modeling for forward scheduling.

Optimal management of deficit irrigation requires: (i) evaluating expected outcomes for alternative management strategies; (ii) testing the feasibility of preferred strategies (iii) translating preferred strategies into detailed, full season irrigation plans; (iv) customizing those plans to accommodate the unique circumstances and constraints of specific fields; (v) tracking implementation of scheduling plans to assure adherence to the overall strategy; and (vi) updating plans as conditions evolve during the season. These capabilities will be illustrated in the present paper.

Advanced technical support is needed for dealing with these analytical challenges, and incremental improvements in current technologies will not meet that need. While current management technologies largely rely on instrumentation to monitor field conditions in real time, deficit irrigation management will require sophisticated modeling -- in future time -- of the whole complex system; irrigation hardware performance; management preferences; operational constraints; the disposition of applied water in heterogeneous fields; and the physiological responses of the crop.

Recognizing the need for a new generation of management modeling, the USDA and other agencies funded development of a practical decision support system for deficit irrigation known as *Irrigation Management Online* (IMO) (Hillyer and Sayde; 2010). This paper reviews the experience and general insights gained from beta testing of IMO for two cases; irrigation with high cost energy in the Columbia Basin, and management of a severely limited water supply for almonds in the California drought.

A DECISION SUPPORT SYSTEM

Development of IMO was based on three key design objectives. The first design objective was that the system should *embrace analytical complexity*. Simple water balance modeling cannot adequately represent the whole complex of dynamic, interacting processes involving soils, climate, crop, water supply, irrigation system and management practices that relate applied water to crop development and yield.

The second design objective was to *streamline the computational process* to facilitate rapid analysis of alternative irrigation strategies. Computationally efficient analytical tools such as linear programming or genetic algorithms have not been capable of dealing with the complexity of deficit irrigation management. Optimization will necessarily involve simulation and iterative search which will entail heavy computational burdens. The analytical software must therefore be designed for maximum speed and efficiency.

The third design objective was to *fully engage the user as a direct participant* in the analytical process. Any seasonal water use plan generated by IMO must align with the objectives, experience and preferences of the farm manager. To account for such subjective factors requires direct input from the client/manager.

These design objectives are reflected in the following key features of the IMO system:

- sophisticated modeling of the disposition and fate of applied water enables more accurate simulation and long-range projections of crop water availability
- modeling of application efficiency coupled with general, ET based yield models can realistically simulate crop response to applied water, an essential capability for optimum management.
- efficient analytical algorithms and advanced software design enable rapid search for optimal strategies and rapid updating of seasonal plans as circumstances change
- an editable calendar enables the farm manager to make short term modifications to irrigation schedules without compromising overall seasonal planning
- record keeping, integrated displays of alternative sources of field data and retrospective analysis of past seasons provide insight into field-specific relationships between water use and crop production, and facilitate system re-calibration for increased analytical precision.

Applications for deficit irrigation management

We will address four analytical tasks to which IMO has been applied for optimum management of limited water:

- Budgeting water; deciding how much water to use for a coming season
- Forward scheduling; planning the seasonal irrigation schedule to make best use of the limited water
- Error detection and recalibration
- Assessment

BUDGETING WATER

Example 1: deciding how much water to use when pumping costs are high

The question of how much water to use for a given field may be moot if the water supply is strictly limited, but when a farm has ample water this can be a challenging economic question. As a general rule the profit maximizing level of water use will be somewhat less than the yield maximizing level. The optimum amount to use may be based on the experience of individual farm managers. Some research leaders have offered general guidelines on the optimum. For two examples, Keller and Bleisner, 1990; English and Raja, 1996) have suggested that when water supplies are limited or costly the economic optimum point will be on the order of 10% or 20% less than full irrigation.

If water has a significant opportunity cost, the optimal level of irrigation may be considerably less (English and Raja, 1996). In that case a production function relating applied water to crop yield may be needed to evaluate alternative water use strategies.

Given the variability of weather, soils, antecedent moisture, distribution uniformity, root distributions and other factors, it is difficult to predict how much applied water will actually be used by a crop, and what potential yield will be. When combined with other factors, such as crop response to chemical use, weather conditions, disease, pests and so on, the yield that will be produced by a given level of applied water is virtually impossible to predict with certainty.

Nevertheless, estimation of yields is important for optimal management of deficit irrigation. It is our position that the analytical engine at the heart of IMO realistically represents the complex relationship between applied water and crop yield on a field scale. Having been derived from first principles, it is sufficiently accurate and robust to guide management decisions. An example follows.

Developing a crop production function

A general relationship between crop consumptive use of water, ET and yield is illustrated in Figure 1 (Raes and Geerts; 2009). Zones (c) and (d) of this function are the economically rational range of interest for deficit irrigation. Yields will increase more or less linearly in zone c. Then, for some crops, yield response rates will decline near maximum potential ET (zone (d)). Beyond that point, zone (e), yields will generally decline with the adverse consequences of excess water use.

Figure 2 indicates how ET relates to applied water (NEEA, 2013). As indicated, applied water tracks ET fairly closely in the range corresponding to zone (c). As applied water approaches the yield maximizing point, progressively increasing losses from surface accumulation and runoff, percolation and surface evaporation will cause the applied water curve to depart progressively farther from the ET curve.

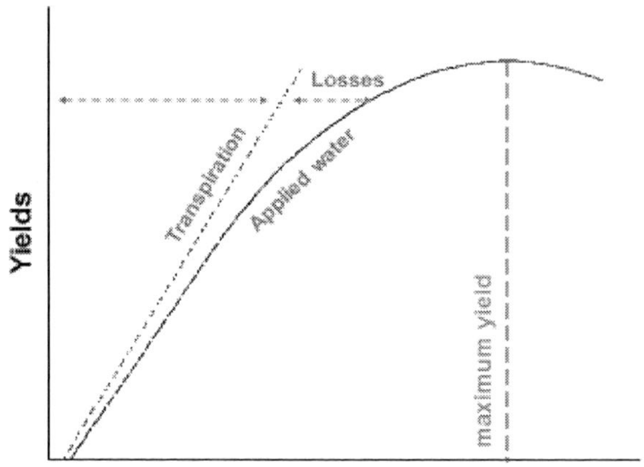

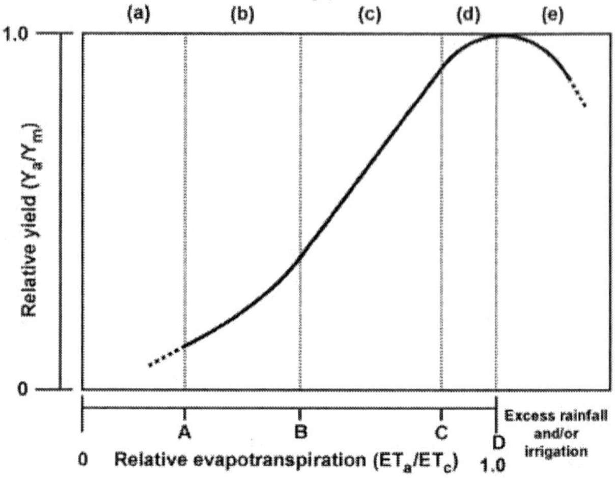

Figure 2: yield response to applied water

Figure 1: yield response to ET

IMO estimates yields by modeling these two relationships in tandem. When an increment of water is applied to the field IMO estimates the resulting pattern of incremental ET that across a heterogeneous field. The physiological response of the crop to incremental ET is then used to estimate yields on a field-wide basis.

The IMO system was used for partial irrigation of a circle of alfalfa on a cooperating farm in the Columbia Basin. The analysis considered alternative levels of water use ranging from 50.5 inches (full irrigation) down to 20 inches (67% below full irrigation). The specific management strategies employed for each level of water use were defined in terms of the following five management parameters, each combination of which would result in a specific level of water use:

i) *Irrigation adequacy*, the percentage of the field to be fully irrigated when water is applied
ii) *Management allowed depletion (MAD);* the amount by which soil water content is allowed to be reduced before an irrigation takes place.
iii) *Target refill level*; the target soil moisture level to which the root zone will be refilled during irrigation (expressed as a percentage of available water holding capacity in the root zone)
iv) *Assumed application efficiency*; the estimated application efficiency to be used in calculating gross irrigation requirements.
v) *Critical growth stage applications*, the seasonal pattern of applying or withholding irrigations according to stages of growth.

Note: in the case of alfalfa the growth stages are associated with the sequence of cuttings, since yields tend to decline with later cuttings. The discontinuity of data points between 27.7 inches and 43.60 inches of applied water derived from elimination of the last cutting.

Nineteen specific combinations of these parameters were used in this analysis to generate paired values of applied water and yields, as summarized in Table 1. The seventh column shows seasonal water use for each instance. The tenth column shows yields as estimated using the FAO 33 algorithm. (However the yield reduction factor derived by calibrating the FAO 33 algorithm (Doorenbos and Kassam, 1979) with water use and yield data from partial irrigation of six fields on the cooperating farm was 1.17, rather than the FAO published factor of 1.10. The derived production function is shown in Figure 3.

Table 1: yield response to applied water

Case	Adqcy	MAD	Refill Target -------	nominal effncy -------	irrigation ending date	Gross applied (inches)	ET	Losses as perc, spray, RO	FAO #33 yields (tons)
1a	87.5	50	100	85	20-Aug	49.9	44.4	6.9	8.99
1b						50.6	44.5	7.4	9.00
1c						50.6	44.5	7.1	9.00
2a	50	50	100	100	20-Aug	49.3	44.3	6.3	8.96
2b						46.1	43.4	5.2	8.91
2c						49.3	44.4	6.3	8.97
4a	nil	50	100	100	20-Aug	48.7	44.2	5.5	8.90
4b						48	44.2	5.3	8.89
4c						48.7	44.2	5.6	8.90
5a	nil	50	80	100	20-Aug	45.5	43.6	4.2	8.80
5b						45.5	43.3	4.3	8.79
5c						44.8	42.9	4.3	8.71
6	nil	60	80	100	20-Aug	43.6	42.2	4.1	8.52
7	nil	70	80	100	20-Aug	42.9	41.2	4	8.00
8	nil	50	80	100	9-Jul	27.7	29.6	2.3	8.08
9	nil	60	80	100	9-Jul	26.4	28.6	2.2	7.83
10	nil	70	80	100	9-Jul	25.1	27.3	2.1	7.32
11	nil	80	80	100	9-Jul	21.3	24.2	1.6	6.35
12	nil	85	60	100	9-Jul	16.8	19.7	1.3	4.66

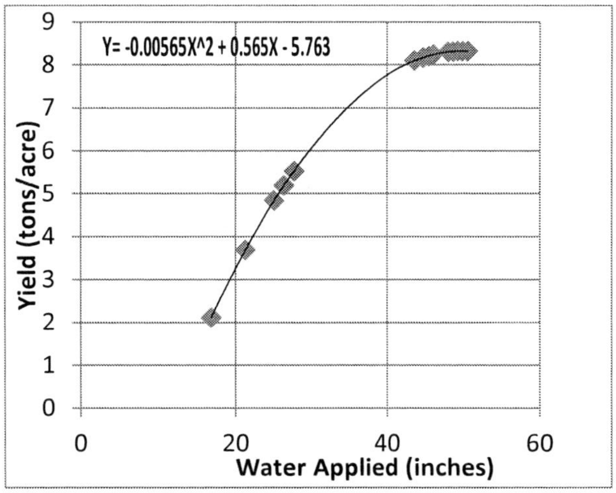

Figure 3: crop response function

Alternative irrigation strategies

This function was used to determine optimum water use for a 125 acre field of alfalfa under a center pivot system on a cooperating farm in the central Columbia Basin with high pumping head (300 ft) and energy costs of \$0.09/kwh. Estimated harvest costs were \$42 per ton, and crop sale price \$220 per ton.

We considered three alternative management objectives; first, to maximize yield per acre; second, to maximize net income per acre; third, to maximize net economic returns to water. Since this farm has more land than water, the water saved by deficit irrigation could be used to increase irrigated acreage, with opportunity costs corresponding to net returns to irrigation. The results are indicated in table 2, below. Net income at full irrigation would be \$143,000. If water and energy use is reduced net income would be increased until water use goes below 42.9 inches.

If water use is reduced below 42.9 inches the net income from the single 125 acre circle would be less than that from full irrigation. However if additional income were derived from sale or use of the water conserved, net farm income could continue to increase. In this case we assumed the opportunity cost of water could be captured by irrigating additional land with the same net economic return to water. At the point of maximum net returns to water (27.7 inches) the profit from cropping on the 125 acre pivot would be reduced by \$43,148, as indicated. But if the conserved water (22.9 inchesx125 ac = 2863 ac-in) were used to irrigate additional land it would yield an additional profit of (2863 ac-in x \$28.93/ac-in = \$82,726). Total farm profits would then be increased by (-\$43,148 + \$82,726 = \$39578).

Table 2: Analysis of alternative optimization strategies

Applied water	Crop yield	Revenue	Energy use (kwh/ac)	Energy Cost ($/ac)	Haying costs ($/ac)	Net income ($/acre)	Net for 125 ac	Change in net farm income	net returns to water ($/ac-iin)
16.8	2.134	470	1260	113	90	$267	$33,314	- $110,001	15.86
21.3	3.708	816	1598	144	156	$516	$64,534	-$78,781	24.24
25.1	4.859	1069	1883	169	204	$695	$86,933	-$56,382	27.71
26.4	5.215	1147	1980	178	219	$750	$93,763	-$49,553	28.41
27.7	5.552	1222	2078	187	233	$801	$100,167	-$43,148	28.93
42.9	8.077	1777	3218	290	339	$1,148	$143,520	$205	26.76
43.6	8.131	1789	3270	294	341	$1,153	$144,118	$802	26.44
44.8	8.209	1806	3360	302	345	$1,159	$144,855	$1,540	25.87
45.5	8.248	1814	3413	307	346	$1,161	$145,118	$1,803	25.52
45.5	8.248	1814	3413	307	346	$1,161	$145,118	$1,803	25.52
46.1	8.276	1821	3458	311	348	$1,162	$145,246	$1,930	25.21
48	8.339	1835	3600	324	350	$1,160	$145,052	$1,736	24.18
48.7	8.352	1838	3653	329	351	$1,158	$144,751	$1,436	23.78
48.7	8.352	1838	3653	329	351	$1,158	$144,751	$1,436	23.78
49.3	8.359	1839	3698	333	351	$1,155	$144,396	$1,081	23.43
49.3	8.359	1839	3698	333	351	$1,155	$144,396	$1,081	23.43
49.9	8.362	1840	3743	337	351	$1,152	$143,950	$635	23.08
50.6	8.360	1839	3795	342	351	$1,147	$143,315	$0	22.66

maximize yield per acre
maximize profit per acre
maximize profit per acre-ft

FORWARD SCHEDULING

Example 2: planning a deficit irrigation schedule for almonds

The second example involves an almond grower in the San Joaquin Valley of California whose available water supply in the fourth year of the recent drought was limited to 250 ac ft for 93.5 acres, or 32 inches, about 60% of full irrigation.

Identifying a research-based strategy

The first step was to review the past practices and experience gained by the producer herself, and to consult with research and extension professionals about recommended general strategies for deficit irrigation.

One primary resource used is shown in Figure 4, from a bulletin prepared by Doll and Shackel (2015) outlining the effect of water stress at various stages of almond development. This provided a general guide to how water should be allocated during the season. A key observation from that bulleting is that deficits can be most easily tolerated during June, the period approaching hull split.

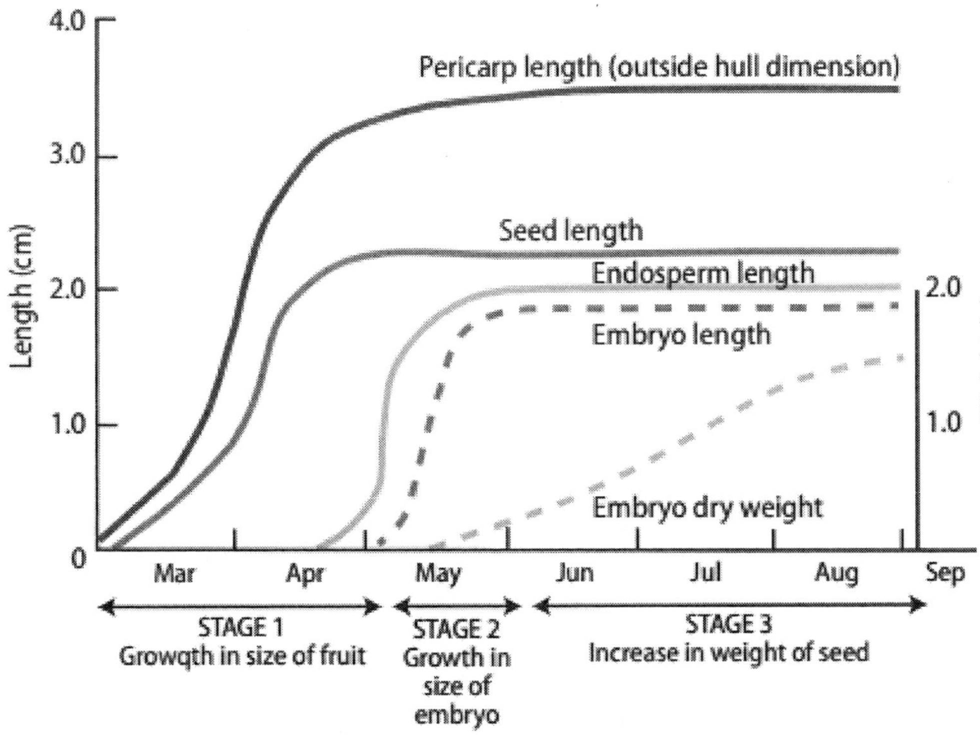

Figure 4: Almond sensitivity to water stress at various stages of development

A second primary source of advice was recommendations for various levels of partial irrigation of almonds, expressed as percentages of PET (PPET) (Goldhammer, IN FAO 66, Steduto, et al., 2012) for each of five phases of crop development. The specific recommendations when available water is 63% of full irrigation, comparable to the cooperating farm situation, are summarized in Table 3.

Table 3: recommended percentages of PET to be applied when the seasonal water supply is 63% of full irrigation

Phase 1: 70% of PET ... the first two months or so following bloom; March and April;

Phase 2: 50% of PET ..... shell hardening, kernel expansion; fruit maturity; May – early June
Phase 3: 25% of PET (later changed to 50%): .... hull split, late June
Phase 4: 100% of PET .... Approaching harvest
Phase 5: 60% of PET .... Post-harvest

This schedule indicates a tolerance for more stress in June, which is consistent with Doll's graph. It also advises more water use in July, approaching harvest.

A third key source, Ken Shackel at UC Davis, advised a generally uniform pattern of stress through the entire season, with the exception of increased stress approaching hull split.

The final strategy was a version of the pattern in Table 3, with two modifications: the first was to allocate less irrigation water to Phase 1, relying on antecedent moisture for a significant fraction of crop water use. The second was to increase water use in Phase 3 to 50% of potential PET.

The strategy thus derived then needed to be translated into specified irrigation dates and set times. The challenge was to allocate 32 inches of water over a seven month season, according to the water use pattern stipulated in Table 1, to maximize crop production, ensure good crop quality and minimize detrimental effects on the following season's crop.

Allocations according to stages of development

IMO was used to downloaded all historical weather data from the California Irrigation Management Information System (CIMIS) Los Banos station and compile a day-by-day profile of average reference ET. These were converted to crop potential ET and Monthly allocations of water were calculated according to the prescribed values in Table 1. The bar chart in Figure 5 shows crop potential ET (blue) and recommended allocations (yellow) for each of the five prescribed stages of crop development.

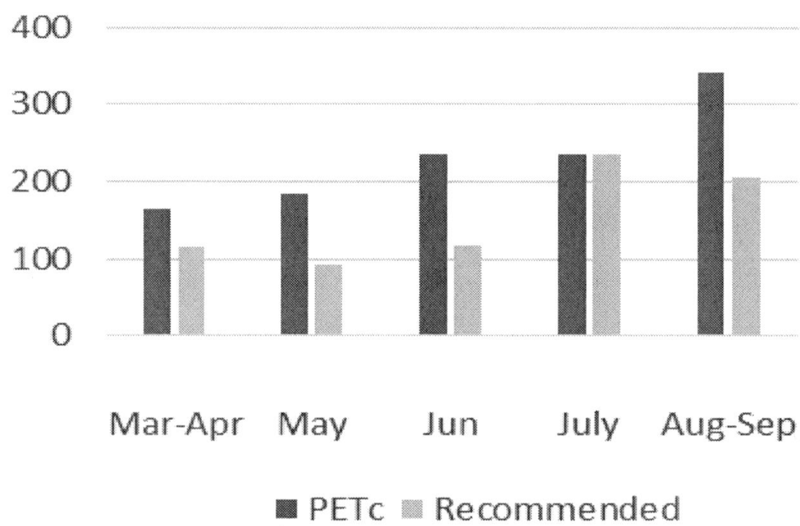

Figure 5: Potential ET and recommended allocations of water water

Forward scheduling

A detailed irrigation schedule was generated using IMO in an iterative search. A preliminary schedule was first generated automatically by IMO to use the 32 inches of available water in a pattern approximating that shown in Figure 5. Successive iterations in a guided search by an analyst until a sequence was found that would adapt the schedule to the specific circumstances of the farm to ensure that it was feasible, practical and consistent with irrigation system capabilities, constraints and normal farm practices.

The resulting schedule is shown in Table 4. A season-long projection of crop available water in a five foot root zone is shown in the accompanying graph (Figure 6). The black data points in upper left are neutron probe measurements to determine antecedent moisture. The red bars represent dates and amounts of irrigation events.

Start Date	Gross Application(Inches)	Set/Block/Rotation (hours)
3/24/2015 6:00 AM	0.72	12.0
4/8/2015 6:00 AM	1.44	24.0
4/22/2015 6:00 AM	1.44	24.0
5/5/2015 6:00 AM	1.44	24.0
5/11/2015 6:00 AM	1.44	24.0
5/20/2015 6:00 AM	1.44	24.0
6/2/2015 6:00 AM	1.44	24.0
6/8/2015 6:00 AM	1.44	24.0
6/17/2015 6:00 AM	1.44	24.0
6/22/2015 6:00 AM	1.44	24.0
7/1/2015 6:00 AM	1.44	24.0
7/6/2015 6:00 AM	1.44	24.0
7/13/2015 6:00 AM	1.44	24.0
7/20/2015 6:00 AM	1.44	24.0
7/28/2015 6:00 AM	1.44	24.0
8/5/2015 6:00 AM	1.44	24.0
8/11/2015 6:00 AM	1.44	24.0
8/24/2015 6:00 AM	1.44	24.0
9/2/2015 6:00 AM	1.44	24.0
9/16/2015 6:00 AM	1.44	24.0
9/29/2015 6:00 AM	1.44	24.0

Table 4: Suggested irrigation dates and set times

Tracking and updating the plan

As the season evolved IMO was used to track water use and field conditions and revise the schedule as needed to ensure adherence to the intended management strategy. The plan was revised during the season to account for weather anomalies and changing forecasts of available water. Figure 7 shows historical daily average PET as derived from the CIMIS station at Los Banos (plotted in red) and specific 2015 daily values (plotted in blue). The 2015 data departed substantially from expected values in May and July. The reduced ET was also compounded by unexpectedly high rainfall early in May. Additionally, the available water supply was increased slightly during the season. Consequently, some of the water originally allotted for May was shifted to July.

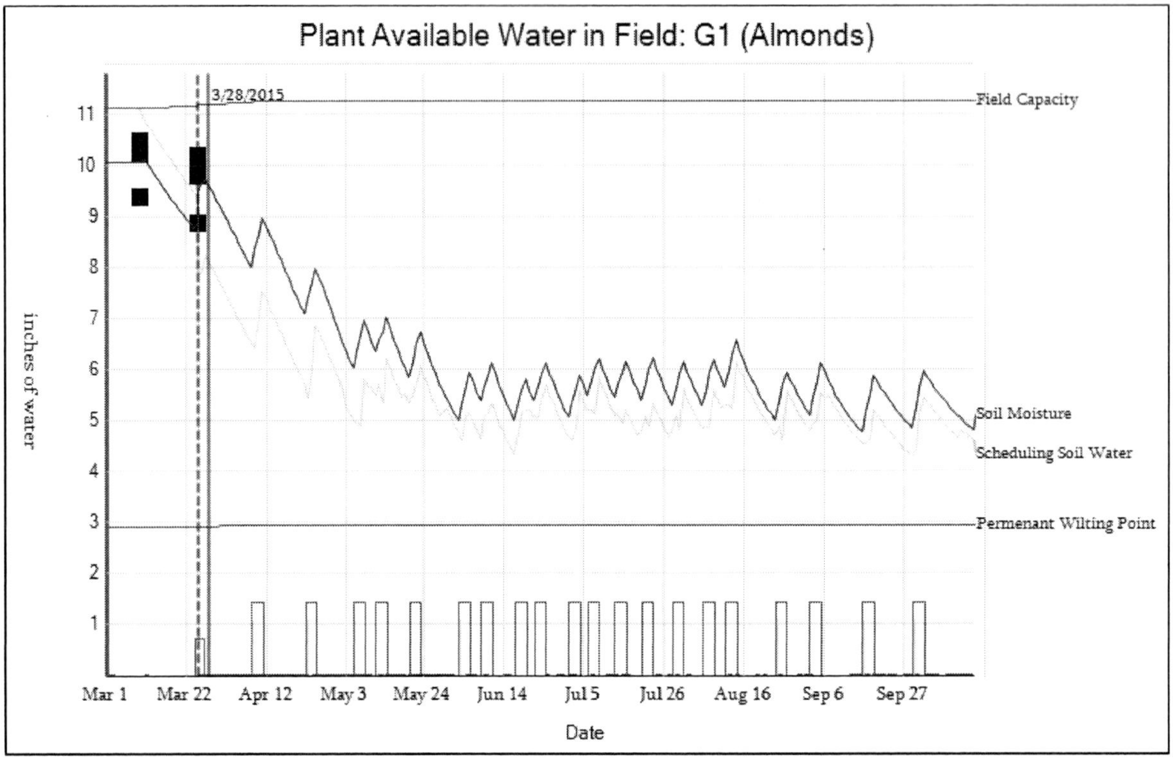

Figure 6: Anticipated seasonal pattern of crop water availability

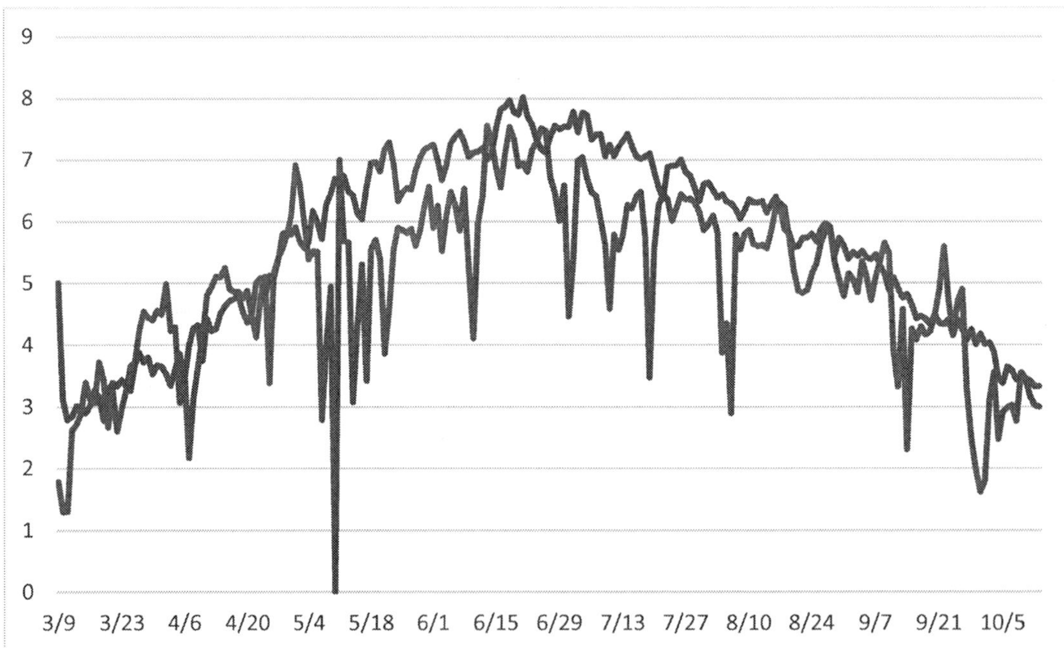

Figure 7: Estimated daily potential ET and observed daily PET

ERROR DETECTION AND RECALIBRATION

Error detection

The long range projections of crop water availability are subject to several sources of error. One factor, antecedent moisture, can often be a significant fraction of a water budget, but how much antecedent moisture will contribute to crop water use over the course of the season can be difficult to predict. Other important uncertain factors are estimates of the *potential* crop ET, upon which the plan is based, particularly due to the variability of k_c, the crop coefficient. Another parameter, K_s, which accounts for the reduction of actual ET when the crop is water-stressed is intrinsically uncertain, and algorithms for estimating K_s are generally linked to soil water holding capacity, which is itself uncertain.

Given these and other elements of uncertainty, deficit irrigation management should include error trapping and recalibrating of the analytical engine as routine operations. The detailed and integrated records of water use, soil moisture conditions and weather produced by IMO, combined with observations of crop development, crop stress and yield, provides an opportunity for systematically processing a mass of potentially valuable information from which a manager can gain insight and refined understanding of optimal water management.

IMO provides two ways to deal with these issues. One is by tracking soil moisture conditions to detect errors in long range projections of actual crop ET. The other is by integrating feedback data from alternative, independent sources. These are illustrated below.

Tracking soil moisture

In the case of the almond producer, it was clear early in the season that pre-season projections of crop available water were significantly lower than indicated by neutron probe measurements, and the error became progressively greater with time. Figure 8 illustrates the cumulative error by mid-May. The error was initially traced to two sources; the crop coefficients for early season ET were too high, and the estimated emitter discharge rate was about 6% low. The crop coefficients were revised in mid-season based on research done separately by Sandon, Ayars and Goldhammer (Goldhammer; IN FAO 66, Steduto et als, 2012). The emitter discharge rate was revised based on District measurements of water deliveries. Subsequently the assumed effective root zone available soil water holding capacity were adjusted. Figure 9 shows how model estimates (blue) would have compared with neutron probe measurements of soil moisture by the end of the season if not calibrated during the season. Figure 10 illustrates the revised soil moisture plot after recalibration. Such recalibration will be an iterative process, with further refinement of model parameters in succeeding years.

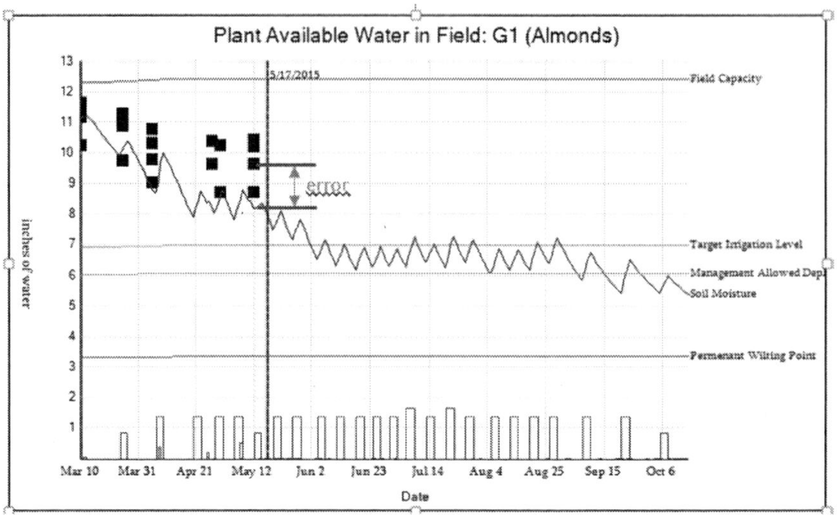

Figure 8: Error in projected soil moisture as of May 17, 2015

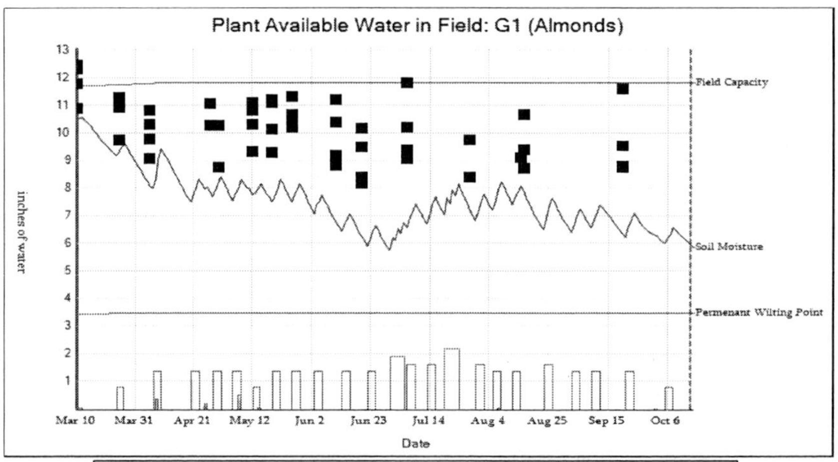

Figure 9: Uncalibrated soil moisture estimates

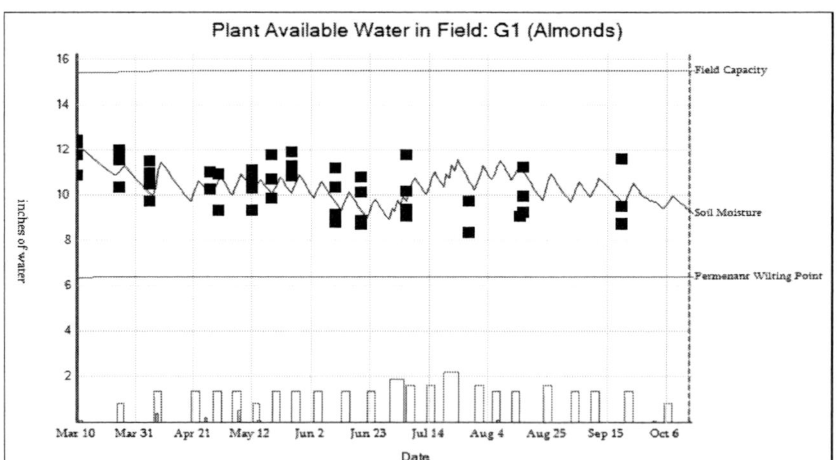

Figure 10: Calibrated projection of season soil moisture

Integrating alternative feedback data

Displaying multiple, independent sources of information about crop water status provides a basis for informed judgement of the quality of each data source. Figure 11 displays three independent data sets in a single graph: soil moisture estimates (derived from ET data), neutron probe readings and, along the bottom of the graph, annotated values of stem water potential.

As an example of the utility of integrated displays, a stem water potential reading of 16.0 in late March indicated incipient stress, indicating that the trees should be irrigated earlier than originally planned. But ET based modeling and neutron probe readings indicated there crop water availability was high. The consistent progression of the soil moisture data was judged to supersede the stem water potential readings and no additional irrigation was called for. (It was later concluded that the SWP readings had not been done correctly.)

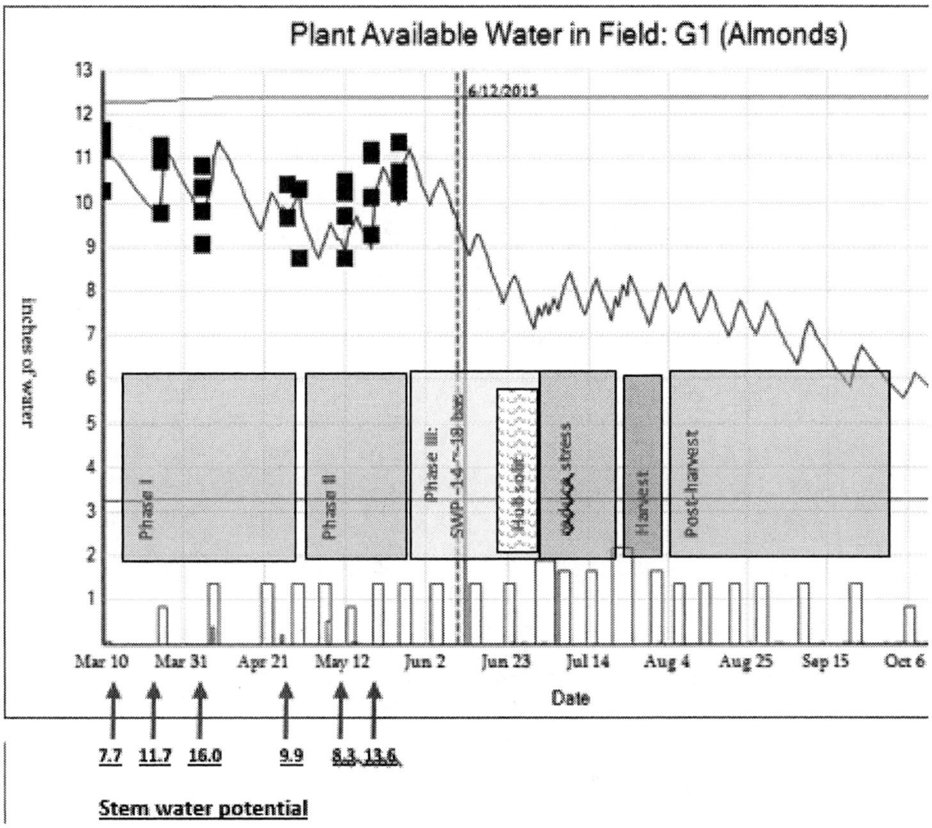

Figure 11: Comparing ET based estimates, neutron probe readings, and stem water potential with the original projection of season soil

While this example involves subjective use of independent data, we have also experimented with a more systematic procedure for combining alternative types of information using Bayesian Decision Theory (English and Sayde, 2008).

ASSESSMENT

It is not possible to quantify the benefits of this decision support system in terms of improved crop production, since there was no 'control' field. Nevertheless we can take note of the farm manager's subjective assessment of the system. Additionally we can describe in detail how well the actual irrigation schedule conformed to the advice of the research and extension community.

The manager's perspective

The value of the forward scheduling with IMO, expressed subjectively by the farm manager, was that it 'takes the guesswork out of it". Early in the season she was concerned that neighbors had begun irrigating and she wondered if she should also. With the seasonal plan in place she delayed starting for about two weeks, which enabled one additional irrigation at a more propitious time later in the season. As the season went on the question of whether to irrigate or delay recurred continuously. Ultimately she was comfortable following the plan precisely for the entire season, with the exception of shifting a day or two on one or two occasions because of conflicts with other activities.

Another important advantage from her perspective was knowing when to order district water.

Conformity with research guidelines

A second question was whether the pattern of water use was aligned with the research-based guidelines from which the irrigation strategy was originally derived. Figure 12 compares the recommended pattern of allocations (blue), the applied irrigation water (red) and the net crop water use (irrigation plus net change in soil water content at each stage, providing a visual indication of how well the pattern of actual allocations tracked the recommended pattern. Total water use for the season was 248 ac ft, almost exactly the original allotment of 250 ac ft.

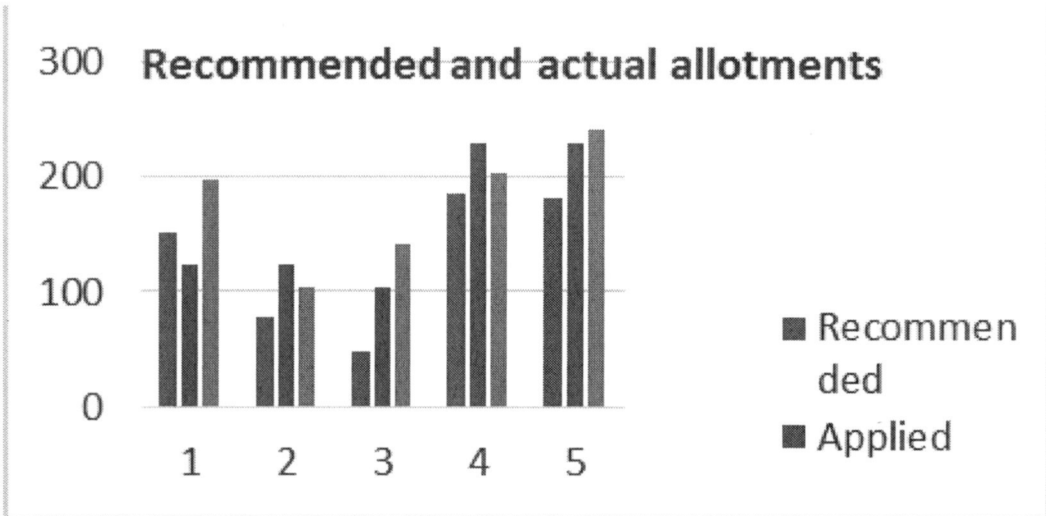

Figure 12: Actual patterns of crop water availability (green) and recommended pattern (blue)

Figure 13 shows a stress index (calculated ratio of actual ET to potential ET) plotted in parallel with Doll and Shackel's graphic of stress sensitivity, providing a general indication of the effectiveness of the stress management strategy. The stress pattern indicates that there was ample water until late May, then increasing stress approaching hull split, and quick recovery approaching harvest.

Yields

Yields in 2015 were 10.5% less than 2014. About 6% less water was applied in 2015 than in 2014, but the lower ET and unanticipated rain in 2015 may have offset that difference.

Our understanding is that the harvest volume was about the same in 2015 as 2014, but kernel weights were slightly lower for nonpareils and significantly lower for two other varieties. One possible explanation for the reduced kernel weights might be that early season (April and May) water supplies were higher than planned relative to late season water use. As a rule, water stress should be more or less balanced throughout the season (Shackel, personal Communication), and some degree of stress early in the season would condition the trees to later stress. But unexpected rainfall and lower than expected potential ET resulted in high levels of crop available water until mid May, followed by significant stress through June. From Doll's graph in Figure 13, early season high crop water availability may have induced full growth of the outer shell early in the season, but the subsequent water supply was unable to support full growth of the kernels later in the season. We did adjust the plan in May to account for the anomalous weather, but perhaps not aggressively enough. There may also have been been an echo effect from stress in the preceeding three years of intensive drought.

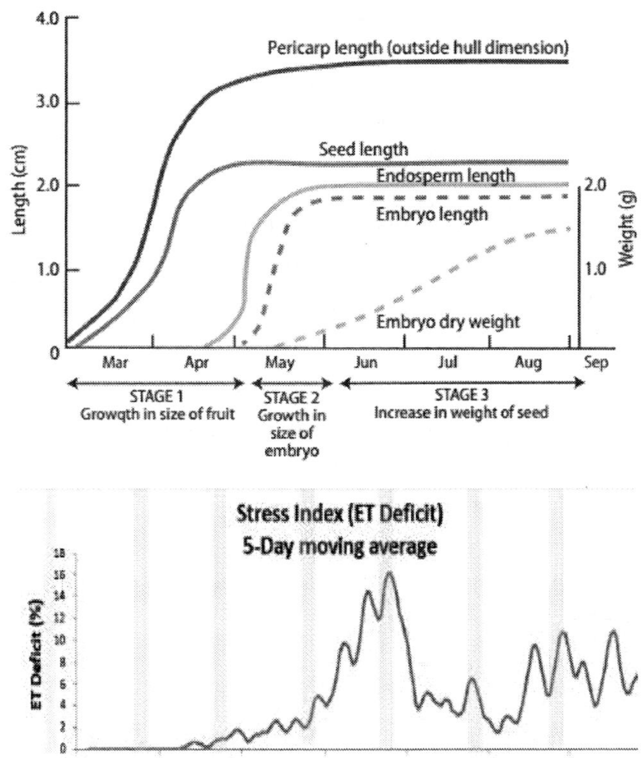

Figure 13: Comparing seasonal sensitivity to water stress to the seasonal pattern of estimated stress moisture with neutron probe data

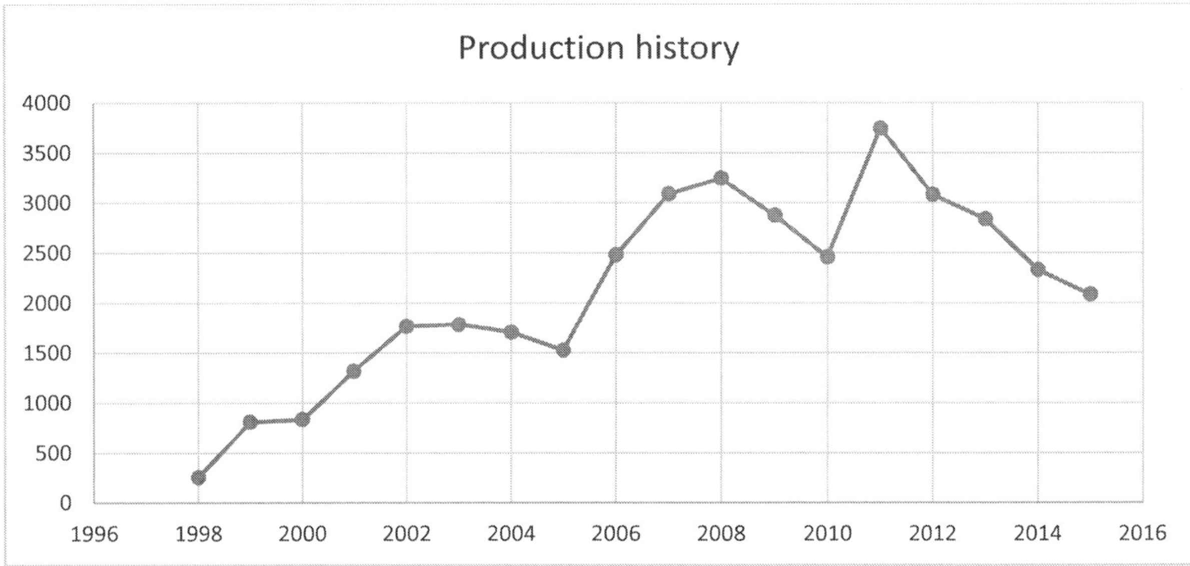

Figure 14: history of orchard yields since planting

CONCLUSIONS

Modeling of crop yields in response to partial irrigation, though intrinsically uncertain, provides science-based guidance for deciding how much water to allot to a particular field when water is limited or expensive. Example 1, illustrated value derived from modeling a field specific relationship between applied water and crop yields in order to examine in more precise detail the benefits of conserving water, and determine well defined optimal levels of water use.

Example 2 also illustrated the process of forward scheduling by which an irrigation manager was able to plan in detail for implementing a recommended irrigation strategy under drought conditions. The planning allowed the farm manager to envision an entire season.

The continuation of example 2 illustrated a necessary element of deficit irrigation management, the systematic and continuous processes of error detection and recalibration of the analytical system. The process of error detection, though predominantly based on the quantities that are modeled (i.e. soil water depletion) can also be enhanced by systematically comparing independent sources of feedback data.

Comparison of uncalibrated and calibrated system analysis indicated that the error in initial estimates of antecedent moisture was about 3.5 inches, or 10% of the anticipated water supply.

The implementation of the intended strategy tracked well with the chosen strategy. The pattern of water use over the season was close to the originally stipulated pattern, after adjusting for the recommended increased water use approaching hull split. Yields were less than in previous years, but it is difficult to ascertain the cause.

REFERENCES

BPA, 2012; Web-based Advisory Program for Energy Conservation in Irrigation; Bonneville Power Administration; Final Report to: Bonneville Power Administration; From: Oregon State University

Doorenbos, J., and Kassam, A.H. 1979. Yield Response to water, FAO Irrigation and Drainage Paper 33, FAO, Rome, Italy, pp. 193.

Doll, David; and Ken Shackel; Drought Management for California Almonds, ANR Publication 8515, February 2015 |

English, M.J. 2010. *"A developing crisis in irrigation advisory services"*; ASABE-IA 5th Decennial International Irrigation Conference, Phoenix, Dec. 2010

English, M.J., and Chadi Sayde, 2008, *"A Feedback System to Optimize Crop Water Use Estimates in Irrigation Scheduling"*, World Environmental and Water Resources Congress, May 13-16, Honolulu

English, M.J., Solomon, K. H., and Hoffman, G.J. 2002. *"A paradigm shift in irrigation management."* Invited paper, Jubilee Year Edition, J. Irrigation and Drainage Engineering, ASCE. Vol. 28, No 5.

English, M. J. and S. N. Raja; 1996; *Perspectives on Deficit Irrigation*; Agricultural Water Management, Vol.32(1996)

FAO, 2002; Deficit Irrigation Practices; FAO Water Reports No. 22; Food and Agriculture Organization of the United Nations, Rome, 2002

Geerts, s> and D. Raes; " Deficit irrigation as an on-farm strategy to maximize crop water productivity in dry areas; Agricultural Water Management, V 96 (9), pp 1275-1284; Sept, 2009

Goldhammer, D., 2012; *Almonds* (pp 358-373); IN; *Crop yield response to applied water*, by: Steduto et als, FAO Irrigation and Drainage Paper 66, Food and Agriculture Organization of the United Nations, Rome

Hillyer, C., Sayde C. (2010) A Web Based Advisory Service For Optimum Irrigation Management. In: 5th National Decennial Irrigation Conference, Phoenix, AZ: ASABE

Keller, Jack, and Ron D. Bliesner; 1990; *Sprinkle and Trickle Irrigation*; Blackburn Press

Kirda, C>, P. Moutonnet, C. Hera and D. R. Nielsen EDS; 1999; Crop yield response to deficit irrigation; Soil and Water Management and Crop Nutrition Section of the Joint FAO/IAEA Division of Nuclar Techniques in Food and Agriculture; Kluwer Academic Publishers.

NEEA; 2013; Preparation of a general guide for improving irrigation efficiency: *A Roadmap for Energy and Water Efficiency in Irrigated Agriculture.* Northwest Energy Efficiency Alliance,

Steduto, P, T. C. Hsiao, E. Fereres and D. Raes, 2012; Crop yield response to applied water, FAO Irrigation and Drainage Paper 66, Food and Agriculture Organization of the United Nations, Rome

UCANR, 2016; Summary of University of California research on irrigation management for almond trees under drought conditions; University of California, Division of Agriculture and Natural Resources; 2016

Effect of RDI on Quality and Economic Yield of Navel Oranges

Natalio Mendez, MS, CCA*
California State University, Fresno. 2415 E. San Ramon Ave MS AS 72. Fresno, CA. 93740.
*Currently at Helena Chemical Company, Fresno, CA.

Govind Seepersad, PhD
The University of the West Indies, St. Augustine, Trinidad & Tobago, West Indies.

Florence Cassel Sharma, PhD
California State University, Fresno. 2415 E. San Ramon Ave MS AS 72. Fresno, CA. 93740.

Dave Goorahoo, PhD
California State University, Fresno. 2415 E. San Ramon Ave MS AS 72. Fresno, CA. 93740-8033

Abstract. *Growing oranges under drought conditions is a challenging task which can result in the adoption of regulated deficit irrigation (RDI). This research was conducted at two locations in California using five irrigation regimes, during the months of July and August, to assess the effect of limited water application on the quality and economic yield of navel oranges. The partial budget modelling approach was used to assess the economic impact of reduced water application on the price of the oranges sold in the U.S. and exported to Japan. At both locations there was a significant difference in fruit Brix index and yield at lower RDI treatments compared to fruit receiving higher doses of water. The implementation of RDI in navel oranges can increase crop quality, yet severe RDI levels can decrease yield. Generally, the price of "free-watered" citrus in the U.S. was lower than citrus in the Japanese market. More importantly, the economic analysis adopted in this study showed the Japanese market was an incentive to: (1) manage water to benefit from that niche market, and (2) produce fruits of lower weight that would ultimately be compensated through higher prices.*

Keywords: RDI, Navel oranges Brix, partial budget modelling, micro-sprinkler fertigation system

Introduction

Citrus crops are a highly profitable commodity in the United States (U.S.) with California and Florida being the two leading States (USDA, 2012). In 2012, the U.S. citrus production totaled 11.2 million tons with a value of $3.2 million of which $2 million was produced in California. The contribution of California citrus crops was 30% of the total U.S. citrus fruit production and 42% of the national value (CDFA. 2014). Citrus crops are one of California's most profitable and essential products to the State economy bringing in over $2 billion annually (California Citrus Mutual, 2013). A vast number of citrus crops are currently in production in California, such as: navels, valencia, minneola tangelos, grapefruit, lemons, and mandarins which are produced across the state and have a large portion based in the central San Joaquin Valley (SJV). The importance of the citrus industry for the sustainability of the local economy is fundamental, yet there are issues that can potentially decrease overall productivity of citrus in the SJV. For example, the recent identification of Asian Citrus Psyllid (ACP) and Huanglongbing (HLB) Disease has become a major problem for citrus growers throughout the State because the potential transmission of citrus greening disease which can destroy the citrus industry having detrimental consequence on the economy (Citrus Research Board, 2010; Grafton-Cardwell and Daugherty, 2013). The yearly potential of frost damage is also an issue growers try to overcome and reduce the severity of the harvest impact they will endure.

Currently, the drought situation in California is probably the issue of most concern to farmers, as the availability of water, or lack thereof, can ultimately affect pestilence and frost management.

Regulated Deficit Irrigation (RDI) is an optimization strategy in which irrigation is applied during drought-sensitive growth stages of a crop (Kriedemann and Goodwin, 2003). Basically, water application is limited to drought-tolerant phenological stages of the crop, often during the vegetative stages and the late ripening period. Total irrigation application throughout the crop cycle is therefore below the crop's evapotranspiration (ETc) requirements (Fereres and Soriano, 2007). While this inevitably results in plant drought stress, overall productivity is maximized by the water supplied through RDI, which is generally the main limiting resource (English, 1990).

RDI has also been used to increase the Brix index and soluble solid levels in various crops (Johnstone et al., 2005). For consumer preference, increasing the Brix index- which is an indicator of the sweetness of the fruit, and reducing the size is essential as consumers tend to prefer a sweet orange with an average size. These two fruit attributes increase the desirability of the product and, thus, RDI can be utilized to potentially impact plant physiological stress in an effort to obtain these desired fruit qualities.

One approach by California citrus growers in dealing with the lack of water allocation in recent years has been the adoption of RDI in many crops, such as: grapes, almonds, pistachios, and citrus throughout the State (Goldhamer and Fereres. 2005). While the implementation of RDI practices in various cropping systems has shown to have beneficial effects, it is vitally important to examine the effects of RDI on the various species of citrus production, plant physiology, fruit quality and economic yield.

Objectives

As part of our ongoing research aimed at optimizing water use efficiency (WUE) for various crops grown in the San Joaquin Valley (SJV), for this study the focus was on navel oranges (*Citrus Sinensis* (L.) Osbeck, cv. Washington 'frost nucellar'). The overall objective was to quantify the effects of RDI on fruit quality and economic yield of navel oranges grown in the central region of the SJV, California. For this presentation, a specific objective was to review the fertigation protocol adopted to ensure that the citrus crop received the equivalent of 50%, 75%, 100% (growers practice/control), 125%, and 150% of the reference evapotranspiration (ET), along with adequate plant nutrition needs.

Materials and Methods

The study was conducted at two separate locations in the city of Woodlake, California (Figure 1). Both locations had similar soil types, rootstock, scion, fertilization, irrigation, and age of crop. Irrigation and fertilizer was applied through a micro-sprinkler fertigation system (fanjet). There were five irrigation levels relative to evapotraspiration (ET) which meant that the crop evapotranspiration (ET_c) requirements were met at 50%, 75%, 100% (growers practice/control), 125%, and 150% of ET.

The fields consisted of mature navel oranges planted in 1964. The citrus crop consisted of propagated cultivars Washington 'Frost Nucellar' scion grafted on Troyer citrange (*Poncirus trifoliata)* rootstock. Troyer rootstock has the characteristics to be reasonably vigorous and resistant to *Phytophthora parasitica*, nematodes, and tristeza virus as well as cold tolerance. The tree spacing was 22ft by 22ft. The research sites, referred to as the North and South locations, were each 43,560ft^2 (i.e. 1 acre) (Figure 1).

Figure 1: Google Earth® imagery of the North and South locations.

The study was conducted from spring 2014 to spring 2015 (traditional citrus growing season) in a commercial citrus farm in Woodlake, California. Both North and South locations contained identical soil, irrigation, crop age and fertilizer programs. Moreover, both fields had a soil type of consisting of a sandy loam with an electrical conductivity of 3.6 dS/m, predominantly calcium driven. The experimental design consisted of 3 blocks and 4 herbicide treatments within each block (0%, 50%, 75%, and 100% (growers practice) of Matrix (PRE) herbicide (main plots) with the five different RDI treatments. Each block was replicated three times and consisted of four rows with one row in between each herbicide treatment as a buffer.

Sampling plots were represented by two trees with a one tree buffer between each sub-treatments. During the active growing season (late-March to November), this citrus block was provided with 25-32 hours of irrigation per week, approximately 0.1-0.15 acre-inches per irrigation event. Irrigation events occurred in two weekly application of 10-15 hours per irrigation event dependent on crop evapotranspiration (ETc). The RDI treatments were calculated in relation to the traditional irrigation practices in place by the farmer. In order to ensure irrigation accuracy, all RDI treatments in all blocks were subjected to a flow and pressure test to ensure irrigation uniformity. Moreover, irrigation was calculated by implementing an irrigation schedule that was crop specific by taking into account: evapotranspiration (ETo), crop coefficient (Kc), and a 95% irrigation efficiency factor (University of California Cooperative Extension. 2015). Climatic data from at the nearby Lindcove, California Irrigation Management Information Systems (CIMIS) station was used to develop the irrigation schedule for conventional treatment (1.00 RDI). Irrigation treatments (RDI) were confirmed by using collection cans to determine the efficacy and distribution of the micro-sprinklers.

All criteria were included in the development of an effective, crop specific irrigation schedule in which in the conventional treatment (1.00 of ET) received the actual amount of water calculated from the various climactic conditions. Irrigation treatments (RDI) were controlled upon the outputs of micro-sprinklers. This research included the use of FanJet Micro-Sprinklers from Bowsmith. RDI treatments were then quantified into the following 5 irrigation categories: 0.50, 0.75, 1.00, 1.25, and

1.50 of evapotranspiration, having into account the previously mentioned irrigation strategies. By altering the emitter sizes to increase or decrease the water application rate, the amount of water applied for each RDI treatment was attained even though the overall run time was the same. In this approach treatment 1.00 RDI (grower standard) was considered to be the control and based on the hours of application, total emitter output (overall water application) and an efficiency factor, the volume of water applied was determined by the following equation:

$$Gallons\ Applied\ =\ Actual\ Hours\ Applied\ x\ Emitter\ Output\ x\ Efficiency\ Factor$$

All emitters were evaluated throughout the growing season at random evaluation events to maintain high efficiency in water application. The Bowsmith micro-sprinkler emitters used for this research had a pressure differential in which irrigation system psi could be used to alter the rate at which fanjets emitted water. In order to ensure maximum efficiency of the irrigation system, the fanjets were maintained at a minimum pressure of 20 psi. There were three random irrigation system events, at each location, when the efficiency and capacity were evaluated. This evaluation included three fanjet sprinklers in which each was analyzed by performing 15 second flows and the pressure (psi) at each. The following equation was used to evaluate the efficiency of each treatment in accordance to evapotranspiration (ET) demand:

$$Avg. Gallons\ Applied\ per\ Emitter\ =\ Hours\ Applied\ x\ Avg. Gallons\ per\ Flow$$

Fertilizers was applied through a fertigation system in which urea-ammonium nitrate (UAN-32), ammonium polyphosphate (10-34-0), magnesium nitrate, potassium thiosulfate (KTS), and sulfuric acid (H_2SO_4) were injected into the irrigation system. Granular Urea (42-0-0) was also applied via broadcast spreader. Foliar applications of micro-nutrient mixes were conducted twice a year to ensure adequate nutritional levels. Such applications included a mixture of elements such as: zinc (Zn), molybdenum (Mo), manganese (Mn), copper (Cu), and iron (Fe) in EDTA form, nitrogen fertilizer it also included in foliar spray mixes. The addition of sulfuric acid was intended for the purpose of reducing the impact of high carbonates and bicarbonates in the irrigation water. The intended pH level for the irrigation water was of 5±0.5 due to high bicarbonate levels. All the fertilizers were applied via irrigation water through a micro-sprinkler (fanjet) irrigation system. Both North and South locations received the following amounts of fertilizer during the growing season: nitrogen (N) – 90-100 lbs/ac, phosphorus (P) – 45 lbs/ac, potassium (K) – 70 lbs/ac, and micros ranged within the rates of 5 lbs/ac (for Mo and Cu) to 20 lbs/ac (for Zn).

Soil moisture profiles were attained using the Diviner 2000 by Sentek™ Technologies. Volumetric water content measurements (θ) were monitored to assess the degree of moisture reduction within the soil profile. The incorporation of this technology allowed for the proper monitoring during a two-month period in which water application was a critical factor. The instrumentation also facilitated the management of RDI levels to ensure that there were no excessive water reductions, thereby, avoiding further crop stress and physiological damage. Soil moisture readings events were conducted on a weekly basis during the two-month period (July and August) time during which was the growing period of most interest. Diviner access tubes were places at approximately 5 feet in depth to represent the active root zone of navel oranges; readings are recorded for every 10 centimeters (3.94 inches) to a maximum depth of 100 centimeters (39.37 inches).

Fruit circumference (cm) were taken and measured via a digital fruit sizing caliper weekly during the months of July and August 2014 to account for fruit circumference growth. Each RDI treatment, contained 25 marked fruits measured weekly with the fruit size caliper. In addition, fruit size was also measured at harvest, along with percentage Brix indices, percent juice, percent solids, and sugar to acid ratio. Protocol for the measurement of these parameters were in accordance with the

Western Australia Department of Agriculture and Food's methodology for fresh market citrus quality (Western Australia Department of Agriculture and Food, 2014). Briefly, juice extracted from oranges collected from two trees was measured for Brix index using a refractometer. For percent juice and solid, five pieces of fruit were squeezed with a manual press to separate juice (ml) and solid fruit pulp (g). Percent acidity was determined by titration with 0.01N NaOH, and the sugar to acid ratio was calculated. At harvest, data was obtained for total yield (lbs), marketable fruit (fresh market), and non-marketable fruit (culls and juice). Fruit were collected from all sub-plots and subjected to general fruit standards (yield weight (lbs) and fresh market standards) and fruit quality standards (fruit weight (g), fruit solids (g), juice (mL), and sugar to acid ratio). The desired parameters were chosen due to their importance in determining fresh market quality standards and relevance in obtaining high economic yield return (California Citrus Mutual. 2013). The data collected for the growing season was subjected to analysis of variance using univariate general linear model used for split block design using SPSS® software (SPSS, 2013).

In addition to fertigation and crop related data, an assessment of the economic yield was conducted by adopting a partial budget modelling approach. Generally, owners of the citrus farms are often asked to make decisions based on resource constraints and market realities. In many cases, decisions are incremental, such as bringing more land into production, expanding or reducing an enterprise or changing how an enterprise is managed. In this case, climatic factors and plant agronomy were brought together to estimate whether water-saving technology applied to the navel oranges will result in lower yields but higher farm income from better prices.

Results & Discussion

Water Application During Irrigation: Actual run time for water applied during the irrigation of the crop in 2014 is provided in Table 1. Throughout the research, two irrigation uniformity tests were conducted at the beginning and middle end of the study to verify the efficiency of the water application system (Tables 2 and 3).

Table 1: Actual irrigation applied based on ETo and Kc for navel oranges in Woodlake, California from July-September 2014.

Week (2014)	ETo	Kc	ETc	Hours Scheduled/ Month	Actual Hours Applied/ Month
21-Jul	1.75	0.65	1.14	29	32
28-Jul	1.52	0.65	0.99	25	29
4-Aug	1.51	0.65	0.98	25	27
11-Aug	1.69	0.65	1.10	28	31
18-Aug	1.69	0.65	1.10	28	26
25-Aug	1.57	0.65	1.02	26	25
1-Sep	1.51	0.65	0.98	25	22
8-Sep	1.33	0.65	0.86	22	23
15-Sep	1.28	0.65	0.83	21	20
22-Sep	1.09	0.65	0.71	18	18
29-Sep	1.08	0.65	0.70	18	20
TOTALS	*16.02*	*0.65*	*8.17*	*262*	*273*

At the North and South Locations, the amount of water applied as a percentage of the required ET were close to the rates established for the five RDI treatments. This was facilitated by the use of Bowsmith fanjets which allowed for the control of the water delivery by adjusting emitter output in response to pressure (psi) in the irrigation lines. For the North Location, the 0.50 ET treatment was calculated to be 52.17% of the actual ET with an average pressure of 19.8 psi, the 0.75 treatment was at 73.81% with an average psi of 19.7, the 1.00 treatment was at 95.64% with an average psi of 18.9, the 1.25 treatment was at 123.51% with an average psi of 19.6, and the 1.50 treatment was at 149.29% with an average psi of 19.8 (Table 2). This was indicative that the irrigation system in place was at high efficacy and uniformity.

For the South Location, it was established that for the 0.50 treatment the actual amount of the crop ET applied was of 52.26% with an average psi of 20.1, the 0.75 treatment was at 74.71% with an average psi of 19.9, the 1.00 treatment was at 97.84% with an average psi of 19.2, the 1.25 treatment was at 125.95% with an average psi of 20.2, and the 1.50 treatment was at 147.31% with average pressure of 18.9 psi (Table 3).

Table 2: Average emitter output for each treatment throughout the growing season in comparison (%) to evapotranspiration (ET_o) for the North Location.

Week	Hours Applied /YR	Grower Standard	50%	75%	100%	125%	150%
21-Jul	32	342.4	178.62	252.736	327.456	422.912	511.168
28-Jul	29	310.3	161.88	229.042	296.757	383.264	463.246
4-Aug	27	288.9	150.71	213.246	276.291	356.832	431.298
11-Aug	31	331.7	173.04	244.838	317.223	409.696	495.194
18-Aug	26	278.2	145.13	205.348	266.058	343.616	415.324
25-Aug	25	267.5	139.55	197.45	255.825	330.4	399.35
1-Sep	22	235.4	122.80	173.756	225.126	290.752	351.428
8-Sep	23	246.1	128.39	181.654	235.359	303.968	367.402
15-Sep	20	214	111.64	157.96	204.66	264.32	319.48
22-Sep	18	192.6	100.48	142.164	184.194	237.888	287.532
29-Sep	20	214	111.64	157.96	204.66	264.32	319.48
Totals	*273*	2921.1	1523.89	2156.15	2793.61	3607.968	4360.902
% of ET Applied			*52.17%*	*73.81%*	*95.64%*	*123.51%*	*149.29%*

Table 3: Average emitter output for each treatment throughout the growing season in comparison (%) to evapotranspiration (ET_o) for the South Location.

Week	Hours Applied /YR	Grower Standard	50%	75%	100%	125%	150%
21-Jul	32	342.40	178.944	255.808	335.008	431.264	504.384
28-Jul	29	310.3	162.168	231.826	303.601	390.833	457.098
4-Aug	27	288.9	150.984	215.838	282.663	363.879	425.574
11-Aug	31	331.7	173.352	247.814	324.539	417.787	488.622
18-Aug	26	278.2	145.392	207.844	272.194	350.402	409.812
25-Aug	25	267.5	139.8	199.85	261.725	336.925	394.05
1-Sep	22	235.4	123.024	175.868	230.318	296.494	346.764
8-Sep	23	246.1	128.616	183.862	240.787	309.971	362.526
15-Sep	20	214	111.84	159.88	209.38	269.54	315.24
22-Sep	18	192.6	100.656	143.892	188.442	242.586	283.716
29-Sep	20	214	111.84	159.88	209.38	269.54	315.24
Totals	*273*	2921.1	1526.62	2182.36	2858.037	3679.221	4303.026
% of ET Applied			*52.26%*	*74.71%*	*97.84%*	*125.95%*	*147.31%*

The inclusion of the efficiency evaluations during the season was essential in an effort to determine if there were overall inadequacies in the irrigation systems. Overall, the distribution uniformity of the drip irrigation systems was excellent and in accordance with the manufacturer's specifications. More importantly, the experiments at both locations had differences in relative irrigation rates that facilitated the comparison of RDI applied at approximately 0.50, 0.75, 1.00, 1.25 and 1.50 the crop evapo-transpiration (ETc) rates calculated from CIMIS data.

Brix Measurements: RDI on citrus crops has been used to regulate fruit sugar content as explained by Garcia-Tejero et al. (2010) and by Gonzalez-Altozano & Castel (2000) in their research with mandarins (cv. Clementina de Nules) in which RDI was a successful practice to increase Brix content. The desirability for high sugar index in navel oranges has been a primary factor for California citrus growers to develop innovative strategies to increase sugar content. Romero et al. (2006) demonstrated that the introduction of RDI had positive effects in the regulation of sugar level in navel oranges. In the current study the collection of Brix data was conducted monthly on all plots in both the North and South Locations to assess the influence of RDI on fruit sugar levels.

At the North Location, RDI had no significant effect to Brix concentrations for fruits collected during the first month. However, from Months 2 through Harvest there was a significant influence (Month 2: P=0.012, Months 3-Harvest: P=0.000) (Table 5; Figures 1 and 2). The average Brix concentration for oranges subjected to the 0.50 and 0.75 RDI treatments attained the greatest sugar content increase throughout the growing season with Brix indices of 3.67±0.19 in Month 1 and 21.74±0.60 at Harvest for 0.50 RDI; and 3.79±0.17 for Month 1 and 19.59±0.31 at Harvest for the 0.75 RDI treatment. More importantly, at harvest the navel oranges subjected to 0.50 RDI were twice as sweet at those receiving 1.50 RDI. Furthermore, the 1.50 RDI treatment resulted in oranges having the lowest Brix with a mean of 4±0.19 in Month 1 and 12.82±0.45 at Harvest (Figure 3).

Table 4: Significant difference of RDI on Brix in the North Location (α=0.05).

Statistical Significance for Fruit Brix (%) Content: North Location

Month	Irrigation
1	0.857
2	0.012
3	0.000
4	0.000
5	0.000
6	0.000
7	0.000
8	0.000
Harvest	0.008

*Significance (α = 0.05).

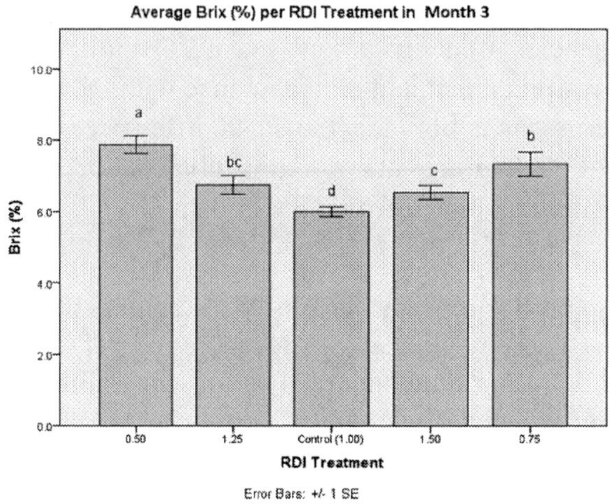

Figure 1: Average Brix index per RDI treatment in the North location in month 3.

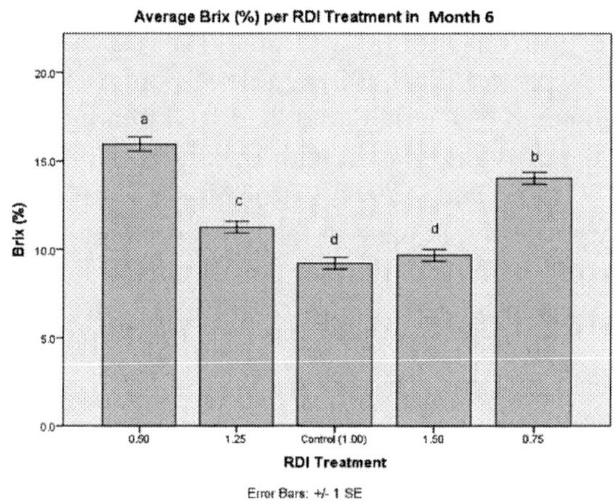

Figure 2: Average Brix index per RDI treatment in the North location in month 6.

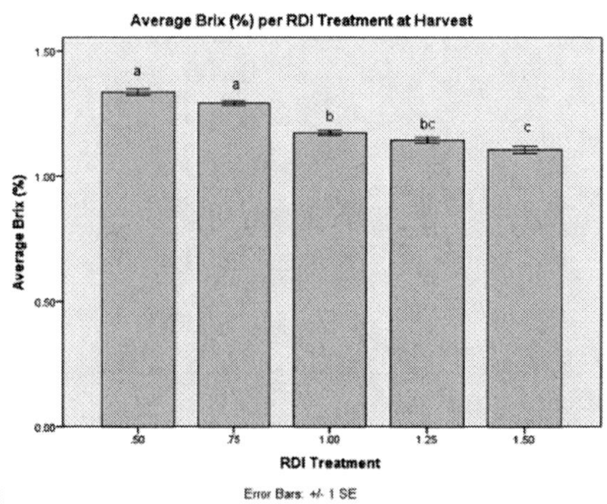

Figure 3: Average Brix index per RDI treatment in the North location in month 6.

As was the case with the North location, there were no significant differences in Brix indices as a function of RDI within the first two sampling events (Table 5; Figures 4 and 5). From Month 3 to Harvest there was a significant effect in Brix levels among the different RDI treatments (Month 3-Harvest all at P=0.000) as the average Brix indices increased from 8.5±0.25 to 20.91±0.45, and from 6.04±0.32 to 12.85±0.39, for oranges receiving 0.50 RDI and 1.50 RDI, respectively. With the exception of Month 4, there was no significant RDI x herbicide interaction effect on Brix levels (Table 5).

Table 5: Significant difference of RDI on Brix in the South Location.

Statistical Significance for Fruit Brix (%)

Content: South Location

Month	Irrigation
1	0.513
2	0.503
3	0.000
4	0.000
5	0.000
6	0.000
7	0.000
8	0.000
Harvest	0.000

*Significance (α = 0.05).

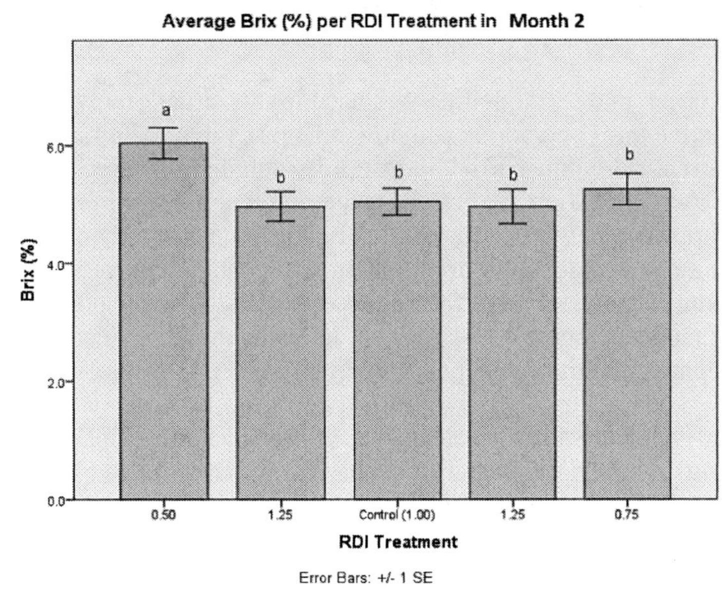

Figure 4: Average Brix index per RDI treatment in the South location in month 2.

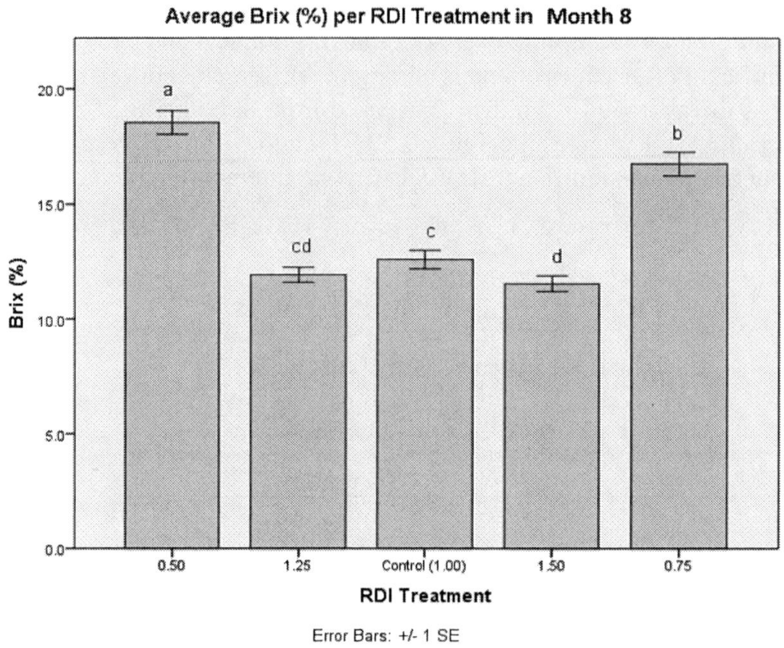

Figure 5: Average Brix index per RDI treatment in the South location in month 8.

These above results are similar to the findings by Romero et al. (2006) in which the sugar content of citrus (cv. Clemenules) increased with a reduction in irrigation. Furthermore, the finding in this study that a reduction in irrigation rates can significantly increase the sugar concentration of navel oranges is consistent with studies conducted on table grapes (Chaves et al., 2007, 2010; Ebel et al., 1993, 1995), pears (Cui et al., 2007), and apples (Ebel et al., 1993). This increase in Brix level in the fruit could be attributed to the induction of enough stress to prevent any further growth but translocation of the photosynthates from the source to the sink (Matthews et al., 1988; Quick et al., 1992). The current study endorses the cultural practice of using RDI to increase Brix levels in the cultivar of navel oranges grown in the Woodlake orchard.

Economic yield: The partial budget modelling was used in this situation given its presentation of incremental changes in the farm operations and how it helps evaluate the financial effect of the intervention. In the current study, the partial budget only included resources that could be changed, such as the reduced water application. Only the change under consideration was evaluated for its potential to vary the farm income. Hence, the partial budgets were based on changes in the following: Increase in income; Reduction or elimination of costs; Increase in costs; and, Reduction or elimination of income. The net impact of the above effects will be the positive financial changes minus the negative financial changes. A positive net indicates that farm income will increase due to the change, while a negative net indicates the change will reduce farm income.

Findings from this study indicate that the price of navel oranges in the Japanese market was an incentive for citrus growers to manage water to benefit from that niche market. In the case of the bulk market, water can be applied up to the point of yield maximization for the bulk market at optimum Brix. There is however, an incentive to reduce water application which can result in a lower yield, measured by fruit weight but a higher brix content. The loss in weight will result in lower revenue using the bulk citrus market model. But, the incentive to produce fruits of lower weight would be compensated through higher prices which more than compensated for the lower yield weight.

Concluding Remarks

- The current study focused on the evaluation of the adoption of RDI as a fertigation strategy for optimizing the sweetness (Brix indices) and economic yield of navel oranges grown in the central SJV, California.
- Firstly, the inclusion of an efficiency assessment of the irrigation system during the season is essential in an effort to determine if there were overall inadequacies in the irrigation systems. Overall, the distribution uniformity of the drip irrigation systems in the current study was excellent and in accordance with the manufacturer's specifications.
- The irrigation system described in the current study comprising micro fertigation sprinklers, commonly referred to as fan jets was suitable comparing the RDI applied at approximately 0.50, 0.75, 1.00, 1.25 and 1.50 the crop evapo-transpiration (ETc) rates calculated from climate data downloadable from California Irrigation Management Information Systems (CIMIS),
- Even though the loss in weight of the navel orange associated with reduced irrigation will result in lower revenue in the local and national bulk citrus market, there is an incentive for growers to produce the navel oranges for international export markets, such as Japan. In the longer term, the lower weight would be compensated through higher prices which can result in a timely return of investment associated with the adoption of the RDI technology.
- The current study endorses the cultural practice of using RDI to increase Brix levels and economic yield for the navel oranges (*Citrus Sinensis* (L.) Osbeck, cv. Washington 'frost nucellar') grown in the Woodlake orchard located in SJV.

References

California Citrus Mutual. (2013). California citrus facts [online]. Available at http://www.cacitrusmutual.com/industry/all-about-citrus/citrus-facts.html. Last accessed: 26 October, 2016.

CDFA. (2014). California Agricultural Statistics Review, 2013-2014. California Department of Food and Agriculture.

Chaves, M. M., T. P. Santos, C. R. Souza, M. F. Ortuno, M. L. Rodrigues, C. M. Lopes, J. P. Maroco, and J. S. Pereira. (2007). Deficit irrigation in grapevine improves water- use efficiency while controlling vigour and production quality. Ann. Appl. Biol. 150, 237-252.

Chaves, M. M., O. Zarrouk, R. Francisco, J. M. Costa., T. Santos, A. P. Regalado, M. L. Rodrigues, and C. M. Lopes. (2010). Grapevine under deficit irrigation: hints from physiological and molecular data. Ann. Bot. 105, 661-676.

Citrus Research Board. (2010). Entomology/pests: Asian citrus psyllid [online]. Available at http://www.citrusresearch.org/links.

Cui, N., T. Du, S. Kang, F. Li, J. Zhang, M. Wang, and Z. Li. (2008). Regulated deficit irrigation improved fruit quality and water use efficiency of pear-jujube trees. Agric. Water Manage. 95, 489-497.

Ebel, R. C., E. L. Proebsting, and M. E. Patterson. (1993). Regulated deficit irrigation may alter apple maturity, quality, and storage life. Hort Sci. 28, 141-143.

Ebel, R. C., E. L. Proebsting, and E. G. Evans. (1995). Deficit irrigation to control vegetative growth in apple and monitoring fruit growth to schedule irrigation. J. Hort. Sci. 30, 1229-1232.

English, M. (1990). "Deficit Irrigation. I: Analytical Framework." J. Irrig. Drain. Eng. 116, 399-412.

Fereres, E., and M. A. Soriano. (2007). Deficit irrigation for reducing agricultural water use. J. Exp. Bot. 58, 147-159.

Garcia-Tejero, I., Romero-Vicente, R., J. A.Jiménez-Bocanegra, G. Martinez-Garcia, V. H. Duran-Zuaro, and J. L. Muriel-Fernandez. (2010a). Response of citrus trees to deficit irrigation during different phenological periods in relation to yield, fruit quality, and water productivity. Agric. Water Manage. 97, 689-699.

García-Tejero, I., J. A.Jiménez-Bocanegra, G. Martínez, R. Romero, V. H. Durán-Zuazo, & J. L. Muriel-Fernández. (2010b). Positive impact of regulated deficit irrigation on yield and fruit quality in a commercial citrus orchard [*Citrus sinensis* (L.) Osbeck, cv. salustiano]. Agric. Water Manage. 97, 614-622.

Goldhamer, D., and Fereres, E. (2005). The Promise of Regulated Deficit Irrigation in California's Orchards and Vineyards. California Department of Water Resources – California Water Plan Update 2005. Available at: http://www.waterplan.water.ca.gov/docs/cwpu2005/vol4/vol4-cropwateruse-promiseofregulateddeficitirrigation.pdf. Last accessed: 2 October, 2016.

Gonzalez-Altozano, P., and J. R. Castel (2000). Castel. Regulated deficit irrigation in 'Clementina de Nules' citrus tress. II: vegetative growth. J. Hort. Sci. Biotech. 75, 388-392.

Grafton-Cardwell, E. E., and M. P.Daugherty. (2013). Asian Citrus Psyllid and Huanglongbing Disease. Pest Notes Publication #74155. University of California Agriculture and Natatural Resources Statewide Integrated Pest Management Program. Available online: http://www.ipm.ucdavis.edu/ PDF/PESTNOTES/pnasiancitruspsyllid.pdf. Last accessed November 1, 2016.

Johnstone, P. R., T. K. Hartz, M. LeStrange, J. J. Nunez, and E. M. Miyao. (2005). Managing fruit soluble solids with late-season deficit irrigation in drip-irrigated tomato production. J. Hort. Sci. 40, 1857-1861.

Kriedemann, P. E., and I. Goodwin. (2003). Regulated Deficit Irrigation and Partial Root Drying. Land & Water, Australia.

Romero, P., J. M. Navarro, J. Perez-Perez, F. Garcia-Sanchez, A. Gomez-Gomez, I. Porras, V. Martinez, and P. Boria. (2006). Deficit irrigation and rootstock: their effects on water relations, vegetative development, yield, fruit quality and mineral nutrition of Clemenules mandarins. Tree Physiol. 26, 1537-1548.

SPSS. 2013. SPSS for Windows Statistical Package. Standard version 23.0

USDA. (2012). Citrus production by typer and State - United States. United States Department of Agriculture. Available at: http://www.nass.usda.gov/ Statistics_by_State/Florida/Publications/Citrus/cit/2011-12/cit0612.pdf. Last accessed: 26 October, 2016.

Western Australia Department of Agriculture and Food. (2014). Measuring internal maturity of citrus. Western Australia Department of Agriculture and Food. Available at: https://www.agric.wa.gov.au/citrus/measuring-internal-maturity-citrus?page=0%2C1. Last accessed: 25 October, 2015.

This page intentionally left blank.

The Northern Colorado Limited Irrigation Research Farm: Past and Future Experiments

Kendall DeJonge, Ph.D., P.E.

USDA-ARS WMSR, 2150 Centre Ave, Bldg D Ste 320, Fort Collins CO 80526, kendall.dejonge@ars.usda.gov

Thomas Trout, Ph.D.

USDA-ARS WMSR, 2150 Centre Ave, Bldg D Ste 320, Fort Collins CO 80526, thomsas.trout@ars.usda.gov

Louise Comas, Ph.D.

USDA-ARS WMSR, 2150 Centre Ave, Bldg D Ste 320, Fort Collins CO 80526, louise.comas@ars.usda.gov

Sean Gleason, Ph.D.

USDA-ARS WMSR, 2150 Centre Ave, Bldg D Ste 320, Fort Collins CO 80526, sean.gleason@ars.usda.gov

Huihui Zhang, Ph.D.

USDA-ARS WMSR, 2150 Centre Ave, Bldg D Ste 320, Fort Collins CO 80526, huihui.zhang@ars.usda.gov

Abstract: *Agricultural water research has been conducted for over 100 years at the USDA-ARS Limited Irrigation Research Farm (LIRF) near Greeley, Colorado. Recent experiments since 2008 have focused on deficit irrigation of commodity crops, primarily corn. Experimental design and facilities are described, as well as water production function (yield vs. evapotranspiration) data obtained from 2008-2016. Water balance is determined using several inputs including on-site reference evapotranspiration, canopy cover, and feedback from soil water content using neutron probe measurements. Physiological measurements, from root length density to plant transpiration via sap-flow, have been taken and example results are also shown. Ground-based remote sensing is an important component of the program, from continuous infrared canopy temperature readings in focused plots to less frequent multispectral and thermal imaging taken from a high clearance tractor with a customized boom. Results and publications from previous projects are briefly highlighted, and the current research focus outlined.*

Keywords. Deficit irrigation, limited irrigation, water productivity, water use efficiency, water stress, infrared thermometry, IRT, CWSI, remote sensing, image processing, root growth, sap flow, plant traits

Introduction

Irrigation water supplies in the Great Plains and much of the western U.S. are declining. Supplies originally developed for irrigated agriculture are being diverted to growing urban areas and for ecosystem restoration. Groundwater use in many areas has exceeded sustainable amounts and must decrease to prevent aquifer depletion. Temperature increases due to climate change will likely reduce mountain snowpack accumulation that is critical to irrigation water supplies and may increase watershed evapotranspiration and crop water requirements. Irrigated agriculture will likely have less water available in the future than it had in the past. Sustaining irrigated agriculture and meeting future food and fiber needs of the growing global population will require increasing productivity per unit of water.

To respond to these needs, the Limited Irrigation Research Farm (LIRF) was established near Greeley, CO in 2008. Managed by the USDA-ARS (Agricultural Research Service) Water Management and Systems Research Unit (formerly Water Management Research Unit), this farm and research group have the overall goal of defining management strategies to sustain irrigated agriculture with limited water supplies. Specific objectives include:

- Determining potential for crop evapotranspiration (ET) savings from growth-stage based strategic deficit irrigation.
- Improving irrigation scheduling by developing crop coefficients from ground-based remote sensing.
- Understanding physiological responses to crop water stress.

This paper provides an overview of the LIRF, the datasets collected and works published since 2008, and goals/aims going forward.

Farm and Experiment Description

The LIRF is located northeast of Greeley, CO (40°26'50" N, 104°38'10" W, 1425 m asl). The 16-ha facility was developed to conduct research on irrigated crop water requirements and crop response. Annual and seasonal average precipitation at this semi-arid site is 340 and 220 mm, respectively. A 4.7-ha experimental field was divided into 4 equal crop sections. From 2008 to 2011, maize (*Zea mays* L., the dominant crop in the region) was grown in rotation with sunflower (*Helianthus annuus*), dry bean (*Phaseolus vulgaris*), and winter wheat (*Triticum aestivum*), as shown in Figure 1. Each field section was divided into 4 replicate blocks, and each block was divided into six 9 x 43 m plots containing 12 N-S oriented crop rows (0.76 m row spacing) on which six irrigation treatments were randomly assigned (randomized block design). From 2012 to 2016, the western two sections and eastern two sections were combined and used to grow maize and sunflower (*H. annuus*) in rotation, with 12 irrigation treatments. Irrigations were applied using surface drip tubing, placed next to each plant row. The east and west edges of each crop section contained a 6 row buffer, with all measurements taking place in the middle 4 to 6 rows. While several crops were evaluated in the overall experiment, the results shown in this paper will be exclusively for maize. All treatments of the same crop were planted at the same population and received the same nitrogen applications. Minimum tillage was used to maintain surface residue from the previous crop and minimize surface evaporation.

Figure 1. Aerial view of the water productivity plots at LIRF on August 1, 2008. Crops from left to right are dry beans, winter wheat (post-harvest), sunflower, and maize. Note visible treatment effects in maize, which was at a more advanced growth stage and has more aboveground biomass than dry bean and sunflower.

Irrigation and Water Balance

The full irrigation treatment was irrigated such that water availability (irrigation plus precipitation plus stored soil water) was adequate to meet crop water requirements, as predicted by the reference evapotranspiration and crop coefficients from FAO-56 methodology (Allen et al., 1998). Adequacy was monitored by ensuring the soil water content remained in the readily available water range. The remaining treatments were irrigated to achieve total water applications (irrigation plus precipitation) that approximated the target treatment amounts. In the 2008-2011 experiment (Trout, 2016), the scaling of the remaining treatments was typically done throughout the season; in the 2012-2016 experiment there were varying levels of water stress and ET reduction imposed from V7 (7 leaves) until VT (tasseling), and from R4 (dough) to maturity. During the sensitive reproductive stages between VT and R4, the crop was irrigated to eliminate water stress. Irrigation applications to each treatment were measured with turbine flow meters (Badger Recordall Turbo 160 with RTR transmitters). Soil water content, SWC, was measured 2 or 3 times each week on the days before and/or after irrigation in the crop row near the center of each plot. Soil water content was measured in 30 cm depth increments between 30 and 150 cm depth, and at 200 cm depth with a neutron soil moisture meter, NMM, (CPN-503 Hydroprobe, InstroTek, San Francisco, CA). The NMM was calibrated gravimetrically at the site. The calibration was used to convert instrument relative counts to volumetric soil water content (SWC). The NMM measures SWC within an approximately 15 cm radius from the measurement point, and was assumed to represent the soil profile within 15 cm of the measurement depth (eg. the 30 cm depth measurement represented the 15 – 45 cm depth). The SWC in the surface 15 cm was measured in the row near the NMM access tube with a portable time domain reflectometer (Minitrase, Soilmoisture Equipment Corp, Santa Barbara, CA) with 15 cm long rods.

By assuming zero runoff due to relatively small field slopes, adequate soil infiltration and surface residue, and drip irrigation, the crop ET (ETc) was estimated as

$$ETc = I + P - \Delta S - DP \qquad \text{Eq. 1}$$

Full details of water balance are described in Trout and DeJonge (2013). An example of soil water trends under two irrigation treatments is shown in Figure 2.

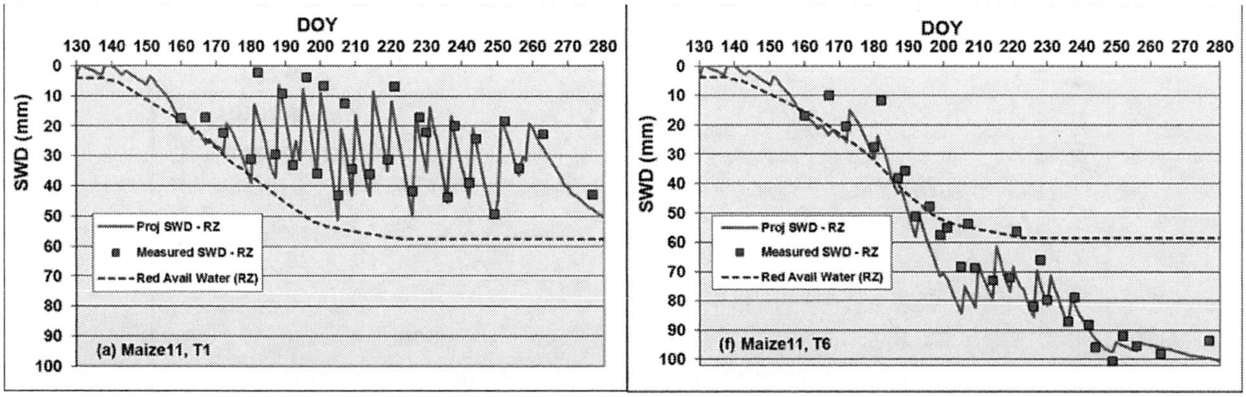

Figure 2. Soil water deficit (SWD, mm) for LIRF 2011 maize fully irrigated treatment (T1, left) and deficit irrigated treatment (T6, right). Squares are the measured SWD in the active root zone, solid line is the modeled SWD, and the dashed line represents the readily available water for the active root zone.

Water Production Function

Crop productivity and its relationship with water is often considered either in terms of yield vs. water applied, or yield vs. ET, the latter of which is called a water production function (WPF). The productivity of maize in the 2008-2011 experiment is shown below (Figure 3). . Results from 2012-2016 have shown similar WPF among treatments with the same deficit target between the two stress periods (late vegetative and maturation periods) with evidence that different deficit targets between these periods (e.g. substantial deficit in either period with near full irrigation in the other) can cause departures (Comas et al., 2014).

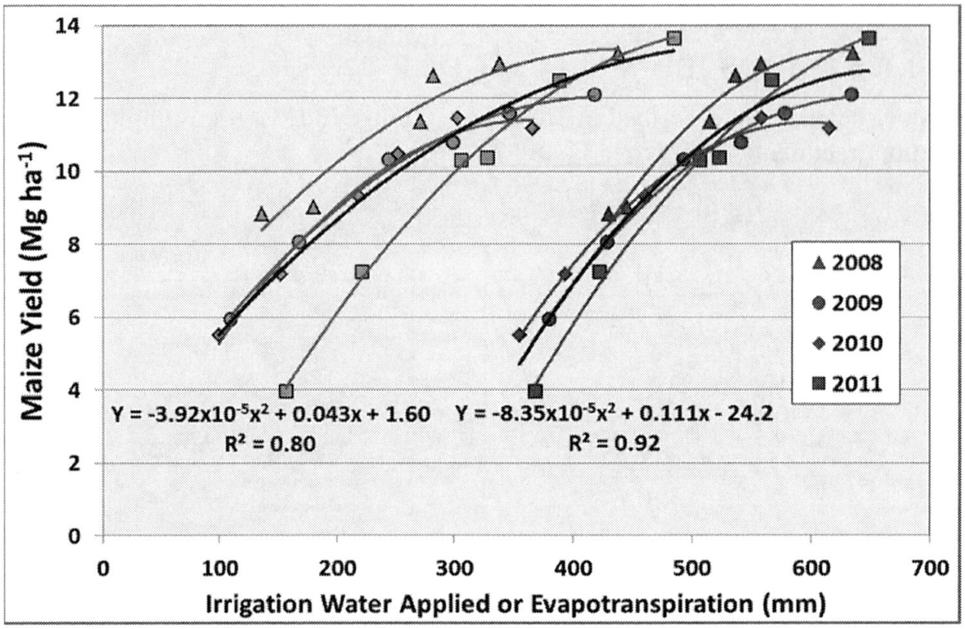

Figure 3. Maize grain yield (@15.5% moisture content) vs. irrigation water applied (left curves, blue symbols) and crop evapotranspiration (right curves, red symbols). Narrow (colored) lines are 2nd degree polynomial regression fits to the mean yield data for each year. Thick (black) lines and equations are regression fits to the combined 4 years of data.

The water production function based on applied irrigation water is fairly flat at full irrigation and curves downward as the water application decreases, showing that the decrease in yield for each unit decrease in water applied is relatively small when the deficit is small, but the rate of yield decrease gets larger as the deficit increases. This means that the marginal productivity of irrigation water (additional yield per unit additional water) is relatively low near full irrigation, showing the potential benefit to the farmer of reducing irrigation and transferring water to higher-valued uses. The water use efficiency, or productivity per unit of irrigation water applied, increases from about 29 kg ha^{-1} mm^{-1} of water applied at full irrigation to about 40 kg ha^{-1} mm^{-1} when irrigation is reduced by 50%. This is because irrigation is more efficient, precipitation is more effectively used by the crop, and the crop extracts more water from the soil.

However, the water production function for grain yield based consumptive use or ET (the right curves in Fig 6) moves to the right and is relatively straight and consistent until at maximum ET, where it levels and spreads some. This implies that the corn is equally efficient in its use of every additional unit of water consumed. The water use efficiency in terms of ET is about 20 kg ha^{-1} mm^{-1} at full irrigation. This is smaller than when based on irrigation water because it also counts precipitation used by the crop. The water use efficiency based on ET stays relatively constant for deficits up to about 15%, and then decreases. Because corn requires about 250-300 mm of water to produce any yield, the water use efficiency declines with deficit irrigation.

Infrared Thermometry

Canopy temperature has been used as an indicator of crop water stress, since a reduction in plant available water results in lower transpiration rates and consequently higher canopy temperatures. The LIRF experiment has provided a framework to collect comprehensive datasets of canopy temperature. Temperature of the corn canopy was acquired on a continuous basis (Figure 4) using infrared thermal radiometer (IRT, model: SI-121, Apogee Instruments Inc., Logan UT) with a 36° field of view and ±0.2 °C accuracy. The IRTs were attached to telescoping posts and angled 23° below horizontal and 45° from north (looking northeast) to ensure viewing primarily crop canopy once canopy cover was nearly complete. Data was typically omitted when canopy was less than 80% total ground cover, to limit view of soil background. Measurements were sampled every 5s and averaged over 30 min intervals.

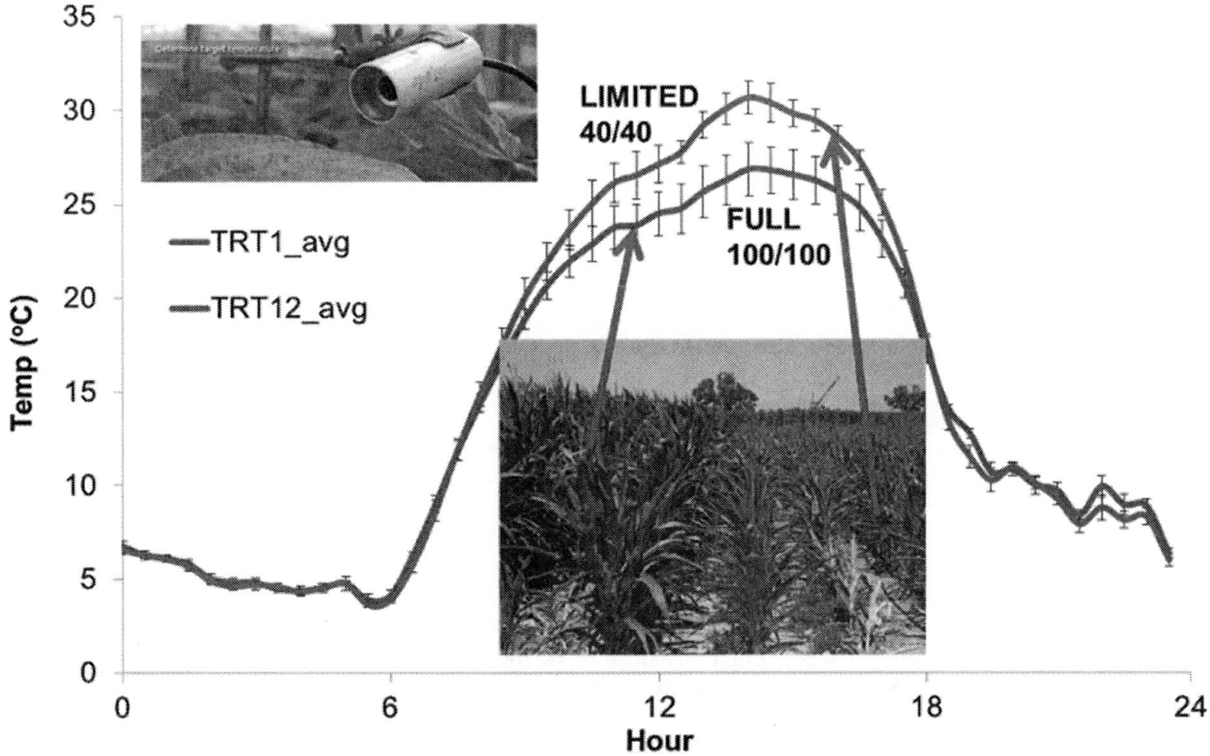

Figure 4. Representative single-day canopy temperature for fully irrigated maize (blue) and limited irrigation maize (red). Error bars indicate standard deviation from four replicates.

Several indices have been developed for monitoring and quantifying water stress from infrared thermometry, using canopy temperature (Tc) as a main driver. The most famous of these is the Crop Water Stress Index (CWSI) established in the early 1980s (Idso et al., 1981; Jackson et al., 1981). While this index is often considered the gold standard, it has additional meteorological requirements which include onsite air temperature and humidity, as well as creation of site or region-specific baselines. Our research recently showed that canopy temperature alone can be strongly correlated with plant physiological measurements of crop water stress such as leaf water potential (DeJonge et al., 2015), which is discussed in more detail later.

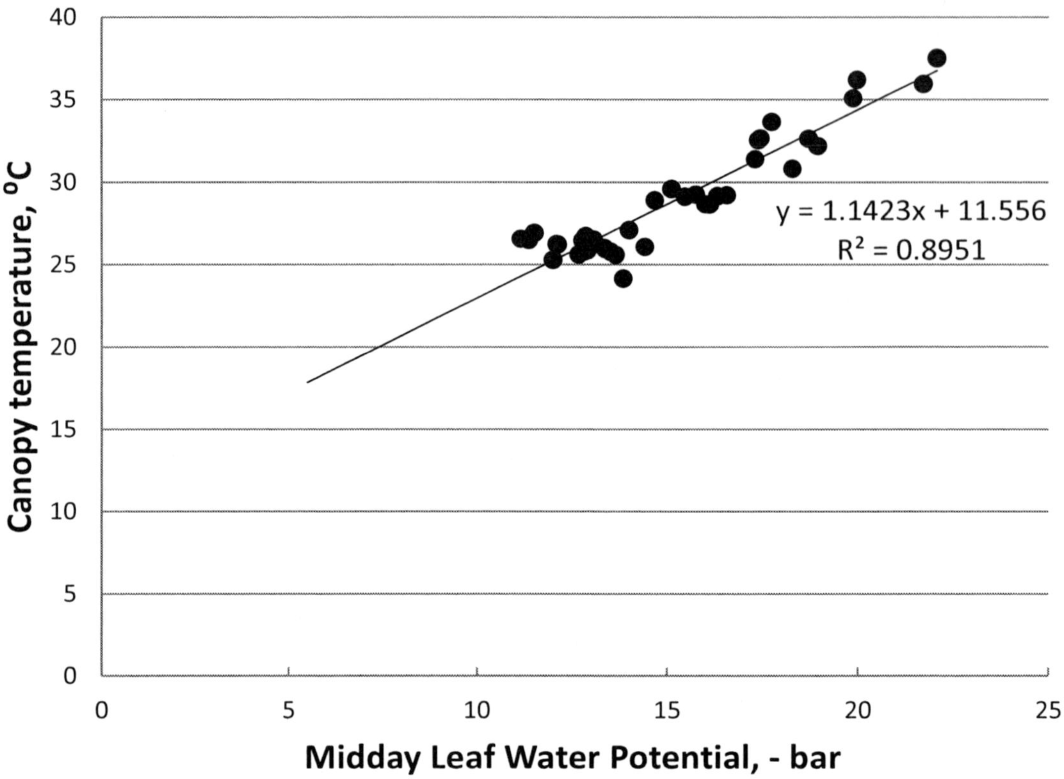

Figure 5. Canopy temperature at 1400 h (Tc) vs. midday leaf water potential (Ψ_L) taken on leaves collected in a two hour window straddling 1400 h on three dates in 2013 (growth stage in parentheses): 23 July (vegetative), 15 August (anthesis), and 3 September (grain-filling). Adapted from DeJonge et al. (2015).

Research conducted at LIRF has resulted in several simpler alternative indices such as the canopy temperature ratio Tcratio (Bausch et al., 2010), degrees above non-stressed canopy temperature DANS (Taghvaeian et al., 2014), and degrees above canopy temperature threshold DACT (DeJonge et al., 2015). These simpler indices may have promise for use by producers given the advantages of less data needs (Table 1), and can be converted to a stress coefficient to reduce crop transpiration under water stress (Kullberg et al., 2016). New methods are being created to process nadir thermal images in new quantifiers of water stress (Han et al., 2016), with more discussion to follow. Additional work is being conducted to create smartphone-based tools for farmers to quantify water stress and ET reduction.

Table 1. Comparison of basic data required for each K_s method tested and associated ET_c estimation RMSE. From Kullberg et al. (2016).

Category	**Requirement**	***K_s* method**			
		CWSI	**DANS**	**DACT**	**T_{cratio}**
Canopy Temperature	Target	X	X	X	X
	Non-Stressed Reference		X		X
Environmental	Relative Humidity	X			
	Air Temperature	X			
	Clear Sky	X	X	X	X
Pre-Calculation	Baselines (locally calibrated)	X			
	Threshold Temperature			X	
	Scaling Coefficient (locally calibrated)		X	X	
	Daily ET_c RMSE (mm/day)	0.77	0.80	0.80	0.83
	Daily ET_c RMSE (%)	14.6	15.2	15.2	15.6

Ground-Based Remote Sensing

Remote sensing data has been collected for crop irrigation management. Ground-based digital RGB, multispectral and thermal imagery were taken with a highboy tractor with a customized boom (Figure 6) using a Canon RGB camera, Tetracam Mini MCA multispectral camera (Tetracam, Inc.) and a FLIR IR camera A645s (FLIR Systems, Inc., Portland, USA). Canopy cover was estimated by separating transpiring (green) from nontranspiring (soil, non-green canopy) using RGB or multispectral images, and near real-time crop coefficient was determined based on crop canopy cover measurements (Trout and Johnson, 2007). We are developing an unmanned aerial system equipped with multispectral and thermal cameras to monitor vegetation, quantify crop water status, and estimate crop coefficient and crop ET at desirable spatial and temporal resolutions.

Figure 6. High clearance tractor ground-based remote sensing data collection platform. The tractor drives in "border rows" between plots where no samples are taken, and captures nadir images of adjacent plots without direct contact or disturbance.

Using high resolution thermal imagery (Figure 7), we also developed a methodology to obtain canopy temperature distribution, and proved that canopy temperature standard deviation (CTSD) could be used for maize water stress detection and be a potential tool for irrigation scheduling (Han et al., 2016).

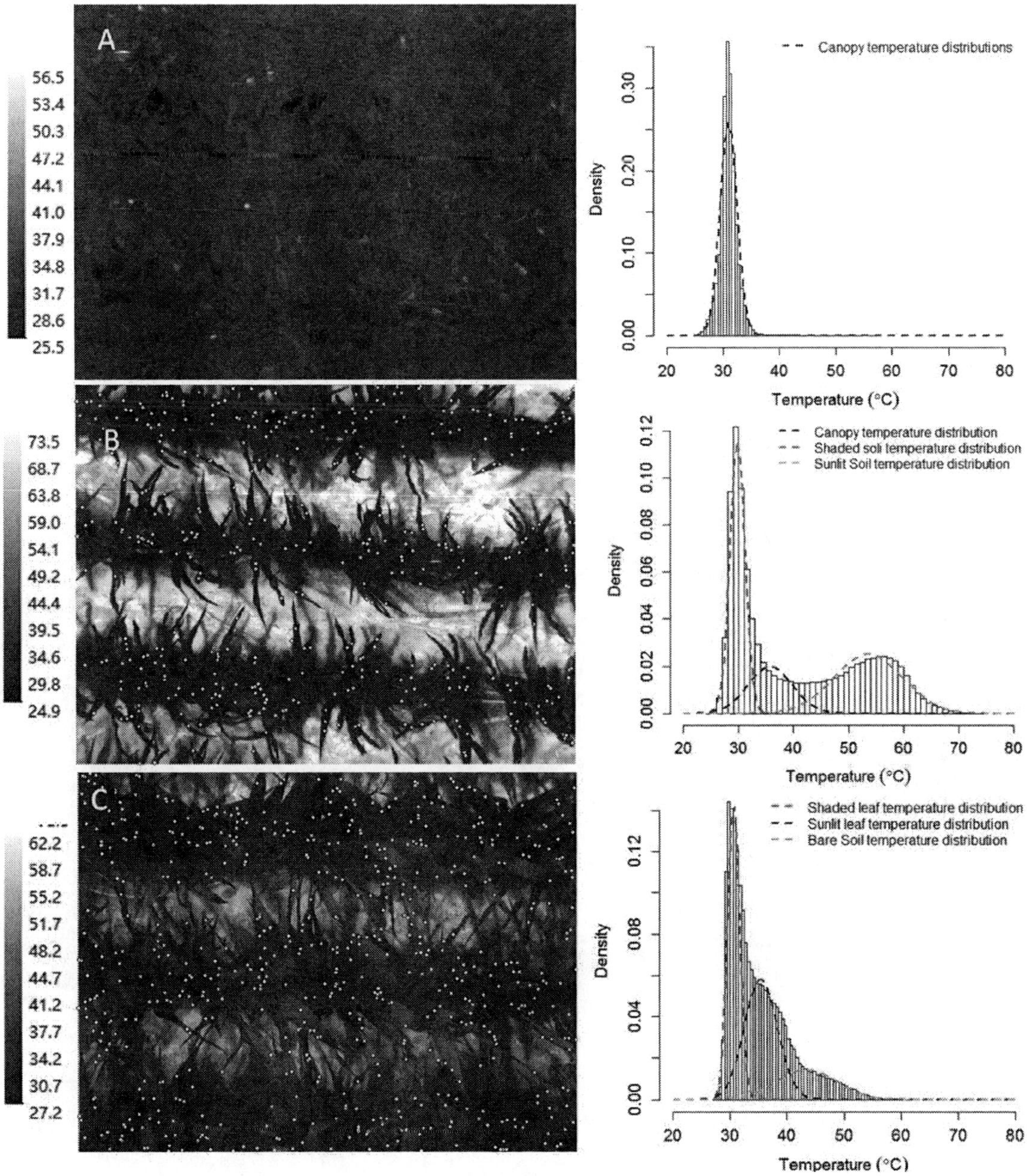

Figure 7. Examples of thermal images of maize taken in 2013 and their temperature distributions (A: DOY 241, Treatment 1: 100% ET; B: DOY 183, Treatment 12: 40% ET; C: DOY 204, Treatment 10: 65% ET). Yellow points were manually selected for calibration of the methods used. From Han et al. (2016).

Physiological Measurements

Physiological measurements are taken to understand plant responses underlying effects of deficit irrigation on yield, document plant stress and water use, and ground-truth the remote sensing of stress. Leaf water potential, Ψ_L, taken with a Scholander pressure chamber, is the standard measure of plant stress (Figure 9). Measurements are typically taken at midday within a two hour window past solar noon to document the maximum daily stress achieved. Measurements of canopy development, root system development (Figure 10 and Figure 11), leaf level transpiration, stomatal conductance, carbon fixation and fluorescence (indicating electron transport through photosystem II) under varying levels of water availability are made to quantify plant development above and below ground, water use and the acclimation of photosynthesis to water limitations. Measurements have documented roots deeper in the soil profile in crops well adapted for drought such as sunflower compared to maize (Figure 11). Measurements have also shown that root growth stimulation under deficit irrigation mainly occurred at middle depths in the rooting profile (30-70 cm) under frequent (4-5 d) surface drip irrigation (Figure 11). Measurements of sap flow with heat-balance type sap flow gages are made to quantify whole plant water use (Figure 12). Crop transpiration determined from water balance is closely aligned with measurements from sap flow (data not shown).

Figure 9. Midday leaf water potential (Ψ_L) measured with a pressure chamber. A distal portion of a leaf is cut, put in a zippered plastic bag, and transported in an insulated cooler to the pressure chamber, where it is trimmed with a razor blade, covered with a damp cloth, and place in the chamber for measurement.

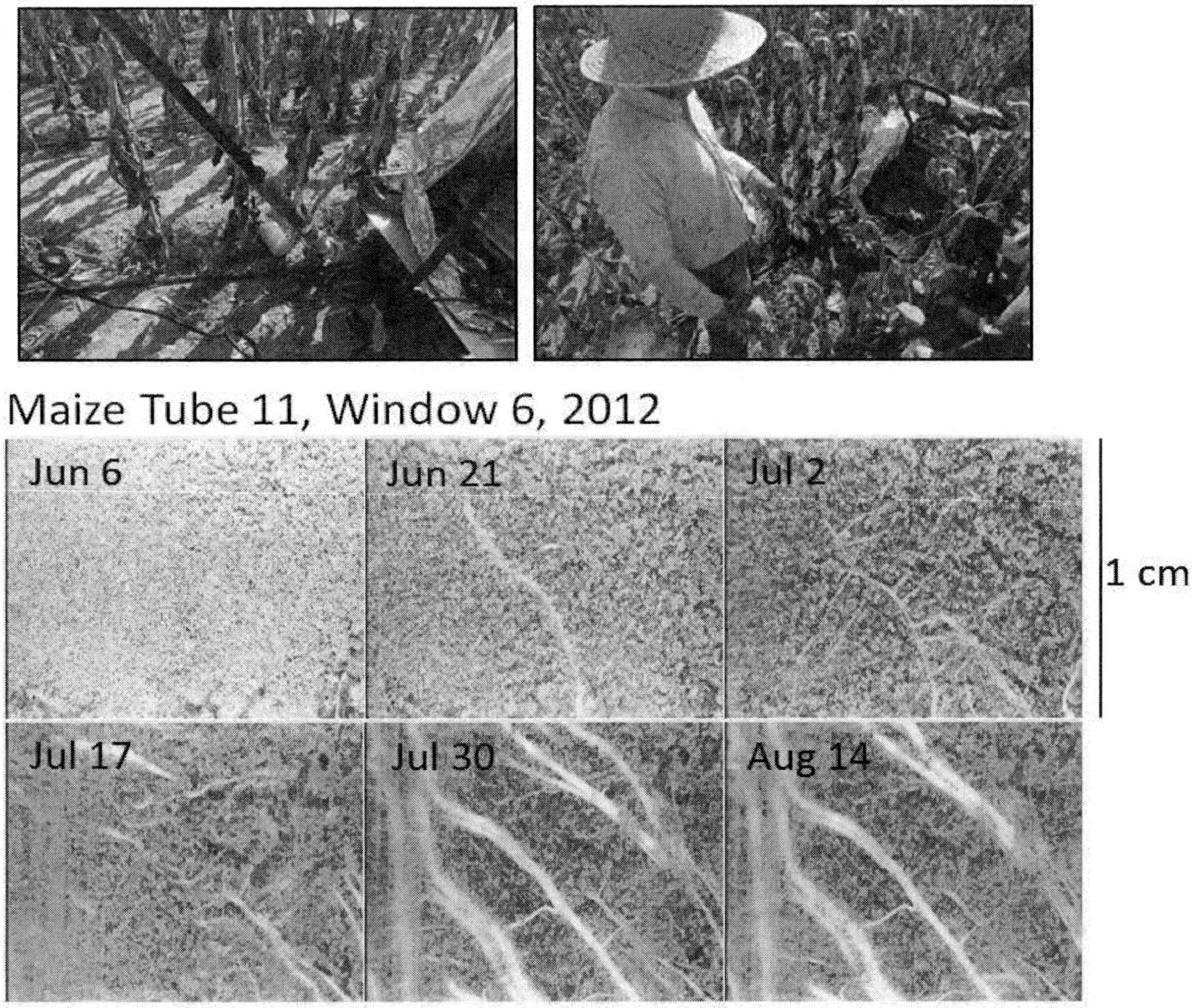

Figure 10. Minirhizotron camera system (Bartz Technology Corporation, Carpinteria, CA, USA) and root images recorded over the season. Upper left picture shows camera inserted into a clear acrylic tube installed in the field in a stand of sunflower. Upper right picture shows the computer system used to record images from a mobile cart. The lower pictures show a series of images with flushes of root growth recorded over the season from one position on the root tube.

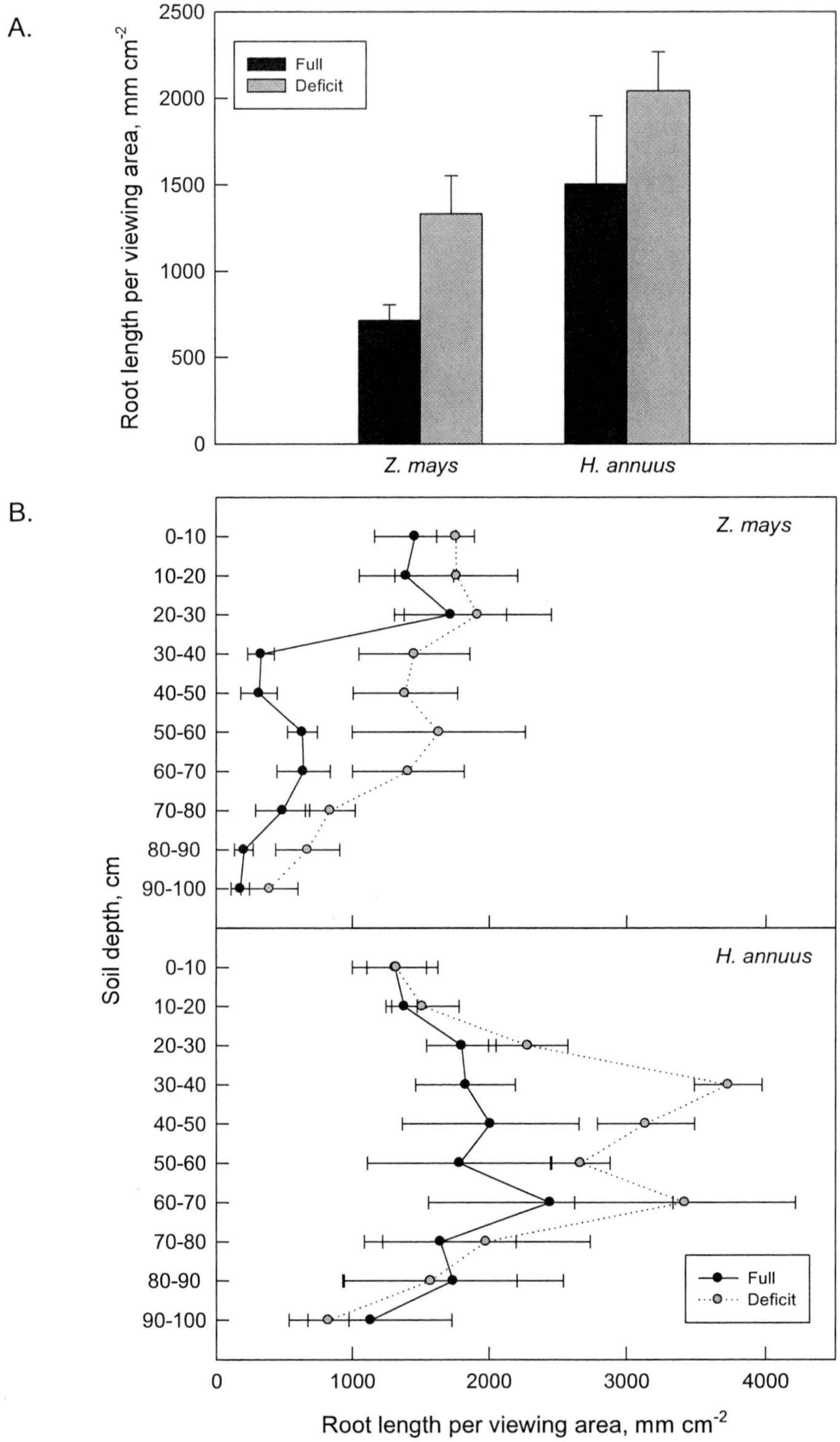

Figure 11. Annual root production (A) and distribution (B) in maize (Zea mays) and sunflower (Helianthus annuus) in 2012 from Comas et al. (2013). Root growth is expressed in terms of root length per viewing area of the minirhizotron window. Each bar and point represents root growth averaged among four minirhizotron tubes per treatment, with each tube installed in a different treatment plot. Error bars represent standard error of the mean.

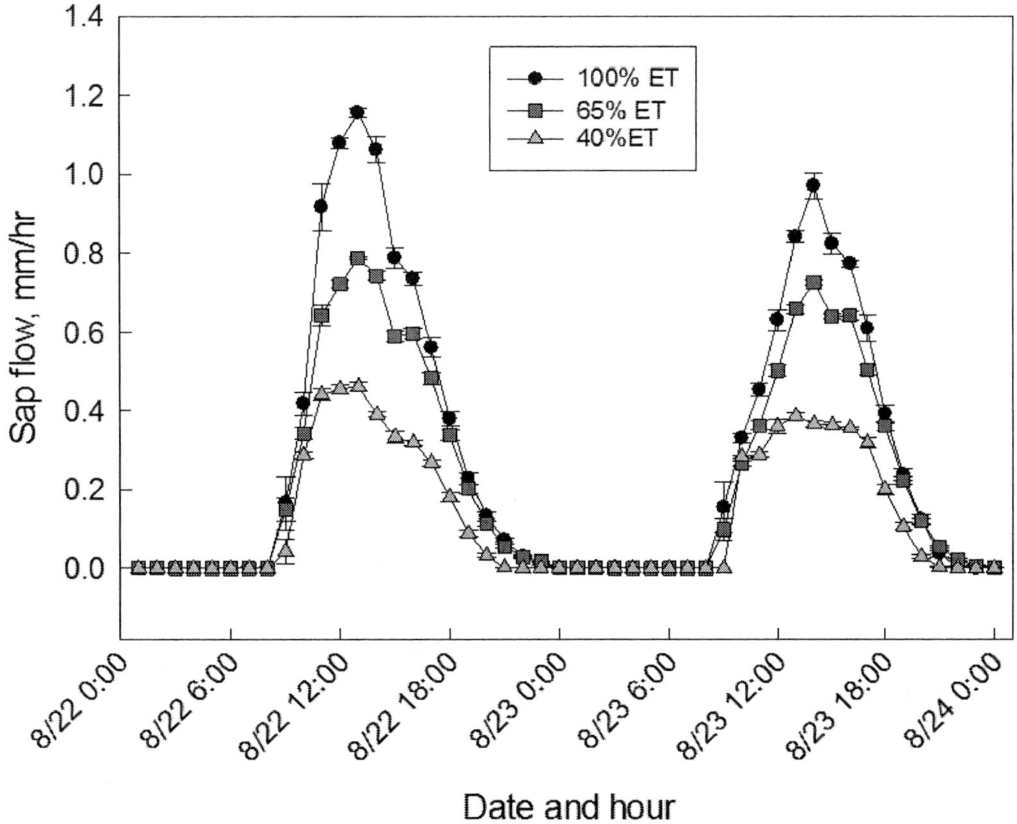

Figure 12. Hourly whole plant transpiration over two days in 2015. Each point is the hourly average of sap flow gages placed in four plots per treatment. Treatments are designated by the target ET. Error bars represent standard error of the mean.

Plant Traits

Drought is defined as a long-term absence of water, whereas deficit irrigation is a strategic management of water supplies based on critical growth stages. Improved production under drought periods of different lengths and timing within the growth season, as well as under regulated deficit irrigation will require different trait characteristics. Plants with higher water use efficiency (WUE), i.e. carbon income per unit water used, are best suited to deficit irrigation environments, whereas plants exhibiting resistance to short but extreme periods of water stress are best suited to dry environments. Greater water use efficiency can be achieved via either higher rates of growth or lower rates of water loss. In contrast with traits conferring higher WUE, the traits associated with improved drought tolerance are still poorly understood, but are likely to include: 1) robust water transport tissues that are less susceptible to dysfunction, 2) deeper or more efficient root systems, and 3) photosynthetic activities and apparatus that are less susceptible to damage during stress. At LIRF we evaluate different genetic varieties of maize to achieve a better understanding of which traits lead to better performance under drought and deficit irrigation. We take both a practical approach comparing commercially available genotypes that are available to farmers, as well as a biologically-informed approach, using an open-source "mapped" population that facilitates understanding performance~trait and trait~genetic relationships.

Published and preliminary data show that photochemical, stomatal, and water transport functioning are closely aligned during water stress in maize (Figure 13), with significant intrinsic variation existing in each of these traits across maize genotypes (Figure 14). This evidence strongly supports the idea that improved performance of maize in under both deficit irrigation as well as rain-fed environments will require improved water transport networks, which currently are not considered a priority for maize improvement (Gleason et al., 2012). Future work at LIRF will dig deeper into this issue.

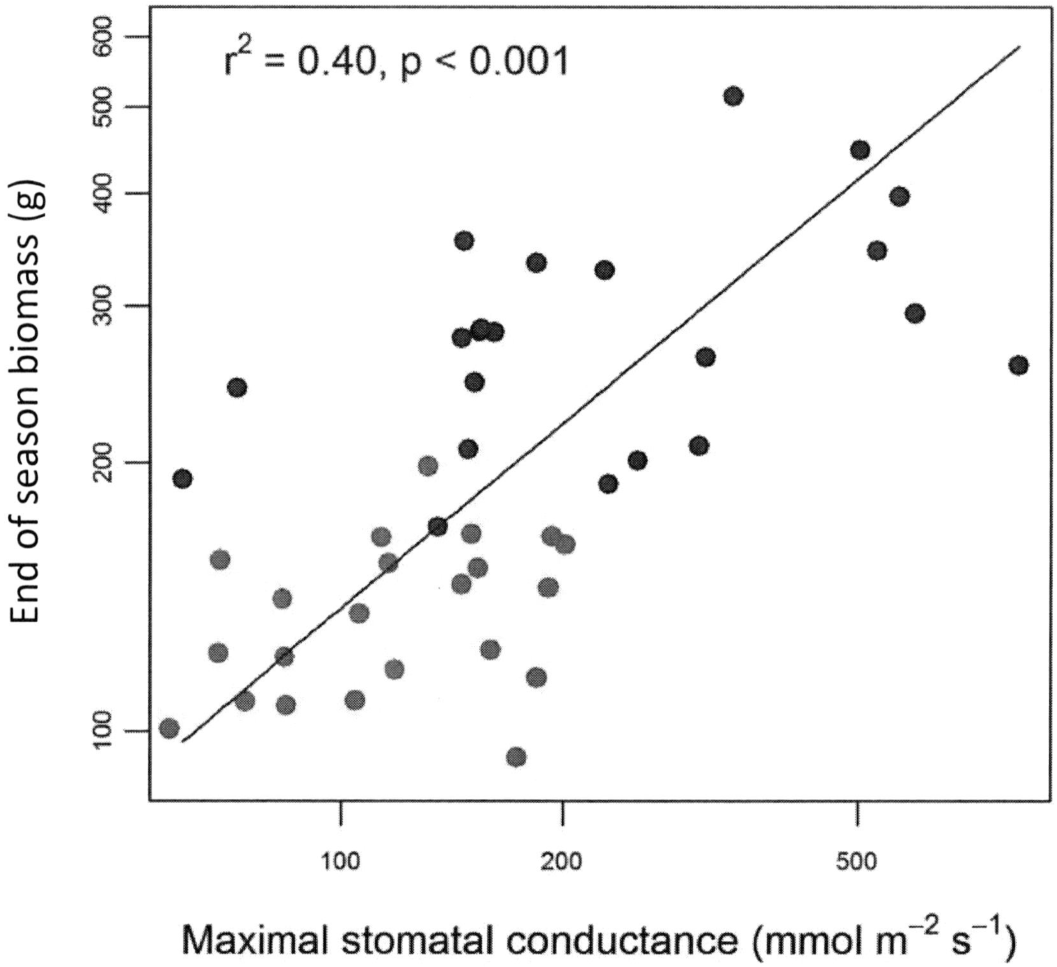

Figure 13. Relationship between maximal stomatal conductance (water loss at a given atmospheric demand) and end of season biomass. Each symbol represents a different maize genotype. Closed red and blue circles denote deficit-irrigated and fully-irrigated treatments, respectively. Note that maximal stomatal conductance explains ~40% of the total variation in end of season biomass across genotypes and treatments. This suggests that research focusing on the traits necessary to maintain high rates of stomatal conductance under water stress will also result in higher yield. It also suggests that much variation in these traits exists among maize varieties, and therefore, better performance can likely be achieved via trait-informed plant breeding and gene editing efforts.

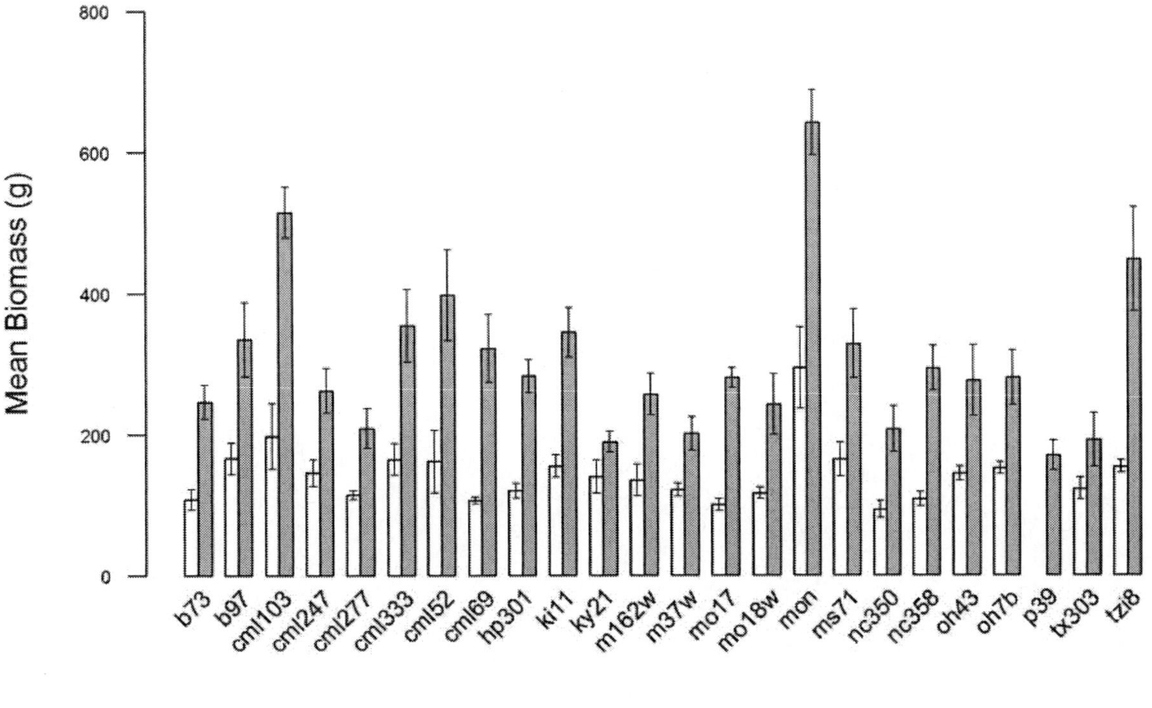

Figure 14. End-of-year biomass increment in fully-irrigated (gray bars) and deficit-irrigated (white bars) treatments across 24 inbred maize genotypes grown in Greeley, Colorado, USA. Note that the response to deficit irrigation (percentage reduction in biomass when grown under deficit vs full irrigation) varies markedly across genotypes. This suggests that response to deficit irrigation depends critically on the variety of maize that is grown.

Future Goals/Aims

Future work at LIRF will continue to explore new questions and objectives, which include:

1. Improve water use efficiency (WUE) by identifying plant traits, mechanisms, and agronomic practices that increase productivity per unit of water used by the crop.

2. Develop simple and accurate methods to quantify evapotranspiration (ET) in agricultural systems under limited water availability to improve the efficiency of irrigation scheduling.

3. Create Water Production Functions (WPF, yield per ET) for alternative crops under limited water availability.

Conclusions

The Limited Irrigation Research Farm (LIRF) near Greeley, CO is an agricultural experiment station that is exploring the management of limited or deficit irrigation, and gaining understanding on how this management affects crops at multiple levels, from satellite and ground-based remote sensing scales, to plant-based scales such as sap flow and root growth. Focus is also placed on quantifying water balance and evapotranspiration under water stress, and identifying plant traits that may be resistant to water limitations. The research Unit and farm has an annual Field Day, and welcomes tours throughout the field season – please contact us if you are interested in a field tour.

Acknowledgements

The authors would like to acknowledge past researchers at the unit and collaborators, and especially the technical and administrative support staff and students who have helped this project continue to yield results. The hard work from these people is imperative to this research, and we are grateful.

References

Allen, R.G., Pereira, L.S., Raes, D., Smith, M., 1998. Crop evapotranspiration: Guidelines for computing crop water requirements, FAO irrigation and drainage paper 56. Food and Agriculture Organization of the United Nations, Rome.

Bausch, W., Trout, T., Buchleiter, G., 2010. Evapotranspiration adjustments for deficit-irrigated corn using canopy temperature: A concept. Irrigation and Drainage 60, 682-693.

Comas, L.H., Becker, S.R., Cruz, V.M.V., Byrne, P.F., Dierig, D.A., 2013. Root traits contributing to plant productivity under drought. Front Plant Sci 4, 442.

Comas, L.H., Wiggans, D.R., Young, J.S., Gleason, S.M., Shaner, D.L., 2014. Adapting irrigation management to water scarcity: Constraints of plant growth, hydraulics and carbon assimilation, ASA, CSSA and SSSA International Annual Meeting, Long Beach, CA, USA.

DeJonge, K.C., Taghvaeian, S., Trout, T.J., Comas, L.H., 2015. Comparison of canopy temperature-based water stress indices for maize. Agricultural Water Management 156, 51-62.

Gleason, S., Butler, D.W., Ziemińska, K., Waryszak, P., Westoby, M., 2012. Stem xylem conductivity is key to plant water balance across australian angiosperm species. Functional Ecology 26, 343-352.

Han, M., Zhang, H., DeJonge, K.C., Comas, L.H., Trout, T.J., 2016. Estimating maize water stress by standard deviation of canopy temperature in thermal imagery. Agricultural Water Management 177, 400-409.

Idso, S.B., Jackson, R.D., Pinter Jr, P.J., Reginato, R.J., Hatfield, J.L., 1981. Normalizing the stress-degree-day parameter for environmental variability. Agricultural Meteorology 24, 45-55.

Jackson, R.D., Idso, S.B., Reginato, R.J., Pinter, P.J., 1981. Canopy temperature as a crop water stress indicator. Water Resources Research 17, 1133-1138.

Kullberg, E.G., DeJonge, K.C., Chávez, J.L., 2016. Evaluation of thermal remote sensing indices to estimate crop evapotranspiration coefficients. Agricultural Water Management In Press.

Taghvaeian, S., Comas, L.H., DeJonge, K.C., Trout, T.J., 2014. Conventional and simplified canopy temperature indices predict water stress in sunflower. Agricultural Water Management 144, 69-80.

Trout, T., 2016. USDA-ARS Colorado maize water productivity dataset 2008-2011, Ag Data Commons.

Trout, T.J., DeJonge, K.C., 2013. Estimating production and crop water use of deficit irrigated crops, in: Wahlin, B.T., Anderson, S.S. (Eds.), Using 21st Century Technology to Better Manage Irrigation Water Supplies, USCID Seventh International Conference on Irrigation and Drainage. USCID, Phoenix, AZ, pp. 259-267.

Trout, T.J., Johnson, L.C., 2007. Estimating crop water use from remotely sensed ndvi, crop models, and reference et, USCID Fourth International Conference on Irrigation and Drainage, Denver, CO, pp. 275-285.

A Smartphone Application for Scheduling Irrigation in Cotton

Wesley M. Porter, Assistant Professor, Ext. Precision Ag and Irrigation Specialist
Crop & Soil Sciences Department, University of Georgia, 2360 Rainwater Rd, Tifton, GA 31793-5766
wporter@uga.edu

Vasileios Liakos, Post-Doctoral Research Scientist
Crop & Soil Sciences Department, University of Georgia, 2360 Rainwater Rd, Tifton, GA 31793-5766

Calvin D. Perry, Director, Stripling Irrigation Research Park
University of Georgia, 8207 GA-37, Camilla, GA 31730

Jose H. Andreis, Applications Programmer
Agricultural and Biological Engineering Department, University of Florida, PO Box 110570, 1741 Museum Road, Gainesville, FL 32611-0570

Edward M. Barnes, Senior Director, Agricultural & Environmental Research
Cotton Incorporated, 6399 Weston Parkway, Cary, NC 27513

Kelly T. Morgan, Associate Professor
IFAS Southwest Research and Education Center, University of Florida, 2685 SR 29 North Immokalee, FL 34142

Clyde Fraisse, Professor
Agricultural and Biological Engineering Department, University of Florida, PO Box 110570, 1741 Museum Road, Gainesville, FL 32611-0570

Katie W. Migliaccio, Professor
Agricultural and Biological Engineering Department, University of Florida, PO Box 110570, 1741 Museum Road, Gainesville, FL 32611-0570

George Vellidis, Professor
Crop & Soil Sciences Department, University of Georgia, 2360 Rainwater Rd, Tifton, GA 31793-5766
yiorgos@uga.edu

Abstract. *The goal of this work was to develop an easy-to-use and engaging irrigation scheduling tool for cotton which operates on a smartphone platform. The Cotton SmartIrrigation App (Cotton App) uses an interactive ET-based soil water balance model. The Cotton App uses meteorological data from weather station networks, soil parameters, crop phenology, crop coefficients, and irrigation applications to estimate root zone soil water deficits (RZSWD) in terms of percent as well as of inches of water. The Cotton App sends notifications to the user when the RZSWD exceeds 40%, when phenological changes occur, and when rain is recorded at the nearest weather station. It operates on both iOS and Android operating systems and was released during March 2014. The Cotton App was evaluated in field trials for three years and performed well when compared to other irrigation scheduling tools. Its geographical footprint is currently limited to the states of Georgia and Florida, United States, because it uses meteorological data only from weather station networks in these states. A new version which will be released in 2017 uses national gridded meteorological data sets and will allow the Cotton App to be used in most cotton growing areas of the United States.*

Keywords: Smartphone, Irrigation, Cotton, Crop coefficient, Evapotranspiration, Phenology

Introduction

Cotton (*Gossypium hirsutum* L.) is the most important fiber crop in the world and one of the most important agronomic crops in the United States where in 2014 it had a production value in excess of USD 5 billion. In the United States, the cotton crop under irrigation has increased steadily over the past two decades because irrigation serves both to reduce risk of crop loss but also to build resiliency and yield stability. Approximately 40% of U.S. cotton is currently irrigated but irrigation water is becoming limited in many cotton growing areas such as the Texas high plains, Arizona, and California and competition for water is increasing rapidly in areas normally associated with plentiful water resources. As a result, the organizations representing growers are investing in the development of irrigation scheduling tools which improve irrigation water use efficiency. In response, a significant amount of research has been conducted on this topic.

Cotton's water needs are a function of phenological stage (Fig. 1). Evapotranspiration (ET) is also an important factor in estimating cotton's daily water use and several cotton irrigation scheduling tools have been developed which use estimated crop ET (ETc) to develop irrigation recommendations. These models typically use a crop coefficient (Kc) which represents the crops phonological stage to calculate ETc from a reference ET (ETo) as shown in equation 1 (Jensen , 1968; Doorenbos and Pruitt 1975, 1977; Burman et al. 1980a, b; Allen et al. 1998).

$$ET_c = ETo \times Kc \qquad (1)$$

Models which use only ETc to estimate irrigation requirements are simple and easy-to-use but they do not consider moisture available in the soil profile which sometimes leads to over-application of irrigation water. Incorporating soil water balance increases accuracy but also increases the number of parameters needed as well as the complexity of the model.

Recent technological advances that allow for widespread internet access through handheld devices such as tablets and smartphones provide a novel platform on which to deliver sophisticated yet easy-to-use ET-based irrigation scheduling tools. Smartphone tools, typically referred to as smartphone applications or apps, are being developed at exponential rates for every imaginable use. The functionality of an app differs from a web tool in that apps are with the user at all times since they reside on the smartphone, are readily accessible, and engage the user through notifications (Migliaccio et al., 2015; 2016). Some apps use notifications, similar to text messages, to prompt users to respond to critical events and eliminate the need to interact with the tool on a daily basis. Migliaccio et al. (2016) presented a suite of SmartIrrigation

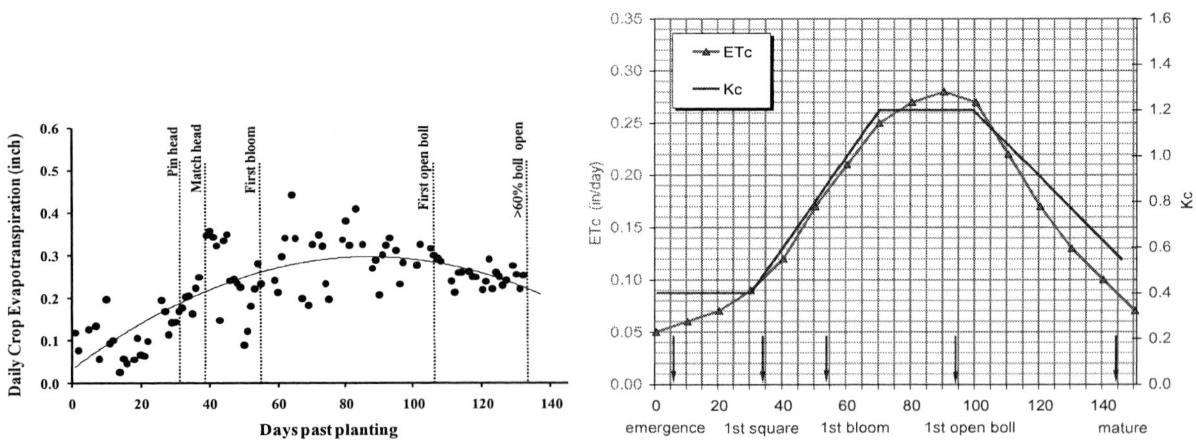

Figure 1. Measured crop water use (ETc) from a cotton field in Louisiana over the growing season (left) and water use and crop coefficient curve for cotton in Stoneville, Mississippi (right) (Perry and Barnes, 2012).

apps which were recently released to provide real-time irrigation schedules for avocado, citrus, cotton, soybean, strawberry, blueberries, turf, and vegetables. Information about and links to download these apps can be found at www.smartirrigationapps.org. This paper describes the Cotton SmartIrrigation App (hereafter referred to as the Cotton App) which was released in 2014. Our objectives were to develop a novel ET-based irrigation scheduling tool for cotton that requires minimal user interaction, is delivered to the user on a smartphone platform, and outperforms many other irrigation scheduling tools.

Materials and Methods

The model which drives the Cotton App is an interactive ET-based soil water balance model. It uses meteorological data, soil parameters, crop phenology, crop coefficients, and irrigation applications to estimate root zone soil water deficits (RZSWD) in terms of percent and inches of water and provides these two pieces of information to the user. The model does not deliver direct irrigation application recommendations. However, the user may utilize the RZSWD information to make appropriate irrigation decisions.

ET and Kc

The model uses meteorological data to calculate ETo using the Penman–Monteith equation (Allen et al. 1998). This method, also known as FAO 56, is widely accepted for irrigation scheduling. The model then uses Kc to estimate ETc as shown in equation 1. For annual crops, Kc changes with phenological stage. Kc typically begins with small values after emergence and increases to 1.0 or above when the crop has the greatest water demand. Kc decreases as crops reach maturity and begin to senesce. We used information from published studies (Perry and Barnes, 2012) to develop a prototype Kc curve for southern Georgia and northern Florida conditions. The curve was calibrated and validated with a series of plot and field studies in 2012 and 2013. Details of the calibration and validation effort are provided by Vellidis et al. (2016b). In the model, changes in phenology and associated changes in Kc are driven by accumulated heat units commonly referred to as growing degree days (GDDs). GDDs are calculated using equation 2.

$$GDD = \frac{T_{max}+T_{min}}{2} - T_{base} \tag{2}$$

For cotton, T_{base} is 60°F. Any temperature below T_{base} is set to T_{base} before calculating the average. Figure 2 presents the relationship between GDDs and Kc, and the corresponding phenological stages as used in the model. GDDs required for phenological stages are derived from Ritchie et al. (2004).

Soil Water Balance Model

ETc is used by the model to estimate daily crop water use. ETc, measured precipitation, and irrigation are then used to estimate the plant available soil water. Plant available soil water is a function of the soil's plant available water holding capacity and current rooting depth. The model allows users to select from one of seven generic soils shown in Table 1. As the plant rooting system grows, the depth of the profile from which the plant can extract water also increases. In the model, the initial rooting zone depth is 0.15 m (6 in) and increases by 7.5 mm day^{-1} (0.3 in day^{-1}) until it reaches a maximum depth of 0.75 m (30 in). At emergence, the soil profile from 0 to 0.75 m is assumed to be at 85% of maximum plant available soil water holding capacity.

Today's plant available soil water is calculated by subtracting yesterday's ETc from yesterday's plant available soil water and adding any precipitation or irrigation measured. The model allows for three types of irrigation – high pressure overhead sprinkler, low-pressure overhead sprinkler, and subsurface drip. It

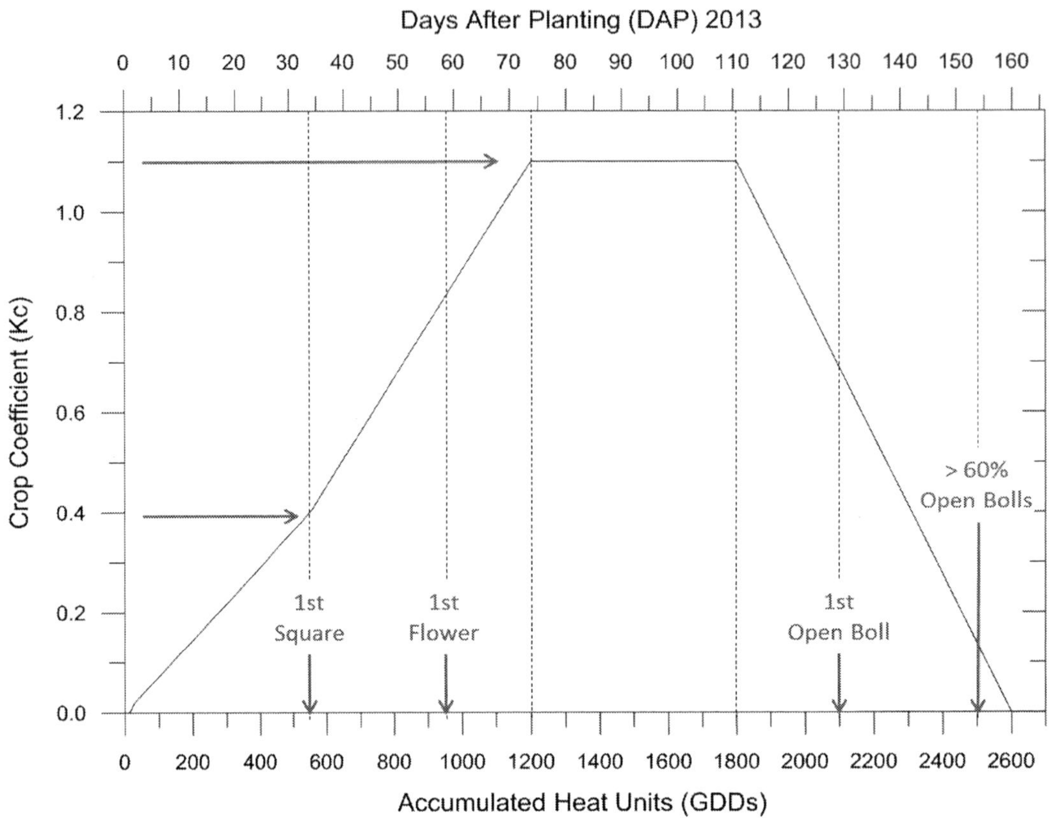

Figure 2. Kc curve used in the model. Maximum Kc is 1.1 which is maintained between 1200 and 1800 GDDs. An inflection point and Kc rate change occurs at 550 GDDs. The top axis indicates how DAP coincided with GDDs in 2013.

uses an efficiency factor of 75% for high pressure sprinkler and 85% for low pressure sprinkler to account for evaporation and drift before the water droplets reach the soil and a 90% efficiency factor for subsurface drip irrigation. The model also assumes that 90% of measured precipitation reaches the soil to account for canopy interception and other losses. A maximum of 25 mm (1 in) and a minimum of 5 mm (0.2 in) in daily precipitation is used in soil water balance calculations. The maximum is used because even if the RZSWD is greater than 25 mm, it is unlikely that more than that amount will infiltrate into the soil profile during a 24 hr period. The minimum is used because less than 5 mm of precipitation in a 24 hr period does not have an appreciable effect on soil moisture. All these parameters are used to calculate root zone soil water deficit (RZSWD) in inches and % RZSWD.

Table 1. Plant available water capacity (AWC), field capacity (FC), and wilting point (WP) of the seven generic soil types used in the Cotton App.

Soil type	AWC ($cm^3\ cm^{-3}$)	FC ($cm^3\ cm^{-3}$)	WP ($cm^3\ cm^{-3}$)
Sand	0.05	0.10	0.05
Loamy sand	0.06	0.12	0.06
Sandy loam	0.10	0.18	0.08
Loam	0.14	0.28	0.14
Silt loam	0.20	0.31	0.11
Clay loam	0.14	0.36	0.22
Clay	0.12	0.42	0.30

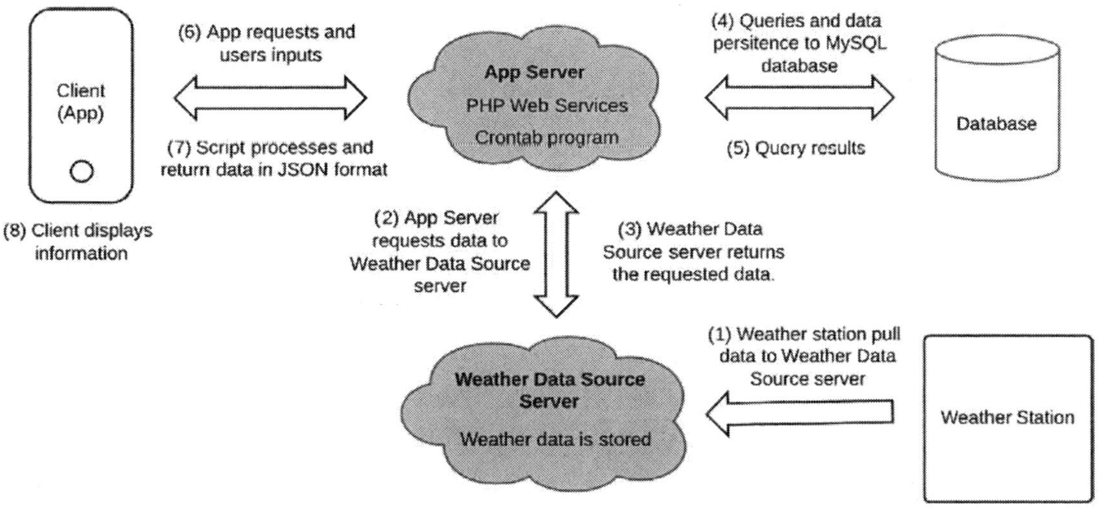

Figure 3. Diagram of interaction among client, server and weather stations (Migliaccio et al., 2015).

Model Calibration and Validation

During 2012 and 2013 we used large plots at the University of Georgia's Stripling Irrigation Research Park (SIRP) located near Camilla, GA to calibrate the model and in 2013 we used five producer fields located in southwestern Georgia to validate the model. In 2013, we used the model adjustments made following the 2012 growing season to schedule irrigation in the plots.

Meteorological Data

Meteorological data, and especially accurate precipitation data, are critical to the Cotton App. In its current version, the Cotton App pulls meteorological data from the Georgia Automated Environmental Monitoring Network (GAEMN) (http://weather.uga.edu) and the Florida Automated Weather Network (FAWN) (http://fawn.ifas.ufl.edu) thus currently limiting the Cotton App's footprint to these two states.

Smartphone App Development

Figure 3 presents the flow of information between the Cotton App, server, and automated weather station networks. Our design principles for the Cotton App were that it should provide the most accurate, site-specific, real-time information we could offer the user. In addition, the Cotton App would require minimum user input which, when necessary, it would solicit from the user by sending notifications. It would not be necessary for the user to check the Cotton App regularly. Finally the Cotton App would provide ready-to-use output and be engaging.

User Interaction

After initial setup, the user is directed to the field setup screen. A user may register multiple fields but only one at a time. Field registration begins with the field location. By default, the Cotton App pins the field on a map at the smartphone's location but the user may reposition the pin by dragging it to the desired location (Fig. 4). Accurately locating the field's position is important because it is used to locate the weather stations nearest to the field. The user then enters a unique field name and planting date. The Cotton App automatically selects the closest weather station but also displays the next four closest weather stations and the user has the option to select any of those. Finally, the user selects soil type from the options presented in Table 1, irrigation system type, and the default irrigation rate. The default irrigation rate is the amount of irrigation the user typically applies during an irrigation event.

The main user interface screen (Fig. 5) is field-specific but the user can move between fields by swiping the screen from left to right or right to left. The circles at the top of the screen indicate the number of

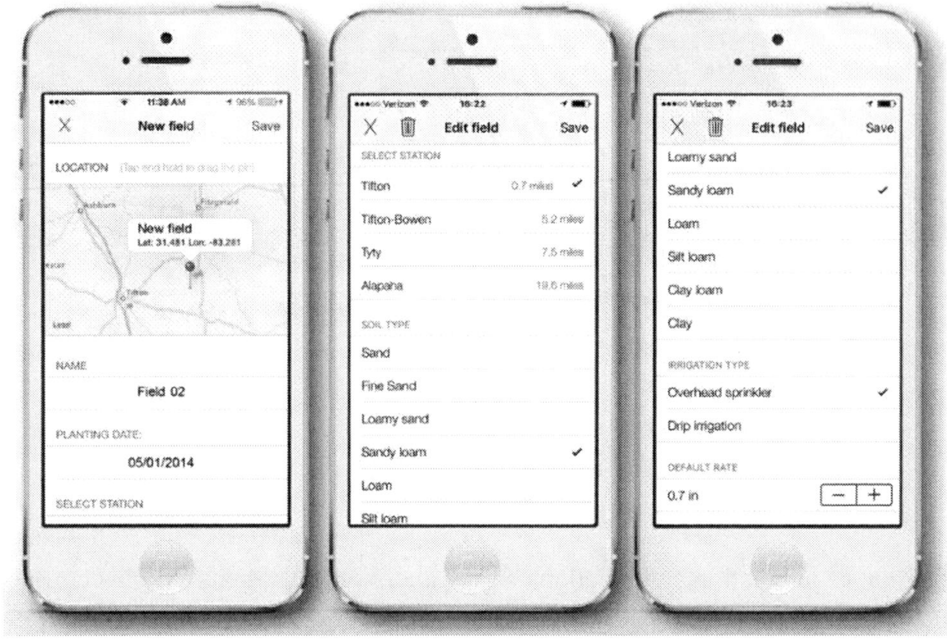

Figure 4. Screenshots of an iPhone running the Cotton App with the new field setup screens. The Cotton App pins the field on a map at the smartphone's location but the user may reposition the field by dragging the pin (left). The Cotton App automatically selects the closest weather station (center) but also displays the next four closest weather stations and the user has the option to select any of those. The user then selects soil type (center), and irrigation system type and default irrigation rate (right).

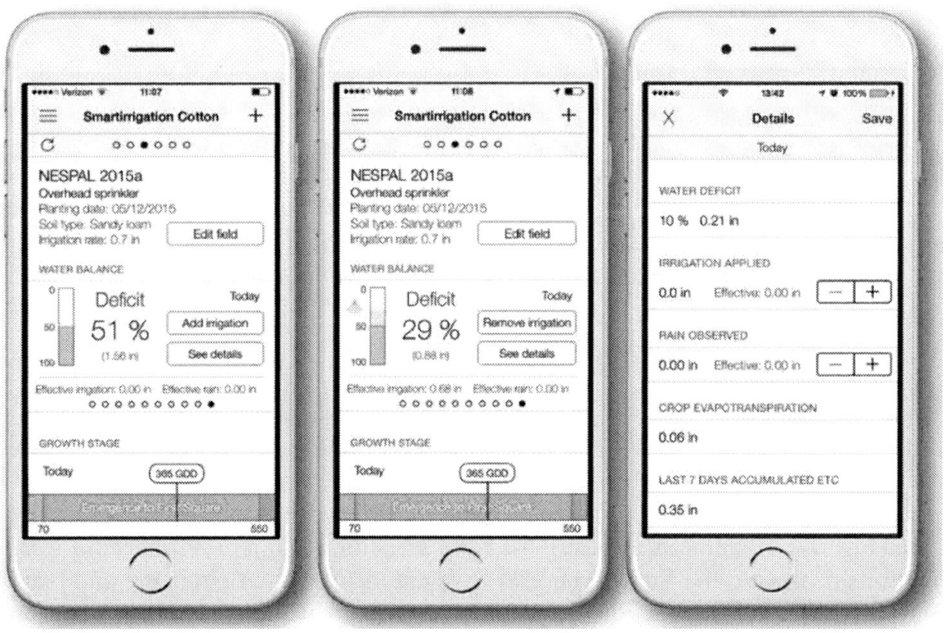

Figure 5. Screenshots showing the main user-interface screen of the Cotton App (left and center). On each of these screenshots, the user can view information about the RZSWD, whether precipitation was recorded or irrigation was applied within the past day, as well as the phenological stage of the crop. Any of this information can be edited by tapping on the "See details" button. If irrigation events were not recorded properly, they can be added or removed.

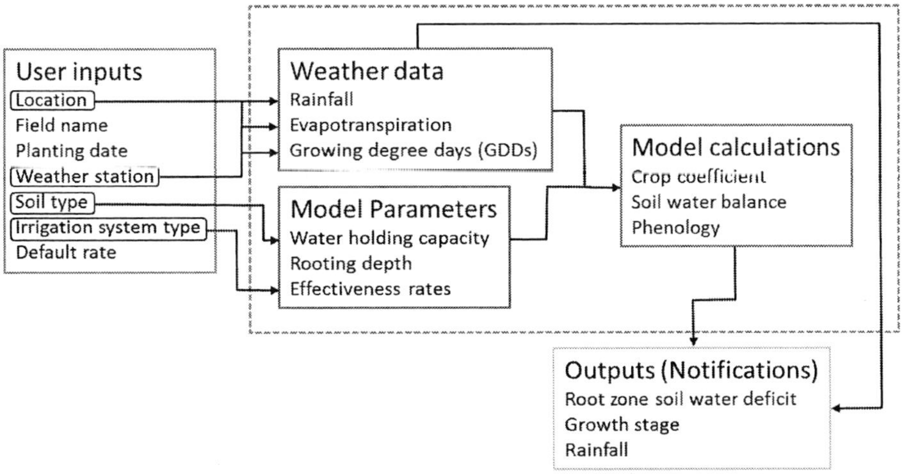

Figure 6. Flow of information in the Cotton App. Components internal to the model are enclosed by the dashed line.

fields registered by the user (in Fig. 5 there are six). The circles are added in the sequence in which fields are registered and the solid circle indicate the field currently being displayed. The *Edit Field* button allows the user to edit any of the information entered during field setup. Below that, the Cotton App displays the current RZSWD. The bar graph on the left is scaled from 0 to 100% RZSWD and moves downwards as soil water is depleted. To the right of the bar, the RZSWD is displayed numerically and below that, in parentheses, is the amount of irrigation water required to refill the profile to 100% capacity. When irrigation is applied, the user *must* record that irrigation by pressing the *Add irrigation* button. The Cotton App then credits the default irrigation amount (multiplied by the efficiency factor) to the soil water balance model. A sprinkler symbol indicates that an irrigation event has been added and the irrigation's effect on RZSWD is shown with a lighter shade of blue on the bar graph (Fig. 5).

Below the bar graph, the screen displays the amount of effective irrigation and effective rain added to the model on this day. If more or less than the default irrigation is added to the field or if the rain amount recorded at the nearest weather station is different from the rain received at the field, the user can adjust the amounts by touching the *See details* button (Fig. 5). Irrigation and rain amounts can be corrected retroactively for the past nine days. The Cotton App will perform best when precipitation data are accurate and the best way to provide these data is to use a local rain gage to adjust rain data recorded at the weather station.

The soil water balance model is run once a day early in the morning after the weather data for the past day are uploaded to the server. The display is updated the first time the user opens the Cotton App after the model run. The model also runs and the display updates if the user adds or removes an irrigation event, corrects rainfall amounts, or changes any of the field parameters (such as soil type) which may affect RZSWD. The Cotton App allows the user to view RZSWD, irrigation, and rain data, and growth stage data for the current day and the past nine days. Past data can be viewed by swiping along the series of ten circles located below the RZSWD display. The current day is represented by the circle at far right.

Estimated phenological development (growth stage) and accumulated GDDs are presented at the bottom of the screen. It is important that the user ground-truth the model's changes in phenological stage as they occur because as described earlier, this is the parameter that forces changes in Kc. If the crop is not progressing at the same rate as predicted by the Cotton App, then the Kc used may be too high or too low and the RZSWD will not reflect field conditions accurately. If the discrepancies are large, use of the Cotton App should be discontinued in this field. At this time, there is no provision for the user to adjust phenological stage. Figure 6 presents a schematic of how the Cotton App interacts with inputs and outputs.

Figure 7. Screenshots showing notifications for RZSWD (left), rain (right) and phenology change (right).

Notifications

Notifications are pushed to the user when a rain event is recorded at a weather station associated with a registered field, when phenological changes occur, and when RZSWD exceeds 40% (Fig. 7). A 50% RZSWD or depletion of 50% of plant available soil water is a commonly accepted irrigation threshold for many agronomic crops. The Cotton App begins to push daily notifications to the user when RZSWD exceeds 40% to allow the user time to trigger the required irrigation event.

Cotton App Performance

For three growing seasons, 2013 - 2015, the Cotton App was a treatment in a cotton irrigation scheduling study conducted at SIRP. Every year, the Cotton App was compared to other scheduling methods some of which changed from year to year. Throughout the three years, only two other treatments were used repeatedly – the University of Georgia Extension Checkbook Method hereafter referred to as the Checkbook Method which was used in 2013, 2014 and 2015 and Watermark® sensors with a 50 kPa irrigation threshold which was used in 2014 and 2015. Only the results from these three treatments will be discussed. Treatment yields were analyzed using an analysis of variance GLM procedure follow by means separation LSD test.

The Checkbook Method tabulates the amount of water a crop needs during each week of its life-cycle. Producers subtract the amount of precipitation received from the weekly requirements and add the remainder via irrigation. The Checkbook Method does not account for environmental conditions and so tends to over-irrigate when ET rates are low.

Results and Discussion

Table 2 summarizes the performance of the Cotton App compared to the Checkbook Method for 2013-2015 and compared to the Watermark® sensors with a 50 kPa irrigation threshold for 2014-2015. 2013 and 2015 were wetter than normal years while 2014 was a drier than normal year. The Cotton App outperformed the Checkbook Method in terms of mean yield regardless of tillage treatment and did this

Table 2. Performance of the Cotton App compared to other irrigation scheduling treatments conducted at the University of Georgia's Stripling Irrigation Research Park. Cotton yield is reported as lint (fiber) yield. Treatment yields were analyzed using an analysis of variance GLM procedure follow by means separation LSD test. Means with the same t Grouping letter are not significantly different (from Vellidis et al., 2016b)

Year		Conventional Tillage			Conservation Tillage		
Rain	Scheduling Method	Lint Yield	Irrigation	WUE[2]	Lint Yield	Irrigation	WUE
(mm)[1]		($kg\ ha^{-1}$)	(mm)	($kg\ ha^{-1}\ mm^{-1}$)	($kg\ ha^{-1}$)	(mm)	($kg\ ha^{-1}\ mm^{-1}$)
2013	Checkbook	1289 [b]	310	4.1	1513 [b]	323	4.6
(696)	Cotton App	1411 [a]	76	18.5	1664 [a]	76	21.8
2014	Checkbook	1915 [b]	388	4.9	1860 [b]	388	4.7
(285)	Cotton App	2067 [a]	231	8.9	2011 [a]	231	8.7
	Watermark 50 kPa Threshold	1974 [b]	315	6.2	1721 [c]	372	4.6
2015	Checkbook	1814 [a]	165	11	1748 [a]	165	10.6
(575)	Cotton App	1926 [a]	146	13.1	1841 [a]	127	14.5
	Watermark 50 kPa Threshold	1849 [a]	108	17.1	1953 [a]	108	18.0

[1] Precipitation in mm during the growing season.

2 WUE = water use efficiency

most effectively during the two wet years. However the differences were statistically significantly different only in 2013 and 2014 because of large intra-treatment variability in yield during 2015 (Vellidis et al., 2016a). The Cotton App also outperformed the Checkbook Method in water use efficiency. This is because the Checkbook Method does not take into account periods with low ET which occur frequently in wet years. The Cotton App outperformed the Watermark® sensors method in 2014 but in 2015, the Watermark® sensors conservation tillage treatment outperformed the Cotton App conservation tillage plots. The yield differences between these two irrigation treatments were statistically significant in 2014.

Expanding the Cotton App's Geographical Footprint

The Cotton App's geographical footprint is currently limited to Georgia and Florida for two reasons. The first is that the project team which developed the suite of SmartIrrigation Apps had already developed the protocols to use data from GAEMN and FAWN. Adding weather networks from other states which provide the meteorological data needed to calculate ETo using the Penman–Monteith equation requires additional resources but is relatively straightforward.

The second reason inhibiting use of the Cotton App in other states is that the Kc curve currently used in the model was calibrated to environmental conditions found in southern Georgia and northern Florida using varieties developed for this environment. Consequently the Kc curve may not be appropriate for the environmental conditions and varieties in other regions. To make the Cotton App useable across the U.S. cotton belt will require a library of Kc curves as well as widespread access to meteorological data.

One solution to the meteorological data problem may be to use national gridded meteorological datasets offered by the U.S. National Oceanic and Atmospheric Administration Weather Service (NOAA NWS). We evaluated the NOAA NWS 2.5km grid Real Time Mesoscale Analysis (RTMA) tool (http://www.nco.ncep.noaa.gov/pmb/products/rtma/) and found that it underestimates precipitation of large events during the summer. Summer precipitation in the southeastern United States is driven by localized convective thunderstorms. As a result, in-field precipitation amounts can be substantially different from those estimated for a 2.5-km grid as well as from precipitation recorded at the nearest meteorological station on any given day.

NOAA NWS also recently released an experimental forecast reference ET (FRET) tool http://1.usa.gov/1Poz2va which we evaluated during the 2015 growing season for 20 locations in Florida, Georgia, and South Carolina. FRET appears to overestimate daily ET when unusually low ET is calculated from weather station data. Overestimating ET during low ET days erodes the advantage that the Cotton App has over irrigation scheduling tools like the Checkbook Method. A trial version of the Cotton App using the NOAA NWS 2.5km grid RTMA precipitation estimation and FRET is currently under development and will be released prior to the 2017 growing season.

Conclusion

Meteorological station-driven precipitation is the Cotton App's weakest feature since in-field precipitation amounts can be significantly different from those recorded at the nearest weather station on any given day. For the Cotton App to be used *most effectively* and to produce the most accurate results, users should correct precipitation recorded at weather stations with data from the field. Because notifications are pushed to the user whenever precipitation is recorded at the weather station, this may be simple to do. A bigger problem may lie with rain received at the field but not recorded at the weather station, because in this case, users will not have knowledge of the event until they visit the field.

Since its release in 2014, the Cotton App has been used by 373 by growers, consultants, and researchers to schedule irrigation in 660 unique fields during the 2014, 2015, and 2016 growing seasons. Twenty updates have been released over this time period – 12 for the Android and eight for iOS platforms, respectively. Reviews from users are positive and the University of Georgia Extension Cooperative Extension Service is now actively promoting the use of the Cotton App in Georgia. An online tutorial is available at http://smartirrigationapps.org/cotton-app-development. Research trials have shown that the Cotton App has the potential to greatly increase water use efficiency when utilized for scheduling irrigation on cotton in the south Georgia/North Florida regions.

Acknowledgements

This work was made possible by funding from the USDA NIFA National Integrated Water Quality Program (awards # 2011-51130-31143 and # 2013-51130-21486), the USDA NRCS Conservation Innovation Grants Program (award # 69-3A75-13-83) and grants from the Georgia Cotton Commission and Cotton Incorporated.

References

Allen, R.G., Pereira, L.S., Raes, D., Smith, M. 1998. Crop Evapotranspiration—Guidelines for Computing Crop Water Requirements. FAO Irrigation and drainage paper 56. Rome, Italy: Food and Agriculture Organization of the United Nations. ISBN 92-5-104219-5.

Burman, R.D., Wright, J.L., Nixon, P.R., Hill, R.W., 1980a. Irrigation management—water requirements and water balance. In: Irrigation, Challenges of the 80's, Proc. of the Second National Irrigation Symposium, Am. Soc. Agric. Engr, St. Joseph, MI, pp. 141–153.

Burman, R.D., Nixon, P.R., Wright, J.L., Pruitt, W.O., 1980b. Water requirements. In: Jensen, M.E. (Ed.), Design of Farm Irrigation Systems, ASAE Mono., Am. Soc. Agric. Eng., St. Joseph, MI, pp. 189–232.

Doorenbos, J., Pruitt, W.O., 1975. Guidelines for predicting crop water requirements. Irrigation and Drainage Paper No. 24, Food and Agricultural Organization, United Nations, Rome, Italy, pp. 168.

Doorenbos, J., Pruitt, W.O., 1977. Guidelines for predicting crop water requirements. Irrigation and Drainage Paper No. 24, 2nd ed., Food and Agricultural Organization, UN, Rome, Italy, pp. 144.

Jensen, M.E., 1968. Water consumption by agricultural plants. In: Kozlowski, T.T. (Ed.), Water Deficits and Plant Growth, Vol. II. Academic Press, Inc., New York, NY, pp. 1–22.

Migliaccio, K.W., K.T. Morgan, C. Fraisse, G. Vellidis, L. Zotarelli, C. Fraisse, B. A. Zurweller, J. H. Andreis, J. H. Crane, and D. Rowland. 2015. Smartphone apps for irrigation scheduling. Trans of the ASABE. 59(1): 291-301, doi:10.13031/trans.59.11158.

Migliaccio, K.W., K.T. Morgan, C. Fraisse, G. Vellidis, L. Zotarelli, C. Fraisse, B. A. Zurweller, J. H. Andreis, J. H. Crane, D. Rowland. 2016. Performance evaluation of urban turf irrigation smartphone app. Computers and Electronics in Agriculture 118, 136–142 doi:10.1016/j.compag.2015.08.015.

Perry, C.D. and E.M. Barnes. 2012. Cotton Irrigation Management for Humid Regions. Cotton Incorporated, Cary, NC. Available at http://www.cottoninc.com/fiber/AgriculturalDisciplines/Engineering/Irrigation-Management/

Ritchie, G.L., C.W. Bednarz, P.H. Jost, and S.M. Brown. 2004. Cotton growth and development. Bulletin 1252, Cooperative Extension Service, University of Georgia, Athens, Georgia, http://www.caes.uga.edu/publications.

Vellidis, G., Liakos, V., C. Perry, W. Porter, and M. Tucker. 2016a. Irrigation scheduling for cotton using soil moisture sensors, smartphone apps, and traditional methods. In S. Boyd, M. Huffman and B. Robertson (eds) Proceedings of the 2016 Beltwide Cotton Conference, New Orleans, LA, National Cotton Council, Memphis, TN (paper 16779, p.772-780).

Vellidis, G., V. Liakos, J.H. Andreis, C.D. Perry, W.M. Porter, E.M. Barnes, K.T. Morgan, C. Fraisse, K.W. Migliaccio. 2016b. Development and assessment of a smartphone application for irrigation scheduling in cotton. *Computers and Electronics in Agriculture* 127:249–259, doi:10.1016/j.compag.2016.06.021.

Proceedings of the 2016 Irrigation Association Technical Conference, Las Vegas, Nevada, December 5-9, Available from the Irrigation Association, Fairfax, Virginia

Irrigation Scheduling Of Field Corn Under Institutional Constraints

Dr. Freddie R. Lamm
Professor and Research Irrigation Engineer
Kansas State University
Northwest Research-Extension Center
Colby, Kansas
Voice: 785-462-6281
Fax: 785-462-2315
flamm@ksu.edu

ABSTRACT

Two pre-anthesis (pre-silking)and two post anthesis (post-silking) deficit sprinkler irrigation strategies for four corn hybrids where total irrigation was constrained to 11.5 inches against a fully irrigated control were compared in terms of grain yield and yield components, water use, and crop water productivity. This study was in response to a voluntary agreement of producers in a region of northwest Kansas (USA) where they agreed to reduce irrigation water application to 55 inches over a 5 year period. This study attempted to determine the best irrigation strategy for these limited applications. Results indicated full irrigation was still relatively efficient but used 30 to 36% more water. When corn prices are greater, managing at the full irrigation level and reducing irrigated land area may be more profitable. Pre-anthesis water stress was more detrimental to grain yield than similar levels of post anthesis stress because of reductions in kernels/ear. When water is greatly restricted, a 50% reduction in irrigation post-anthesis might fare reasonably well by relying on stored soil water and precipitation for grain filling. These results might not repeat on less productive soils or under harsher environmental conditions.

INTRODUCTION

In the semi-arid Central Great Plains and particularly northwest Kansas, soils are generally productive deep silt loam soils but precipitation is limited and sporadic with mean annual precipitation ranging from 16 to 20 inches across the region, which is only 60-80% of the seasonal water use for corn. Irrigation is often used to mitigate these water stress effects but at the expense of the continued decline of the Ogallala Aquifer.

In 2012, the Kansas legislature passed new water laws that allowed creation of a new water management structure known as a Locally Enhanced Management Area (LEMA). It allows stakeholder groups of various sizes to locally come together and design a management strategy to reduce overdraft of the Ogallala Aquifer in their area subject to approval by the Kansas Division of Water Resources. The first LEMA to be approved known as Sheridan High Priority Area 6 became a reality within Sheridan and Thomas Counties in northwest Kansas in 2013. The stakeholders in a 100 square mile area voluntary agreed to reduce their average water right to 11 inches/year for the next 5 year period. This area is centered approximately 30 miles east of the KSU Northwest Research-Extension Center at Colby, Kansas. In Kansas, annual rainfall decreases approximately 1 inch for every 18 miles moving east to west and greatest annual rainfall in western Kansas is in the months of May, June,and July, so a similar appropriate restriction at Colby to the Sheridan HPA #6 LEMA might be approximately 12 inches instead of

11 inches. Corn is the major irrigated crop in the region and producers in this LEMA would prefer to continue growing corn due to the availability of good local markets that include two large cattle feeding operations as well as a nearby dairy. The LEMA reduction of water right to 11 inches represents about a 27% reduction in water from the 80% chance Net Irrigation Requirement for Sheridan County (15 inches). The producers within the LEMA have the flexibility to apply their 5-year allocation of water as they so determine, but could benefit from research that determines when water can be restricted without large corn yield penalty.

ET-based irrigation scheduling has been promoted in the Central Great Plains for many years (Rogers, 1995). As producers move to deficit irrigation strategies this method of scheduling can still be useful in alerting the producer to soil water conditions and can help the producer decide when to allocate their limited supply (Lamm and Rogers, 2015). Management Allowable Depletion (MAD) values have been established as a means of helping producers know when to irrigate, but these established values have been questioned as too harsh for modern corn production (Lamm and Aboukheira, 2011; 2012).

Sprinkler irrigation does not allow for large amounts of water to be timed to a specific growth stage without incurring runoff, so strategies must be employed that can slowly restrict or slowly increase water available to the crop and to soil water storage for later usage. Preliminary computer simulation indicated that on average, approximately 40% of the seasonal irrigation amount is required prior to anthesis (Figure 1), so an imposed reduction of 50% during the pre-anthesis period might be acceptable most years, yet not be excessive in the drier years. However, this does not fully reflect the ability of the soil profile to be a "bank", so examining a higher irrigation regime is also warranted.

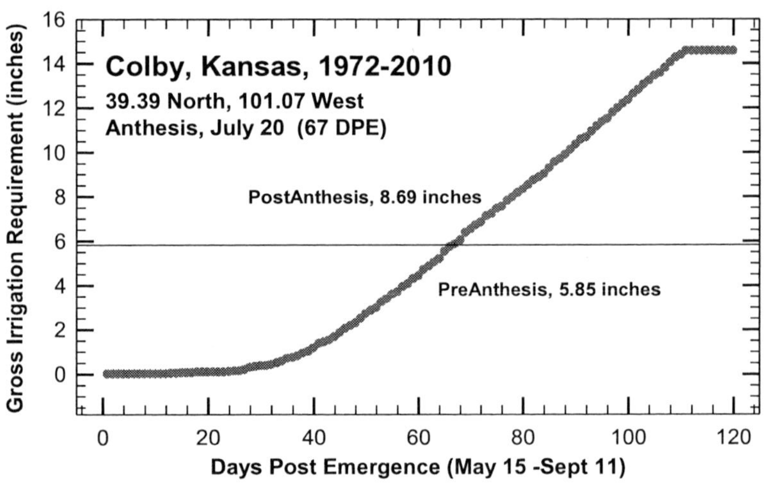

Figure 1. Seasonal gross irrigation requirements for field corn at Colby, Kansas.

A 4-year field study was conducted to examine restriction of irrigation to approximately to 50 or 75% of the ET-Rain value for either the pre-anthesis period or during the post-anthesis period. Since grain filling (post anthesis) is important, intuitively, one might surmise that those strategies restricting water during the pre-anthesis stages would always be preferable, but the pre-anthesis period is also when the number of kernels/acre is being potentially set and also the soil water storage allows for "banked" water to be used later by a deep rooted crop such as corn. These deficit strategies were compared to a fully-irrigated control treatment.

PROCEDURES

Four different commercial corn hybrids (two specifically marketed as drought tolerant) were compared under five different irrigation regimes in a three year (2013-2015) field study at the KSU Northwest Research-Extension Center at Colby, Kansas. For brevity only the average datas from the four hybrids will be discussed here. The irrigation regimes were: 1) Full irrigation (100% ET) with no restriction on total irrigation; 2) Irrigation restricted pre-anthesis to 50% of ET, 100% of ET thereafter with 11.5 inches total restriction; 3) Irrigation restricted pre-anthesis to 75% of ET, 100% of ET thereafter with 11.5 inches total restriction; 4) Irrigation restricted post-anthesis to 50% of ET with 11.5 inches total restriction; and 5) Irrigation restricted post-anthesis to 75% of ET with 11.5 inches total restriction. Irrigation amounts of 1 inch/event were scheduled according to water budget weather-based irrigation scheduling procedures only as needed subject to the specific treatment limitations. As an example, during the pre-anthesis stage Irrigation Trt 3 would only receive 75% ET, but after anthesis would receive irrigation at 100% until such time that the total irrigation is 11.5 inches. Soil water was monitored periodically (approximately 2 to 3 times/month) to a depth of 8 ft. in 1 ft. increments with neutron moderation techniques. This data was used to assess MAD values as well as to determine total water use throughout the season. Corn yield and yield components were determined through hand harvesting a representative sample at physiological maturity. Crop water productivity was calculated as grain yield/crop water use. The 5 irrigation treatments (whole plot, 6 reps) were in a RCB design with irrigation applied using a lateral move sprinkler and the 4 corn hybrid treatments superimposed as split plots. The data were analyzed using standard PC-SAS procedures.

RESULTS AND DISCUSSION

Weather Conditions and Irrigation Requirements

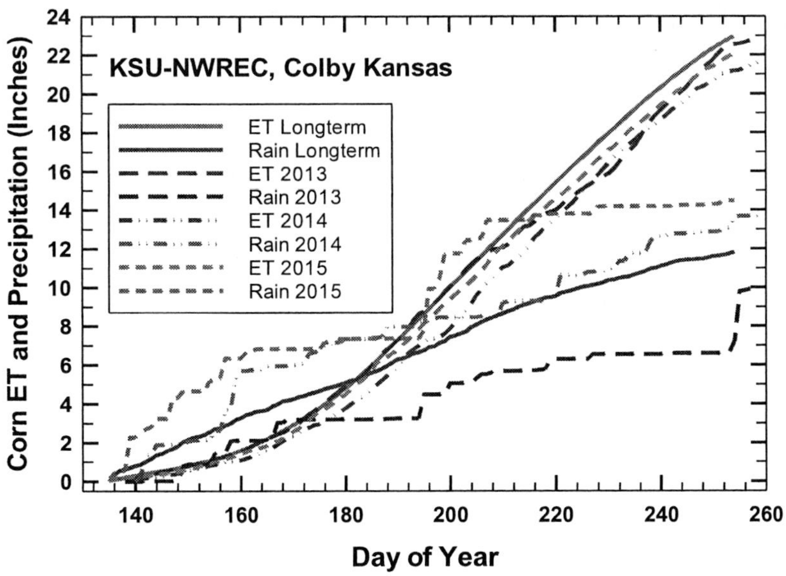

Figure 2. Cumulative calculated crop ET and precipitation during the growing season for Colby, Kansas, 2013 to 2015.

Overall weather conditions for the three years were favorable for excellent corn production during the study. Calculated crop ET for 2013 through 2015 was slightly lower than long term values and seasonal precipitation was 2 to 3 inches greater than normal in 2014 and 2015 and 2 inches less than normal in 2013 (Figure 2).

Full irrigation amounts varied from 12.48 inches in 2014 to 15.36 inches in 2013 (Figure 3 and Table 1). The treatments with pre-anthesis water restrictions (Trt 2, 50% ET pre-anthesis and Trt 3, 75% ET pre-anthesis) reached their water limitation (11.5 inches) in two of the three years (2013 and 2015) as did the post anthesis deficit irrigated treatment that was irrigated with 75% of ET during the post anthesis period. The irrigation treatment using the least amount of water during the three years of the study was the treatment where irrigation was restricted to 50% of ET during post-anthesis period (Trt 4).

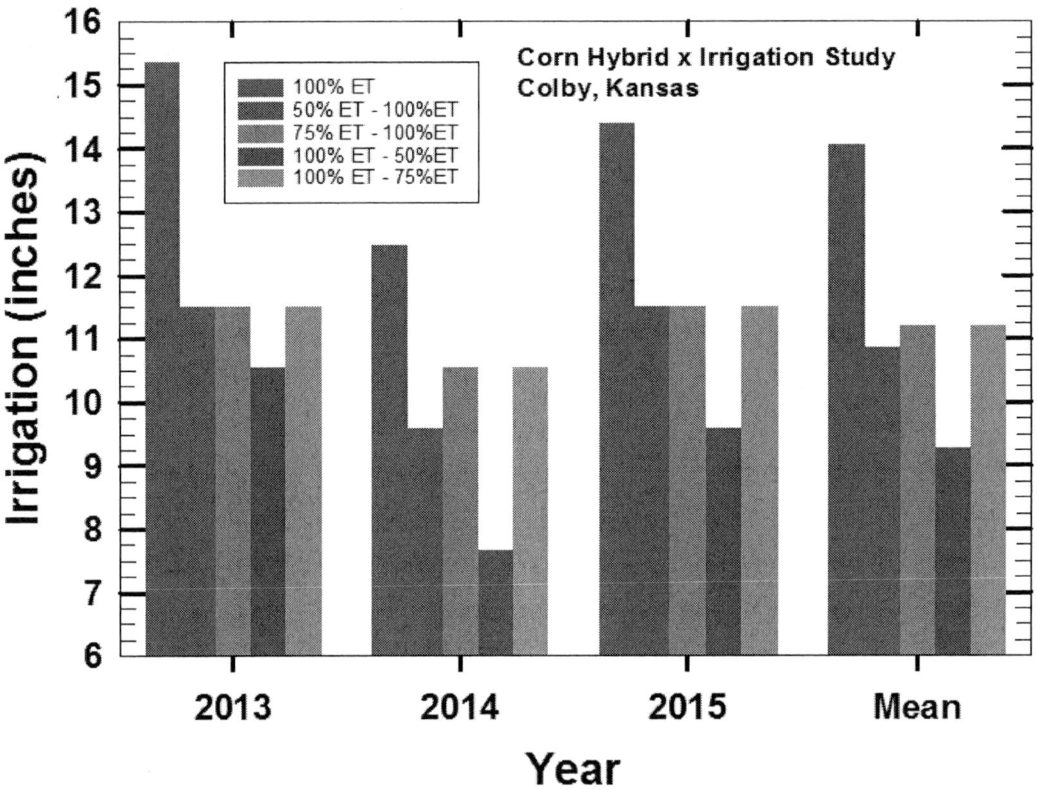

Figure 3. Irrigation amounts for the five irrigated corn treatments during the three years of the study.

Crop Yield and Water Use Parameters

Corn grain yield was greatest in 2014 and was lowest in 2013, the year with the greatest irrigation need (Figure 4 and Table 1). Fully irrigated corn grain yields ranged annually from 241 to 251 bushels/acre with the deficit-irrigated lowest yields ranging from 215 to 237 bushels/acre. Corn yield was greatest for unrestricted irrigation (Trt 1) but required 30 to 36% more irrigation, but was still very efficient with only a 2 to 4% reduction in water productivity (WP) (Figure 4 and 5 and Table 1). Lower yields occurred for pre-anthesis water restrictions (Trt 2 and 3) than for similar post-anthesis restrictions (Trt 4 and 5). These results suggests that obtaining sufficient kernel set was more important than saving irrigation for grain filling in this study. When irrigation is greatly restricted, a 50% reduction post-anthesis appears as a promising alternative, relying more heavily on stored soil water and precipitation for grain filling.

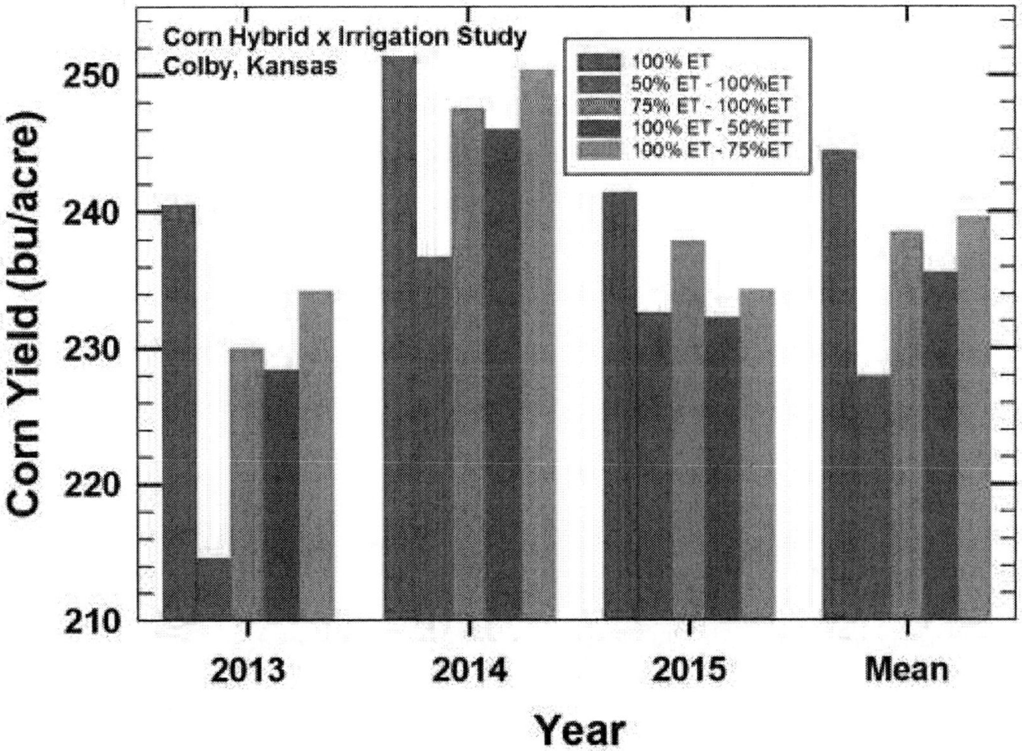

Figure 4. Corn yields for the five irrigation treatments during the three years of the study.

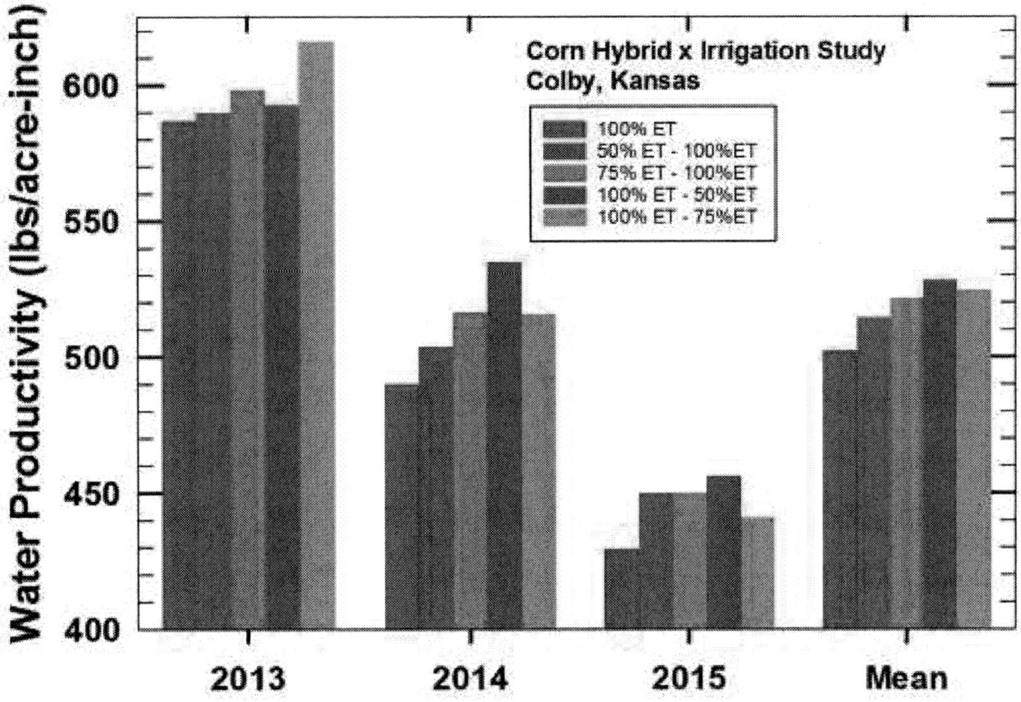

Figure 5. Water productivity for the five irrigation treatments during the three years of the study.

Table 1. Corn yield, yield component, and water use parameters in an irrigated corn study at Colby, Kansas, 2013-2015.

Irr Trt.	Irr. Amount	Yield, bu/a		Plant density, p/a		Ears/ plant		Kernels/ ear		Kernel mass, mg		Water use, inches		WP, lbs/acre-in	
Year 2013															
1. 100% ET	15.36	241	A	32452	A	1.00	A	542	A	349	A	23.0	A	587	B
2. 50/100% ET	11.52	215	C	32779	A	0.99	A	483	C	349	A	20.5	C	590	B
3. 75/100% ET	11.52	230	B	32634	A	0.99	A	522	B	347	A	21.6	B	598	AB
4. 100/50 % ET	10.56	228	B	32561	A	0.99	A	524	B	344	A	21.7	B	593	B
5. 100/75% ET	11.52	234	B	32561	A	1.00	A	527	AB	349	A	21.4	B	616	A
Prob > F		<0.0001		0.8328		0.3872		<0.0001		0.3976		0.0001		<0.0001	
Year 2014															
1. 100% ET	12.48	251	A	33215	A	1.00	A	566	A	339	A	28.76	A	490	C
2. 50/100% ET	9.60	237	B	33360	A	1.00	A	539	B	336	A	26.34	D	504	B
3. 75/100% ET	10.56	248	A	33251	A	1.01	A	557	A	337	A	26.89	C	516	B
4. 100/50 % ET	7.68	246	A	33069	A	1.00	A	558	A	338	A	25.82	E	535	A
5. 100/75% ET	10.56	250	A	33215	A	1.00	A	566	A	338	A	27.22	B	516	B
Prob > F		0.0010		0.6060		0.1034		0.0059		0.9002		<0.0001		<0.0001	
Year 2015															
1. 100% ET	14.40	241	A	32380	A	1.00	A	575	A	330	A	31.50	A	429	A
2. 50/100% ET	11.52	233	A	32525	A	1.00	A	563	A	323	A	28.98	A	450	B
3. 75/100% ET	11.52	238	A	32597	A	1.00	A	574	A	324	A	29.65	A	450	B
4. 100/50 % ET	9.60	232	A	32452	A	0.99	A	574	A	320	A	28.59	A	456	C
5. 100/75% ET	11.52	234	A	32670	A	0.99	A	573	A	322	A	29.78	A	441	B
Prob > F		0.0786		0.6613		0.0900		0.8987		0.6180		0.5629		<0.0001	
All Years															
1. 100% ET	14.08	244	A	32682	A	1.00	A	561	A	339	A	27.75	A	502	C
2. 50/100% ET	10.88	228	C	32888	A	1.00	A	529	B	336	A	25.26	D	515	C
3. 75/100% ET	11.20	239	B	32827	A	1.00	A	551	A	336	A	26.05	C	522	B
4. 100/50 % ET	9.28	236	B	32694	A	1.00	A	552	A	334	A	25.36	E	528	A
5. 100/75% ET	11.20	240	B	32815	A	1.00	A	556	A	336	A	26.14	B	524	A
Prob > F		<0.0001		0.5298		0.3079		<0.0001		0.4560		<0.0001		<0.0001	

Examination of Yield Components

Yield can be calculated as:

$$Yield = \frac{Plants}{Area} \times \frac{Ears}{Plant} \times \frac{Kernels}{Ear} \times \frac{Mass}{Kernel} \qquad \textbf{Eq. 1.}$$

The first two terms are typically determined by the cropping practices and generally are not affected by irrigation practices later in the season. Water stresses during the mid-vegetative period through about 2 weeks after anthesis can greatly reduce kernels/ear. Kernel mass, through greater grain filling, can partially compensate when insufficient kernels/ear are set, but may be limited by late season water stress or hastened senescence caused by weather conditions.

In this study, the yield component most strongly affected (as much as 6% corn yield variation) by irrigation practices was kernels/ear and was significantly affected (Pr F<0.05) in two years and also for the average of all years (Table 1 and Figure 6). Full irrigation (Trt 1) had the greatest number of kernels/ear while the 50% ET pre-anthesis treatment (Trt 2) consistently had the smallest value. These results suggest that pre-anthesis water stresses must be limited so that sufficient kernels/ear (i.e. sinks) can be set for modern corn hybrids.

Because all the yields components combine directly through multiplication to calculate yield, their effect on yield can be easily compared in Figure 6. The numbers on the lines refer to the 5 irrigation trts and the lines just connect similar data (i.e., the lines are not showing any pattern of results from one trt. to the next). A variation of 1% in any yield component would affect yield by the same 1%. It can be observed that there is much greater horizontal dispersion for kernels/ear than for all the other yield components which vary less than approximately 1%. Thus, irrigation treatment had a much greater effect on kernels/ear and the fully irrigated 100%ET, Trt 1 and the pre-anthesis 50% ET, Trt 2 were affected the greatest.

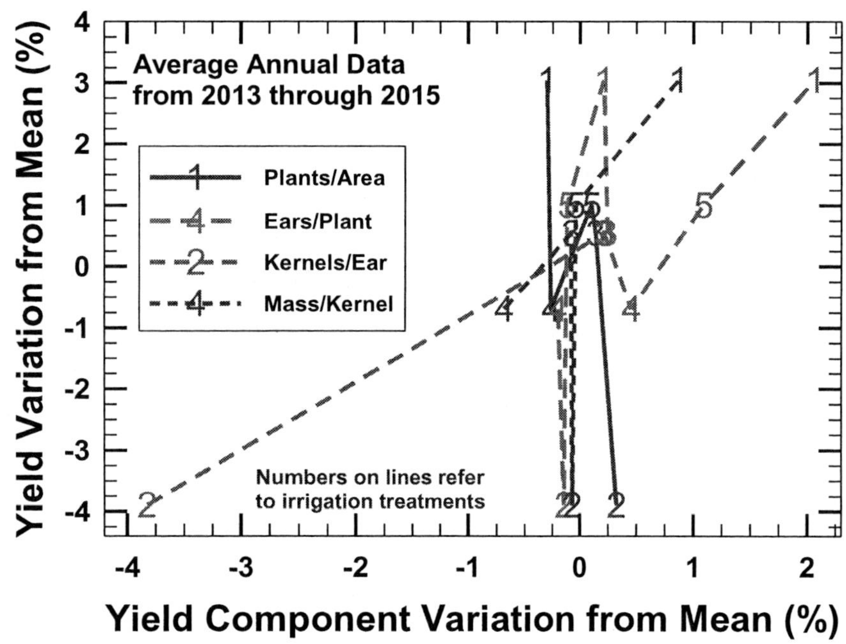

Figure 6. Yield variation as affected by variation in the yield components for the 5 different irrigation treatments.

Although Trt 4 (50% ET post-anthesis) averaged using 1.6 inches less irrigation than Trt 2 (50% ET pre-anthesis), its average corn yield was 8 bushels/acre greater (Table 1). Treatment 4 also had the greatest water productivity of all five treatments although all water productivities were respectable. It can be seen in Figure 6 that the major difference between Trt 4 and 2 is that Trt 4 was able to set a kernels/ear value much closer to the mean value than Trt 2.

CLOSING THOUGHTS AND CONCLUSIONS

- **Full irrigation was still relatively efficient but used 30 to 36% more water.**

When irrigation is not severely restricted, corn prices are greater, and/or irrigation costs are lower, managing irrigation at this level and reducing irrigated land area may be more profitable.

- **Pre-anthesis water stress was more detrimental to grain yield than similar levels of post-anthesis water stress because of reductions in kernels/ear.**

This result is somewhat counter to typical older guidelines which indicated that moderate stress during the vegetative stage for corn may not be detrimental. This may be indicating that kernel set on modern hybrids is a greater factor in determining final yields.

- **When water is greatly restricted, a 50% reduction post-anthesis might fare reasonably well by relying on stored soil water and precipitation for grain filling.**

The rationale behind this comment is that it is important to establish a sufficient number of kernels/ear (i.e., sinks) that potentially can be filled if soil water and weather conditions permit.

- **These results might not repeat on less productive soils or under harsher environmental conditions.**

On coarser soils (e.g. sandy soils), stored soil water and sporadic precipitation might not be sufficient to "carry" the crop through the post-anthesis period as well as in this study. However, it can be noted that the 50% ET post anthesis treatment (Trt 4) still performed better than the 50% pre-anthesis treatment (Trt 2) in 2013, the year with the greatest irrigation need.

ACKNOWLEDGEMENTS

This research was supported in part by the Ogallala Aquifer Program, a consortium between USDA Agricultural Research Service, Kansas State University, Texas AgriLife Research, Texas AgriLife Extension Service, Texas Tech University, and West Texas A&M University.

REFERENCES

Lamm, F. R. and A. A. Aboukheira. 2012. Effect of late season water stress on corn in northwest Kansas. ASABE paper no. 121337206. Available from ASABE, St. Joseph, MI. 10 pp.

Lamm, F. R. and A. A. Aboukheira. 2011. Effect of early season water stress on corn in northwest Kansas. ASABE paper no. 1111338. Available from ASABE, St. Joseph, MI. 11 pp.

Lamm, F. R. and D. H. Rogers. 2015. The importance of irrigation scheduling for marginal capacity systems growing corn. Applied Engineering in Agric. 31(2): 261-265.

Rogers, D. H. 1995. Using Evapotranspiration Reports for Center Pivot Irrigation Scheduling. KSU Coop. Extension Service, Manhattan, Kansas, Irrigation Management Series L-915. 6 pp

This page intentionally left blank.

Evapotranspiration and Crop Coefficient for Pecan Trees in El Paso, Texas

Shalamu Abudu[1], Zhuping Sheng[1], Ari Michelsen[1], Olga Rodriguez[1], James King[2]
1. Texas A&M AgriLife Research Center at El Paso, 1380 A&M Circle, El Paso, TX 79927-5020; PH (915) 859-9111; FAX (915) 859-1078 ; e-mail: shalamu.abudu@ag.tamu.edu

2. Department of Civil Engineering, New Mexico State University, Las Cruces, NM 88003-8001; PH (575) 646-5377; FAX (575) 646-6049; e-mail: jpking@nmsu.edu

Abstract. *In this study three-year daily actual pecan evapotranspiration (ET) was measured using an Open Path Eddy Covariance (OPEC) system for improved pecan irrigation scheduling in the west Texas. To monitor the amount and timing of water consumption of pecan trees, a monitoring network for measuring evapotranspiration with ET tower with an OPEC system, soil moisture and groundwater levels was installed at a Pecan Orchard in Tornillo, El Paso County, Texas. Three-year data (2012-2014) were analyzed and summarized in order to quantify the pecan ET at site. The results showed that the actual pecan ETs range from 1054mm to 1167mm for the growing season of March-October as compared to reference ET ranging from 1528 mm to 1635mm calculated by using ASCE standardized method. Maximum daily evapotranspiration range from 8.7mm to 9.4mm and all the maximum values occurred at the end of June and the beginning of July. The daily and monthly pecan tree crop coefficients were determined. The results from this study provide guidelines for precision irrigation for achieving better water conservation and improved production of pecans in the region.*

Key words: Pecan; Evapotranspiration; Irrigation Scheduling; Soil Moisture; Crop Coefficients;

Introduction

Irrigation water management is a major concern for pecan growers in the Southwest United States, particularly in the El Paso area because of its arid climate, with an average annual precipitation of 216 mm and an average annual evaporation of over 2,000 mm. Texas ranked second nationally in pecan production, and El Paso County has been one of the leading eight counties in pecan production with an approximately 3,500 hectares of irrigated pecan orchards. There is a great need for estimating pecan evapotranspiration (ET) accurately for improved irrigation scheduling of the pecan trees in the region. Numerous studies have been undertaken to quantify pecan evapotranspiration over past decades in both Las Cruces and El Paso pecan farms. Thomson (1974) reported that pecan consumptive water use in the El Paso-Las Cruces area ranged from 680 to 1000 mm per season depending on tree size. Miyamoto (1983) estimated that close-spaced and full-grown pecan tree ET was in the range of 1000 to 1300 mm per growing season in El Paso, Texas. Sammis et al. (2004) reported that 21-year-old trees with spacing of 9.7 x 9.7 m and a diameter of 30 cm had 1210 mm of seasonal ET averaged over two years through OPEC system measurements in the Las Cruces area, New Mexico. Liu and Sheng (2013) reported that mature trees with spacing of 9.1 x 9.1 m and a diameter of 37 cm consumed 1085 mm of water (ET) in 2007 through soil moisture measurements in the El Paso area, Texas.

The pecan water use in El Paso area is met by surface water from the Rio Grande and supplemented by the groundwater during drought years. Early spring irrigation on pecan farms in the El Paso area also uses return flow and groundwater based on the availability of Rio Grande water in early spring. Current irrigation scheduling depended on the availability of water from the Rio Grande through canals and groundwater pumping. At pecan farm in Tornillo, El Paso County, Texas, the irrigation events are about 13-16 times a year with the varying irrigation application amount from 100mm to 150 mm for each irrigation. This study aims at gaining better understanding of the evapotranspiration of the pecan trees and developing strategies for conserving water through improved irrigation scheduling in the El Paso area. To measure actual pecan ET, an OPEC system was established to monitor carbon dioxide flux, latent heat flux, sonic sensible heat flux, momentum flux, a computed sensible heat flux, temperature, humidity, horizontal wind speed and wind direction, net radiation, soil heat flux, soil temperature and soil water content in a pecan farm in El Paso, Texas since June 2010. This paper presents an estimate of evapotranspiration of mature pecan trees based on three-year (2012-2014) actual ET measurement using OPEC system at the study site. Daily and monthly crop coefficients were then developed for each year, which can be used to improve irrigation scheduling for water conservation in the pecan farms in the El Paso Area.

Materials and Methods

To monitor water consumption of pecan trees, a monitoring network was established at a Pecan orchard in Tornillo, El Paso County, Texas. It is located about 65 kilometers southeast from the city of El Paso. The soil profile at this site includes loam to a depth of 0.38 m, silty fine sand from 0.38 m to 0.94 m, silty clay from 0.94 m to 1.09 m, loam from 1.09 m to 1.47 m, silty clay from 1.47 m to 1.75 m, loam from 1.75 m to 2.39 m, clay from 2.39 m to 2.72 m, and fine sand (saturated) from 2.72 m to 2.89 m (not through) from a hand-augured borehole. For loam in El Paso, field capacity is 27 to 35%. The permanent wilting point for loam in El Paso ranges from 12 to 20% (Miyamoto, 1983). The pecans are approximately 10.55 m high and 0.32 m in diameter with spacing of 9.1x 9.1m. The number of pecan trees per hectare near the ET tower is 103 on average. Its canopy area measured at noon is 61.4 m^2. One tree occupied a surface area of 82.8m^2. Maximum pecan root depth was observed to be 1.62 to 2.29 m from four different dug holes with a depth to 2.44 m in this farm. An average maximum pecan root zone depth was determined to be 1.83 m. The observed groundwater depth varied from 2.21 to 2.73 m below land surface during the period from February 2008 to May 2009.

An 18-meter high ET tower with an Open Path Eddy Covariance (OPEC) system manufactured by Campbell Scientific, Inc. has been in operation in this field covering over two thousand acres since June 2010. The OPEC system methodology is not described here for brevity. Blanford and Gay (1992) presented a complete description of the equipment and theory behind OPEC. In addition, a standard Campbell weather station located a mile away from the ET tower for measuring climate variables and daily evaporation. The weather station data were used to calculate grass reference ET using ASCE's standardized equation (ASCE, 2005) and infilling the missing data at ET tower using regression method. Daily and monthly pecan crop coefficients were determined from the grass reference ET and actual ET measured using the OPEC system.

Results and Discussion

Soil Moisture and Irrigation

Based on the three-year (2012-2014) recorded soil moisture data in 30cm top soil in the pecan farm (as shown in Figure 1), the number of irrigation events and timing of irrigation in each year are almost same. Most of the irrigation events for this pecan orchard occurred when the soil moisture approached the minimum tolerance line. This indicates that pecan irrigation practice follows to some degree the behavior of crop and hydrological conditions in the soil. Generally, the irrigation started on March and last irrigation had taken place in the end of October each year. As shown in Figure 1, the total numbers of irrigation events were 14, 14 and 15 times during the growing season from March to October in the year of 2012, 2013 and 2014, respectively. Soil moisture observation shows that similar irrigation amounts were applied in 2012 and 2013 with 14 irrigation events. However, the irrigation in 2014 is slightly different from previous years with 15 times of irrigation events and the application rates in summer were somewhat smaller than previous years, particularly in July and August they are much smaller per irrigation as demonstrated by lower soil moisture (Figure 1).

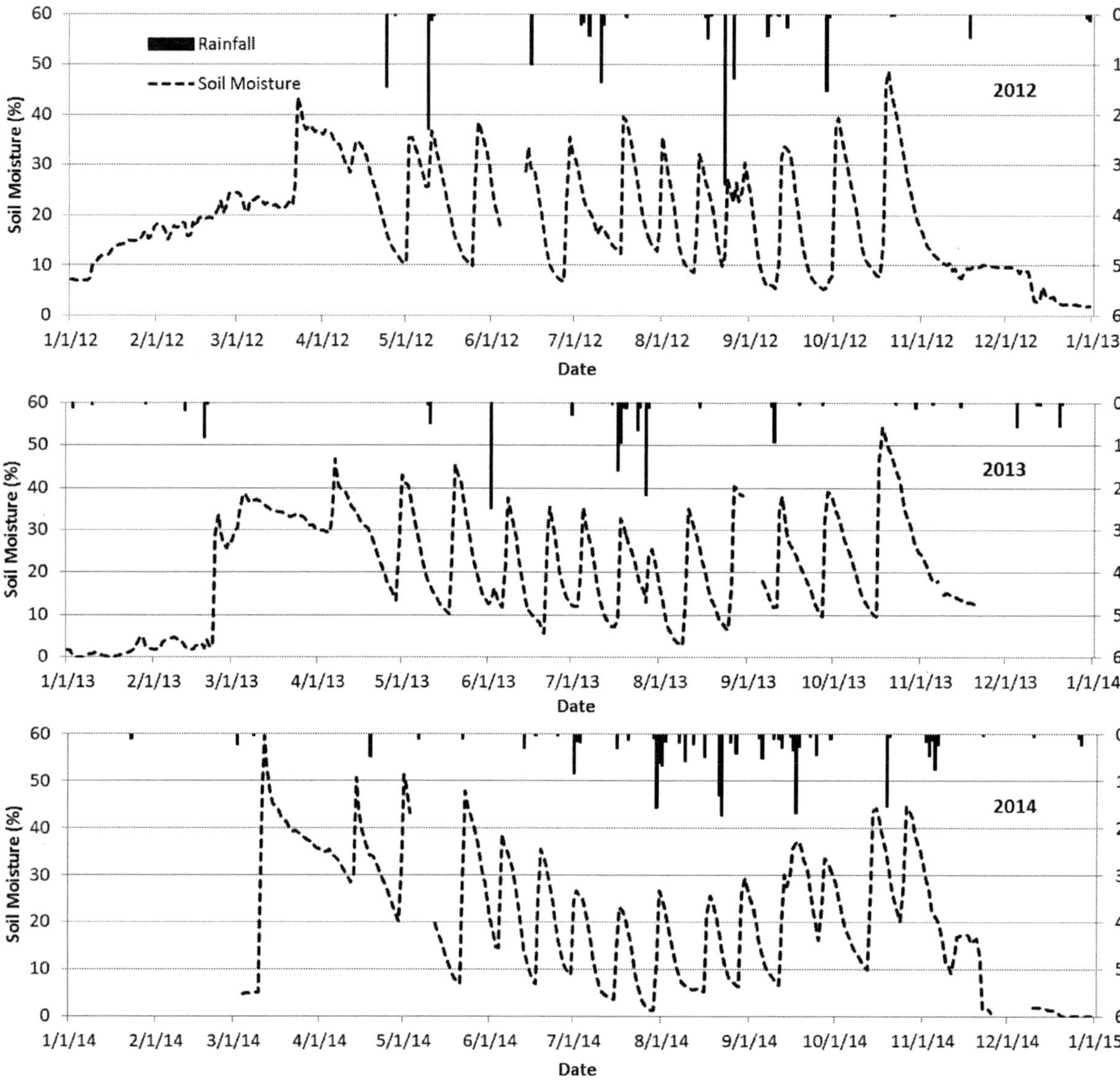

Figure 1. Soil moisture status and precipitation events for the years of 2012, 2013, 2014 at the pecan farm

Pecan Evapotranspiration

The ASCE standardized reference evapotranspiration equation (ASCE, 2005) was employed to derive reference ET from climate variables that measured in the standard weather station located one mile apart from the ET tower. The calculated daily reference ET values and measured daily pecan ET from the OPEC system observations were shown in Figure 2. As shown in Figure 2, the maximum ET of 8.7 mm/day, 9.4mm/day and 9.2mm/day were observed on 7/1/2012, 6/22/2013 and 7/1/2014, respectively. Average pecan ET is 4.5, 4.8 and 4.3 mm/day during the growing season from March 1 to October 31, which is the similar growing season as in previous studies (Miyamoto, 1983; Sammis et al., 2004; and Liu and Sheng, 2013).

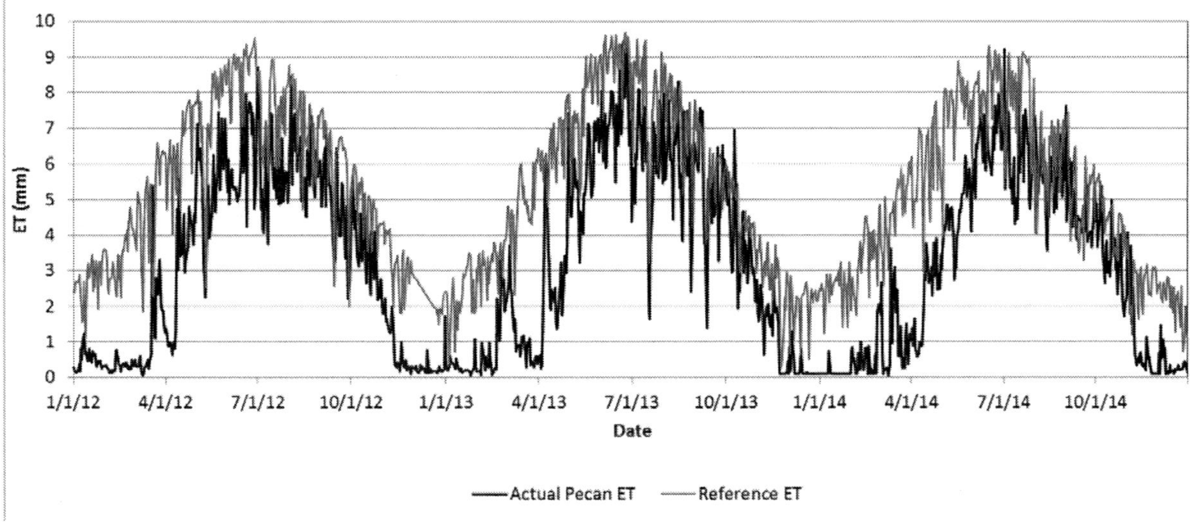

Figure 2. Measured actual pecan ET by OPEC system and calculated reference ET using ASCE standardized equation for the year of 2012-2014 at the study site.

The comparison of monthly measured pecan ET in this study with other studies (Miyamoto, 1983; Sammis et al., 2004; and Liu and Sheng, 2013) is shown in Table 1. The recent studies (Miyamoto, 1983; Sammis et al., 2004; and Liu and Sheng, 2013) showed the annual pecan ET value is in range of 1000 to 1460 mm for mature trees in the El Paso-Las Cruces area. The annual pecan ET value that was measured by OPEC system in this study from 2012 to 2014 were 1153mm, 1248mm and 1112mm, respectively and they are in this reported range. The total seasonal evapotranspiration measured with the OPEC system in the study site were 1097, 1167 and 1054mm for 2012, 2013 and 2014 respectively. The pecan growing season was identified as March 1 to October 31 in this pecan farm based on the previous studies and specific conditions of the study site. Miyamoto (1983) estimated a mature pecan orchard's consumptive use as 1310 mm for the growing season of April 1 through October 15. Miyamoto (1983)'s study involved orchards that were 8 to 35 years old and ranged in trunk diameter from 13 to 45 cm and heights from 7.4 m to 18.8 m located in El Paso, Texas, and Las Cruces, New Mexico areas. Sammis et al. (2004) reported the seasonal pecan ET values of 1260mm for 2001 and 1170mm for 2002, and an average of 1210mm for the growing season of from April to November (Table 1). Sammis et al. (2004)'s study was conducted for the mature pecan farm with the tree spacing of 9.7 m x 9.7 m, average orchard height of 12.8 m and with an average tree diameter at breast height of 30 cm. In this study, the pecan trees are approximately 10.6 m high and 32 cm in diameter with spacing of 9.1m x 9.1m. Hence, the results from this study are comparable to the other studies (Miyamoto, 1983; Sammis et al., 2004; and Liu and Sheng, 2013) since the pecan orchards of the study area contain mature pecans trees and similar in grass cover, tree size and irrigation conditions.

The monthly ET reached its maximum value in August at the rates of 188mm/month in 2012, 220mm/month in June in 2013, and 201mm/month in 2014 (Table 1). Except for Liu and Sheng (2013)'s study, the maximum monthly ET values are all occurred either July or/and August in all years in all studies (Miyamoto, 1983; Sammis et al., 2004). From Table 1 it can be concluded that the monthly ET pattern of this study is close to Sammis et al. (2004) in terms of pattern and magnitude of the monthly ET values. The three years average seasonal pecan ET in this study was 1106mm, is about 8% lower than Sammis et al (2004) reported value in average and 15% lower than the Miyamoto(1983) reported value in average. Considering all the other factors that affect the pecan ET are similar, the main reason for the lower pecan ET values in the Tornillo Farm, El Paso is attributed to the fact that there is insufficient irrigation in this study site due to surface water shortage in the farm. The evapotranspiration rate under water stress condition tends to be lower than the evapotranspiration rate in full irrigated crops without water shortage.

Table 1. Comparison of monthly evapotranspiration (ET) in this study with other studies from the literature (mm)

Studies	Growing season	Mar	Apr	May	Jun	Jul	Aug	Sep	Oct	Nov	Seasonal
Sammis (2004), 2001	Apr1-Nov20		88	177	202	**221**	**210**	185	136	40	1260
Sammis (2004), 2002	Apr1-Nov6		136	176	218	**199**	**198**	170	73		1170
Miyamoto (1983)	Apr1-Oct15		70	119	225	278	**290**	239	86		1307
Liu & Sheng (2013)	Mar12-Nov11	7	58	86	101	156	194	228	236	21	1086
This Study 2012	March1-Oct31	37	88	168	184	180	**188**	141	110		1097
This Study 2013	March1-Oct31	34	88	176	**220**	189	189	157	113		1167
This Study 2014	March1-Oct31	34	67	145	**201**	187	177	137	106		1054

Notes: Underlined values are the highest monthly ET values

Pecan Crop Coefficients for Irrigation Scheduling

Improved irrigation water management requires accurate scheduling of irrigations which in turn requires an accurate calculation of daily crop evapotranspiration. Crop coefficients are the basic parameters in estimating evapotranspiration on a daily basis for the irrigation scheduling. The crop coefficient (K_c) is defined as the ratio of measured evapotranspiration (ET)/potential evapotranspiration (ET_0) referenced to grass. As shown in Figure 3, three daily crop coefficient equations were developed for pecan trees based on the day of year (DY) as the independent variable for each year. The fourth-order polynomials were fit to the calculated Kc data for 2012, 2013 and 2014 with the coefficient of determination of 0.80, 0.83 and 0.81, respectively. The polynomial equations are as follow:

$$Kc = 0.0000000006DY^4 - 0.0000006DY^3 + 0.0001DY^2 - 0.0072DY + 0.1842 \quad \text{For 2012} \quad (1)$$
$$Kc = 0.0000000004DY^4 - 0.0000004DY^3 + 0.0001DY^2 - 0.0068DY + 0.1962 \quad \text{For 2013} \quad (2)$$
$$Kc = 0.0000000005DY^4 - 0.0000005DY^3 + 0.0001DY^2 - 0.0063DY + 0.1294 \quad \text{For 2014} \quad (3)$$

As can be seen in Figure 3 and equations (1), (2) and (3) that the coefficients of developed daily K_c equation polynomial are consistent and essentially the same value. The crop coefficient and evapotranspiration increased until maximum leaf area occurred in end of July and the start of August. The daily K_c equations developed in this study showed the similar tendency as the developed daily crop coefficients equations by Sammis et al. (2004) for pecan using both day of year and growing degree days as the base using 2001 and 2002 OPEC system measured ET data in Las Cruces, New Mexico, about 125 kilometers north of our study site. Our equations developed for three years are consistent between the years and the general tendency of all polynomials are the same (as in Figure 3), and can be used to estimate daily K_c values with higher accuracy in El Paso area.

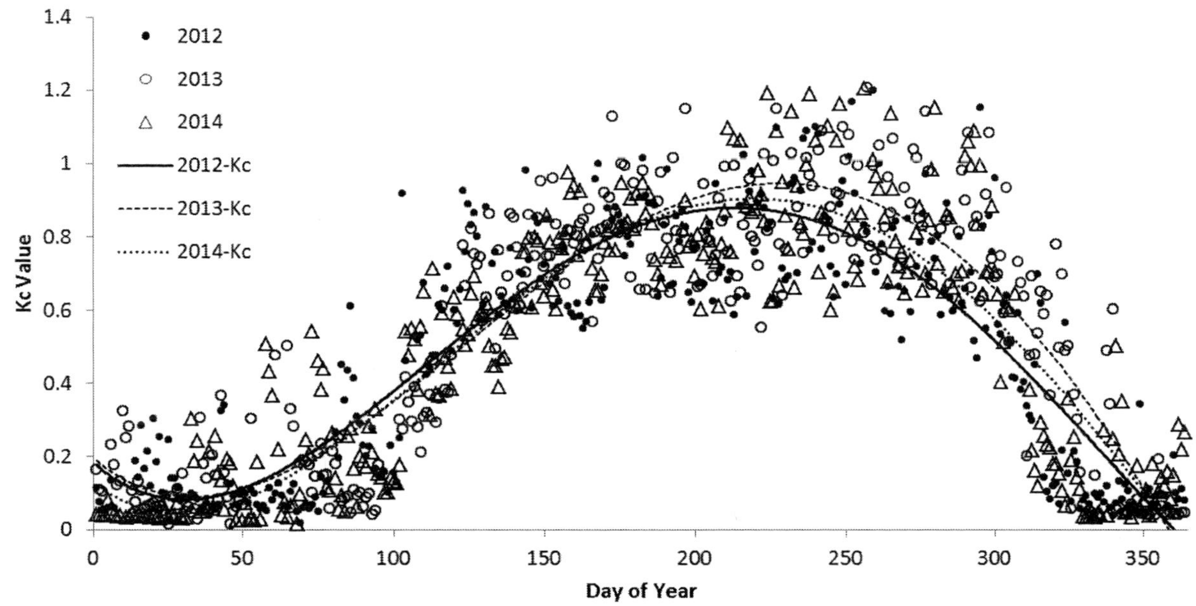

Figure 3. Daily crop Coefficient for pecan trees using day of the year as a time base for 2012-2014

Based on the measured monthly pecan ET values and calculated reference ET by ASCE's standardized equation (ASCE, 2005); the monthly crop coefficients for pecan are estimated for 2012, 2013 and 2014 (Figure 4). Each month K_c values are consistent with little discrepancies among the study years and the monthly K_c values for all year are smaller than 1.0 with an average highest monthly K_c values of 0.89occurred in August. According to other studies in the region (Miyamoto, 1983; Sammis et al., 2004; and Liu and Sheng, 2013), the K_c values derived from this study are smaller than others, indicating that the deficit irrigation that inhibits the evapotranspiration process and possible yield lost under water stress conditions in the study site.

Figure 4 shows the comparison of monthly K_c values to other studies in the region. The general tendency and values from this study is similar to Sammis et al (2004)'s study, except that their reported values are consistently higher than the values derived from this study with exception of October. Miyamoto's pecan K_c values for the months of April and May were lower than Sammis et al. (2004)'s values and the values reported in this study for all three years. During the middle and later growing season, the Miyamoto (1983)'s K_c values are much higher than both Sammis's values and values from this study with values over 1.0 for July, August and September. One possible cause for such a difference is that the Miyamoto's K_c values were derived under more ideal environment that facilitated by the experimental farm of Texas AgriLife Research Center with sufficient irrigation and possible high pecan evapotranspiration under favorable conditions. This indicates that there is still potential pecan yield increase in the Tornillo pecan farm in this study. In general, the developed daily and monthly pecan tree crop coefficients from this study can be used in pecan irrigation scheduling in similar pecan farms with the similar climate and water management conditions as the El Paso area, Texas.

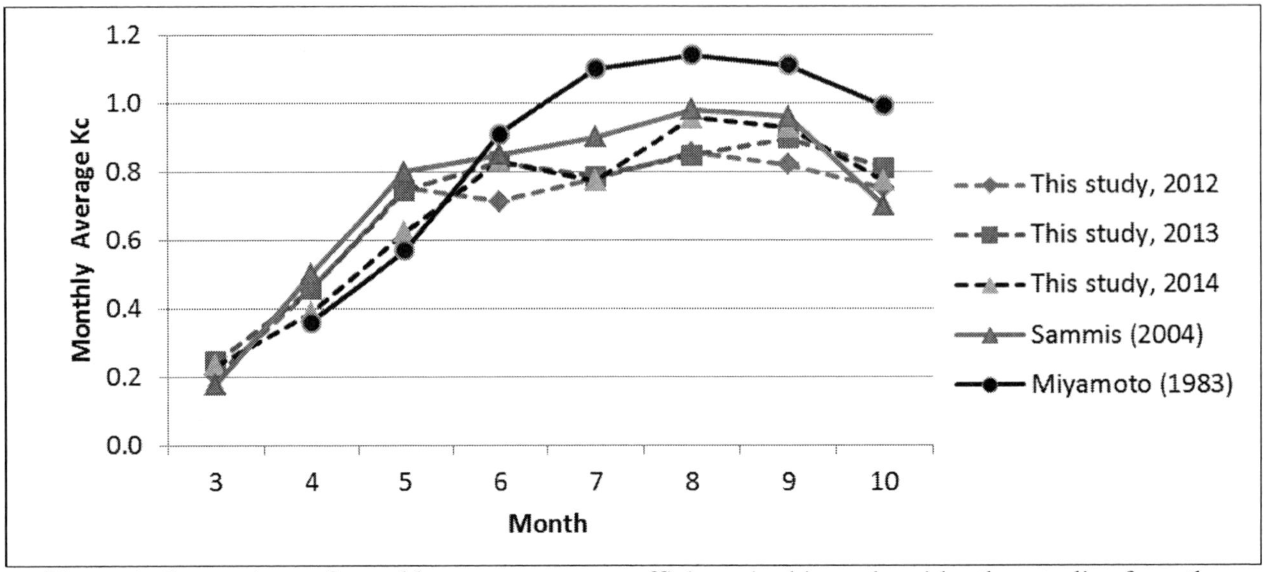

Figure 4. Comparison of monthly average crop coefficients in this study with other studies from the literature for the same region

Conclusions

Pecan irrigation in El Paso, Texas is surface water from the upper Rio Grande in a normal year and the combination of surface water of the upper Rio Grande and local groundwater in drought year. Pecan growers in this region have concerns about irrigation water management and improvement of pecan production. An OPEC system was established to monitor and quantify the consumptive water use of pecan in El Paso, Texas. Daily reference ET was derived through the ASCE standardized reference evapotranspiration equation using measured climate parameters from the standard weather station installed in the study site. Actual pecan ET was measured by OPEC system that is installed in the middle of pecan farm from 2010. A daily pecan crop coefficient equations and monthly crop coefficient values were derived from using actual pecan ET and calculated reference ET. Both ET values and K_c values were compared with other studies that reported in the literature for the same region with similar conditions.

The results from the study indicate that the actual pecan ETs was 1097mm, 1167mm and 1054mm for growing season of March-October. Maximum daily evapotranspiration were 8.7mm, 9.4mm and 9.2mm for 2012, 2013, 2014 and all the maximum values were observed at the end of June and the beginning of July. The maximum K_c value is 0.89 on August, indicating that there is still potential for increasing pecan yield in the Tornillo pecan orchard through improving irrigation scheduling practices or increasing water use efficiency. The crop coefficients developed in this study can be used in a water balance irrigation scheduling model which could be used in conjunction with other irrigation scheduling techniques to improve irrigation management of pecans.

Acknowledgements

This project is supported in part by National Institute of Food and Agriculture (NIFA) Hatch Project (1007968) of the United States Department of Agriculture (USDA), Texas AgriLife Research and Five R Enterprises Inc. Additional support provided by the NIFA, USDA, under award number 2015-68007-23130 helped to complete this article.

Reference

ASCE. (2005). ASCE Standardized Reference Evapotranspiration Equations, Edited by Richard G. Allen, IvanA. Walter, Ronald L. Elliott, Terry A. Howell, Daniel Itenfisu, Marvin E. Jensen, and Richard L. Snyder, Reston,VA, ASCE.

Blanford J. H., Gay L. W. (1992). Test of a robust eddy correlation system for sensible heat flux, Theor. Appl. Climatol. 46:53-60

Liu Y., Sheng Z. (2013). Soil moisture status in an irrigated pecan field, J. of Irrigation and Drainage Engineering 139(1):26-40.

Miyamoto S. (1983). Consumptive water use of irrigated pecan J. Amer. Soc. Hort. Sci. 108(5):676-681.

Sammis T.W., Mexal J.G., Miller D. (2004). Evapotranspiration of flood-irrigated pecans, Agricultural Water Management 69:179-190.

Thomson W.C. (1974). Scheduling pecan irrigations by computer, p.7-18. In: M.D. Bryant (ed.). Proc. Western Pecan Conf., New Mexico State Univ., Las Cruces.

Adapting the Kansas Crop Water Allocator (CWA) to Multi-year Use

Danny H. Rogers
Extension Agricultural Engineer
Kansas State University
Biological and Agricultural Engineering
Manhattan, Kansas
Voice: 785-532-2933
Email: drogers@ksu.edu

Jonathan Aguilar
Extension Agricultural Engineer
Kansas State University
Southwest Research-Extension Center
Garden City, Kansas
Voice: 620-275-9164
Email:jaguilar@ksu.edu

Isaya Kisekka
Research Agricultural Engineer
Kansas State University
Southwest Research-Extension Center
Garden City, Kansas
Voice: 620-275-9164
Email: ikisekka@ksu.edu

Introduction

Water supply for irrigation from the Ogallala in Kansas continues to become more limited, mostly due to loss of well capacity associated with declining aquifer thickness. Irrigation water use in Kansas is also constrained by an annual appropriation of water which includes, among other designations, the maximum total volume of water that can be diverted and the land area to which it can be applied. This annual appropriation of water to a user is through a permit system that once completed is referred to as a water right and can be maintained indefinitely if the terms of the water right are followed. The allowable volume of water as determined by the water right for most water rights in western Kansas is seldom the limiting factor today as most of the water rights were established before the occurrence of severe declines of the Ogallala and higher efficiency irrigation systems. However, whatever limits water availability, the irrigation producer must adjust the irrigation management strategy to the water availability. A tool to help in this decision making process for an annual allocation of water is the Crop Water Allocator (CWA). The original CWA was a planning tool that could help producers find the optimum combination of crop mix and irrigation amount for a given land area and fixed water volume in terms of net return per acre (Klocke et al., 2006).

Annual water allocations, as established by the 1945 Kansas Water Appropriation Act (K.S.A. 82a-701, et seq.), work reasonably well when allocations match long term water supply availability but impose little conservation incentive, especially as supplies become limited and irrigation practices use deficit irrigation management strategies. Institutional reductions of water allocation in areas where allocations are now known to exceed long term availability are problematic since the water allocation process results in an allocation that is defined as a real property right. In 1978, the Kansas legislature enacted the Groundwater Management District Act which contained provisions for the initiation of Intensive Groundwater Use Control Areas (IGUCA) (K.S.A. 8 82a – 1036-1038). IGUCAs allow for the implementation of additional corrective control provisions in areas of excessive deterioration of water supplies. While a number of localized IGUCAs have been established to address localized groundwater issues, the act, to date, has not been used to address the regional decline of the Ogallala. In several of the established IGUGAs , the total volume of water allocations were reduced but several new allocation concepts were allowed in lue of the annually based allocation to an authorized location, such as a multi-year water allocation and relocation of water allocations between points of diversions and/or authorized acreages.

Several other options have also been enacted by the Kansas Legislature that can be used to modify an individual water right at least temporarily, including the Localized Enhanced Management Area (LEMA)(S.B. 310) act and Water Conservation Area (WCA)(S.B. 275) act. LEMA's might be described as a voluntary IGUCA. The formation of an IGUCA involves a public hearing process in which the Chief Engineer (CE) from the Kansas Division of Water Resources takes input on water issues of a designated area and proposed control options. While producers have input to the process, the CE ultimately determines the final outcome of any new restrictions and management options available to water right holders in the IGUCA. IGUCAs do have periodic review and can be altered but the ultimate decision still lies with the CE. The process to form a LEMA, which can be formed within a Groundwater Management District, goes through the public hearing process with the CE to receive input on the LEMA management proposals and the CE can offer suggestions for changes but these changes must be acceptable to the LEMA originators. Once the CE accepts the LEMA, the proposal becomes the water policy for the region for the time period of the LEMA. A WCA is similar to a LEMA but has a streamlined process to allow any water right owner or group of owners an opportunity to develop a water management plan to allow for increased management flexibility with the ultimate goal of reducing withdrawals in an area in an effort to extend the useful life of the Ogallala aquifer. One LEMA and several WCAs have been formed and include as part of the water management scheme, a multi-year water allocation instead of an annual allocation.

Since multi-year water allocation is a potential option to irrigation water right owners, the question of what is the best allocation of the water resource relative to the crop and land resources available. Since management program discussed above are targeted to areas with declining water resources, the water allocation amounts must be reduced from current usage values, resulting in allocations that will be deficit as compared to full irrigation. Many of the current multi-year allocations use a 5 year base. The amount is dependent on the target area. The current LEMA set the new allocation to be an approximately 20 percent reduction of the 10 year average use in the area prior to LEMA establishment, in this case, the prior average annual use was 14 inches per acre, the LEMA allocation was set to 55 inches in 5 years (an average of 11 inches on an annual basis). To help producers and water managers consider impacts of multi-year allocations, and evaluate crop selection options, the CWA program was modified to accommodate multi-year allocations.

Description of CWA

The Multi-Yr CWA allows program operators to customize the inputs to their specific conditions but loads with default values that represent typical costs, yields, etc. in the same fashion as CWA. Figures 1 and 2 show the two pages of input for the program. Many input requirements contain default values. The program operator can customize the model by clicking on each input box and either selecting an input option from the dropdown menu or entering the desired value. Boxes with a question mark provide additional background information on the input as a help to the user. Crops of interest to a producer would be checked by clicking on the crop box next to the name. The land split selection determines how the acreage can be divided between crops or irrigation amount. A 50-50 selection means one half of the field can be of one crop that receives a certain irrigation amount and the one-half another crop or amount. The same crop could be selected but with different irrigation amounts. The total amount of irrigation application however cannot exceed the annual gross irrigation amount specified, although one split could receive the total amount and the other split(s), a reduced amount or none. The applied irrigation input limits the maximum amount of water that can be applied in a single year.

For each crop selected for consideration, the user should select current or projected crop price and the maximum yield that might be expected for each crop if grown under well watered conditions. Embedded into CWA are yield-water relationship curves (production functions) for each crop, an example curve is shown in figure 3. Crop yield are determined from the applied irrigation. The relationships used have been developed from irrigated field research conducted in the high plains region of western Kansas. The data from this research was then used as input to a crop simulation model that was executed to develop the applied irrigation and annual precipitation range. These curves are site specific to the annual rainfall, so the results are customized to the production conditions of western Kansas. All inputs including crop-specific production costs can also be customized by the operator of the program.

The original CWA calculates the net economic return from all possible combinations of crops and irrigation allocations among crops for each acreage allocation as determined by the land split and then ranks the net returns starting with the maximum. Net economic return is calculated by subtracting the production costs and irrigation costs from the total return, calculated by multiplying crop yield by the crop price. Net return does not include costs associated with land and equipment investments. The multi-year CWA uses a similar approach, however since the number of possible combinations become astronomically large quickly, statistical sorting of some options occurs.

The multi-year water allocation is set on the "Field and Irrigation" input page. The total number of inches of water for the allocation period is entered in the Total Water Allocation box and the number of years of the allocation, limited to 6 years, is entered into the Total Years box. The simulation run is started once the Calculate button is clicked at the bottom of either entry page. The top 100 crop selection combinations from the simulation are displayed for user's review.

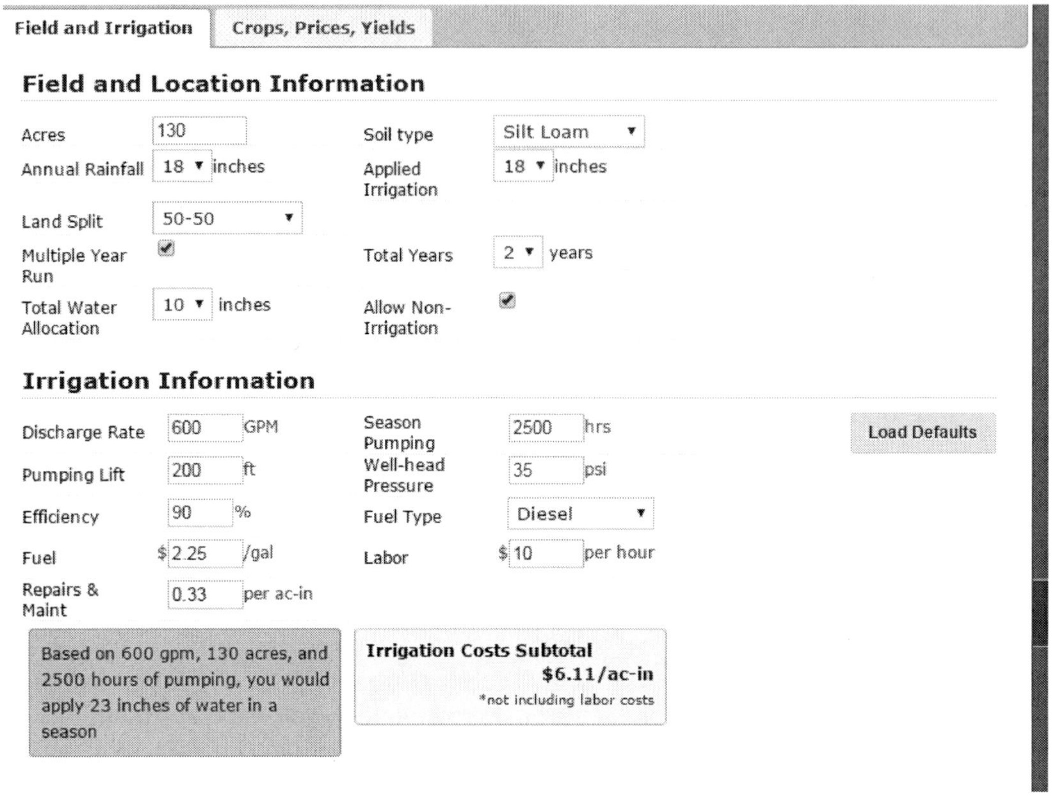

Figure 1: "Field and Irrigation" input page of the Multi-Yr CWA.

Figure 2: " Crops, Prices, Yields" input page of the Multi-Yr CWA. The illustration shows all of the available crop options as marked for consideration.

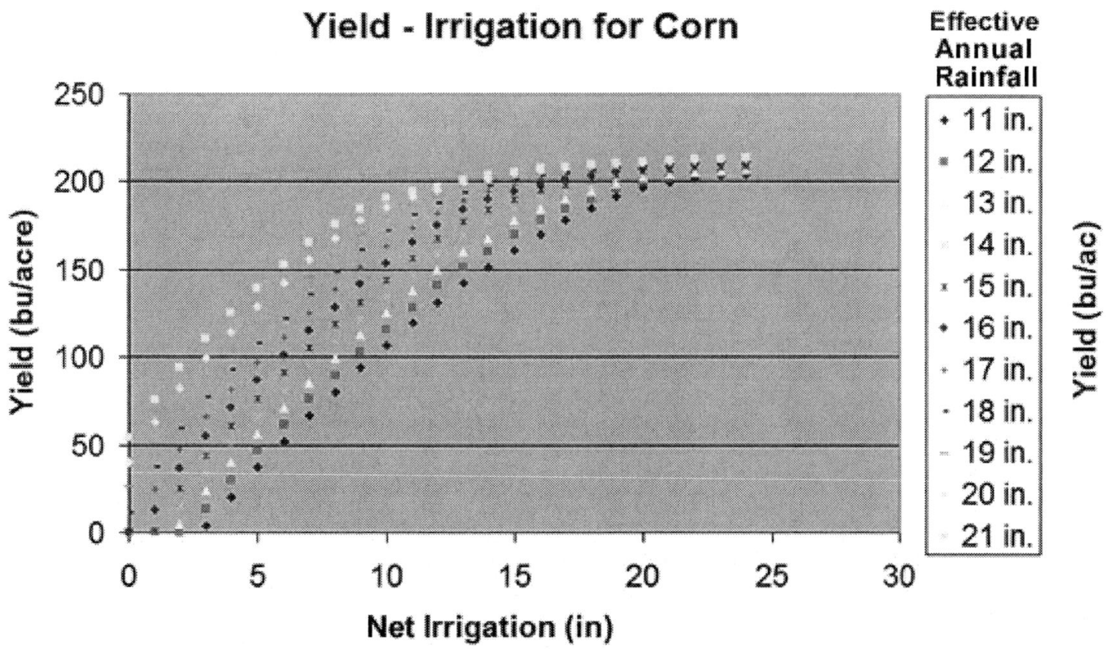

Figure 3: A Yield-irrigation relationship curve used in CWA. The example shown is for corn.

Results from Multi-Year CWA

Multi-year CWA begins evaluating the possible combination when the "Calculate" button is clicked. An example run is shown in Figure 4 (the two input pages) and Figure 5, which shows the first three options of the simulation run. For the input conditions in this example, corn was selected for both halves of the irrigated field and irrigated with the same amount of water. For years 2 and 3, sorghum was selected with equal irrigation amounts but at a lesser level than the corn of year 1, and finally sorghum years 4 and 5 at still a lesser amount than years 2 and 3. The final column shows the average return for this 5-year period was $252/acre. The next best option substituted soybean for sorghum in year 5 with a slight reduction in net return. Rank 3 option substituted corn for soybean in year 5. The ranking of other options are not shown. No time value of money, water or change in other costs or crop prices occur during the simulation period.

Sensitivity changes could be made by altering an input and generating new output. It is best to change only one input at a time. For example, figure 6 shows the results of changing the maximum yield potential of corn from 220 bu/ac to 240 bu/a for the yield-irrigation curve shown in Figure 3. The maximum yield potential can be altered by producer input based on their experience with the production capability of a particular field for non-water limited growing conditions. This single adjustment resulted in corn being selected as the first option (shown as rank 2 in Figure 6) for the entire five year period with the irrigation being divided equaled between the years. In the second option (rank 3), sorghum was a substitute for corn. In Figure 6, rank 1 is the first option selected from the figure 5 example; notice the pin on the righthand side of the chart has been activated. This saved the results from that simulation so that it could be easily compared to the change made in the next simulation.

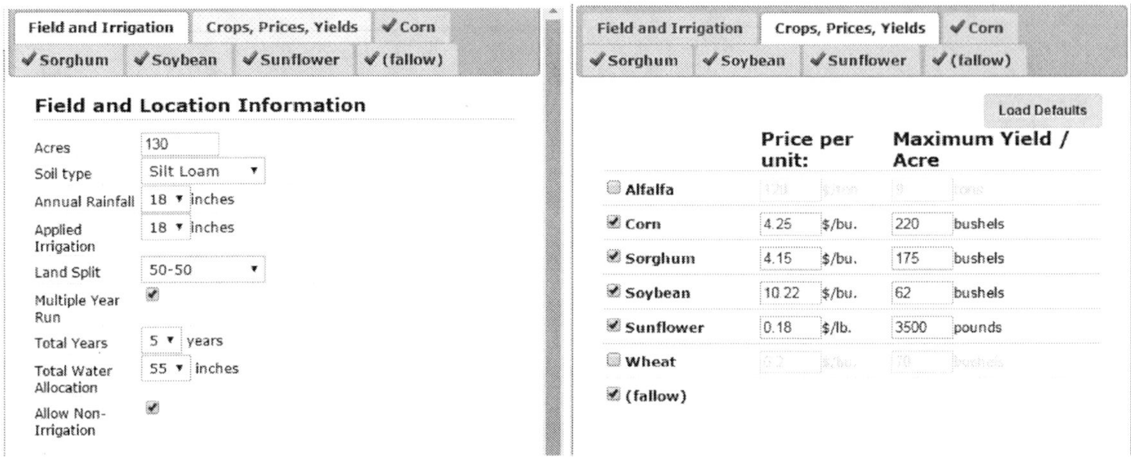

Figure 4: Input values for an example multi-year CWA simulation. The output chart for this example is shown in Figure 5.

Rank	Year	Acres	Crop	Yield /acre	Irrig. applied inches	Op. Costs $/acre	Returns $/acre	Annual Net RTN $/acre	Multi-year Ave. Net RTN $/ac
1	1	65.0	Corn	194.5 bu	13.0	$561	$827	$266	$252/ac
		65.0	Corn	194.5 bu	13.0	$561	$827		
	2	65.0	Sorghum	155.8 bu	11.0	$394	$646	$253	
		65.0	Sorghum	155.8 bu	11.0	$394	$646		
	3	65.0	Sorghum	155.8 bu	11.0	$394	$646	$253	
		65.0	Sorghum	155.8 bu	11.0	$394	$646		
	4	65.0	Sorghum	149.9 bu	10.0	$380	$622	$243	
		65.0	Sorghum	149.9 bu	10.0	$380	$622		
	5	65.0	Sorghum	149.9 bu	10.0	$380	$622	$243	
		65.0	Sorghum	149.9 bu	10.0	$380	$622		
2	1	65.0	Corn	194.5 bu	13.0	$561	$827	$266	$251/ac
		65.0	Corn	194.5 bu	13.0	$561	$827		
	2	65.0	Sorghum	155.8 bu	11.0	$394	$646	$253	
		65.0	Sorghum	155.8 bu	11.0	$394	$646		
	3	65.0	Sorghum	155.8 bu	11.0	$394	$646	$253	
		65.0	Sorghum	155.8 bu	11.0	$394	$646		
	4	65.0	Sorghum	149.9 bu	10.0	$380	$622	$243	
		65.0	Sorghum	149.9 bu	10.0	$380	$622		
	5	65.0	Sorghum	149.9 bu	10.0	$380	$622	$238	
		65.0	Soybean	48.4 bu	10.0	$263	$495		
3	1	65.0	Corn	194.5 bu	13.0	$561	$827	$266	$250/ac
		65.0	Corn	194.5 bu	13.0	$561	$827		
	2	65.0	Sorghum	155.8 bu	11.0	$394	$646	$253	
		65.0	Sorghum	155.8 bu	11.0	$394	$646		
	3	65.0	Sorghum	155.8 bu	11.0	$394	$646	$253	
		65.0	Sorghum	155.8 bu	11.0	$394	$646		
	4	65.0	Sorghum	149.9 bu	10.0	$380	$622	$243	
		65.0	Sorghum	149.9 bu	10.0	$380	$622		
	5	65.0	Corn	164.4 bu	10.0	$476	$699	$233	
		65.0	Sorghum	149.9 bu	10.0	$380	$622		

Figure 5: Top three example results from Multi-year CWA using the input pages of Figure 4

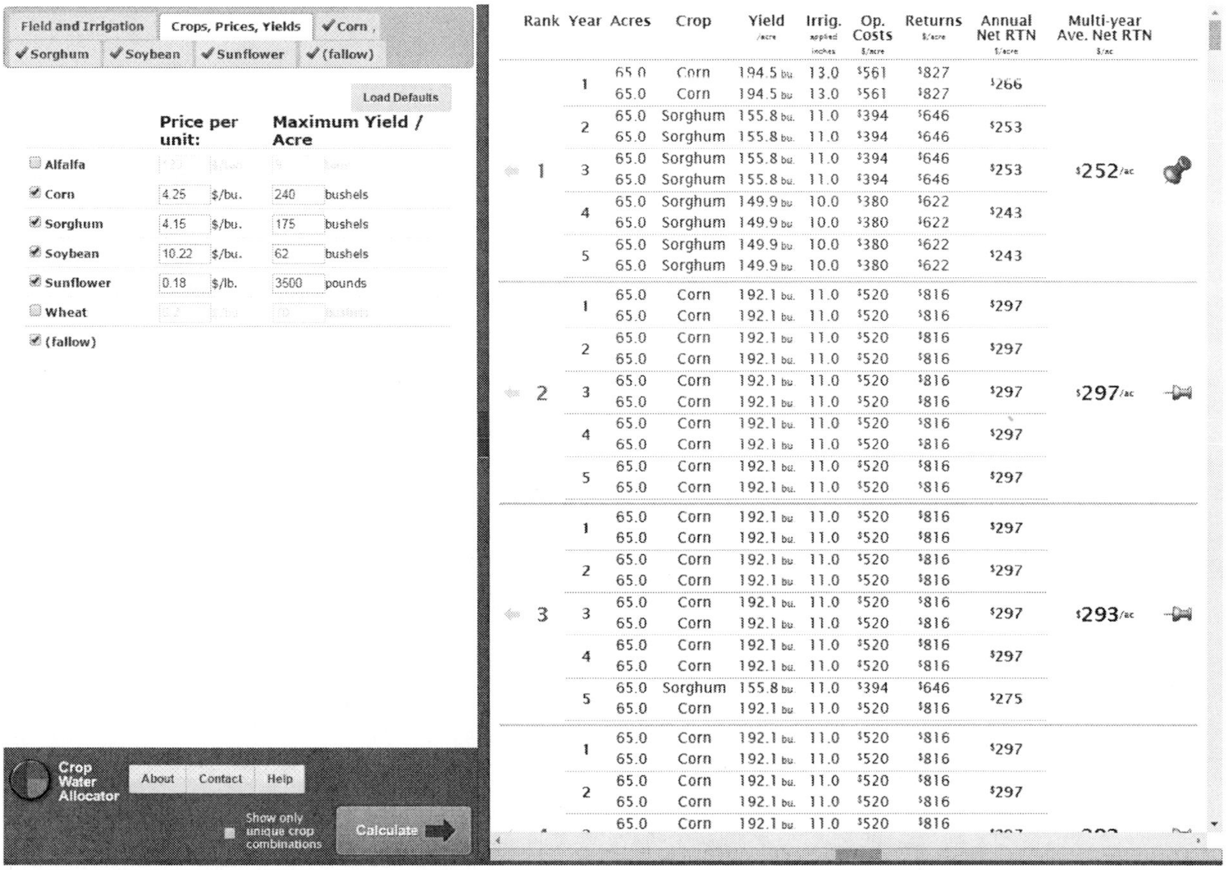

Rank	Year	Acres	Crop	Yield /acre	Irrig. applied inches	Op. Costs $/acre	Returns $/acre	Annual Net RTN $/acre	Multi-year Ave. Net RTN $/ac
1	1	65.0	Corn	194.5 bu.	13.0	$561	$827	$266	$252/ac
		65.0	Corn	194.5 bu.	13.0	$561	$827		
	2	65.0	Sorghum	155.8 bu.	11.0	$394	$646	$253	
		65.0	Sorghum	155.8 bu.	11.0	$394	$646		
	3	65.0	Sorghum	155.8 bu.	11.0	$394	$646	$253	
		65.0	Sorghum	155.8 bu.	11.0	$394	$646		
	4	65.0	Sorghum	149.9 bu.	10.0	$380	$622	$243	
		65.0	Sorghum	149.9 bu.	10.0	$380	$622		
	5	65.0	Sorghum	149.9 bu.	10.0	$380	$622	$243	
		65.0	Sorghum	149.9 bu.	10.0	$380	$622		
2	1	65.0	Corn	192.1 bu.	11.0	$520	$816	$297	$297/ac
		65.0	Corn	192.1 bu.	11.0	$520	$816		
	2	65.0	Corn	192.1 bu.	11.0	$520	$816	$297	
		65.0	Corn	192.1 bu.	11.0	$520	$816		
	3	65.0	Corn	192.1 bu.	11.0	$520	$816	$297	
		65.0	Corn	192.1 bu.	11.0	$520	$816		
	4	65.0	Corn	192.1 bu.	11.0	$520	$816	$297	
		65.0	Corn	192.1 bu.	11.0	$520	$816		
	5	65.0	Corn	192.1 bu.	11.0	$520	$816	$297	
		65.0	Corn	192.1 bu.	11.0	$520	$816		
3	1	65.0	Corn	192.1 bu.	11.0	$520	$816	$297	$293/ac
		65.0	Corn	192.1 bu.	11.0	$520	$816		
	2	65.0	Corn	192.1 bu.	11.0	$520	$816	$297	
		65.0	Corn	192.1 bu.	11.0	$520	$816		
	3	65.0	Corn	192.1 bu.	11.0	$520	$816	$297	
		65.0	Corn	192.1 bu.	11.0	$520	$816		
	4	65.0	Corn	192.1 bu.	11.0	$520	$816	$297	
		65.0	Corn	192.1 bu.	11.0	$520	$816		
	5	65.0	Sorghum	155.8 bu.	11.0	$394	$646	$275	
		65.0	Corn	192.1 bu.	11.0	$520	$816		
	1	65.0	Corn	192.1 bu.	11.0	$520	$816	$297	
		65.0	Corn	192.1 bu.	11.0	$520	$816		
	2	65.0	Corn	192.1 bu.	11.0	$520	$816	$297	
		65.0	Corn	192.1 bu.	11.0	$520	$816		
		65.0	Corn	192.1 bu.	11.0	$520	$816		

Figure 6: Results of Figure 4 example with the single change of input for corn maximum yield increase from 220 bu/ac to 240 bu/ac.

Figures 7 and 8 show the results of the last two combinations displayed; the only difference is figure 7 results show the 100 top ranking results based solely on the net return. Figure 8 display results show the top combination were sorted to display only the unique crop combinations. This latter display has a broader range of crop combinations, so less profitable crop options might be viewed.

The CWA is a long range planning tool, therefore the selected irrigation amount indicated is based on long term averages and the selected precipitation value. The irrigation amount applied during any given year should be based on growing conditions of that year, since large variations can occur (Rogers et al., 2015, Kisekka et al., 2015). Once the first growing season is completed, a new evaluation could be completed using updated crop prices, production costs, and remaining irrigation amount for the remainder of the years from the initial simulation.

Rank	Year	Acres	Crop	Yield /acre	Irrig. applied inches	Op. Costs $/acre	Returns $/acre	Annual Net RTN $/acre	Multi-year Ave. Net RTN $/ac
100	1	65.0	Corn	192.1 bu	11.0	$520	$816	$297	$281/ac
		65.0	Corn	192.1 bu	11.0	$520	$816		
	2	65.0	Corn	192.1 bu	11.0	$520	$816	$297	
		65.0	Corn	192.1 bu	11.0	$520	$816		
	3	65.0	Corn	202.3 bu	12.0	$565	$860	$294	
		65.0	Corn	202.3 bu	12.0	$565	$860		
	4	65.0	Corn	179.3 bu	10.0	$496	$762	$266	
		65.0	Corn	179.3 bu	10.0	$496	$762		
	5	65.0	Soybean	51.1 bu	11.0	$272	$522	$251	
		65.0	Soybean	51.1 bu	11.0	$272	$522		
101	1	65.0	Corn	192.1 bu	11.0	$520	$816	$297	$281/ac
		65.0	Corn	192.1 bu	11.0	$520	$816		
	2	65.0	Corn	192.1 bu	11.0	$520	$816	$297	
		65.0	Corn	192.1 bu	11.0	$520	$816		
	3	65.0	Corn	192.1 bu	11.0	$520	$816	$297	
		65.0	Corn	192.1 bu	11.0	$520	$816		
	4	65.0	Sorghum	155.8 bu	11.0	$394	$646	$275	
		65.0	Corn	192.1 bu	11.0	$520	$816		
	5	65.0	Corn	192.1 bu	11.0	$520	$816	$239	
		65.0	Sunflower	2956.9 lb	11.0	$352	$532		

Figure 7: Results of simulation for the top 100 results from the Figure 4 simulation run.

Rank	Year	Acres	Crop	Yield /acre	Irrig. applied inches	Op. Costs $/acre	Returns $/acre	Annual Net RTN $/acre	Multi-year Ave. Net RTN $/ac
100	1	65.0	Corn	220.5 bu	14.0	$602	$937	$335	$257/ac
		65.0	Corn	220.5 bu	14.0	$602	$937		
	2	65.0	Corn	212.2 bu	13.0	$585	$902	$317	
		65.0	Corn	212.2 bu	13.0	$585	$902		
	3	65.0	Corn	192.1 bu	11.0	$520	$816	$297	
		65.0	Corn	192.1 bu	11.0	$520	$816		
	4	65.0	Corn	192.1 bu	11.0	$520	$816	$239	
		65.0	Sunflower	2956.9 lb	11.0	$352	$532		
	5	65.0	Sunflower	2258.4 lb	6.0	$312	$407	$95	
		65.0	Sunflower	2258.4 lb	6.0	$312	$407		
101	1	65.0	Corn	212.2 bu	13.0	$585	$902	$317	$256/ac
		65.0	Corn	212.2 bu	13.0	$585	$902		
	2	65.0	Soybean	57.0 bu	14.0	$310	$583	$304	
		65.0	Corn	220.5 bu	14.0	$602	$937		
	3	65.0	Soybean	55.5 bu	13.0	$303	$567	$291	
		65.0	Corn	212.2 bu	13.0	$585	$902		
	4	65.0	Soybean	51.1 bu	11.0	$272	$522	$274	
		65.0	Corn	192.1 bu	11.0	$520	$816		
	5	65.0	Sorghum	53.2 bu	0.0	$182	$221	$95	
		65.0	Corn	150.7 bu	8.0	$489	$640		

Figure 8: Results of simulation for the top 100 results from the Figure 4 simulation run but sorted to only show unique crop combinations.

Conclusions

New irrigation water management options have become available to Kansas producers that face limited irrigation water supplies. One new management option is the allocation of water resources on a multi-year basis rather than the traditional annual water allocation. To help producers make decision on how to use the available land and irrigation water resources that result in the optimal economic returns, the planning tool, Crop Water Allocator, was modified to accommodate a multi-year water allocation. While many factors influence the outcome, the Multi-year CWA program may be a tool to help them determine the best crop acreage mix of the increasingly limited water resources.

References

Kisekka, I., J. Aguilar, and D.H. Rogers. 2015. Year to Year Variations in Crop Water Use Functions. In: Proc. 27th Annual Central Plains Irrigation Conference, Feb. 17-8, 2015, Colby, KS. Available from CPIA, 760 N. Thompson, Colby, KS. pp. 44-49.

Klocke, N.L., L.R. Stone, T.J. Dumler, G.A. Clark, and S. Briggeman. 2006. Crop Selection and Water Allocations for Limited Irrigation. In: Proc. 18th Annual Central Plains Irrigation Conference, Feb. 21-22, 2006, Colby, Kansas. Available from CPIA, 760 N. Thompson, Colby, Kansas. pp. 157-161.

KANSAS WATER APPROPRIATION ACT. K.S.A. 82a-701 through 82a-737 and 82a-740 through 82a-745 and K.S.A. 42-303 and 42-313 Revised December 2015

Intensive Groundwater Use Control Areas. K.S.A. 82a-1036, K.S.A. 82a-1037, and K.S.A. 82a-1038. Also see: IGUCA Rules and Regulations, Division of Water Resources, Kansas Department of Agriculture. Sept. 18, 2009.

Rogers, D.H., J. Aguilar, I. Kisekka, P.L. Barnes, and F.R. Lamm. 2015. Agricultural Crop Water Use. In: Proc. 27th Annual Central Plains Irrigation Conference, Feb. 17-8, 2015, Colby, KS. Available from CPIA, 760 N. Thompson, Colby, KS. pp. 1-18.

Water Conservation Areas. Kansas Senate Bill No. 275. April 215.

Precision Ag Irrigation Language (PAIL) Project

Diganta Adhikari, PhD,
Irrometer Co., Inc., Riverside, CA

Aaron Berger,
AgSense, LLC, Huron, SD

Dan Berne,
Next Chapter Marketing, Portland, OR

R. Andres Ferreyra, PhD,
Ag Connections, Inc., Murray, KY, United States

Charles Hillyer, PhD
Texas A&M Agrilife Extension, TX, United States

Steve Melvin,
Lindsay, Omaha, NE

Bart Nef,
Campbell Scientific Inc, Logan, UT

Tom Penning,
Irrometer Co., Inc., Riverside, CA

Joseph M. Russo, PhD,
ZedX, Inc., Bellefonte, PA

Terry Schiltz,
AgSense, LLC, Huron, SD

Abstract. Agriculture is entering the era of "big data" which will be the basis for evidence-based management in precision agriculture. The collection, storage, and streaming of big data requires technical tools that must be integrated into a farm enterprise. However, these tools are incompatible with each other due to different designs, data formats, and transfer protocols. Furthermore, data delivered by a tool represents only one part of a set of information required by a grower to make a management decision. Growers have been reluctant to adopt these tools because of their incompatibility in design and their inability to be integrated into a holistic solution for decision making. Only through the implementation of data exchange standards will these disparate tools be adopted by growers and their supporting cast in the agricultural industry.

The Precision Ag Irrigation Language (PAIL) project is part of an industry-wide effort under the AgGateway business consortium to create open data exchange standards for agriculture. The focus of the PAIL project is on irrigation data exchange standards. PAIL is a collaborative effort of 20+ companies; it was chartered by AgGateway's Precision Agriculture Council in 2013 following preliminary work organized by the Northwest Energy Efficiency Alliance (NEEA). The PAIL team is nearing submission of a draft open standard to ASABE with the goal of it becoming an international standard in ISO. This paper describes the PAIL project, including its scope and primary deliverables (process models, Core Documents, and data exchange schemas for Core, Operations, and Observations). It discusses how these deliverables can be used in a farm enterprise.

Keywords. information management. irrigation. irrigation technology. precision irrigation. standards.

Introduction

The United Nation's Food and Agriculture Organization (FAO) has projected the earth's population will exceed 9 billion by 2050 (Alexandratos and Bruinsma, 2012). This increase will require an additional billion tons of cereal produce alone; nearly a 33% increase over current levels. The FAO expects most of the gains to come from increased yield and increased land in production. However, in developed countries, where FAO projects an 8% decrease in land for production, cereal gains must come from an increase in yield. Even with the 33% increase in production, the FAO believes that water demand will increase by only 11%. The reduced rate of increase is expected to come from improvements in water use efficiency and a reduction in rice production. Most of the increase in efficiency will come from improvements in stress tolerance and reduced water needs in new varieties (Baulcombe, 2010). In developed nations, some of the increase in water use efficiency will be from improved management practices. Regardless of the projections, farmers in the future will be pressured to increase production on less land and with reduce water use due to competition with other sectors in society.

It is not necessary to look beyond the United States to find evidence of pressure on irrigated farms. Irrigated agriculture in the United States (U.S.) accounts for 80-90% of the consumptive water use and approximately 40% of the value of agricultural production (Schaible and Aillery, 2012). This value, totaling nearly $118 billion US dollars, is produced on 57 million acres. According to the most recent Farm and Ranch Irrigation Survey (USDA, 2012) , 25,853 out of 296,303 irrigated farms reported reduction in yields due to a shortage of ground or surface water. This reduction is in addition to yield losses due to 6,011 farms discontinuing irrigation. The number of farms discontinuing irrigation is up more than 30% from the last survey.

The need for a standard

Agriculture has become a data-driven endeavor. New sources of information about soil, weather, crop status, machine operation, marketing, and economics all facilitate the evidence-based decision-making that defines precision agriculture. Using these new data streams requires tools and the evidence of this is found in the proliferation of new applications (apps) for mobile devices. A search of the Google Play store for the words "Agriculture" or "irrigation" yields 92 and 82 results, respectively. Even though these apps improve accessibility to data, growers are still responsible for relating the data to decisions in a farm enterprise. Furthermore, accessed data can be from diverse sources representing different scales, formats, and units. Consequently, the exercise of relating data can involve one or more tasks, such as combining data from multiple sources into a single output; performing calculations that transform data into specific recommendations; or using data as input into models to predict some potential outcome. Each of these tasks requires moving and transforming data. Tasks working together can be considered integration. The integration produces decision-making power that is greater than the sum of the individual tasks and data streams. It provides the evidence needed for evidence-based management.

There are many approaches for managing irrigation as shown in Figure 1, which is Table 22 of "Methods used in Deciding When to Irrigate" section of the Farm and Ranch Irrigation Survey (USDA, 2012). As can be seen in the figure, most approaches do not utilize technical tools, which are necessary for evidence-based management. In fact, technical tools, such as an irrigation schedule resulting from a computer simulation model, represent only 64,037 out of 369,917 approaches. This imbalance in favor of non-technical approaches has persisted over the last seven surveys dating back to 1988. (Smith et al., 2010)

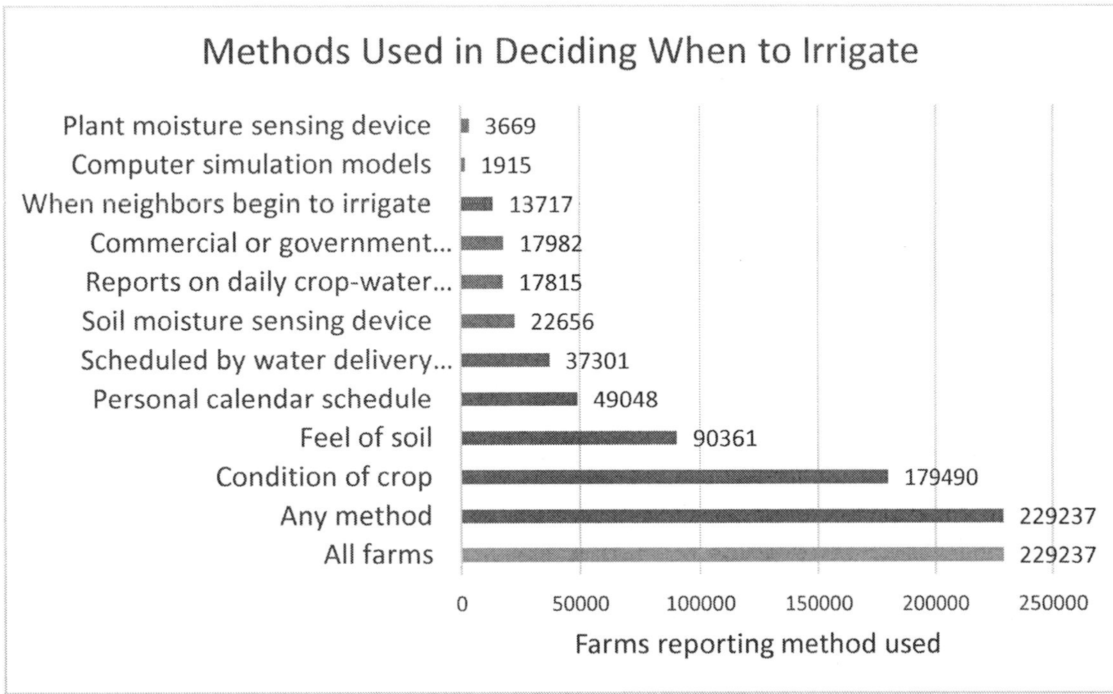

Figure 1. Table 22, Farm and Ranch Irrigation Survey, 2013 Census of Agriculture.

There are several potential explanations for the poor adoption of technical tools. First, there may be no incentive to change if things are working, even if there is an unforeseen benefit. Second, it takes more effort to install and maintain a soil moisture sensor than to just "feel the soil." Third, a technical tool may only provide some of the data or information to support a decision. For example, when maximizing the value of water, a soil moisture sensor only tells part of the story. Growers also need to know how much water was applied, how much the crop has used, what the weather has done, and the condition of the crop. Sensors exist for each of these information sources, but only as separate tools. Maximizing the value of water requires integrating all these tools. Therein lies the problem, the tools do not communicate to each other and as such they are not Integrated. The integration of tools is currently the responsibility of a grower, who may or may not have the know-how, people, funds, or time to do it. No matter the reason, the effort required can be discouraging.

Farm Management Information Systems (FMIS) are an obvious point of integration for technical tools. By facilitating integration, an FMIS alleviates some of the grower's burden. Implementing this integration requires an FMIS to have special code to interoperate between tools, and ultimately between sources of data. As new tools emerge, an FMIS must continue to expand. If each of the different tools could produce data in the same format, integration would be simpler and cheaper. A common data format would not only facilitate integration, but likely lead to a proliferation of new and more comprehensive FMIS solutions. This proliferation would in turn lead to increased adoption of more efficient technical approaches for deciding when to irrigate.

The multitude of technical tools to mine new data sources, the availability of cheaper telemetry, and the expanding role of FMIS all portend an important opportunity to improve irrigation management. However, there is no established framework for integrating these disparate tools and incorporating telemetry. This lack of a framework creates an immediate need for an irrigation-related, data exchange standard. Without a standard for data exchange, irrigation will miss the "Big Data" revolution and will instead remain a "manual" management activity as evidenced by the choice of approaches shown in Figure **1**.

The PAIL Project

In 2011, a group of companies, representing the irrigation segment of agriculture, was brought together by the Northwestern Energy Efficiency Alliance (NEEA) to explore the development of data exchange standards for irrigation. In late 2013, the development effort was moved into AgGateway (www.aggateway.org), a nonprofit consortium of about 240 companies dedicated to the implementation of standards for Agriculture. This move led to the chartering of the PAIL project by AgGateway's Precision Agriculture Council in early 2014. The companies participating in the project became known as the PAIL team.

The goal of the PAIL project is to develop industry-wide standards that will enable the exchange and use of data from different irrigation management systems. Data are currently stored in a variety of proprietary formats and each company is responsible to bear the cost and effort for making an exchange. The PAIL project seeks to develop a common language that can enable data exchange and, in the process, begin addressing the integration of technical tools for evidence-based, irrigation management.

The PAIL project covers a wide range of data topics, which can be organized into two broad categories: operations and observations.

- **Observations** are the field, atmospheric, plant, or other in situ measurements that apply to irrigation management. Data collection tools include weather stations, soil moisture sensors, or crop-related sensing. This work is based on, and extends, the ISO19156 standard for observations and measurements (International Organization for Standardization, 2011).
- **Operations** are all activities associated with the application of water with an irrigation system. Activities include, but are not restricted to, management-level communications and record-keeping. The operations data set is based around a "Recommendation", which describes a suggested course of action; a "Work Order," which describes a desired course of action; and a "Work Record," which describes the action that occurred. This work is based on, and extends, the ISO11783-10 standard for communications between agricultural machinery and FMIS (International Organization for Standardization, 2015).

There are several deliverables that will come from the PAIL project. Of those, five are important for this paper.

- **User Stories** (Jeffries, 2001) and Use Cases (Jacobson, 1992) that describe, in a semi-structured way, the typical management scenarios involving the exchange of data. User stories and use cases effectively define the scope of the standard.
- **Process Models / BPMN Diagram**s (von Rosing et al., 2015) that represent the different processes performed by actors in irrigation field operations. Explaining Business Process Modeling Notation (BPMN) is beyond the scope of this paper, but there are two aspects relevant to PAIL. The first is that BPMNs are based on a business process, that is, the management process as seen from the perspective of a farmer, whose goal is to operate as a profitable enterprise. The second element is that the process of building the BPMN results in identification of a set of messages (and data thereof) that define the communications that occur during irrigation management.
- **A field trial** (or "beta-test") that serves to expose potential conflicts or shortcomings of the standard. The trial also serves as a demonstration of the standard's value to potential adopters. The PAIL team conducted a trial during 2015 and is performing a second in 2016.
- **The XML Schema** (Fallside and Walmsley, 2004) is the primary technical deliverable. The schema contains a structured and unambiguous definition of data and its format.
- **A U.S. National Standard**, submitted to ASABE. A standards project, X632, is already in progress in the ASABE irrigation management committee, NRES-244. Drafting of this standard is underway and submission for balloting is expected in late 2016. This ASABE standard will subsequently be submitted to the International Organization for Standardization (ISO) as a new work item proposal.

Design Goals

The PAIL team applied several guiding principles during the design of a data exchange standard. These principles reflected the needs of individual companies in PAIL and project goals as a whole. At each point during the development process, where critical design decisions emerged and multiple solutions were available, the design principles guided the team's decisions. The principles were not set in stone from the start of the project. Instead, they emerged as each member contributed to the development and expressed their individual needs. Each guiding principle is described below.

Simple Beats Clever

On the surface this may seem like a different flavor of "KISS," but the intention is subtle. When formatting data, it is often possible to express the same thing in multiple ways. Some ways may be more practical for one domain than another. There is a temptation to find a clever way to include both ways in the same data format. However, having more than one way to express the same thing creates added burden for consumers of data. Wherever possible, PAIL chose simple solutions over those that are ingeniously comprehensive.

Small Packets

Data relevant to irrigation move through a variety of transport systems. Cell modem, sat-phone modem, mesh network radios, spread spectrum, radios, and direct machine-to-machine communications are all relevant. Some of these mediums (e.g. machine-to-machine via internet) have robust bandwidth capability, but many do not (e.g. sat-phone service billed by the byte). The low-bandwidth systems are just as important as the high-bandwidth, so the PAIL standard must be suitable for bandwidth-constrained applications. To that end, the schemas strive to minimize the size of the data packets to the greatest extent possible.

Make It Useful for Consumers of Data

It is often convenient for producers of data to send "everything" to data consumers, especially if data is sent electronically over the Internet. However, the consumers can be overwhelmed and miss key data they need, or spend unnecessary time looking for it. When transferring data to a consumer, the producer should include only reference data that is necessary for a consumer to complete a desired transaction.

JSON Friendly

The PAIL schemas are expressed as XML Schema Definition documents. This implies that all PAIL documents will be XML documents. However, while XML is a mature language, it is not the only document formatting language available. RESTful APIs have become the mechanism of choice for many web-based platforms. XML and JSON are, in general, compatible formats. However, there are some ambiguities regarding how to interpret certain XML schema structures into JSON. AgGateway has established some guidelines to prevent these ambiguities when translating XML to JSON. PAIL has followed these guidelines wherever possible.

Use Compound Identifiers

The Compound Identifier is a construct originally developed in AgGateway's ADAPT group (AgGateway, 2016). These objects provide a locally-scoped unique identifier that enables the use of objects by reference. More detail on compound identifiers is provided in the Identity section below.

Paper Overview

In this paper, we present the core elements of the PAIL project, the business processes those elements were derived from, and an introduction to the data structures defined in the standard. The intended audience is both engineering research professionals who will review the standard, and practitioners who will ultimately implement the standard. This paper will enable interested persons to decide if the PAIL standard can help their organizations serve the irrigation industry and, ultimately, the irrigators themselves.

Actors, User Stories, and Core Documents

Development of the PAIL data standards began by eliciting knowledge about the needs of various "actors" in irrigation: growers, their farm staff, consultants, and service providers. The PAIL team initially represented the various actors' needs and perspectives using "user stories" (Jeffries, 2001). The team also represented the data they record and exchange during irrigation operations through a set of "core documents."

Actors

The planning, executing and recording of irrigation events typically involve several people. Of course, an individual can assume multiple responsibilities, so the actors are best seen as *persons* occupying one or more *roles*. The PAIL standard identifies these actors in Table **1** below.

Table 1. Actors in the PAIL Data Flow

Actor	Description
Grower	Has authority to make decisions for all aspects of the farm. Develops a *Crop Plan* (core document) to convey what crops will be grown, and when, on which fields. Creates *Work Orders* (core documents) out of *Recommendations* (core documents) received from the Consultant.
Consultant	Has expertise to recommend how fields should be irrigated throughout the growing season, or over multiple seasons. Reviews the Grower's *Crop Plan*. Uses data from field equipment, such as soil sensors and field weather stations, to support the recommendation process. Requests and receives data from offsite Data Providers. Integrates all relevant data to create an irrigation *Recommendation* (core document) for the Grower.
Irrigator	Performs tasks related to irrigating one or more fields; i.e., performs the actual irrigation field operation. Uses a *Work Order* (core document) received from the Grower or Consultant to initiate, run, and end an irrigation operation. May make a preemptive change in a work order; for example, if a rain event occurs the irrigator may suspend or halt an irrigation operation.
Data Provider	Collects, stores and makes available various forms of *Observations and Measurements* (O&M, core document) data. Collects and stores proprietary irrigation operation event data. Derives *Work Records* (core documents) from the irrigation operations event data, and makes them available to the Grower Note: The tasks described above could be performed by more than one Data Provider. For example, the irrigation operations data could be handled by one provider, the weather data sourced by another, and the soil water data by yet another.

User Stories

User stories provide the PAIL team a high-level set of development requirements.

Table 2. PAIL User stories

Phase	As a/an	I want to ...	So that I can ...
Planning	Grower	create a Crop Plan.	communicate my intentions for one or more growing seasons.
	Consultant	review the Crop Plan to know what crops will be planted and how they will be grown.	make irrigation recommendations based on the grower's goals.
	Consultant	retrieve soil moisture, field weather and other field scouting data.	integrate it into my data analysis and recommendation to the grower.
	Data Provider	retrieve, store and organize field, weather and other relevant data.	send requested data to an authorized user.
	Consultant	retrieve derived weather data from a weather data service provider.	integrate it into my data analysis and recommendation to the grower.
	Consultant	create a Recommendation.	can advise the grower with a seasonal irrigation work plan.
	Grower	review the Recommendation from my consultant.	ensure it is consistent with my farm practices and current conditions.
Execution	Grower	create an irrigation Work Order.	be sure the Irrigator knows how much water to apply and where to apply it.

	Irrigator	use the irrigation Work Order to send a command to the irrigation system controller.	begin and end the irrigation as planned, or modify as field conditions change.
	Data Provider	store a Work Record of what happened during an irrigation event.	provide a record as requested from an authorized user.
Reporting	Consultant	retrieve a Work Record of an irrigation event.	use the data as input for the next irrigation Recommendation.
	Grower	store and retrieve a Work Record.	use it as input for planning next season's crops and field operations, and provide reports, as necessary, to regulators and/or insurance providers.

AgGateway's Core Documents for Field Operations

Growers currently face increasing pressure to document their field operations (e.g., irrigation, crop nutrition, crop protection), both for regulatory and commercial reasons. AgGateway's Core Documents for Field Operations support these activities and provide a common set of communications among Growers, Irrigators, Consultants, and Data Providers. In summary, the grower plans how to grow a crop, and then enters a cycle where observations and measurements are made about the state of the crop, an expert recommends a course of action, the grower (or an agent thereof) decide what course of action to take, the action is taken, the results are recorded, and the cycle begins anew. A grower may have a similar interest for the purposes of establishing production costs and the cost-effectiveness of specific agricultural practices.

More formally, the Core Documents (enumerated in Table 3) define data that can be exchanged during specific processes associated with a field operation. The definitions are quite flexible because of the myriad of ways growers implement their record-keeping in response to regionally-specific regulatory requirements, market characteristics or farming operations, and personal preference.

Table 3. Core Documents

Document Name	Abbr.	Type	What It Conveys	Actor Involved
Crop Plan	Plan	Strategic	A high-level document describing how a crop will be grown on a given piece of land during a crop season. "This is how we're going to grow this crop this season."	Grower, or other actor involved in the strategic planning for the field operations.
Observations and Measurements	O&M	Tactical/ Predictive	A document containing data measured/observed in the field. "This is what's happening (or what we think might happen) in the field."	Crop scout, remote observation or a person tasked with monitoring conditions in the field.
Recommendation	REC	Tactical	"This is what I recommend we should do" This document is not always acted upon; it is acted upon via a work order, upon approval.	An individual, such as a consultant or agronomist, with the expertise / licensing necessary to recommend a course of action.
Work Order	WO	Tactical	"This is what we are going to do."	An individual with authority to order the work done.

Work Record	WR	Tactical/ Historical	"This is what we actually did in the field."	May be automatically generated; otherwise, an operator that performed the task.
Supporting Documents				
Reference Data and Setup File		All	"This is the common information we need to set up and support accurate and efficient data exchange."	Grower, or other actor involved in managing the grower's production data.

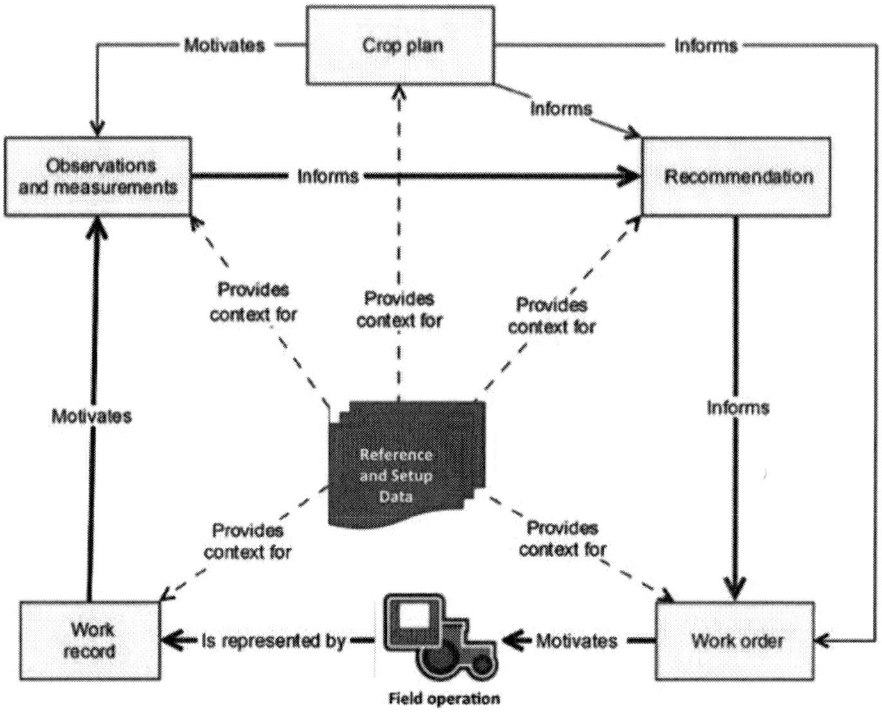

Figure 2. AgGateway Core Documents for Field Operations. The diagram in Figure 2 shows the relationships among the core documents.

- The Crop Plan informs or motivates the other documents.
  - Example: a crop plan defines an irrigation water quota available to a given field; this quota informs the Recommendation of whether to irrigate or not on a given day.
- Observations and Measurements inform Recommendations.
  - Example: soil water content measurements indicating the need to irrigate.
- Recommendations inform Work Orders.
  - Example: a consultant recommends irrigating because a corn crop's anthesis will happen soon.
- Work Orders motivate Field Operations.
  - Example: A grower purchases crop protection products from a retailer and requests their application.
  - Example: A grower communicates to an operator (irrigator actor) that a field must be irrigated with a certain depth of water over a certain period.
- Field operations are represented by Work Records.
  - Example: A telemetry system installed on a center pivot summarizes and reports data about the application of water on the field on a given day.
- Work Records motivate Observations and Measurements.

- Example: A crop scout goes out to the field to determine whether there are still symptoms of water stress in a crop following an irrigation operation.

Core Documents Flow

The previous section described the Core Documents and the relationships among them. In this section, an example is provided of the exchange of core documents as part of a Grower's business processes (Figure 3).

- The Grower shares the Crop Plan with an Agronomist and an Irrigation (O&M) service.
- The Grower shares a historical record of Work Records and O&M with the Agronomist.
- The Agronomist makes a recommendation ("Irrigation Plan") informed by the Crop Plan, the historical record, and fresh O&M.
- The Grower, informed by the Recommendation, orders a course of action through a work order ("Irrigation prescription") sent to the pivot panel, which executes the field operation.
- The Work Record ("Irrigation record") is returned to the grower (e.g. through a web service associated with the pivot's telemetry system.)
- The Grower processes the Work Record, creating a report shared with a regulator or value partner (e.g. a banker).

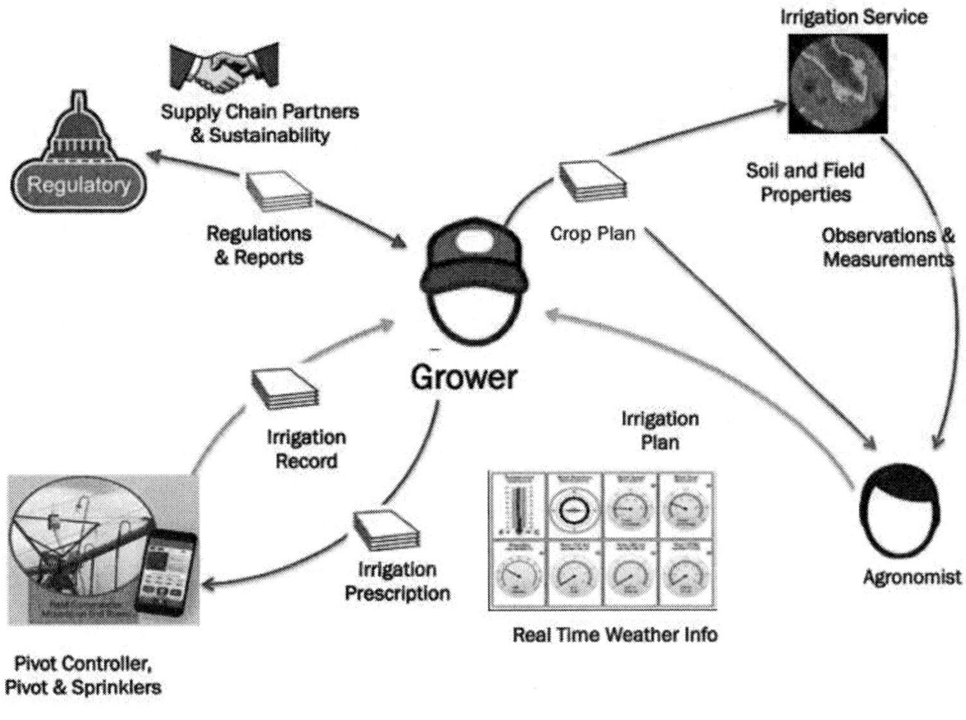

Figure 3. An example of the Exchange of Core Documents as part of a Grower's business processes.

Business Process Models

Figures 4 and 5 formalize the ideas shown above, bringing actors, Core Documents and relationships together in the context of formal processes. For clarity, Operations (Creation of Work Orders and Work Records) have been placed in Figure 4, and Observations (Procurement and use of O&M, Creation of Recommendations) have been placed in Figure 5.

As mentioned earlier, a detailed description of BPMN is out of the scope for this paper. A quick introduction supported by the key in Figure 7 should be sufficient to understand the following diagrams.

Different actors are represented by the rectangular horizontal *pools* in the diagram.

- The processes carried out by each actor are contained in the corresponding actor's pool.
- Processes begin, end, and sometimes are paused by *events*, shown as circles in the diagrams.
- There are different kinds of events, triggered by time (shown with a clock-face icon), receiving a *message* (shown with an envelope icon), or by a *rule* being met.
- Communication among pools happens through *messages*. Note that some of those messages correspond to Core Documents.
- The flow of a process can fork, depending on the outcome of an activity. The places where flow diverges (and converges) is shown with *gateways* (diamond shapes). The PAIL diagrams of Figures 5 and 6 only show a kind of gateway called "Exclusive-OR", where the divergent outcomes are mutually exclusive (i.e. only one outcome is possible in any given situation).

Figure 5 shows five different processes involved in irrigation operations.

- Grower creating a work order (from a received Recommendation) and sending it to the Irrigator.
- Grower requesting work records from a Data Provider and storing them in an FMIS.
- Irrigator executing a Work Order received from the Grower.
- Data Provider storing event data received during the execution of the field operation.
- Data Provider assembling work records from stored event data, and sending them to the Grower upon request.

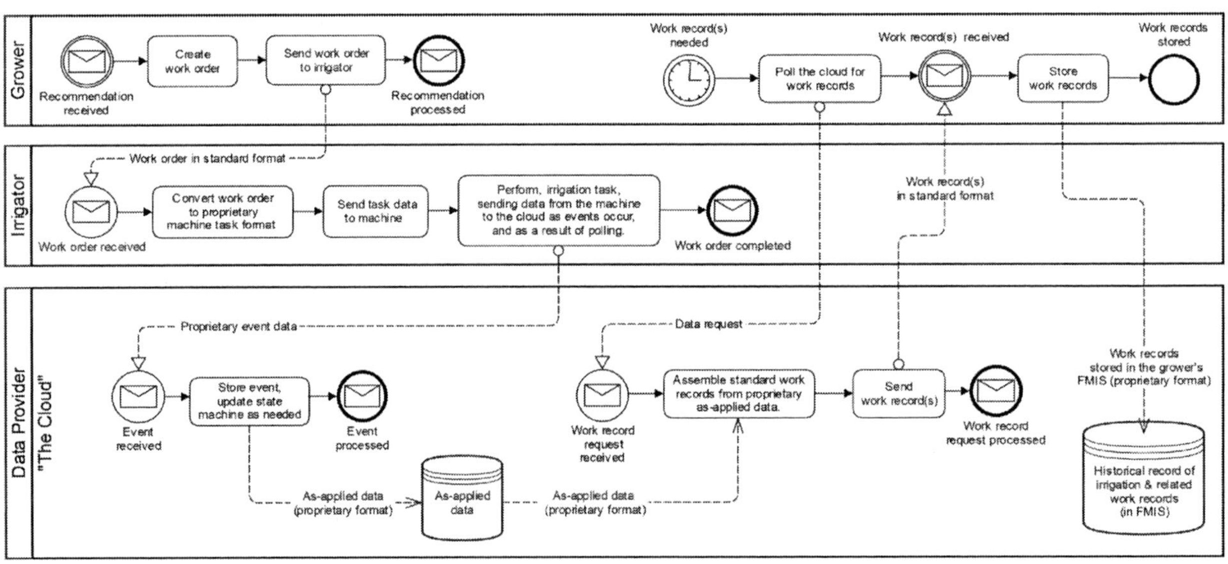

Figure 5: BPMN diagram for Operations.

Figure 6 shows three different processes involved in irrigation observations (in addition to the repeated first process above).

- Grower shares Crop Plan with Consultant, kicking off the Recommendation-creation process.
- Consultant starts season upon receipt of Crop Plan, enters a loop of requesting data from Service Provider(s), using it to create a Recommendation, and sending that to the Grower loop executes until end of season.
- Data Provider honors requests for Observations & Measurements data.

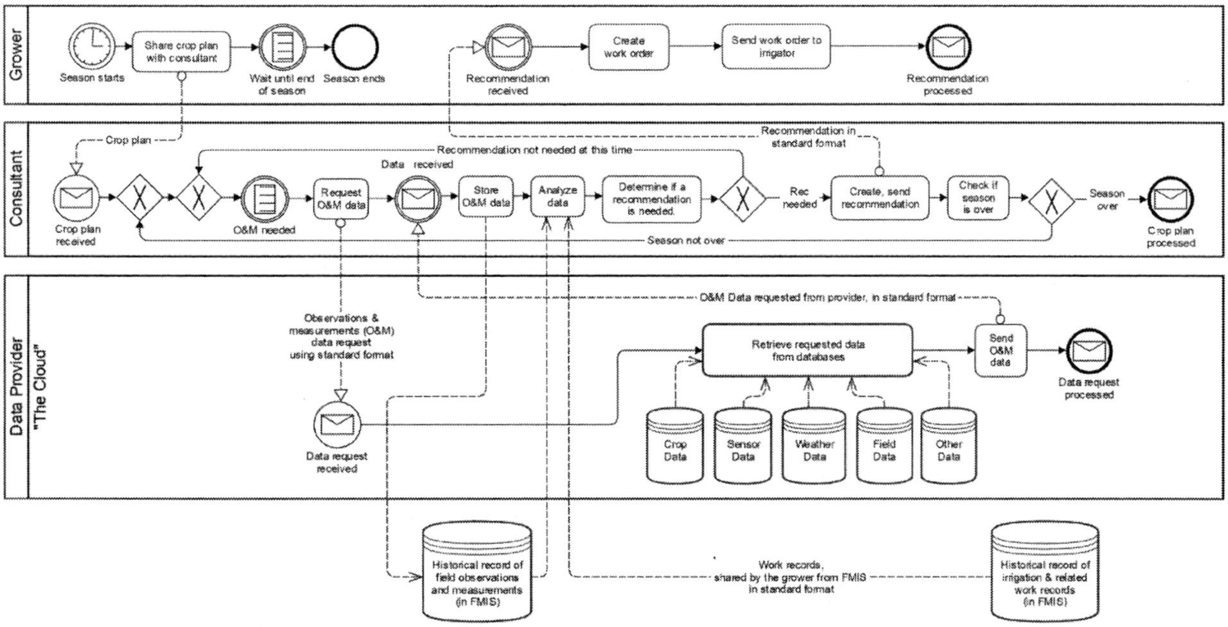

Figure 6: BPMN diagram for Observations.

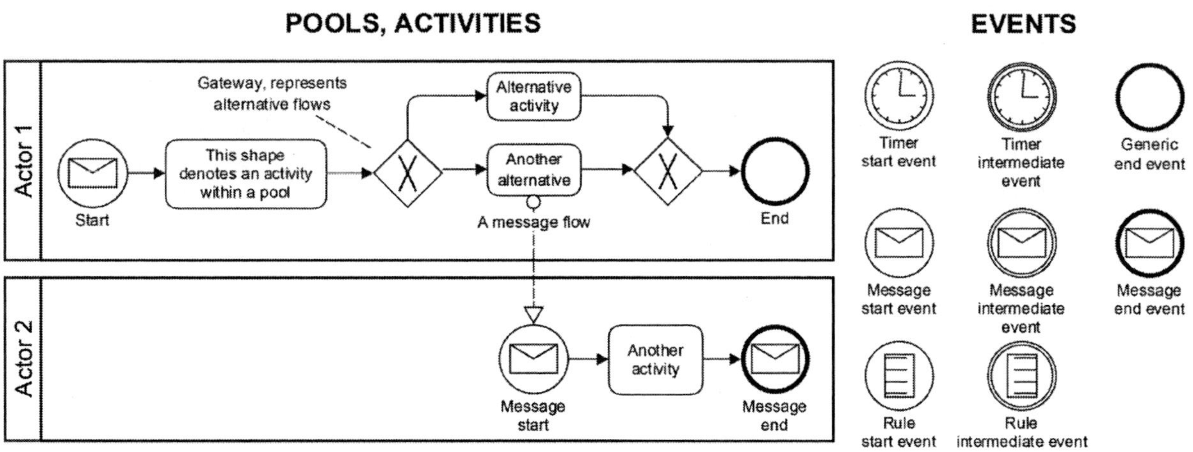

Figure 7: Key to interpret the symbols used in the BPMN diagrams shown in Figures 5 and 6.

PAIL Data: Basic Concepts

Identity

Many objects specified by the proposed PAIL standard are used *by reference* in other objects (for example, a grower, farm and field may be referenced in a work order) and thus need identifiers that can be used by the referencing object. A Unified Modeling Language (UML) class diagram (International Organization for Standardization, 2005) is the mechanism used by PAIL (and other AgGateway precision agriculture-themed standards work) to do this referencing among objects (Figure 8). It centers on an object class called CompoundIdentifier, which provides objects with a simple integer identifier (the ReferenceIdentifier) for use in the local scope of any instance of a data model, and allows associating an arbitrary number of (optional) unique identifiers (the list of UniqueIds) to that ReferenceId.

Each UniqueId, in turn, can be of four different types:

- A Universally Unique Identifier, or UUID (Leach et al., 2005).
- An arbitrary string (to accommodate proprietary alphanumeric identifiers)
- A long integer (to accommodate proprietary integer identifiers)
- A uniform resource identifier, or URI (W3C/IETF, 2001).

Time

Accurately capturing the time at which various events happen is an important part of agricultural record keeping. This is particularly true in irrigation, where water volumes are frequently calculated as a flow rate (e.g., in gallons per minute) multiplied by a duration. The documentation of an event time uses a simplification of the TimeScope used in AgGateway's ADAPT toolkit (AgGateway, 2016). The simplification consists of two timestamps, a required Context attribute that specifies the meaning of the TimeScope through an enumerated vocabulary (not shown), and an optional human-readable Description.

Reference, Setup, and Configuration Data

Reference and Setup Data as providing context to the Core Documents is shown in Figure 2. Their role is explained in greater detail in Figure 8.

Reference data refers to information that a manufacturer makes available for the purchase, setup and/or use of their products, and pertains to *all instances* of a manufacturer's equipment and/or product and product components; i.e., reference data is not grower-specific or specific to an individual sale or single instance of a thing. For example, the product name, EPA number and active ingredients are reference data for a crop protection product, but a lot number is not. In another example, the model and series number are reference data for a center pivot irrigation machine, but the serial number is not.

The intent is to share reference data sets across the whole industry so that different stakeholders can interpret shared documents the same way. This includes names and identifiers of seed varieties, crop protection products, active ingredients, etc. AgGateway has several teams working to create reference data sourcing infrastructure for the industry. (AgGateway, 2015A).

Setup data provides information needed to set up data exchange between the grower and machinery or other actors (e.g., crop advisors.) It refers to two categories of information. Unlike Reference Data, Setup data is grower-specific. The two types of setup data include grower data and configuration data.

Grower Data represents basic information about the grower, farm, fields, and actors. This may include farm names, field boundaries, the specific products the grower has a permit to use, etc.

Configuration Data specifies the state of specific instances of things such as farm equipment and instruments (e.g. soil sensors, irrigation pivots, combines, etc.) This may include their location, what they are connected to, who installed them, etc.

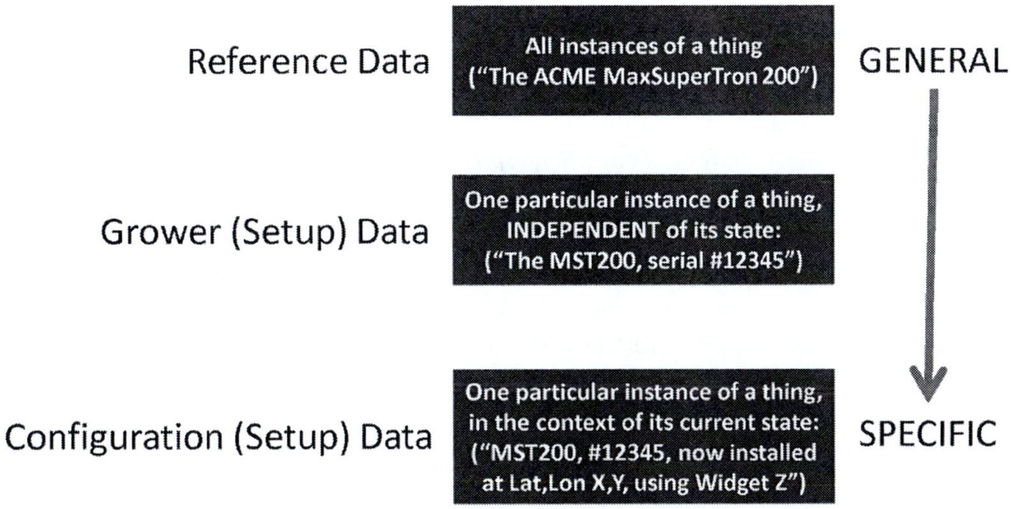

Figure 8: Reference, Setup, and Configuration Data

Data Pedigree

In support of the interpretation of represented data, the PAIL team included functionality for specifying the origin of critical information such as time and location, as well as to specify how the system handles setup data represented in a data file.

- LocationDataSource: Was the location GPS-derived? Was it obtained mechanically (e.g., through an encoder) or estimated? This is important when interpreting data from irrigation equipment such as a center pivot, where the quality assurance procedures to use for different sources of position data might vary (e.g., ensuring accurate GPS-derived positions may require trimming / removing trees that may obscure the sky near the edge of a field, whereas ensuring accurate mechanically-derived positions may require ground-truthing the accuracy of the reported azimuth of a pivot.)
- TimeDataSource: Were the recorded times derived from a GPS? Were they server-mediated when an event was uploaded by the telemetry system? This knowledge is important because the latter option is susceptible to introducing event timing errors under conditions of telemetry system communication errors, whereas the former is not.
- SetupDataPedigree: Is the system keeping track of changes in setup data and reporting the corresponding time series of setup information along with the communicated data, or is it only keeping track of the latest setup? This has important implications for a user: in the latter case, the user would need to access data often to keep accurate track of changes in setup (such as the length of a center pivot) that may affect the meaning of reported data.

The intent of recording this information is not in any way prescriptive; while it is undoubtedly more convenient for a user to have the most accurate and complete options available for these kind of data, there are many legacy systems installed that produce valuable data; the purpose of the pedigree data is to provide the consumer of PAIL data files with valuable information for interpreting exchanged data.

Documents

As architected by AgGateway's SPADE (AgGateway, 2015) and ADAPT (AgGateway, 2016) teams, the five Core Documents mentioned earlier share most of their attributes. Specific details about the different attributes in the Document-derived classes are outside the scope of this document; for the moment, it is enough to note that they answer the following questions:

- **What**: The products or services being applied, or the data being reported.
- **Where**: Grower / Farms / Fields / Cropzones / GPS locations.
- **Who**: People involved and their role: operator, agronomist, trucker, etc.
- **When**: When should / did the operation happen?
- **How**: Product rates, equipment settings, etc.
- **With What**: What equipment is involved?
- **Context items**: A generic system to encode geopolitical-context-dependent information such as (for the US) FSA, EPA, DOT numbers, harvested commodity codes and other geography-specific data that growers must track for insurance and other purposes.

It should be noted that the actual PAIL implementation does not include the abstract Document class. Consultation with developers on the team suggested that implementing the individual child classes separately in the schema (as opposed to extending a Document data type) was in line with the "simple beats clever" approach discussed earlier, and desirable for their production environments.

The Draft Standard and the Schemas

The draft standard (ASAE X632) being proposed by the PAIL team has three parts (a fourth, pertaining to pumps, is in development). Each part includes an annex with a data schema covering the data presented in that part of the standard, as follows:

Part 1: **Common elements** is meant to be used throughout the rest of the standard. They include definitions, business process models, core concepts, product reference data, and setup data.

Part 2: **Operations** include Recommendations, Work orders and Work Records (Plan is out of scope in the first version), and irrigation-equipment-specific, reference data.

Part 3: **Observations** include Observations & Measurements (O&M) and their corresponding Reference (e.g., sensors, loggers, codes for features of interest), and setup data.

Discussion

Development philosophy

The PAIL project has sought to develop a common language that enables integration of multiple, disparate technical tools and sources of water management data. Working with a large variety of tools and data sources requires more expertise than any one discipline or entity can provide; it requires a collaborative approach. In "The Cathedral and the Bazaar," Raymond (2001) describes two philosophies of software development:

- The Cathedral is essentially the traditional academic approach where a group of experts and thinkers apply their substantial knowledge to a problem, test it, and deliver it to the expected consumers via publications, seminars, and classes. This approach has its benefits. The cathedral can produce solutions that are cohesive, clearly scoped, and well-founded in research. The disadvantage is that these solutions do not always accommodate the practical realities of the practitioners. This problem usually emerges from a desire to avoid complexities that would complicate an otherwise simple conceptual framework or when the complexities are caused by issues unrelated to the application domain. Those omissions are often perceived by practitioners as a lack of understanding of real-world conditions and leads to the "Ivory Tower" perception of academic solutions.

- The Bazaar is an open approach where anyone can participate (within bounds of reason). Participants are expected to contribute and the major impacts come from those who do most of the work. The Bazaar approach is messy, slow, and often contentious. However, the Bazaar has a significant advantage. The result is a product the practitioners need. The nature of participatory development means that, by the end of the development cycle, practitioners have already adopted the new system. This contrasts with the Cathedral approach, where motivating adoption is the critical and last step of the development process.

PAIL's development has followed the bazaar model. Any corporation or individual can join AgGateway and participate in the development of a standard. As of this writing, the development process has gone on for nearly three years and by the time the standard is released, several companies will have already adopted an earlier version. Those companies are the same ones that helped develop the standard.

A vehicle for research

PAIL can also provide value to the research community. Many decision-support system (DSS) tools are developed by researchers with the intention of providing growers an easier way to implement robust management practices. These DSS tools incorporate advanced analytical methods and often include field validation that demonstrates their potential for resource conservation, improved efficiency, or greater profitability. A problem is that the tool itself, however, is typically developed by a graduate student whose field of study is not interface design or software engineering, and who does not necessarily use robust industry standard practices for software development. This lack of standard practices in development becomes an obstacle to industry adoption. Additionally, when the student graduates, development stops and does not continue unless the principal investigators can find additional funding. The end result is that the DSS tool will "sit on a shelf collecting dust", be perceived as no longer in active development, and be abandoned by users.

Grant-driven research is not an optimal framework for developing and maintaining applied, production oriented technical tools. These tools require customer support, continual debugging, and a commitment to evolving software for customer's needs. Commercial development is geared towards those needs and software companies are successful because they provide those services effectively.

Standardized interoperability provides a means for researchers to deliver research products, in the form of DSS tools, without the burdens associated with maintenance and customer support. The DSS tools can be written to interact with the interfaces or data formats defined by the PAIL standard, freeing the researchers (and the graduate students) from the need to build, maintain and support a "user friendly" interface. Instead, companies can integrate the DSS tools into their products and focus on providing the user interface and customer support. Thus, the PAIL standard is a means to deliver the benefits of research to growers without the burden of continually requiring funding to support maintenance.

There is another research-oriented aspect that is an indirect consequence of the bazaar model of development. Companies that drive PAIL's development are focused on providing services that are needed now or in the near-future. To be useful, the standard must be relevant to current practices. Research, on the other hand, is focused on developing new tools, which may require data or concepts not yet in use by industry. Because the standard is focused on current practices, it could conceivably not have sufficient constructs to support new tools. Significant effort was made to develop a standard that is generic enough to avoid these conflicts but no standard can account for every eventuality. When a researcher encounters a situation where a new tool cannot be expressed in PAIL, this is an indirect indication of a significant incompatibility with current practices. Such an incompatibility will motivate a researcher to educate the industry, propose a change in a practice, or suggest a modification to the standard itself.

A framework where irrigation is integrated with other field operations

The ISO11783 data format (International Organization for Standardization, 2015) is commonly used to represent planned tasks (i.e. work orders) and actual tasks (i.e. work records) for the field operations of planting, tillage, crop protection, crop nutrition and harvest. It is not commonly used in irrigation for reasons that can be found in the format documentation. For example, ISO11783 cannot easily accommodate the radial geometries inherent in center-pivot systems. Also, the ISO11783 format is complex to understand and pervaded by tradeoffs (such as avoiding the representation of floating-point arithmetic). The ISO11783 format, while appropriate at its time of conception, is not necessarily useful today. The PAIL standard, on the other hand, is highly aligned with the new ADAPT object model (AgGateway 2016) which retains backward compatibility with ISO (for the benefit of the previously-mentioned field operations). It provides a richer, business-process-oriented semantics. The ADAPT design allows irrigation to coexist with the other operations as part of a grower's business process. In the context where growers must comply with complex regulatory requirements as a cost of doing business (e.g. reporting on crop nutrition products applied through irrigation), this alignment with ADAPT is likely to be very advantageous for growers.

Conclusion

The goal of the PAIL project is a data exchange standard that creates a "common language" for irrigation technology. The standard will promote adoption of evidence-based management practices by making technical tools easier to integrate into a farm enterprise.

The PAIL team developed the standard by first creating process models, using Use Cases, User Stories, and BPMN diagrams, to describe irrigation management. The process models were created from a grower's perspective and represent management practices as they are now rather than an idealized version. Based on these process models, the team designed a robust data model that incorporates all relevant data flows and messages. The data model is rendered as an XML Schema, which will be available publicly with the publication of the standard. Two field trials have been undertaken to validate the efficacy of the standard and to demonstrate its utility in actual irrigation settings. Trial results can contribute to the documentation of the standard and be the basis for training materials.

The PAIL data exchange standard will be submitted to ASABE for balloting in Q3/Q4 of 2016. Once accepted by ASABE, the PAIL standard will also become an ISO standard. Development on the PAIL data exchange standard will continue even after it is recognized by an international standards organization. The PAIL team is currently working on additional sections that cover drip irrigation, pumping systems, and testing. Any person interested in participating on the PAIL team should contact member.services@aggateway.com.

Literature Cited

AgGateway. (2015). "The Spade Project." *The Spade Project*, PDF, , <http://s3.amazonaws.com/aggateway_public/AgGatewayWeb/About%20Us/CommunicationsKit/AgGatewaySPADE3_11415.pdf> (Jun. 13, 2016).

AgGateway. (2016). "The ADAPT Toolkit: Implementing Interoperability in Precision Agriculture." *The ADAPT Toolkit: Implementing Interoperability in Precision Agriculture*, PDF, , <http://s3.amazonaws.com/aggateway_public/AgGatewayWeb/About%20Us/AgGateway_ADAPT_Toolkit_5616.pdf> (Jun. 13, 2016).

Alexandratos, N., and Bruinsma, J. (2012). *World agriculture towards 2030/2050: the 2012 revision*. ESA Working paper, FAO.

Baulcombe, D. (2010). "Reaping benefits of crop research." *Science*, 327(5967), 761–761.

Colorado State University. (n.d.). "Agricultural Water Conservation Clearinghouse." <http://www.agwaterconservation.colostate.edu/ToolsWeather.aspx> (Jun. 6, 2016).

Fallside, D. C., and Walmsley, P. (2004). "XML schema part 0: primer second edition." *W3C recommendation*, 16.

GS1. (2014). "Global Location Number (GLN)." *Global Location Number (GLN) | GS1*, Text, , <http://www.gs1.org/gln> (Jun. 13, 2016).

Hillyer, C., and Robinson, P. (2010). "Envisioning The Next Generation Of Irrigation Schedulers." *5th National Decennial Irrigation Conference Proceedings*, ASABE, St. Joseph, Michigan, Phoenix, Arizona.

International Organization for Standardization. (2005). *ISO/IEC 19501:2005 Information technology -- Open Distributed Processing -- Unified Modeling Language (UML) Version 1.4.2*. International Organization for Standardization, Geneva, Switzerland.

International Organization for Standardization. (2011). *ISO 19156:2011 Geographic information -- Observations and measurements*. International Organization for Standardization, Geneva, Switzerland.

International Organization for Standardization. (2015). *ISO 11783-10:2015 Tractors and machinery for agriculture and forestry -- Serial control and communications data network -- Part 10: Task controller and management information system data interchange*. International Organization for Standardization, Geneva, Switzerland.

Jacobson, I. (1992). *Object-oriented software engineering: a use case driven approach*. ACM Press ; Addison-Wesley Pub, [New York] : Wokingham, Eng. ; Reading, Mass.

Leach, P. J., Mealling, M., and Salz, R. (2005). "RFC 4122 - A Universally Unique IDentifier (UUID) URN Namespace." *IETF Tools*, <https://tools.ietf.org/html/rfc4122> (Jun. 13, 2016).

Raymond, E. S. (2001). *The cathedral and the bazaar: musings on Linux and Open Source by an accidental revolutionary*. O'Reilly, Beijing; Cambridge, Mass.

von Rosing, M., White, S., Cummins, F., and de Man, H. (2015). "Business Process Model and Notation—BPMN." *The Complete Business Process Handbook*, Elsevier, 429–453.

Schaible, G., and Aillery, M. (2012). "Water conservation in irrigated agriculture: trends and challenges in the face of emerging demands." *USDA-ERS Economic Information Bulletin*, (99).

Smith, R., Baillie, J., McCarthy, A., Raine, S., and Baillie, C. (2010). *Review of precision irrigation technologies and their application*. A Report for National Program for Sustainable Irrigation, University of Southern Queensland, National Centre for Engineering in Agriculture.

Taylor, K. (2009). "Final Report of the 'Scheduling irrigation for commercial agricultural growers within the El Dorado Irrigation District using permanently placed soil moisture sensors' Project."

TruePoint Solutions. (2008). "TrueISM." <http://www.truepointsolutions.com/trueism.htm> (May. 28, 2010).

USDA. (2012). *Farm And Ranch Irrigation Survey (2013)*. National Agricultural Statistics Service, Washington, D.C.

W3C/IETF. (2001). "URIs, URLs, and URNs: Clarifications and Recommendations 1.0." *World Wide Web Consortium (W3C)*, <https://www.w3.org/TR/uri-clarification/> (Jun. 13, 2016).

Wang, D., and Cai, X. (2009). "Irrigation Scheduling---Role of Weather Forecasting and Farmers' Behavior." *Journal of Water Resources Planning and Management*, 135(5), 364–372.

Managing Variable-Rate Irrigation Using NDVI

K. C. Stone, P. J. Bauer, and G. C. Sigua

Coastal Plains Soil, Water, and Plant Research Center,
US Department of Agriculture – Agricultural Research Service, Florence, SC, USA
** Corresponding author. Email: ken.stone@ars.usda.gov*

Abstract. *Variable rate irrigation (VRI) systems are capable of spatially allocating limited water resources while potentially increasing profits and conserving water. However, compared to traditional irrigation systems, VRI systems require a higher level of management. Delineation of management zones for spatial irrigation applications typically have been static through the growing season and has been based on grower's their past experience and knowledge of variability in their fields (soil types or soil EC). In this research, we investigated the use of static management zones and the potential use of dynamic management zones based on remotely sensed crop vegetative indices. The static zones were managed using soil properties and using an expert system. The dynamic zones were managed by using remotely sensing crop vegetative indices using Crop Circle NDVI sensors to calculate spatial crop coefficients. Initial results indicate that the vegetative indices varied throughout the field and could be used to spatially allocate water differentially.*

Introduction

Variable-rate irrigation (VRI) systems are irrigation systems that are capable of applying different water depths both in the direction of travel and along the length of the irrigation system. Spatial water applications attempt to overcome site-specific problems that include spatial variability in topography, soil type, and soil water availability. Irrigation management in some areas of the southeastern U.S. could benefit VRI because of the highly variable soils with low water holding capacities.

A widely used method of estimating irrigation requirements is the FAO-56 method (Allen et al., 1998), in which crop coefficients are used for determining the irrigation requirement of a crop over the growing season using reference evapotranspiration (ET_o) measurements. The FAO-56 method provides standard generalized estimates of the crop coefficients that may not be appropriate for every location, and it does not readily lend itself to VRI management. A potential method to estimate spatial crop coefficients than can be used in VRI systems is using remotely sensed canopy reflectance. Bausch and Neale (1987) proposed a concept for deriving crop coefficients from reflected canopy radiation. They plotted the seasonal normalized difference vegetation index (NDVI) and found that it resembled the seasonal basal crop coefficient curve. They reported that crop coefficients derived from NDVI were independent of time-based parameters such as planting date and effective cover date that are usually associated with traditional crop coefficients and that basal spectral crop coefficients were a real-time crop coefficient that permitted the crop to express its response to weather, management practices, and stresses. In a summary of vegetation index-based remote sensing for estimating crop coefficients, Glenn et al. (2011) reported that remotely sensed NDVI-based crop coefficients can help reduce agricultural water use by matching irrigation rates to the actual water needs of a crop as it grows instead of to a modeled crop growing under optimal conditions. These NDVI-based crop coefficients could also be used as a method of estimating spatial crop coefficients for scheduling spatial irrigation using a VRI system.

In this research, our objective was to evaluate and compare three irrigation management methods for their potential in managing VRI systems. The three irrigation management methods were (1) using remotely sensed crop vegetative indices to estimate crop coefficients, (2) using the Irrigator Pro for Corn expert system, and (3) using measured soil water potentials.

Materials and Methods

From 2012 to 2014, corn (*Zea mays*) was grown under conservation tillage on a 6 ha site under a VRI system near Florence, South Carolina. The soils (figure 1) under the center-pivot irrigation system are highly variable. Three irrigation treatments were evaluated for their potential utilization for spatial irrigation management using the VRI system. The first treatment was based on remotely sensing the crop normalized difference vegetative index (NDVI treatment) combined with a 7-day water balance, and irrigations were initiated when the SWP fell below -30 kPa. The NDVI treatment was used to estimate crop coefficients using methods similar to those used by Bausch (1993), Hunsaker et al. (2003), and Glenn et al. (2011). These estimated crop coefficients were used in the FAO 56 dual crop coefficient method for estimating crop evapotranspiration (ET_c) and irrigation requirements. Initially in 2012, the crop coefficients were based on the FAO 56 crop coefficients for field corn ($K_{cb\ ini} = 0.15$, $K_{cb\ mid} = 1.15$, and $K_{cb\ end} = 0.5$). After crop establishment and NDVI measurements were collected, the crop coefficients were updated and estimated by multiplying the NDVI measurement by a slope of 1.5. The second irrigation treatment was based on the Irrigator Pro for Corn expert system that was developed by the USDA-ARS National Peanut Research Laboratory (Davidson et al., 1998; Lamb et al., 2004, 2007). In this research, Irrigator Pro for Corn was implemented using spatial management zones corresponding to variable soil types. Irrigator Pro uses soil texture and soil water potential (SWP) measurements to estimate the soil water holding capacity in the root zone for water balance calculations. The third and more traditional irrigation treatment (SWP treatment) was based on using SWP sensors to maintain SWP values above -30 kPa (approx. 50% depletion of available water) in the top 30 cm of soils.

The irrigation system was a 137 m center-pivot irrigation system modified to permit variable application depths to individual areas 9.1 × 9.1 m in size (Omary et al., 1997; Camp et al., 1998). The center-pivot length was divided into 13 segments, each 9.1 m in length. For this experiment, the outer nine segments (segments 5 to 13) of each pivot quadrant were used for the three irrigation treatments in a randomized block design with three replicates per quadrant (with a total of 12 replicates for the entire pivot). A more detailed description of the water delivery system may be found in Omary et al. (1997) and for the control system in Camp et al. (1998).

Crop Management: No-til corn was planted in 76 cm rows in a circular pattern with a planting population of 79,000 seeds ha^{-1}. All nitrogen fertilizer, except preplant granular applications (25 kg ha^{-1} N), was applied via fertigation. Nitrogen (225 kg ha^{-1}) was applied through the pivot via fertigation in 2012 on 25 May (90 kg ha^{-1}), 31 May and 4 June (67 kg ha^{-1}). In 2014, fertigation applications were on 19 May (90 kg ha^{-1}), 4 June and 9 June (67 kg ha^{-1}). In 2013, fertigation applications were applied on 25 May (90 kg ha^{-1}) and 17 June (67 kg ha^{-1}). The total N applied via fertigation in 2013 was reduced to approximately 157 kg ha^{-1} due to pumping plant repairs and a management oversight.

NDVI and SWP Measurements: The NDVI measurements were collected over center rows for each pivot segment throughout the growing season at approximately two-week intervals until tasseling using a Crop Circle ACS-430 active crop canopy sensor and GeoSCOUT GLS-400 datalogger (Holland Scientific, Inc., Lincoln, Neb.). The mean NDVI values were calculated from the collected reflectance measurements and crop coefficients were calculated by multiplying by a slope of 1.5. In 2012, mean calculated crop coefficients were 0.41 (2 May), 1.01 (15 May), 1.08 (24 May), 1.19 (1 June), and 1.16 (8 June, post-tassel). For 2013, mean crop coefficients were 0.38 (14 May) and 1.03 (31 May). Due to a malfunction of the GPS, no additional NDVI readings were available in 2013; therefore, the FAO K_{cb} value of 1.15 was used. In 2014, mean crop coefficients were 0.30 (6 May), 0.42 (14 May), 0.92 (27 May), 1.07 (4 June), and 1.16 (12 June). Since we did not collect NDVI readings after tasseling, the last calculated crop coefficient was the midpoint crop coefficient ($K_{cb\ mid}$) until the late-season stage ($K_{cb\ end}$).

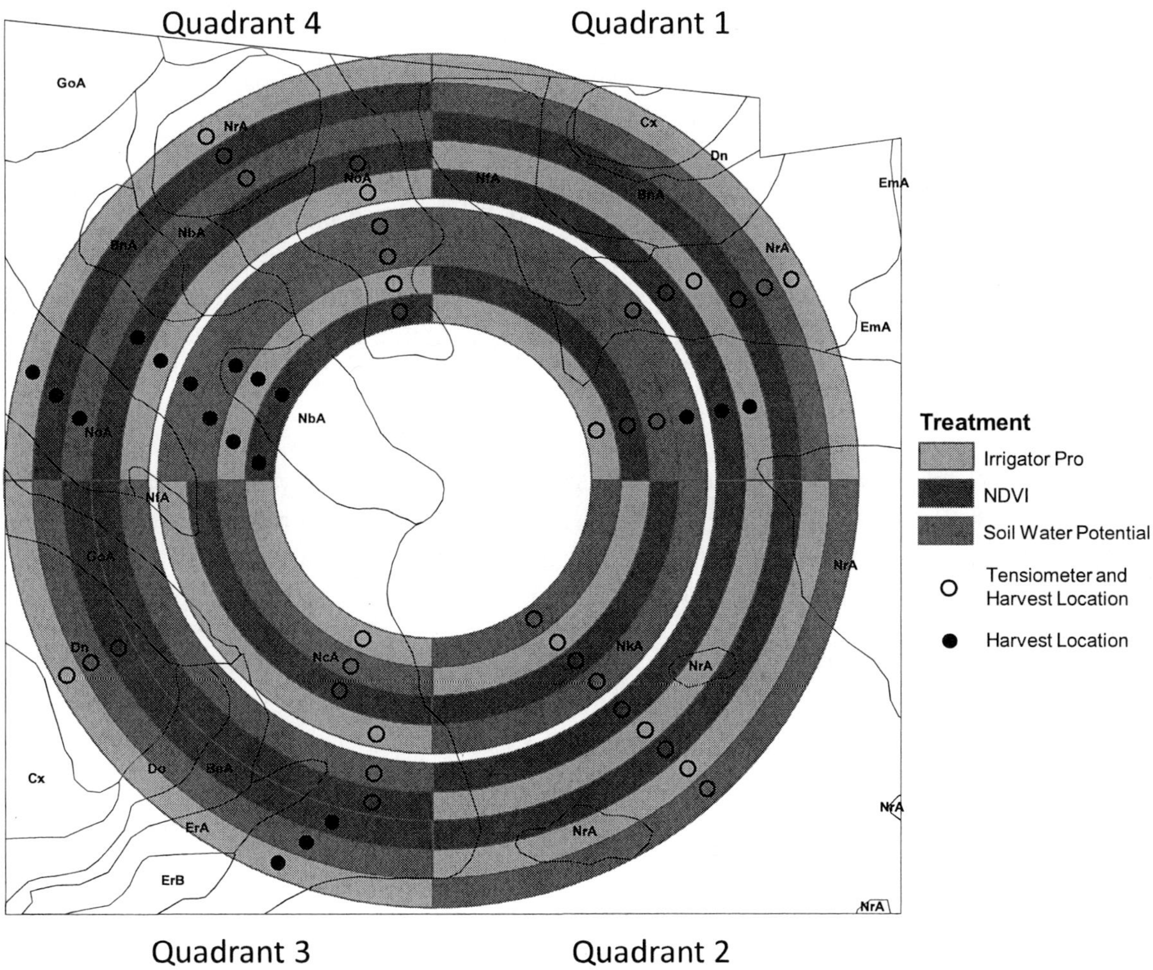

Figure 1. Plot map for the 2012-2014 irrigation study.

Soil water potentials were manually measured and tabulated at 36 locations (fig. 1) within the experiment. In each treatment and replication, tensiometers were installed in the predominate soil type within each plot at two depths (0.30 and 0.60 m). The predominate soil type in each plot was used to manage irrigation for the entire SWP treatment plot. Measurements were recorded at least three times each week. The 0.30 m tensiometers in the SWP and NDVI treatments were used to initiate irrigation applications. When the SWP of the SWP treatments decreased below -30 kPa, a 12.5 mm irrigation application was applied to that plot. Additionally, if the SWP decreased below -50 kPa, an additional 12.5 mm of irrigation was applied if the rainfall forecast was less than 50%. For the NDVI treatment plots, when the SWP decreased below -30 kPa and the 7-day calculated water balance (ET – rainfall) exceeded 12.5 mm, a 12.5 mm irrigation application was initiated. Irrigation for the Irrigator Pro for Corn expert system was initiated when the calculated available water in the soil was about 50% depleted. All irrigations were halted when the corn reached black layer each year.

Harvest Details: Corn grain yields were determined by weighing the grain harvested from a 6.1 m length of two rows near the center of each plot using a plot combine. A total of 54 yield samples were collected near the 36 tensiometer monitoring sites. Subsamples were collected from the plots and air-dried to obtain seed moisture content. Grain yields were corrected to 15.5% moisture. After yields and total water applied to each treatment were determined, the water use efficiency (WUE) was calculated by dividing the mean plot yield by the total water applied (irrigation + rainfall). The WUE values were reported in units of kg grain ha^{-1} mm^{-1} of water applied.

Statistical Analyses: All data were statistically analyzed in SAS (SAS Institute, Inc., Cary, N.C.) using Proc GLM. The experimental design was a randomized block design with twelve replicates. An initial analysis combined over all years indicated that the years were significantly different, so analysis was conducted on each year individually for yield and total water usage. Treatment means were separated using the Waller-Duncan k-ratio and Fisher's least significant tests.

Results and Discussion

Rainfall: For the three-year study, the growing season (April to August) rainfalls were 468 mm in 2012, 620 mm in 2013, and 414 mm in 2014. In 2013 only two irrigations were required. In 2012, two to nine irrigations were required depending on treatment in late June and early July. The 2014 season required the greatest number of irrigation events (7 to 21 depending on treatment), and had greatest total irrigation depth.

Corn Yields: An overall analysis of variance for corn yield indicated that the growing year was the only significantly different variable. The average corn yields for the three-year study across the three irrigation treatments ranged from 10.3 to 16.2 Mg ha^{-1}. The 2012 overall yield (15.6 Mg ha^{-1}) was significantly greater than the overall yields of the other two years (table 1). Even though 2013 had the highest rainfall, it had a significantly lower yield (10.5 Mg ha^{-1}) and may be attributed to reduced nitrogen application in that year.

Because the corn yields for the three years were significantly different, we analyzed them individually. In 2012, the mean corn yield across all irrigation treatments was 15.6 Mg ha^{-1} and treatments were not significantly different (table 1). This indicated that all three irrigation treatments adequately provided enough water for the corn crop. In 2013, the mean treatment corn yields were not significantly difference and averaged 10.5 Mg ha^{-1}. It also had the greatest rainfall during the growing season with the least number of irrigation events (0 to 2 depending on treatment), yet it had the lowest mean yield of the three-year study. In 2014, the mean corn yields across irrigation treatments were not significantly different and had an overall mean yield of 13.5 Mg ha^{-1},. Year 2014 had the least cumulative rainfall for the three-year study and the greatest number of irrigation events.

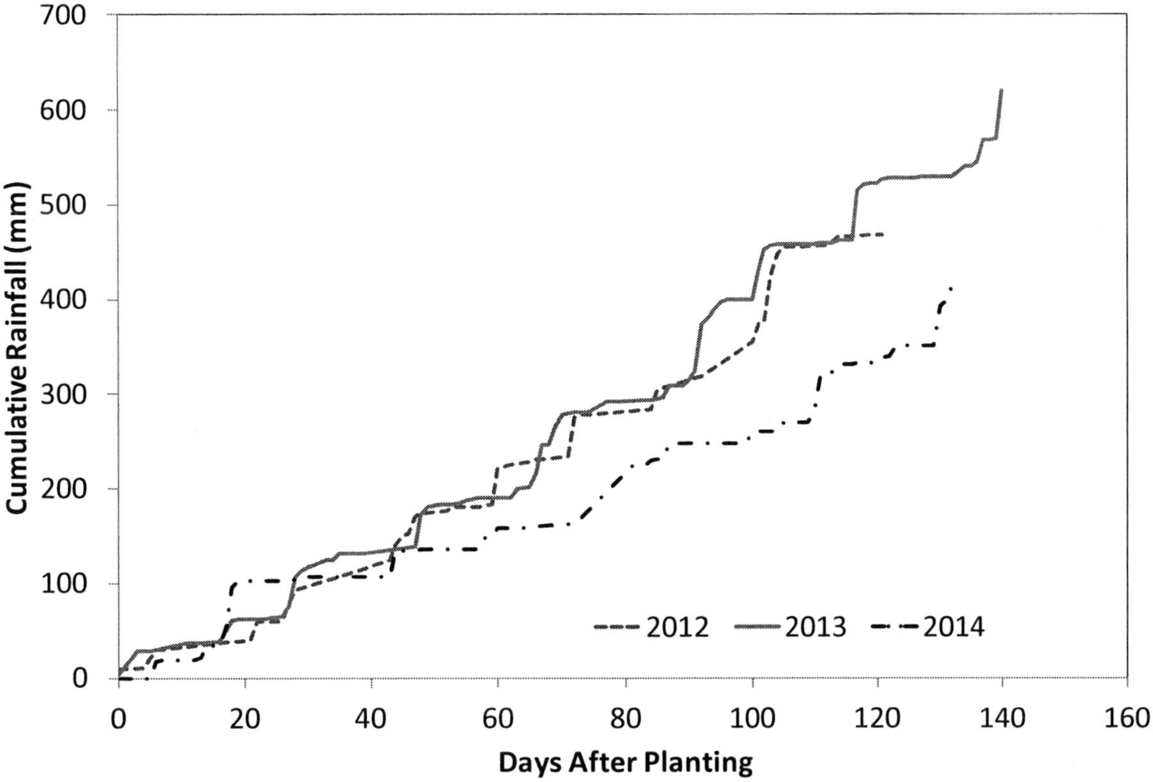

Figure 2. Growing season cumulative rainfall.

Based on the three-year study with different rainfall distributions throughout the growing seasons, each of the three full irrigation scheduling methods provided adequate supplemental irrigation to produce good to excellent corn yields for the region (Wiatrak, 2010).

Total Water, Irrigation, and WUE: The total water (rain + irrigation) received by the corn crop varied over the three years by irrigation treatment. The total water the corn crops received from 2012 to 2014 was 526, 627, and 570 mm, respectively, with 2012 to 2014 yearly mean irrigation water applied to the corn crop was 57, 7, and 156 mm, respectively (table 1). In only year 2012 were the total water and irrigation treatment mean depths significantly different. The Irrigator Pro treatment had significantly higher total water and irrigation water depth than the other treatments. The Irrigator Pro treatments typically called for more early-season irrigation events. The 2012-2014 WUE across irrigation treatments was 29.8, 16.8, and 23.8 kg ha^{-1} mm^{-1}, respectively and were significantly different.

Table 1. Mean corn yields, irrigation depths, and water use efficiencies for the three irrigation treatments.

Treatment	Yield [a] ($Mg\ ha^{-1}$)			Irrigation (mm)			Water Use Efficiency (kg grain $ha^{-1}\ mm^{-1}$)		
	N	Mean	SD	*N*	Mean	SD	*N*	Mean	SD
2012									
Irrigator Pro	18	16.2 a	1.2	18	76.9 a	27.3	18	29.7 a	2.9
NDVI	18	15.6 a	1.9	18	53.3 b	28.2	18	30.1 a	4.7
SWP	18	15.1 a	1.9	18	41.3 b	18.5	18	29.6 a	3.7
Year mean	54	15.6 A	1.7	54	57.2 B	28.8	54	29.8 A	3.8
2013									
Irrigator Pro	18	10.8 a	1.3	18	6.4 a	10.0	18	17.3 a	2.1
NDVI	18	10.5 a	2.6	18	7.8 a	6.9	18	16.7 a	4.2
SWP	17	10.3 a	1.7	18	7.8 a	9.9	17	16.5 a	2.7
Year mean	53	10.5 C	1.9	54	7.3 C	8.9	53	16.8 C	3.1
2014									
Irrigator Pro	18	13.3 a	1.7	18	163.7 a	29.8	18	23.0 a	3.0
NDVI	18	13.8 a	1.8	18	152.2 a	27.8	18	24.4 a	3.1
SWP	18	13.5 a	1.5	18	152.4 a	35.4	18	24.0 a	3.0
Year mean	54	13.5 B	1.7	54	156.1 A	31.1	54	23.8 B	3.0
Overall mean	161	13.3	2.7	162	73.5	66.8	161	23.5	6.2

[a] Year means across treatments followed by the same uppercase letter are not significantly different at the 5% level. Treatment means within a year followed by the same lowercase letter are not significantly different at the 5% level.

Conclusions

Corn was grown under variable-rate center-pivot irrigation for three years (2012-2014) to evaluate the potential of using vegetative indices and an expert system for managing spatial irrigations. These two methods were compared with irrigation management using soil water potentials. Rainfall during the three growing seasons varied widely. In 2013, only two irrigation events were required, while the 2014 growing season required 7 to 21 irrigation events depending on the plot.

The 2012 corn crop had the highest overall yield (15.6 Mg ha^{-1}) and was significantly greater than the other two years. In 2014, the overall mean yield was 13.6 Mg ha^{-1}, and even though 2013 had the highest rainfall, it had a significantly lower yield (10.5 Mg ha^{-1}).

The crop irrigation depths for the three years were significantly different and varied from an average of 156 mm in 2014, to 75 mm in 2012, to 7 mm in 2013. In 2012, the Irrigator Pro required significantly greater irrigation than the SWP or NDVI treatments. In 2013 and 2014, there were no significant differences in irrigation depth between the irrigation treatments. The WUE for the three irrigation treatments was significantly different for the three-year study, ranging from 29.8 kg ha^{-1} mm^{-1} in 2012, to 23.8 kg ha^{-1} mm^{-1} in 2014, and to 16.8 kg ha^{-1} mm^{-1} in 2013. However, for individual years, there were no significant differences in WUE among the irrigation treatments.

Overall, the NDVI and Irrigator Pro for Corn treatments managed irrigations as well as the traditional SWP-based treatment. Each of these irrigation treatments was able to adequately manage irrigation and produce adequate crop yields for the region and could be used effectively to manage irrigation under a variable-rate irrigation system.

References

Allen, R., Pereira, L., Rais, D., & Smith, M. (1998). Crop evapotranspiration: Guidelines for computing crop water requirements. FAO Irrigation and Drainage Paper 56. Rome, Italy: United Nations FAO.

Bausch, W. C. (1993). Soil background effects on reflectance-based crop coefficients for corn. *Remote Sens. Environ., 46*(2), 213-222. http://dx.doi.org/10.1016/0034-4257(93)90096-G

Bausch, W. C., & Neale, C. M. U. (1987). Crop coefficients derived from reflected canopy radiation: A concept. *Trans. ASAE, 30*(3), 703-709. http://dx.doi.org/10.13031/2013.30463

Camp, C. R., Sadler, E. J., Evans, D. E., Usrey, L. J., & Omary, M. (1998). Modified center-pivot system for precision management of water and nutrients. *Appl. Eng. Agric., 14*(1), 23-31. http://dx.doi.org/10.13031/2013.19362

Davidson Jr., J. I., Bennett, C. T., Tyson, T. W., Baldwin, J. A., Beasley, J. P., Bader, M. J., & Tyson, A. W. (1998). Peanut irrigation management using EXNUT and MOISNUT computer programs. *Peanut Sci., 25*(2), 103-110. http://dx.doi.org/10.3146/i0095-3679-25-2-9

Glenn, E. P., Neale, C. M., Hunsaker, D. J., & Nagler, P. L. (2011). Vegetation index-based crop coefficients to estimate evapotranspiration by remote sensing in agricultural and natural ecosystems. *Hydrol. Proc., 25*(26), 4050-4062. http://dx.doi.org/10.1002/hyp.8392

Hunsaker, D. J., Pinter, J. P., Barnes, E. M., & Kimball, B. A. (2003). Estimating cotton evapotranspiration crop coefficients with a multispectral vegetation index. *Irrig. Sci., 22*(2), 95-104. http://dx.doi.org/10.1007/s00271-003-0074-6

Lamb, M. C., Masters, M. H., Rowland, D., Sorensen, R. B., Zhu, H., Blankenship, R. D., & Butts, C. L. (2004). Impact of sprinkler irrigation amount and rotation on peanut yield. *Peanut Sci., 31*(2), 108-113. http://dx.doi.org/10.3146/pnut.31.2.0009

Lamb, M. C., Rowland, D. L., Sorensen, R. B., Butts, C. L., Faircloth, W. H., & Nuti, R. C. (2007). Economic returns of irrigated and non-irrigated peanut based cropping systems. *Peanut Sci., 34*(1), 10-16. http://dx.doi.org/10.3146/0095-3679(2007)34[10:EROIAN]2.0.CO;2

Omary, M., Camp, C. R., & Sadler, E. J. (1997). Center-pivot irrigation system modification to provide variable water application depths. *Appl. Eng. Agric., 13*(2), 235-239. http://dx.doi.org/10.13031/2013.21604

Wiatrak, P. (2010). Corn production guide. Clemson, S.C.: Clemson University Cooperative Extension Service. Retrieved from www.clemson.edu/extension/rowcrops/corn/guide/

Sensor Strategies for Scheduling Irrigation in Louisiana

Stacia L. Davis, Ph.D. E.I.T.

Assistant Professor, LSU AgCenter, Red River Research Station, 262 Research Station Drive, Bossier City, LA 71112, (318) 741-7430 ext 1105, sdavis@agcenter.lsu.edu

Abstract. The objective of this study was to evaluate multiple types of soil moisture sensors to determine their applicability for producers in Louisiana agriculture. Irrigation treatments were determined using: A) soil matric potential sensor system, B) volumetric water content sensor system, or C) weekly irrigation depending on rainfall. Overall, both soil moisture sensor systems were capable of limiting irrigation events compared to weekly irrigation during dry periods with 70% water savings without yield reduction occurring at one location. Though accuracy in sensor readings declined over time, they were still helpful in determining trends in soil moisture. However, using a static threshold to trigger irrigation events was not advisable for either sensor system due to their inaccuracy. Proper implementation requires that the producer has the knowledge to interpret the soil moisture data in reference to the physical system for best management practices.

Keywords. Crops, irrigation, scheduling, sensors, soil moisture

Introduction

The competition for agricultural water supplies continues to increase in Louisiana due to many factors including short-term and long-term drought, increasing irrigated acreage, and demand from other sectors like industry, power generation, public supply, and aquaculture. One approach to addressing agricultural water challenges is to improve irrigation efficiency of functioning irrigation systems.

The primary method of agricultural irrigation in Louisiana, estimated as 80% of the irrigated acreage, is furrow irrigation. Using this irrigation method, efficiency can be improved by either addressing infiltration as the water moves across the field or by using tools to determine when irrigation is necessary. Various tools exist to provide feedback to the grower concerning field conditions. Soil moisture sensors are one such tool that can be cost effective; however, there are several brands of sensors on the market, which differ in accuracy, capabilities, and price. The overall objective of this study was to evaluate multiple types of soil moisture sensors to determine their applicability for producers in Louisiana agriculture.

Materials and Methods

In Louisiana, typical existing agricultural irrigation infrastructure includes a groundwater well with an electric, diesel, or propane fueled pump and an underground pipe network with periodic access via aboveground risers. Irrigation is distributed to each furrow using temporary thin walled lay-flat tubing with manually punched holes resulting in overland flow from the edge of the field having the highest elevation toward the tail end (Fig. 1).

Figure 1. Example of surface irrigation using lay-flat tubing in soybeans.

Generally, there is very little control in the applied volume per event when it comes to furrow irrigation. Irrigation volumes depend on pump efficiency, available head pressure, pipe-riser system design, hole size selection in the lay-flat tubing, and infiltration characteristics of the soil. Most of these dependencies require considerable investment and effort to change, which is only likely to occur by producers when required (such as replacing an end-of-life pump) and not just for improving irrigation efficiency. However, using soil moisture to determine when to apply irrigation can delay an application by a few days from the typical producer schedule of once per week, eventually skipping an irrigation event. Measurements of soil moisture can also aid a producer in determining if rain or irrigation infiltrated the soil surface and will be effective for the crop.

Soil moisture sensors generally fall into one of two categories depending on the type of produced data. The soil matric potential sensors, also commonly referred to as granular matrix sensors, estimate the tension or suction required by the plant material to remove the water from the soil. These sensors report soil moisture in centibars where 0 cb represents saturation, -10 to -30 cb represents field capacity, and -1,500 cb represents permanent wilting point. Volumetric water content sensors estimate soil moisture as the volume of water per volume of soil column and is typically reported as a percentage. Thresholds for saturation, field capacity, and permanent wilting point are dependent on the soil type and must be determined prior to application in the field.

A study was designed to measure irrigation application based on the following treatments: A) soil matric potential sensor system, B) volumetric water content sensor system, and C) weekly irrigated treatment. Each treatment was replicated three times with at least six rows on 40 inch spacing and a minimum row length of 300 ft. Irrigation application was measured using volumetric flow meters (McCrometer, Inc., Hemet, CA) assuming equal application across treatments when more than one treatment received irrigation. Yield was used as the primary response variable to determine whether differences in irrigation application resulted in negative impacts to the crop. Yield was harvested from the two middle rows of each plot in a 100 ft portion of the row. Ideally, irrigation was triggered for the sensors at either 80 cb or around 30% (based on soil conditions) depending on measurement type.

The soil matric potential sensor chosen for the study was the Watermark (Irrometer Company, Riverside, CA). It is the most popular soil matric potential sensor on the market due to a comparatively low price point and simplicity (Spaans and Baker 1992, Thompson et al. 2006).

The sensors were coupled with Aqua Trac (AgSense, Huron, SD) telemetry units for logging and transmitting the soil moisture data. Each Aqua Trac unit was capable of managing four sensors; sensors were installed at 6 inches, 12 inches, 18 inches and 24 inches. Each replication received a sensor system.

The GS-1 (Decagon Devices, Pullman, WA) was chosen as the volumetric water content sensor. It was new to the market and meant for agricultural situations. It was chosen primarily due to its comparability to the Watermark based on size and installation style. Decagon RM50G telemetry loggers were used to access the soil moisture data. Each logger can support five sensors thus these sensors were installed every 6 inches, similar to the Watermarks, with one additional sensor at the 30-inch depth.

The study was conducted in 2015 and 2016 at three Louisiana Agricultural Experiment Stations across northern Louisiana to test the treatments on three distinct soil types. The field at the Red River Research Station (Bossier City, LA) is located on sandy clay loam, part of the Red River Alluvial soils inherent to the region. The field at the Macon Ridge Research Station (Winnsboro, LA) is predominantly silt loam representing the soils of the Macon Ridge. The final location, at the Northeast Research Station (St. Joseph, LA), has cracking clay soils that are known to dominate the Mississippi Delta region. Soybeans were grown at the Macon Ridge and Northeast Research Stations whereas cotton was grown at the Red River Research Station. All sensors were installed after planting (and fertilization for cotton) and removed prior to harvest.

Results and Discussion

Due to many uncontrollable circumstances, the goal of determining irrigation application based on yield differences was not successfully completed each year at all research locations. However, soil moisture was collected at all locations and was instrumental in identifying secondary objective information such as deficit conditions, infiltration of various irrigation and rainfall events, and identifying sensor performance. The most informative results are presented here.

2015 Results

The GS-1 sensors provided the most accurate data in the sandy clay loam soil. Due to the perceived accuracy, the changes in soil moisture across all sensors within the measurement area were quantified as crop evapotranspiration (ET_C). The ET_C was used to calculate the ratio of ET_C to reference evapotranspiration (ET_O), also known as crop coefficients (Fig. 2). These estimations were compared to crop coefficients developed for cotton in Louisiana using weighing lysimeters (Kumar et al. 2015). The dips in crop coefficient correspond with two missed irrigation events indicating that a stress coefficient was introduced during the growing season.

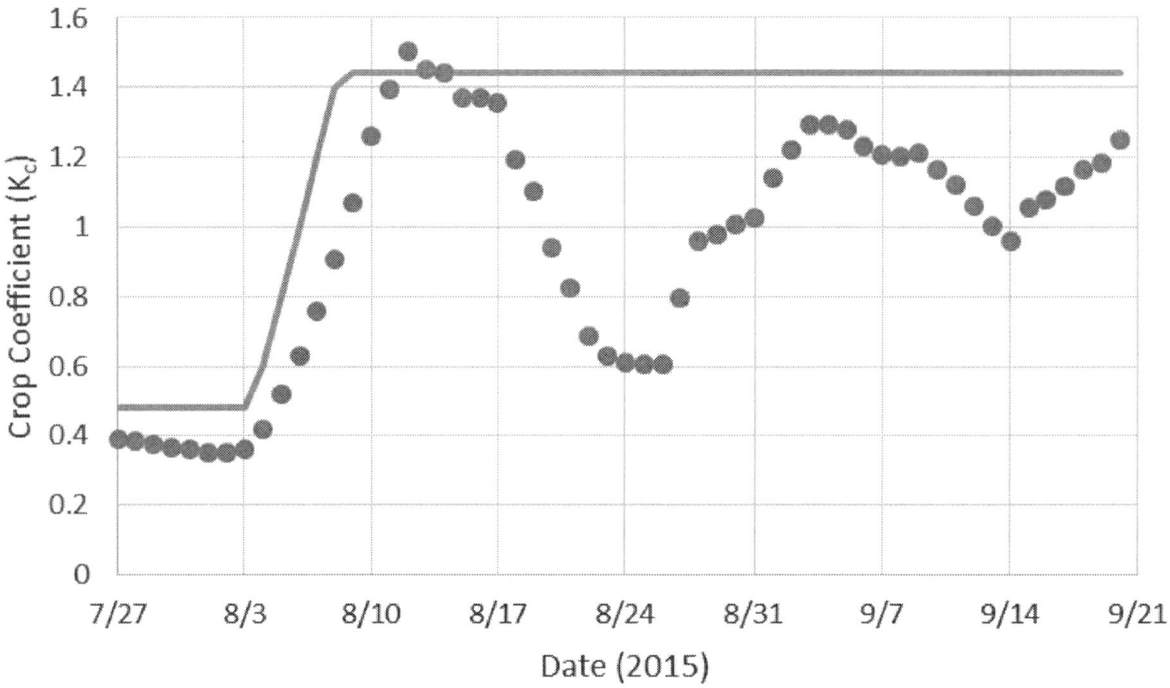

Figure 2. Rolling average crop coefficients (blue dots) were determined using soil moisture sensors for cotton in 2015. Calculated coefficients were compared to published coefficients for Louisiana determined using weighing lysimeters (gray line).

The Watermark sensors did not respond to changes in soil moisture in the sandy clay loam. Though it is possible that infiltration was an issue in these plots, it's unlikely considering that the plots with GS-1 sensors were irrigated at the same time and had distinct and immediate responses. This was the lightest textured soil of the three locations, further strengthening the previous research suggesting that these sensors do not perform well in coarse soils, such as those with significant sand content (Varble and Chavez 2011, Cepuder et al. 2008).

The opposite result occurred of the cracking clay location, having the heaviest soil in the study. The Watermark sensors showed immediate response to irrigation with significant drawdown between events. However, the GS-1 sensors showed very little change in soil moisture over the season and only if they were able to provide data at all. Both sensor types rely on good contact between the sensor and soil for accuracy. It is hypothesized that the cracking soil pulled away from the metal rods of the GS-1 sensors, resulting in inaccurate soil moisture estimations, whereas the mesh construction of the Watermarks provided continued contact with the soil despite cracking.

In general, there were good responses to soil moisture changes in the silt loam soil by both sensor types. However, the GS-1 sensors did not have enough accuracy to utilize the method for determining crop coefficients using the sensor data. The Watermarks did not always respond to irrigation at all sensor depths, indicating that accuracy was questionable as well.

2016 Results

In most locations, equipment for both sensor types were reused from the previous year. As a result, sensor data was much less accurate than expected in the second year. Indications of inaccuracy were obvious with the GS-1 sensors where soil moisture data began to oscillate to below permanent wilting point and above saturation. In some instances, the readings dropped from an acceptable soil moisture to below permanent wilting point suddenly and without reason. The Watermarks were not as obvious, but showed a distinct decline in the range of soil moisture within the wet portion of the drying curve, resulting in inappropriately delayed irrigation. It is hypothesized that the manual labor required to install and remove these sensors under extreme weather conditions during Louisiana summers reduced the amount of care in handling the sensors.

The location with the most complete results occurred at the Northeast Research Station on cracking clay soil. This location received 19.7 inches of rainfall from May through September (Table 1). Irrigation for the weekly treatment occurred during weeks where rainfall was not abundant, resulting in five irrigation events, totaling 29.8 inches of irrigation, that occurred during reproductive growth in the dry period of the growing season. Only two irrigation events were scheduled for the sensor treatments resulting in 8.8 inches of irrigation. Despite the difference in irrigation application, there were no differences in yield between the three treatments, averaging 65.4 bushels per acre. Thus, 70% water savings were achieved by using the sensors despite the previously mentioned inaccuracies. The production soybeans surrounding the plots were unirrigated and of the same variety. Randomly sampled for comparison, the unirrigated soybeans had lower average yield of 40.8 bushels per acre compared to the irrigated soybeans, indicating that irrigation was beneficial in 2016.

Table 1. Summary of irrigation and yield results of soybean for the cracking clay research site located at the Northeast Research Station.

Treatment	Number of Irrigation Events	Cumulative Irrigation (in)	Cumulative Rainfall (in)	Yield Weight (bu/ac)*
Watermark	2	8.8	19.7	63.2 a
Decagon	2	8.8	19.7	64.8 a
Weekly	5	29.8	19.7	68.2 a
Unirrigated	0	0	19.7	40.8 b

*Treatments with significant differences ($\alpha = 0.05$) were represented with different lowercase letters.

Treatment implementation and data collection was inconsistent for the silt loam soil at the Macon Ridge Research Station. As a result, there were no differences in irrigation application across treatments at this location (Table 2). Additionally, most of the plots were accidentally harvested as production; harvest occurred for one plot each of the Watermark, Decagon, and unirrigated treatments. As a result, yield could not be statistically evaluated.

Despite these issues, there was one important outcome observed using the soil moisture sensors. Neither sensor type consistently responded to irrigation after an event. Though it would be plausible that sensor inaccuracies were at play, the anecdotal yield data indicated that there was no difference between irrigated and unirrigated soybeans at this location. Considering the relatively acceptable sensor performance last year, it is likely that irrigation water failed to infiltrate while moving across the field. It is hypothesized that the velocity of the water across the soil surface combined with unideal soil conditions prohibited infiltration into the root zone. This hypothesis was supported by individuals in the agricultural industry that work in this region and on this soil type.

Table 2. Summary of irrigation and yield results of soybean for the silt loam research site located at the Macon Ridge Research Station.

Treatment	Number of Irrigation Events	Cumulative Irrigation (in)	Cumulative Rainfall (in)	Yield Weight (bu/ac)*
Watermark	3	9.0	15.2	46.0
Decagon	3	9.0	15.2	43.8
Weekly	3	9.0	15.2	--
Unirrigated	0	0	15.2	43.7

*Treatments with significant differences (α = 0.05) were represented with different lowercase letters.

There were no differences in yield for the cotton grown on the sandy clay loam (Table 3). Irrigation occurred four times during the dry period for the weekly and Watermark treatments, but only two irrigation events occurred for the Decagon treatment. There were difficulties in maintaining pressure within the lay-flat tubing as the season progressed, so irrigation events were not always typical. Also, replications were significant in the model indicating that growing conditions were variable across the plots. This variation has been an issue at all three research locations.

Table 3. Summary of irrigation and yield results of cotton for the sandy clay loam research site located at the Red River Research Station.

Treatment	Number of Irrigation Events	Cumulative Irrigation (in)	Cumulative Rainfall (in)	Yield Weight (bale/ac)*
Watermark	4	8.8	20.2	2.5 a
Decagon	2	5.3	20.2	2.6 a
Weekly	4	8.8	20.2	2.3 a

*Harvest data was not available at the time of publication.

Conclusion

Overall, both soil moisture sensor systems were capable of limiting irrigation events compared to weekly irrigation during dry periods. There were quantifiable water savings at locations where infiltration occurred. At one location, there was no reduction in yield despite 70% water savings. Though accuracy in sensor readings declined over time, they were still helpful in determining trends in soil moisture. However, using a static threshold to trigger irrigation events was not advisable for either sensor system due to their inaccuracy. Also, it's likely that repeated installation and removal increased the opportunity for sensor failure.

The combination of inefficiencies in irrigation conveyance, lack of trained labor, and aging infrastructure across research stations have led to inconsistencies in conducting irrigation-specific research across the state. Future work in this area will be localized in a more controlled setting to improve irrigation application, data collection, and overall management of the project. Sensors will be checked for quality prior to installation in 2017 and failures will be noted to determine life expectancy of the equipment.

Though the level of measurement inaccuracy is important to know when using a soil moisture sensor, this issue can be overcome when considering that they are just one tool that can provide the producer with information about management. There are many other considerations; for example, a soil moisture sensor tends to provide point measurements used to represent conditions over a large acreage that may have variability in soil type, plant type, or irrigation requirements based on evapotranspiration. As in any technology, the provided information requires knowledge of the physical system for interpretation. Proper implementation requires that the producer understands the data and can make an educated decision based on the provided information and on-site observations.

References

Cepuder, P., S. Evett, L. K. Heng, C. Hignett, J. P. Laurent, and P. Ruelle. In: Evett, S. R., L. K, Heng, P. Moutonnet, and M. L. Nguyen (Eds.), Field Estimation of Soil Water Content: A Practical Guide to Methods, Instrumentation and Sensor Technology. IAEA-TCS-30. Interational Atomic Energy Agency, Vienna, Austria, ISSN 1018-5518.

Kumar, V., T. K. Udeigwe, E. L. Clawson, R. V. Rohli, and D. K. Miller. 2015. Crop water use and stage-specific crop coefficients for irrigated cotton in the mid-south, United States. Agric. Water Manage. 156(2015) 63-69.

Spaans, E. J. A. and J. M. Baker. 1992. Calibration of Watermark soil moisture sensors for soil matric potential and temperature. Plant and Soil. 143: 213-217.

Thompson, R. B., M. Gallardo, T. Aguera, L. C. Valdez, and M. D. Fernandez. 2006. Evaluation of the Watermark sensor for use with drip irrigated vegetable crops. Irrigation Science. (2006) 24: 185-202.

Varble, J. L. and J. L. Chavez. 2011. Performance evaluation and calibration of soil water content and potential sensors for agricultural soils in eastern Colorado. Agricultural Water Management. 101(2011) 93-106.

Calibration of Low cost infrared thermometer to measure Crop Water Stress index

Theodore W. Sammis[1], David R Miller[2] Vincent P. Gutschick[3]

1. Department of Environmental Science P.O. Box 3Q, Las Cruces NM 88001
2. Department of Natural Resources and Environment, University of Connecticut, Storrs, CT, (860) 377-3790
3. Global Change Consulting Consortium, INC. 4804 Calabazilla Drive Las Cruces NM 88001

Written for presentation the IA Irrigation symposium Los Vegas Nevada December 5-9, 2016

Abstract.
The advent of low-cost handheld infrared (IR) thermometers has led to a proliferation of non-contact surface temperature measurements that can be used in many applications from food processing to measuring the water stress of plants. In order to use the IR thermometers in crop water stress measurements (CWSI), the instruments must be calibrated in the plant surface temperature range of use. The low–cost infrared thermometers measure the infrared temperature by using uncooled thermopile detectors that adsorb radiation in the 8 um to 14 um spectral range. These detectors are uncooled; therefore radiation emitted by the detector itself must be considered in the calibration process. When measuring the non water stressed CWSI using the IR thermometers, correction for reflected radiation from the sky is not necessary because the CWSI calculation is the relative difference between the canopy and surrounding air temperature. However, if the CWSI measurements are to be used to calculate the actual transpiration rate of the crop, then the IR thermometer reading must be corrected for the reflected sky radiation and change in emissivity.

Keywords. *Crop Water Stress index, canopy resistance, aerodynamic resistance, evapotranspiration*

1. Introduction

The advent of low-cost handheld infrared (IR) thermometers ($25-$50 U.S.) has led to a proliferation of non-contact surface temperature measurements that can be used in many applications from food processing to measuring the water stress of plants in the field. Plant leaf temperature increases with plant water stress (Howell 1996) and this temperature measurenment can be used to calculate the Crop Water Stress Index (CWSI).

We use the CWSI as a water management tool to maintain optimal water stess levels throughout the growing season; because seasonal maintainence of some water stress, depending on the plant, can increase water use efficiency while not affecting yield. (Chai, et al., 2016). Irrigation amounts restricted to maintain the desired water stress for a paticular crop (Moller, et al., 2007) should be monitored for impacts on the plant water stress level.

To use the IR thermometers to measure plant water stress level, the instruments must be calibrated in the temperature range of use. Low cost infrared thermometers measure the infrared temperature by using thermopile detectors that detect radiation in the 8 um to 14 um spectral range. Radiation emitted by the detector itself must be considered in the calibration process, because these detectors are uncooled. The emissivity setting on the thermometer can either be adjustable as a setting on the IR thermometer or a fixed value of 0.95. The thermometers that have a fixed emissivity value are lower in cost.

1.1 CWSI

An index of crop water stress (CWSI) is defined for sunlit canopy surface temperatures, collected by a hand held infrared thermometer as:

$$CWSI= 1- ET_a/ET_{ns} \quad (1)$$

where The ratio ET_a/ET_{ns} is the relative ET. ET_a is the actual ET and ET_{ns} is the non-stressed ET of the plants.

The CWSI can be calculated from canopy temperature (T_c), air temperature (T_a), and vapor pressure deficit (VPD) (from air temperature and relative humidity), and from a knowledge of the upper and lower surface temperatures at the possible extremes of ET, represented graphically by upper and lower base lines in Figure 1.

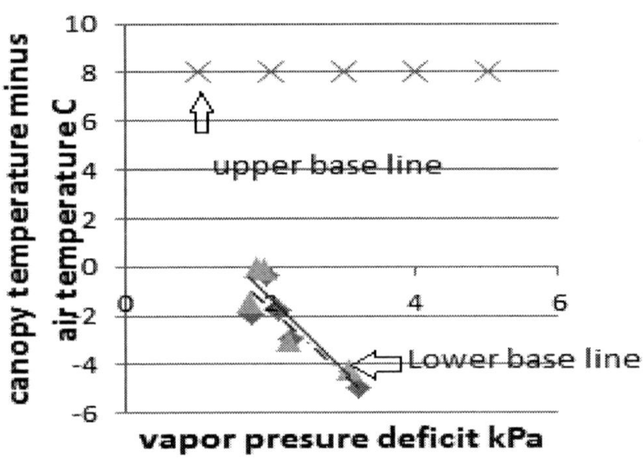

Figure 1. Upper and lower base lines for the CWSI for grapes in Napa California measured with a low cost infrared thermometer.

The upper base line represents complete stress where ET is zero. The lower base line represents the no water stress condition. Measurement of leaf – canopy temperature difference and VPD determines the relative distance between these two extremes and the relative ET in equation 1. The lower base line must be measured when there is no moisture, fertilizer or insect stress. Nitrogen stress along with water stress can cause stomatal closure with a resulting increase in T_c-T_a. (Rudnick, and Irmak , 2014)

The graphical solution of the CWSI was developed by Tanner (1963), Jackson et al. (1981), and Idso (1982). CWSI base lines have been developed for soybeans (Candogan et al, 2013), grapes (Bellvert et al. (2013), corn (Payero and Irmak, 2006), Broccoli (Gültaş 2010) and potatoes (Erdem et al., 2005) and many other crops including ornamentals (Sammis and Jerrigan, 1992).

2.0 Theory of IR thermometers measurements.

All objects emit radiation in the form of electromagnetic waves distributed across the electromagnetic spectrum. The distribution and intensity of the radiation emitted is determined by the surface temperature of the object according to Planck's law (Fowler, 1998). Lcaf surfaces near air temperature emit radiation contained within the infrared part of the electromagnetic spectrum, at wavelengths ranging from 8–14 μm. The IR thermometer detector adsorbs the radiation received over this wavelength range, which increases its temperature, and in turn, it provides a voltage, or current, in proportion to intensity of the radiation load it receives. The signal strength (S) is a nonlinear function of the canopy temperature and is described by the Sakuma–Hattori interpolation equation [Sakuma and Kobayashi 1997],

$$S(T)= C/ [\exp (C_2/ (AT+ B))-1] \qquad (2)$$

Where: A, B, and C are calibration constants related to the properties of the IR thermometer.
C_2 is the second Planck function constant equal to 14,388 um K
T is the surface temperature in degree K

The microprocessor contained in the infrared thermometer solves equation 2 for T and displays as degree C or degree F on the IR thermometer output screen. The constant C is set to one in equation 2 and A and B are calculated as a function of the central wavelength of the sensor (λ_o) in microns and width of the wave length range ($\Delta \lambda$) expressed by Eq. 3 and Eq. 4.

$$A= \lambda_o [1- (\Delta \lambda^2/2 \lambda_o^2)] \qquad (3)$$

$$B= C_2 \Delta \lambda^2/ 24 \lambda_o^2 \qquad (4)$$

The sources of radiation received by the IR sensor are IR emitted by the canopy and reflected sky radiation. These depend on the temperatures of the canopy, its emissivity (ε_S), and the air temperature. Meanwhile, the sensor emits IR as a function of the sensor temperature.

A black body has an emissivity of one and the canopy of a crop has an emissivity of ~0.98 (Chen, 2015). Consequently if the low cost IR thermometer has an emissivity set internally to 0.95 then the calibrated temperature must be corrected to an emissivity of 0.98 if the absolute temperature is required. The infrared thermometers are factory calibrated in a constant controlled temperature indoor environment. Therefore the infrared thermometer temperature must also be corrected for the radiation (Q) the infrared thermometer receives from the sky when using the thermometer outdoors to measure CWSI.

$$Q(received) =0.98*Q(crop) + 0.02*Q(sky) \quad \text{or} \qquad (5)$$

$$Q(crop)=[Q(received)-0.02*Q(sky)]/0.98$$

Where the emissivity of the crop is 0.98

A more detailed description of the measurement errors associated with using low cost, low temperature IR thermometers is given by Saunder (2009).

2.1 *Field of View*

Target size and distance are critical to the accuracy for most IR thermometers. Every IR instrument has a field of view (FOV), that is, a family of angles of vision over which it averages the radiation received. IR thermometers have fixed focus optics, the minimum measurement spot occurs at the specified focal distance that can range for general purpose IR Thermometers from 50 to 150 cm. The FOV can range from 12:1 to 10:1 or 8:1 . When using the infrared thermometer to measure CWSI, it is important to be close enough to measure canopy temperature and not include the temperature of the soil, or the stakes supporting the canopy in the case of grape vines.

2.2 *Calibrating the infrared thermometer to measure CWSI*

The manufactures assume that the temperature of the surrounding and the detector along with the emissivity of the instrument and the emissivity of the canopy are the same, so the only error would be the reflected sky radiation. The IR thermometer factory calibration needs to be improved over the desired temperature range of crops by using a black body calibration source with an emissivity of one discussed in the methods section. The factory calibration is over a wide range of temperature (20 degree C to 520 degree C) and consequently, its resolution is too low in the canopy range of temperature of 20 degree C to 40 degree C.

3. Methods

Twelve infrared thermometers (Sun model EM520B) with a fixed emissivity of 0.95 and a field of view of 8:1 were purchased and calibrated by putting them in a greenhouse where the temperature ranged from 18 to 39 degree C throughout the day. Specification of the infrared sensor is given in Table 1.

Specifications	**Range**
Temperature Range	-20 C to 520 C
Repeatability	+- 2 C
Response time	500 mSec, to reach 95 % of reading
Spectral Response	7-18 um
Emissivity	0.95
Relative humidity operation range	10-95%Rh
Power	9V
FOV	8:1

Table 1 Specifications of All Sun Model Em520B infrared thermometer (AllSun, 2016)

The thermometers were used to measure the temperature in a compactor cup (ThermoWorks compactor cup, 2016) which consisted of an aluminum cup painted black to establish an emissivity of 1.0. An access hole in the cups top was used for the radiation measurements. A second horizontal hole on the cup bottom side was for the precision placement of a thermocouple (type t) connected to a fluke thermometer (model 52-2) with an accuracy of

-+0.3 C (Fluke 2016). Four infrared thermometer readings of the compact cup temperature were taken. Each reading took 3 -5 seconds. During the infrared temperature measurements the fluke thermometer was read continually to make sure that the compact cup was as a constant temperature and in equilibrium with the surrounding air temperature measured with an additional thermocouple. Both the infrared thermometer and the compactor cup temperatures were recorded by hand and the data plotted in an excel spreadsheet. Care was taken to make sure the compacter cup, infrared thermometer and the air thermocouple were not in direct sunlight and consequently, the infrared thermometer case was at air temperature. Besides using a greenhouse, similar measurements were made in an auto interior on a clear day when the interior of the car heated up due to direct sunlight falling on the car but not the compactor cup and infrared thermometer.

One of the thermometers was placed in direct sunlight for 10 minutes to determine the impact of heating the infrared thermometer above air temperature.

A second brand (Cen-Tech) fixed emissivity infrared thermometer (number 13) was purchased to compare to the 12 All Sun thermometers. An additional infrared thermometer [Thermoworks infrared thermometer (2016) Ir-Gun-S (Number 14)] was purchase. It had a variable setting for emissivity and was set to 1.0, which was the same emissivity as the compactor cup. The identical calibration procedure was conducted with these thermometers.

The factory calibration was used for the RH-temperature sensor model -PYLE PTHM20 when making CWSI measurements of grapes (Pyle 2016) using the low cost infrared thermometer.

3. **Results**

The mean black body temperature minus the infrared temperature of the 12 infrared thermometer instrument measurements plotted against the blackbody temperature showed a linear increase in measurement error with increasing temperature (coefficient of determination 0.95) when averaged over all the sensors (Figure 2). However, the large difference between the average and max and min values shows the need for individual calibration of each infrared thermometer. A single calibration function cannot be used to correct the infrared thermometer to the correct black body temperature. Also the difference between the black body temperature and the infrared temperature can be, on average, as much as 2 degrees C ; it is essential to calibrate the low cost infrared thermometers before using them to measure the CWSI. The difference between canopy minus air temperature in the CWSI measurements usually varies from +2 for small leaves to +8 for large leaves to minus 6 degrees C for both size leaves. An error of 2 degrees in the canopy measurement would result in a 20 percent error in the calculation of the CWSI.

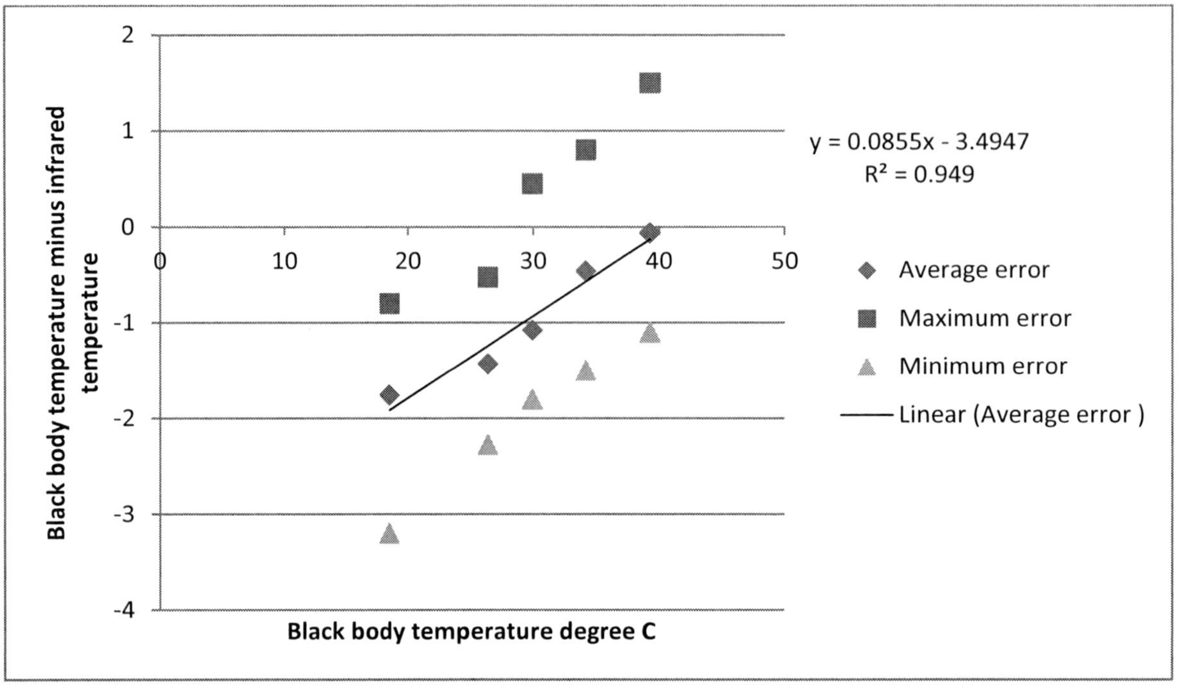

Figure 2. Average and Max and Min difference between the correct surface temperature of a black body and that measured by a low cost infrared thermometer with a fixed emissivity.

The linear calibration of the infrared individual thermometers (back body temperature- infrared thermometer temperature) must be added to the infrared measured temperature as shown for thermometer 3 (Figure 3). The coefficient of determinations of the linear individual infrared thermometers calibration functions ranged from 0.91 to 0.99 for an infrared thermometer with a fixed emissivity of 0.95 (Table 2).

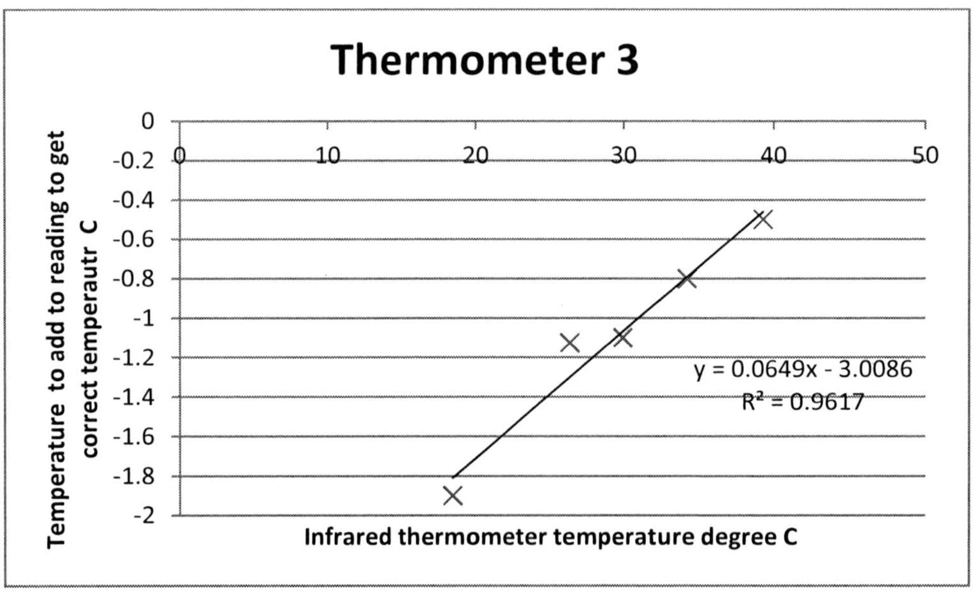

Figure 3. Linear calibration function for infrared thermometer 3.

Table 2 Individual calibration of infrared thermometers with compared to a black body temperature.

Infrared thermometer number	Linear calibration function, temperature added to reading to correct to black body temperature y=ax +b		Coefficient of determination	Emissivity of infrared thermometer
	a	B		
1	0.1018	-2.8219	0.91	0.95
2	0.0747	-3.152	0.96	0.95
3	0.0649	-3.0086	0.96	0.95
4	0.0442	-2.8815	0.91	0.95
5	0.0299	-2.5884	0.96	0.95
6	0.0380	-2.1648	0.92	0.95
7	0.0651	-3.0907	0.91	0.95
8	0.0652	-3.2743	0.94	0.95
9	0.1189	-3.2124	0.99	0.95
10	0.2345	-8.3782	0.95	0.95
11	0.0651	-3.0907	0.91	0.95
12	0.2119	-7.2205	0.91	0.95
13	0.032	-1.879	0.73	0.95
14	No linear calibration function			1.0

When the calibration was conducted on the infrared thermometers that had adjustable emissivity settings, no linear calibration function (Figure 3) could be determined for thermometer 14. A low coefficient of determination (0.73, Table 2) was determined for the other fixed emissivity infrared thermometer (number 13).

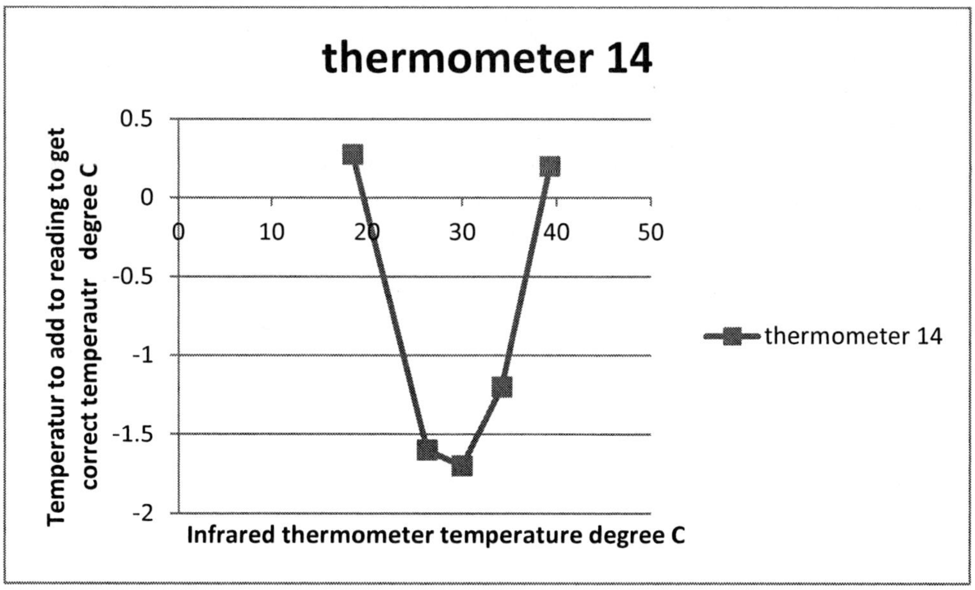

Figure 3. Difference between the correct surface temperature of a black body and that measured by a low cost infrared thermometers with variable emissivity, set to one.

Figure 4 shows that error in canopy temperature measured with a low cost infrared thermometer calibrated for an emissivity of one when not corrected for clear sky radiation and crop canopy emissivity of 0.98.

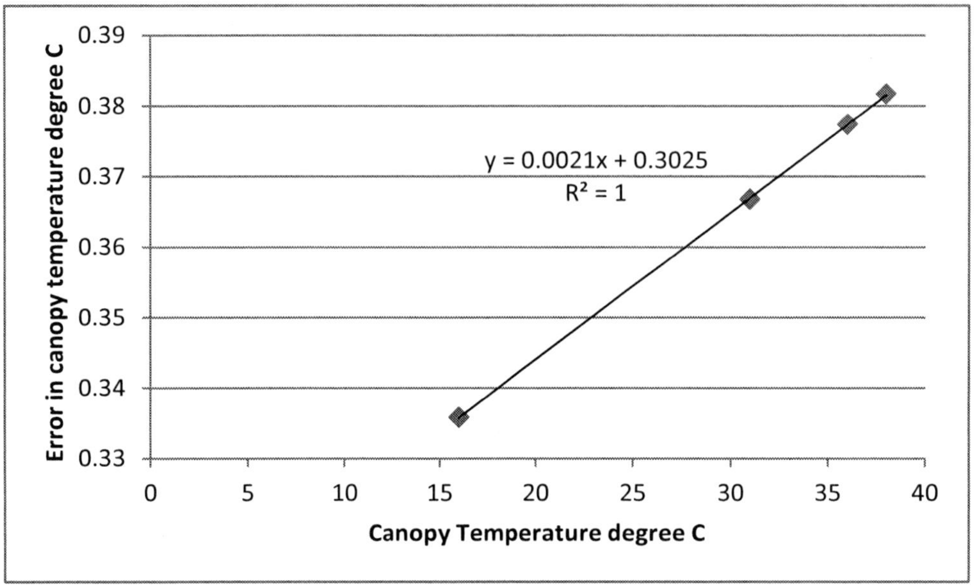

Figure 4. Error in canopy temperature measured with a low cost infrared thermometer calibrated for an emissivity of one when not corrected for clear sky radiation and crop canopy emissivity of 0.98

The infrared thermometer left in the sun for 10 minutes before measurements became sufficiently hot that the temperature display increased 4 to 6 degrees C.

4 Discussion

If the infrared thermometer is not calibrated then an error of up to 20 percent can occur in calculating the CWSI. The canopy and aerodynamic resistance and evapotranspiration rate of the canopy can be calculated using the lower base line slope and intercept of the CWSI and the O'Toole equation (O'Toole and Real 1986) but the error can be significant (10%) if the canopy temperature is not also corrected for emissivity and clear sky in addition to the infrared thermometer calibration. However, the maximum error of not making the emissivity and clear sky correction in the CWSI graphical calculation by Idso and Jackson (1969) is only 2 percent for VPD near zero because the correction has to be applied to both the lower and upper base line temperature measurements. The error decreases below 0.2% as the vapor pressure deficit increases from 0 to 4 MPa at measurement time.

If the infrared thermometer is left in the sun before taking canopy temperature measurements a 35 % error or more can occur in the calculated CWSI. Consequently, it is imperative that the IR thermometer be place in the shade for 15 minutes to equilibrate to air temperature before taking measurements. It is best to shade the instrument from direct sunlight when taking the measurements. In the calibration of the infrared thermometer, the thermometer temperature must also be in equilibrium with the air temperature as must the compactor cup used in the calibration process.

More accurate infrared thermometers that correct for the body temperature of the sensor do not need calibration but the cost is considerable higher (greater than $600) than the low cost infrared thermometers ($25-$50). The more expensive infrared thermometers can be connected to a data logger and the measurements taken automatically from a tractor or all-terrain vehicle as it moves through the field.

However, because of the cost of the of high end infrared thermometers, the calibrated low cost infrared thermometers can be purchased to calculate the CWSI and experience gained by the grower to determine if monitoring of irrigation management by use of the CWSI is desirable before spending the additional money.

5 Conclusions

The low cost infrared thermometers measure the infrared temperature by using uncooled thermopile detectors that detect radiation in the 8 um to 14 um spectral range. Because these detectors are uncooled, radiation emitted by the detector itself must be considered in the calibration process.

Linear calibration to correct the infrared thermometer to a black body temperature in the range of 18 to 39 degree C resulted in coefficients of determination for fixed emissivity thermometers ranging from 0.91 to 0.99. The variable emissivity infrared thermometers had no linear calibration function.

Each individual low cost infrared thermometer must be calibrated.

In order to use low cost IR thermometers in crop water stress measurements (CWSI), the instruments must be calibrated in the temperature range of use otherwise the CWSI error can be as high as 20 percent. Shading the thermometer is important in both the calibration and field measurement procedures. If the infrared thermometer is left in the sun before taking canopy temperature measurements a 35 % error or more can occur in the calculated CWSI.

When measuring the CWSI using the IR thermometers, correction for reflected radiation from the sky and emissivity that is not a black body is not necessary when using the CWSI graphical calculation by Idso (1982) because the error is small, less than 2 percent.

However, if the CWSI measurements are used to calculate the canopy and aerodynamic resistance and the transpiration rate of the crop, then the IR thermometer reading after calibration must be corrected for the reflected sky radiation and change in emissivity or the error can be as high as 10% in the canopy and aerodynamic resistance values.

Reference

AllSun (2016). Model Em520B infrared Thermometer. http://www.all-sun.com/EN/d.aspx?pht=727)

Bellvert ,J., P. J. Zarco-Tejada, J. Girona, E. Fereres (2013). Mapping crop water stress index in a 'Pinot-noir' vineyard: comparing ground measurements with thermal remote sensing imagery from an unmanned aerial vehicle. Precision Agric
DOI 10.1007/s11119-013-9334-5

Candogan, Burak Nazmi, Mehmet Sincik, Hakan Buyukcangaz, Cigdem Demirtas, Abdurrahim Tanju Goksoy, and Senih Yazgan (2013). Yield, quality and crop water stress index relationships for deficit-irrigated soybean [Glycine max (L.) Merr.] in sub-humid climatic conditions Agricultural Water Management. 118,112-121.

Fluke (2016). Fluke 52-2 Dual Input Digital Thermometer. http://en-us.fluke.com/products/thermometers/fluke-52-ii-thermometer.html

Chai Qiang, Yantai Gan, Cai Zhao, Hui-Lian Xu, Reagan M. Waskom, Yining Niu, Kadambot H. M. Siddique (2016). Regulated deficit irrigation for crop production under drought stress. A review Agron. Sustain. Dev. 36: 3

Chen, Chiachung (2015). Determining the leaf emissivity of three crops by infrared thermometry. Sensor, 15, 11387-11401

Erdem, Tolga, A. Halim Orta, Yeşim Erdem, Hakan Okursoy (2005). Crop water stress index for potato under furrow and drip irrigation systems Potato Research. 48(1-2), 49-58.

Fowler, Michael (1998). Black Body Radiation University of Virginia
http://galileo.phys.virginia.edu/classes/252/black_body_radiation.html

Gültaş. T. (2010). Crop water stress index for assessing irrigation scheduling of drip irrigated broccoli (*Brassica oleracea* L. var. italica). Agricultural Water Management 98(1), 48-156.

Howell. T.A. (1996). Irrigation scheduling research and its impact on water use. In: C.R. Camp, E.J. Sadler, and R.E. Yoder (eds.) Evapotranspiration and Irrigation Scheduling, Proceedings of the International Conference, Nov. 3-6, 1996, San Antonio, TX, American Society of Agricultural Engineers, St. Joseph, MI.

Idso, S. B. (1982). Non-water stressed baselines: a key to measuring and interpreting plant water stress. Agric. Meteorol. 27, 59-70.

Idso, S. B., Jackson R. D. (1969). Thermal radiation from the atmosphere. Journal of Geophysical Research 74, 5397-5403

Jackson , R. D. , S. B. Idso, R. J. Reginato and P. J. Printer, Jr. (1981). Canopy temperature as a crop water stress indicator. Water Res. 17,1133-1138.

O'Toole, J. C. and J. G. Real (1986). Estimation of aerodynamic and crop resistance from canopy temperature. Agron. J. 78,305-310.

Payero and Irmak (2006). Variable upper and lower crop water stress index baselines for corn and soybean irrigation science 25, 21-32

Pyle. (2016). Temperature and Humidity meter with dew point and wet bulb temperature product details. http://www.pyleaudio.com/sku/PTHM20/Temperature-and-Humidity-Meter-With-Dew-Point-and-Wet-Bulb-Temperature

Rudnick, D. and Irmak, S. (2014). Impact of Nitrogen Fertilizer on Maize Evapotranspiration Crop Coefficients under Fully Irrigated, Limited Irrigation, and Rainfed Settings." J. Irrig. Drain Eng., 140(12), 04014039.

Saha, S, S. Nadiga, C. Thiaw, J. Wang, W. Wang, Q. Zhang, H. M. van den Dool, H.-L. Pan, S. Moorthi, D. Behringer, D. Stokes, M. Pena, S. Lord, G. White, W. Ebisuzaki, P. Peng, P. Xie (2006). The NCEP Climate Forecast System. Journal of Climate, Vol. 19, No. 15, 3483-3517

Sakuma F and Kobayashi M (1997) Interpolation equations of scales of radiation thermometers *Proc. TEMPMEKO '96, 6th Int. Symp. on Temperature and Thermal Measurements in Industry and Science* ed P Marcarino (Torino, Italy: Levrotto & Bella) pp 305–10

Sammis, Theodore W. , Kara Maraden, Fritz Westover, Junming Wang, David R. Miller. (2015). Improving Irrigation Efficiency and Wine Quality monitoring using Crop Water Stress Index from Canopy Temperature; A Review and Proof of Concept. Paper Number: 2147596 Emerging Technologies for Sustainable Irrigation A joint ASABE / IA Irrigation Symposium Long Beach, California November 10 – 12, 2015

Sammis, T.W. and D. Jernigan (1992). Crop water stress index of ornamental plants. American Soc. of Agric. Eng., Vol.8, No. 2

Saunder, Peter (2009). Calibration and use of low-temperature direct-reading radiation thermometers. Meas. Sci. Technol. 20 025104

Tanner, C. B. (1963). Plant temperatures. Agron J. 55, 210-211.

ThermoWorks compactor cup (2016). Description of compactor cup. http://www.thermoworks.com/IR-Comparator-Cup

ThermoWorks infrared thermometer (2016). Description of infrared thermometer http://www.thermoworks.com/Infrareds

Presented at 2016 Irrigation Show & Education Conference, Dec 5-9, 2016,
Las Vegas Convention Center, Las Vegas, Nevada, USA

Use of Soil Moisture Sensors for Irrigation Scheduling

Ruixiu Sui

USDA-ARS Crop Production Systems Research Unit, Stoneville, Mississippi, USA

Abstract. *Soil moisture sensors were evaluated and used for irrigation scheduling in humid region. Soil moisture sensors were installed in soil at depths of 15cm, 30cm, and 61cm belowground. Soil volumetric water content was automatically measured by the sensors in a time interval of an hour during the crop growing season. Soil moisture data were wirelessly transferred onto internet through a wireless sensor network (WSN) so that the data could be remotely accessed online. Soil water content measured at the 3 depths were interpreted using a weighted average method to reflect the status of soil water in plant root zone. A threshold to trigger an irrigation event was determined with sensor-measured soil water content. An antenna mounting device was developed for operation of the WSN. Using the antenna mounting device, the soil moisture measurement was not be interrupted by crop field management practices. The soil moisture sensor-based irrigation scheduling method has been used for irrigation scheduling in a USDA-ARS Research Farm in Stoneville, MS.*

Keywords. Soil moisture sensor, irrigation scheduling, wireless sensor

Introduction

The Mid-South region is a major area for crop production in the United States. The main crops grown in the region are soybean, corn, cotton, and rice. Though annual precipitation in the Mid-South is approximately 1300 mm, only about 18% of the precipitation occurs during June to August when crops require a large quantity of water to grow. Furthermore, changes in precipitation patterns have made both drought and excessive rainfall more frequent (UCS, 2011; Earth Gauge, 2011). Heavy rainfall causes extensive amounts of runoff from cropland, resulting in only a small amount of the precipitation infiltrating into the soil for crop use. Uncertainty in the amount and timing of precipitation is one of the most serious risks to crop production in the region. To reduce the risk and increase farming profit, the producers increasingly pump large amounts of groundwater from Mississippi River Valley alluvial aquifer each year to irrigate crops during the growing season (Vories and Evett, 2010). Due to the large withdrawals from the aquifer, groundwater levels in the region have declined significantly. Common method used in irrigation scheduling in this region is based on visual assessment of crop response and a "feel" for soil water status. There is a great need for the producers to have objective, reliable, and easy-to-use water management technologies that work for the Mid-South crop/soil environments.

Irrigation scheduling determines the time and amount of water to apply. One of the most popular methods for irrigation scheduling is to measure soil moisture levels in the plant root zone and apply water if there is water shortage for plants. Soil moisture content can be directly determined using manual gravimetric soil sampling by weighing and drying the soil sample. The gravimetric method is simple. However, it is time consuming and expensive as frequent measurements are required. Soil moisture sensors are able to measure soil moisture content indirectly. In recent years, various types of soil moisture sensing devices have been developed and made commercially available for water management applications. Some of these devices are capable of wirelessly transferring the data collected from their sensors.

Evaluations have shown that each type of the sensing devices has its advantages and shortcomings in terms of accuracy, reliability, and cost (Basinger et al., 2003; Chanzy et al., 1998; Evett and Parkin, 2005; Seyfried and Murdock, 2004; Yao et al., 2004). The neutron probe has been shown to be a reliable tool for determining soil water content. However, its use of radioactive source, the maintenance requirement, and the cost have restricted its application in recent years. Meanwhile, electromagnetic (EM) sensors, such as electrical capacitance and resistance type sensors, and time-domain reflectometer (TDR) devices have been rapidly developed and adopted for soil-water measurement (Dukes and Scholberg, 2004; Fares and Alva, 2000; Miranda et al., 2005; Seyfried and Murdock, 2001; Vellidis et al., 2008). Previous research indicated that the EM sensors could be useful tools to determine soil moisture status. However, the sensors must be well-calibrated under specific operation conditions including soil type and temperature (Yoder et al., 1997; Leib et al., 2003; Evett et al., 2006; Sui et al., 2013).

The objective of this study was to develop a practical method to use soil moisture sensor for irrigation scheduling in humid region.

Procedures

Site and Devices. Soil moisture sensing system was implemented at a Research Farm of USDA-ARS Crop Production Systems Research Unit at Stoneville, MS. The system included soil volumetric water content sensors (EC5, GS1, Decagon Devices, Pullman, WA), two models of data loggers (Em50R and Em50G, Decagon Devices) and a data station (ECH2O 900 MHz, Decagon Devices) for data acquisition, and a measurement and control data logger (MCDL) (CR1000, Campbell Scientific, Logan, UT) coupled with a wireless modem (RavenXTG, Sierra Wireless, Carlsbad, CA) for data collection and transmission. An omni-directional antenna was selected to work with the modem. The antenna was mounted on the top of a 3-m tower. The data station, MCDL, and wireless modem were housed together in a water-proof fiberglass box. Em50G and Em50R data loggers used batteries for power supply while the data station and MCDL were powered by a solar power supply (Sui and Baggard, 2015). The system were installed across three fields under coverage of a center pivot irrigation system (Figure 1). The soil type of the fields varied from silt to silt loam. Cotton, corn, and soybean crops were grown in the fields. Field size was approximately 7.5, 6.6, and 6.6 ha for cotton, corn, and soybean, respectively. In growing season of 2015 and 2016, six Em50G loggers were installed in cotton field, while five Em50R and one Em50G in corn field, and five Em50R and one Em50G in soybean field. Soil types were considered in selecting locations to install the sensors. Since soil type relates to soil EC, EC maps of the fields were used to identify the sensing locations in the three fields. At least one sensing location was chosen for each soil type so that the soil-water variability within a field could be observed.

EC5 and GS1 sensors were used with the data logger Em50R and Em50G to measure soil volumetric water content. The Em50R logger used a 900-MHz frequency radio to transmit data to the data station. After the data were automatically downloaded from the data station to the MCDL, the wireless modem automatically transmitted all data through a commercial wireless network to make the data accessible on the internet using a LoggerNet data logger support software (Campbell Scientific, Logan, UT). In the Em50G logger system, data were transmitted through a cellular communication network to a service which made data available on the internet. A software form Decagon was used to download the data and display the soil moisture graphics. Both Em50G and Em50R logger had the capacity to collect data from up to 5 sensors. In this study three soil moisture sensors were used with one data logger.

Sensor Calibration. The EC5 sensors were tested with various Mississippi soils. Six 183cm x 183cm x 71cm wooden compartments were built inside a greenhouse, and each compartment was filled with one type of soil from the Mississippi Delta. The sensors with the data loggers were installed in the soil compartments to measure soil water content. Using a sprinkler water was applied to the soils. Soil samples were periodically collected from the compartments to determine the soil water content using gravimetric method. The soil water content measured by the sensor readings were compared with the soil water content determined by the gravimetric method for sensor calibration.

The sensors were also evaluated using another weighing method. Five 2-Gallon pots were filled with the soil from the field where the sensors were installed. Dry matter and bulk density of the soil in the pot was determined. One EC5 sensor was installed in each pot. After the pot was saturated with water, the pot with the sensor was placed on an electronic scale to continuously measure the weight. Meanwhile, the sensor measured the soil water content. Soil water content in the pot was calculated using the scale-measured weight and the known soil dry matter and bulk density. Scale-measured water content was then compared with the sensor-measured water content for the calibration.

Sensor and logger Installation. The sensors were installed at depths of 15cm, 30cm, and 61cm, respectively. To install the sensors, a hole was drilled at the center of the crop row using a soil auger. The soil moisture sensors were inserted horizontally into the soil at the designated depths. All Em50R and Em50G data loggers were set up to continuously make one measurement of soil water content in every minute and calculated the hourly average of the measurement. Then, the readings of soil water content were wirelessly transmitted to the data station at a time interval of one hour.

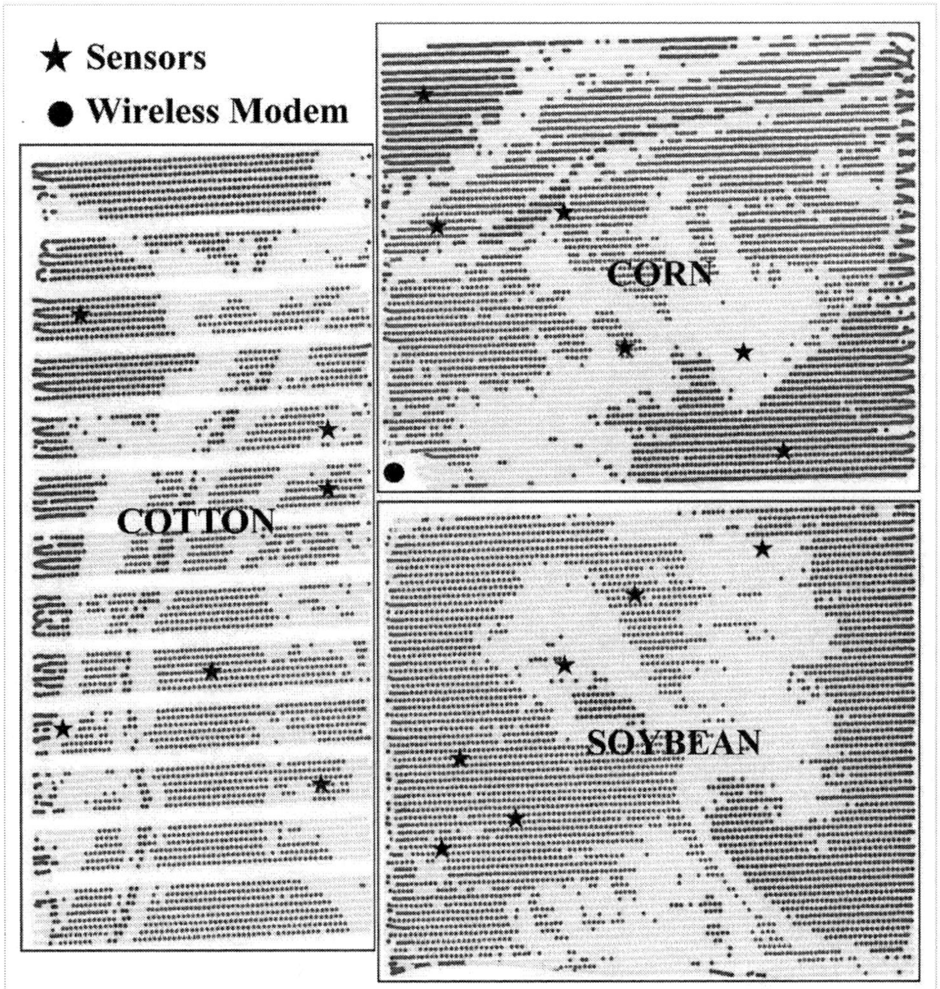

Figure 1. Field layout of the soil moisture sensors. The background is soil electric conductivity maps.

It was found that data transmission of Em50R logger was affected by plant canopy. Data could not properly transmitted when plant canopy was higher than the antenna of Em50R logger. To use a pole to place the antenna to a higher position could eliminate this issue. However, the antenna poles in field will obstruct normal operation of the equipment such as sprayers. To make the wireless sensor more practical for field use, an antenna mount was developed for Em50R data logger. The antenna mount includes a spring, a U-shaped metal base, and a PVC pipe. The spring was mounted between the center of the base and one end of the pipe. A hole was drilled in the pipe at about 30cm from the spring-pipe joint for pulling antenna cable inside the pipe for cable protection. The antenna connected to the cable was installed inside the top end of the pipe. In use of the mount in field, the U-shaped base was inserted into the soil, and the antenna cable was connected to the wireless device (Figure 2). As the agricultural equipment passed over the wireless device and impacted the PVC pipe, the spring in the mount would be bent and the antenna inside the PVC pipe would be protected from damage (Sui and Baggard, 2015).

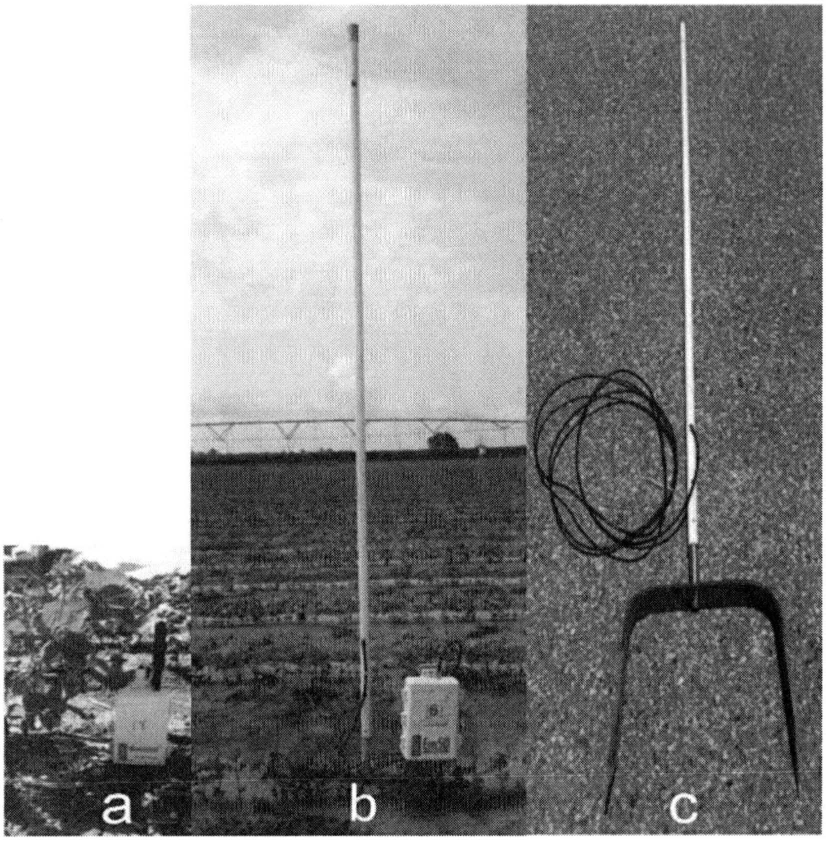

Figure 2. (a) Em50G logger installed in cotton field; (b) Em50R logger with the antenna mount in soybean field; (c) Antenna mount.

Results and Discussion

Sensor Calibration. Readings from the sensors have a linear correlation with the gravimetric soil water content. The sensors over-estimated the water content when the manufacture's "mineral soil" calibration was used. The sensors were capable of detecting general trend of soil-water changes. However, their measurements varied among the sensors and were influenced by soil texture. To obtain accurate measurements, the sensors require soil-specific calibration.

Data Processing and Application. Soil water content measured at the 3 depths were interpreted using a weighted average method to better reflect the status of soil-water in plant root zone. Weighted average of soil water content was used for irrigation scheduling. A weight was assigned to each sensor measurement based on the sensor depth. The weighted average measured by the sensors at 48 hours after the soil is saturated was used as sensor-measured field capacity (FC). Irrigation was triggered when soil moisture content was dropped close to the level approximately 50% of plant available water.

This soil moisture sensor-based irrigation scheduling method has been used in cotton, corn, and soybean crops. In general, it worked fairly well. However, it has been observed that the weights assigned to the measurement at different depths and the threshold to trigger an irrigation event should be adjusted according to crop type and crop growth stages due to the difference of crop root distribution patterns in the soil profile.

Installation and maintenance. The sensor installation and maintenance are critical in application of soil moisture sensors. The sensors should be installed in a representative area of the field. In installation, it needs to make the sensor prongs contact the soil well, and minimize the disturbance of the soil profile. The sensor should be installed on crop row and the plants near the sensing location should not be damaged during sensor installation and maintenance. After growing season, the loggers can be disconnected from the sensors and removed from field. The sensor can remain in the soil for use in next season. However, attention should be given to prevent the sensor from being damaged in field practices such as subsoil tillage. The section of sensor cable above the ground could be damaged by wild animals in field. It needs to be protected using physical or chemical means (e.g., flexible aluminum conduit or spray).

Conclusion

The soil moisture sensors were evaluated for use in irrigation scheduling in a humid region. The sensors with wireless data loggers were able to monitor soil-water status, and the sensor measurements could be used as a guidance for irrigation scheduling. For better results, the sensors require the calibration with soils from the field where the sensors will be installed. In one sensing location, it was suggested to install 3 sensors at different depths across crop root zone. Crop root distributions across the root zone and crop growth stages should be considered when the sensor readings at different measurement depths are used to determine a threshold to trigger irrigation events.

Disclaimer. Mention of a trade name, proprietary product, or specific equipment does not constitute a guarantee or warranty by the U. S. Department of Agriculture and does not imply approval of the product to the exclusion of others that may be available.

References

Basinger, J.M., G.J. Kluitenberg, J.M. Ham, J.M. Frank, P.L. Barnes, and M.B. Kirkham. 2003. Laboratory evaluation of the dual-probe heat pulse method for measuring soil water content. Vadose Zone J., 2:389–399.

Chanzy, A., J. Chadoeuf, J.C. Gaudu, D. Mohrath, G. Richard, and L. Bruckler. 1998. Soil moisture monitoring at the field scale using automatic capacitance probes. Eur. J. Soil Sci., 49:637–648.

Dukes, M.D. and J.M. Scholberg. 2004. Soil moisture controlled subsurface drip irrigation on sandy soils. Applied Engineering in Agriculture, 21(1):89-101.

Earth Gauge. 2011. http://www.earthgauge.net/wp-content/fact_sheets/CF_Drought.pdf. Accessed in May 28, 2011.

Evett, S.R., and G.W. Parkin. 2005. Advances in soil water content sensing: The continuing maturation of technology and theory, Vadose Zone J., 4:986–991, doi:10.2136/vzj2005.0099.

Evett, S.R., J.A. Tolk, and T.A. Howell. 2006. Soil profile water content determination: sensor accuracy, axial response, calibration, temperature dependence, and precision. Vadose Zone J., 5:894-907

Fares, A. and A.K. Alva. 2000. Evaluation of capacitance probes for optimal irrigation of citrus through soil moisture monitoring in an entisol profile. Irrig. Sci., 19:57-64

Leib, B.G., J.D. Jabro, G.R. Matthews. 2003. Field evaluation and performance comparison of soil moisture sensors. Soil Science, 168(6):396-408.

Miranda, F.R., R.E. Yoder, J.B. Wilkerson, and L.O. Odhiambo. 2005. An autonomous controller for site-specific management of fixed irrigation systems. Computer and Electronics in Agriculture, 48(3)183-197.

Seyfried, M.S., and M.D. Murdock. 2001. Response of a new soil water sensor to variable soil, water content, and temperature. Soil Sci. Soc. Am. J., 65:28–34.

Seyfried, M.S., and M.D. Murdock. 2004. Measurement of soil water content with a 50-MHz soil dielectric sensor. Soil Sci. Soc. Am. J., 68:394–403.

Sui, R. and J. Baggard. 2015. Wireless sensor network for monitoring soil moisture and weather conditions. *Appl. Eng. in Agric.*, 31(2):193-200. DOI 10.13031/aea.31.10694.

Sui, R., H.C. Pringle, and E.M. Barnes. 2013. Evaluation of soil moisture sensors. In Proc. Irrigation Show and Education Conference. Falls Church, VA. : Irrigation Association.

UCS. 2011. http://www.ucsusa.org/gulf/gcstatemis_cli.html. Accessed in May 28, 2011.

Vellidis, G., M. Tucker, C. Perry, C. Kvien, C. Bednarz. 2008. A real-time wireless smart sensor array for scheduling irrigation. Computer and Electronics in Agriculture, 61:44-50.

Vories, E.D. and S.R. Evett. 2010. Irrigation research needs in the USA Mid-South and Southeast, humid and sub-humid regions. 5th National Decennial Irrigation Conference Proceedings, 5-8 December 2010, Phoenix Convention Center, Phoenix, Arizona USA IRR10-8679.

Yao, T., P.J. Wierenga, A.R. Graham, and S.P. Neuman. 2004. Neutron probe calibration in a vertically stratified vadose zone. Vadose Zone J., 3:1400–1406.

Yoder, R.E., D.L. Johnson, J.B. Wilkerson, D.C. Yoder. 1997. Soil water sensor performance. Applied Engineering in Agriculture, 14(2):121-133.

This page intentionally left blank.

Relating soil available water fraction to water stress indices

Steven R. Evett, Research Soil Scientist
USDA-ARS Conservation & Production Research Laboratory, Bushland, Texas,
steve.evett@ars.usda.gov

Susan A. O'Shaughnessy, Research Agricultural Engineer
USDA-ARS Conservation & Production Research Laboratory, Bushland, Texas,
susan.o'shaughnessy@ars.usda.gov

Paul D. Colaizzi, Research Agricultural Engineer
USDA-ARS Conservation & Production Research Laboratory, Bushland, Texas,
paul.colaizzi@ars.usda.gov

Robert C. Schwartz, Research Soil Scientist
USDA-ARS Conservation & Production Research Laboratory, Bushland, Texas,
robert.schwartz@ars.usda.gov

Abstract. *Thermometric infrared sensing of production agricultural fields is becoming more commonplace using satellite, aerial and proximal remote sensing platforms, including arrays of infrared thermometers mounted on moving irrigation systems. Using plant canopy temperature data for irrigation management has been thoroughly explored and working solutions have been implemented to trigger irrigations when plant water stress reaches a given threshold value. Using spatial canopy temperature data, maps of plant stress indices, such as the crop water stress index (CWSI), can be produced and used to trigger irrigation in those parts of a field in which the stress index exceeds the threshold, thus enabling variable rate irrigation (VRI). However, knowing the degree of plant water stress does not clearly translate into knowing the amount of irrigation to apply. Past research has shown a strong correlation between crop water stress index (CWSI) and stem water potential. The stem water potential is strongly correlated to the soil water potential in the root zone, although an exact analytical expression for the relationship is lacking, largely due to the dynamic nature of root growth and soil water redistribution in response to irrigation and precipitation. A strong relationship between CWSI and root zone soil water storage has been demonstrated during the latter part of the growing season in crops that were irrigated at different levels throughout the season, but the relationship earlier in the season was not strong in that earlier research, likely because the crop had not yet been severely stressed. At any rate, determining the soil water status before the onset of severe stress is the real objective and one not met by focusing only on soil water storage. Idso and Jackson demonstrated a relationship between the CWSI and the fraction of plant available soil water storage, not just the entire soil water storage. The present work involves a preliminary analysis of the relationship between fractional plant available soil water and CWSI measured at three levels of deficit irrigation of corn in 2016. The strength of correlations depended on the deficit level and the period of the season for which data were analyzed, and correlations were sometimes linear and sometimes nonlinear, particularly for stronger deficits later in the season. It appears that the effective soil water potential over the root zone will have to be computed in order to develop a stronger relationship with CWSI.*

Keywords. Fractional available soil water, variable rate irrigation, accuracy, management allowed depletion, canopy temperature, crop water stress index

The U.S. Department of Agriculture (USDA) prohibits discrimination in all its programs and activities on the basis of race, color, national origin, age, disability, and where applicable, sex, marital status, familial status, parental status, religion, sexual orientation, genetic information, political beliefs, reprisal, or because all or part of an individual's income is derived from any public assistance program. (Not all prohibited bases apply to all programs.) Persons with disabilities who require alternative means for communication of program information (Braille, large print, audiotape, etc.) should contact USDA's TARGET Center at (202) 720-2600 (voice and TDD). To file a complaint of discrimination, write to USDA, Director, Office of Civil Rights, 1400 Independence Avenue, S.W., Washington, D.C. 20250-9410, or call (800) 795-3272 (voice) or (202) 720-6382 (TDD). USDA is an equal opportunity provider and employer.

The mention of trade names of commercial products in this article is solely for the purpose of providing specific information and does not imply recommendation or endorsement by the U.S. Department of Agriculture.

Soil Water Criteria for Irrigation Scheduling

Irrigation scheduling using soil water sensors is an exercise in maintaining the water content of the crop root zone soil above a lower limit defined by the management allowed depletion (MAD) for that soil and crop, but not so wet that too much water is lost to deep percolation, evaporation and runoff. The management allowed depletion for a corn crop on a clay loam soil is only about 0.06 inch/inch. To be useful for managing water to prevent over filling the soil or allowing it to dry so much that the crop yield is compromised more than acceptable, soil water sensors must be accurate. The accuracies needed are on the order of 0.02 to 0.04 inch/inch fractional soil water content (Table 1), which is better than many commercial soil water sensors are able to provide. This, plus the cost and practical problems associated with installing and maintaining soil water sensor networks in the field, motivates alternative approaches to irrigation scheduling based on canopy temperature sensing of crop water stress.

Values of field capacity and permanent wilting point for a particular field (needed for determining the available water holding capacity and MAD values) may be found from NRCS soil maps, at least to a close approximation. The values are, however, likely to change with depth in the soil and with position in the field, meaning that irrigation management should be site specific to be most effective, and to do that requires sensors be installed in the different soils of the field or some other means of mapping crop water stress be employed. NRCS soil maps are available on the Internet: http://websoilsurvey.sc.egov.usda.gov/App/HomePage.htm.

Plant root zones deepen during the growing season, often extending to five foot depth and sometimes deeper if the soil does not have a restrictive horizon and soil water content at depth is large enough to encourage root penetration. Since soil water sensors typically are sensitive only to the soil immediately around them, and since most sensors are small, it is typical that two or more sensors must be installed at different depths in order to gain acceptable understanding of how soil water content is changing in response to irrigation and crop water uptake. Depths of six and 18 inches or six and 24 inches are common. Seeing that the soil is above field capacity at 24 inches may indicate that deep percolation losses are occurring.

Table 1. Example calculation of available water holding capacity†, θ_{AWHC}, which is defined as the difference in water content between field capacity (θ_{FC}) and permanent wilting point (θ_{PWP}), in three soils with widely different textures. Also shown is the management allowed depletion (MAD, inch/inch or $m^3\ m^{-3}$). The small range of MAD severely tests the abilities of most soil water sensors, particularly for the loamy sand soil.

Horizon	θ_{PWP}	θ_{FC}	θ_{AWHC}		MAD		MAD
	----------- $m^3\ m^{-3}$ or in/inch ---------				fraction		$m^3\ m^{-3}$
silt loam	0.086	0.295	0.209	×	0.6	=	0.126
loamy sand	0.066	0.103	0.037	×	0.6	=	0.022
clay	0.190	0.332	0.142	×	0.6	=	0.085

† θ_{FC}, θ_{PWP}, and θ_{AWHC} are soil water contents at field capacity, at the permanent wilting point, and the plant-available water holding capacity (designated as AWHC).

Spatial Variability of Soil Properties

Center pivots are sometimes placed on sloping land and on land that changes from one soil type to another across the area covered by the pivot. Sometimes soil type changes are unrelated to slope and aspect, for example in glacial till soils, flood plains, salt affected soils, etc. Figure 2 illustrates a situation with both slope and soil type variations. There are four soil types irrigated by this pivot, the Lazbuddie clay and Loften clay (LcA and LoA) are in irrigation capability class 2 due to their small slopes and deep profiles. They represent the margins of a playa. The Pullman clay loam (PuB) under the pivot is in class 3 due to its greater slope and potential for runoff. The Pep clay loam has slopes of 3 to 5% and so is in class 4 due to very high runoff potential. Site-specific, variable rate irrigation could be used to reduce irrigation rates on the areas with high runoff potential.

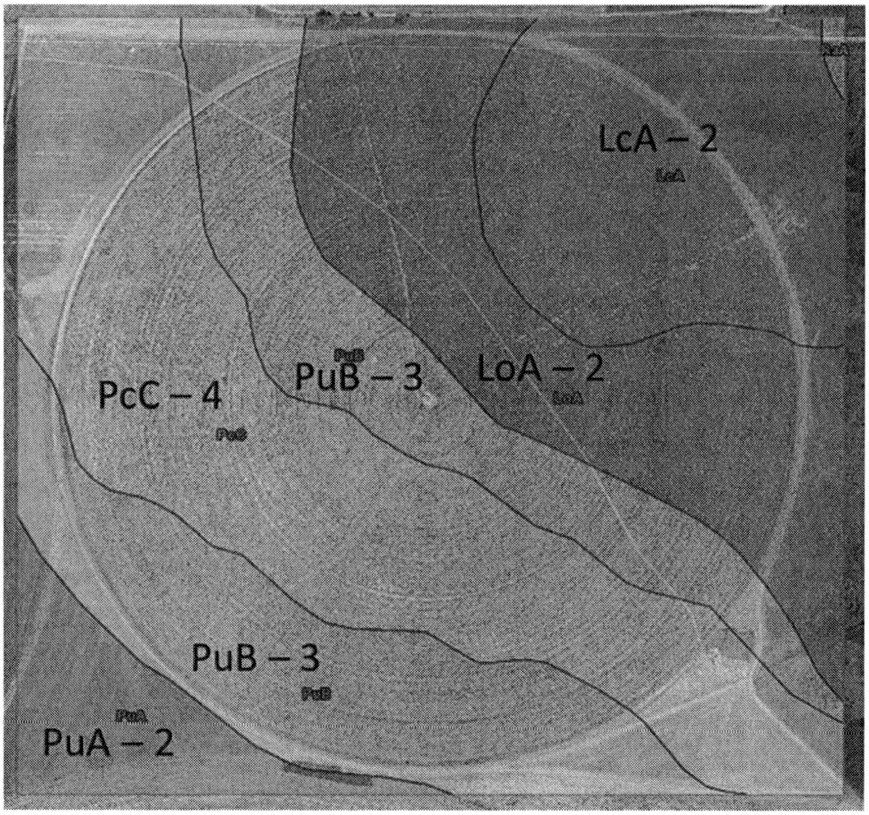

Figure 1. Soil and irrigation capability classification map from the NRCS soil survey web site for a center pivot in the Texas Panhandle. Letter codes indicate soil type; numbers indicate irrigation capability class. See Table 2 for details.

The soils illustrated in Figure 1 are all clays or clay loams, but do differ somewhat in available water holding capacity (Table 2). The most important difference between them is their slope, in particular the greater slopes of the Pep clay loam. One could lump the Lofton, Lazbuddie and Pullman soils together in terms of soil water sensing, leaving only the Pep soil to be sampled separately. However, the interpretation of soil water content data should be viewed in light of the FC and PWP values for each soil type, which means that a given water content will have different meaning in different soils. Depending on the degree of lumping, sensors in eight locations may be necessary to guarantee that soil water content variations are adequately captured.

Table 2. Summary by Map Unit of Classifications in the Area of Interest (AOI) in Figure 2.

Map unit symbol	Map unit name	Rating	AWHC* (in/in)	Acres in AOI	Percent of AOI
LcA	Lazbuddie clay, 0 to 1 percent slopes	2	0.161	40.4	17.1%
LoA	Lofton clay loam, 0 to 1 percent slopes	2	0.140	46.5	19.7%
PcC	Pep clay loam, 3 to 5 percent slopes	4	0.170	68.7	29.1%
PuA	Pullman clay loam, 0 to 1 percent slopes	2	0.165	14.3	6.1%
PuB	Pullman clay loam, 1 to 3 percent slopes	3	0.158	65.3	27.7%
RaA	Randall clay, 0 to 1 percent slopes, frequently ponded			0.5	0.2%
Totals for Area of Interest (the square, not the circle)				235.7	100.0%

***AWHC is available water holding capacity, the water that the soil holds between field capacity and permanent wilting point. In this case it is given in inch per inch for the top 40 inches of soil.**

Canopy Temperature and Soil Water Sensing – Relationship to Crop Water Stress

Factors Influencing Crop Spatial Variability

Crop variations in space are influenced by other factors in addition to soil type, texture, salinity, depth and depth to restricting layers. Slope and aspect affect runoff and evaporative demand. In hilly terrain, evaporative demand is typically greater on south facing slopes than on north facing slopes. Disease and insect pressure can create field variability, as can temporary ponding due to runoff, or lack of sufficient infiltration of applied irrigation due to runoff from steeper slopes. Of course, agronomic mistakes in planting, spraying and fertilization can also create variability in the crop, which will translate into variability in crop water uptake rates and soil water content variability. Several of these factors cannot be ameliorated by irrigation, but irrigation can be varied in response to save water. For example, irrigation of areas of a field hard hit by disease or insect pressure may no longer be economically viable, in which case a Site-Specific VRI (SSVRI) system prescription can be written to stop irrigation in those areas. Irrigation can be reduced on field areas in which slope is causing runoff problems, thus ameliorating parts of the field prone to ponding and water logging. Soil water sensors placed in these two areas (sloping and prone to water logging) will detect problems of lack of soil water on slopes and excess of soil water on areas that pond. Reducing irrigation rates on sloping areas will, however, likely lead to crop water stress there, which can only be addressed by extra irrigations on those areas. While an SSVRI system may allow this site-specific irrigation to occur, there may not be time in the irrigation schedule to allow these extra irrigations on sloped areas. Also, it should be recognized that evaporative loss is a greater fraction of smaller irrigations than of larger irrigations (Tolk et al., 2014).

Plant Water Status Mapping – Connection to ET & Soil Water Status

In dry climates, greater crop canopy temperatures (Fig. 2) indicate greater crop water stress because crops with sufficient soil water availability are cooled by transpiration. Irrigation scheduling based on crop water stress can be accomplished automatically (or manually) using crop water stress data from infrared temperature sensors mounted on moving irrigation systems (Peters and Evett, 2008). An empirical crop water stress index (eCWSI) based on georeferenced data from sensors on a center pivot lateral can be mapped to show spatial changes in crop water stress that develop over time. Crop leaf temperature data from infrared thermometers (IRTs) can be combined with on-site measured weather data from inexpensive weather stations to calculate a crop water stress index that is integrated over the daylight hours to improve stability, resulting in maps of the integrated Crop Water Stress Index (iCWSI) for an entire field (O'Shaughnessy et al., 2010). The iCWSI is well correlated with plant stem water potential, which is a direct indicator of plant water stress. Automated VRI irrigation using a supervisory control and data acquisition (SCADA) system has been demonstrated to produce yields and crop water use efficiencies as good as or better than those resulting from irrigation scheduling using the best scientific irrigation scheduling method – the neutron probe used weekly in many access tubes spread over a field (Evett et al., 2006; O'Shaughnessy et al., 2012a), and has been recently patented (Evett et al., 2014a). These methods are being transferred to commercial center pivot irrigation systems for eventual sale to producers. The wireless IRTs eliminate initial and maintenance costs of wiring (O'Shaughnessy et al., 2012b, 2013); and this IRT technology has been transferred to a manufacturer who offers it for sale (model SapIP-IRT, Dynamax, Inc., Houston, TX).

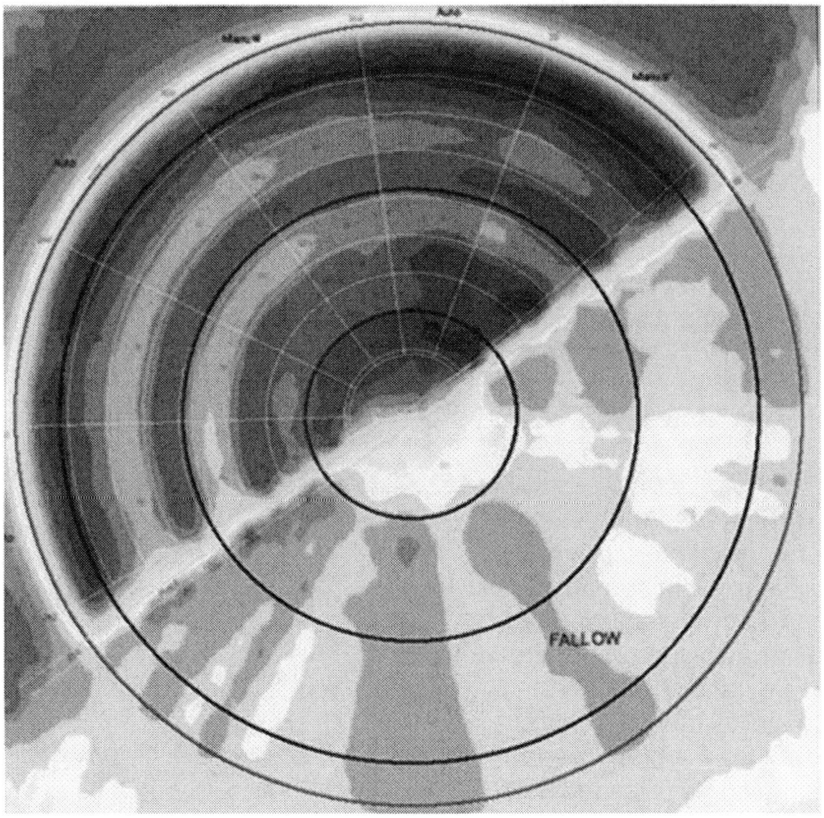

Figure 2. False color image of a center pivot irrigated field showing increasing canopy temperature as irrigation deficit increased from full irrigation (coolest temperature, dark blue) to more deficit irrigation (lighter blue) and to a non-irrigated (dryland) treatment (yellow, brown and red, hottest temperature).

Planting a crop can be seen as the installation of many thousands of sensitive biological soil water sensors per acre. Crop water stress is well correlated with leaf water potential, which in turn is correlated with soil profile water content.

Knowing how much water a soil can accept can be determined from soil water sensors at the locations where those sensors are installed, but is unknown elsewhere in the field. Soil water content is spatially variable horizontally and vertically in all soils and temporally variable over the irrigation season (e.g., Padhi et al., 2011). Spatial variation in soil water content occurs even in level fields (e.g., Longchamps et al., 2015). Spatio-temporal dependence of soil water content can vary with respect to soil types, but the variable that is important for irrigation management is the fractional plant available soil water (fPASW) in the root zone. For example,

$$fPASW = (SWC-PWP)/(FC-PWP)$$

Where SWC is the soil water content, FC is soil water content at field capacity and PWP is the soil water content at permanent wilting point. Fractional PASW has a range of 0 to 1. Rab et al. (2009) found that PASW was strongly correlated with wheat yield in Australia.

While in-situ soil water sensors provide frequent soil water content measurements, it often is not feasible to install an adequate number of sensors to accurately represent soil water content variability in producers' irrigated fields (Hedley and Yule, 2009). However, it is possible to map the CWSI (Fig. 3, A) using data from a plant feedback ISSCADA system moving across the field (Peters and Evett, 2008; O'Shaughnessy et al., 2016). Since a strong relationship exists between the CWSI (range of 0 to 1) and the fractional PASW (0 to 1) (Fig. 3, B) (Jackson et al., 1981), data on the fractional PASW in the root zone obtained at a few locations could be cokriged with a map of

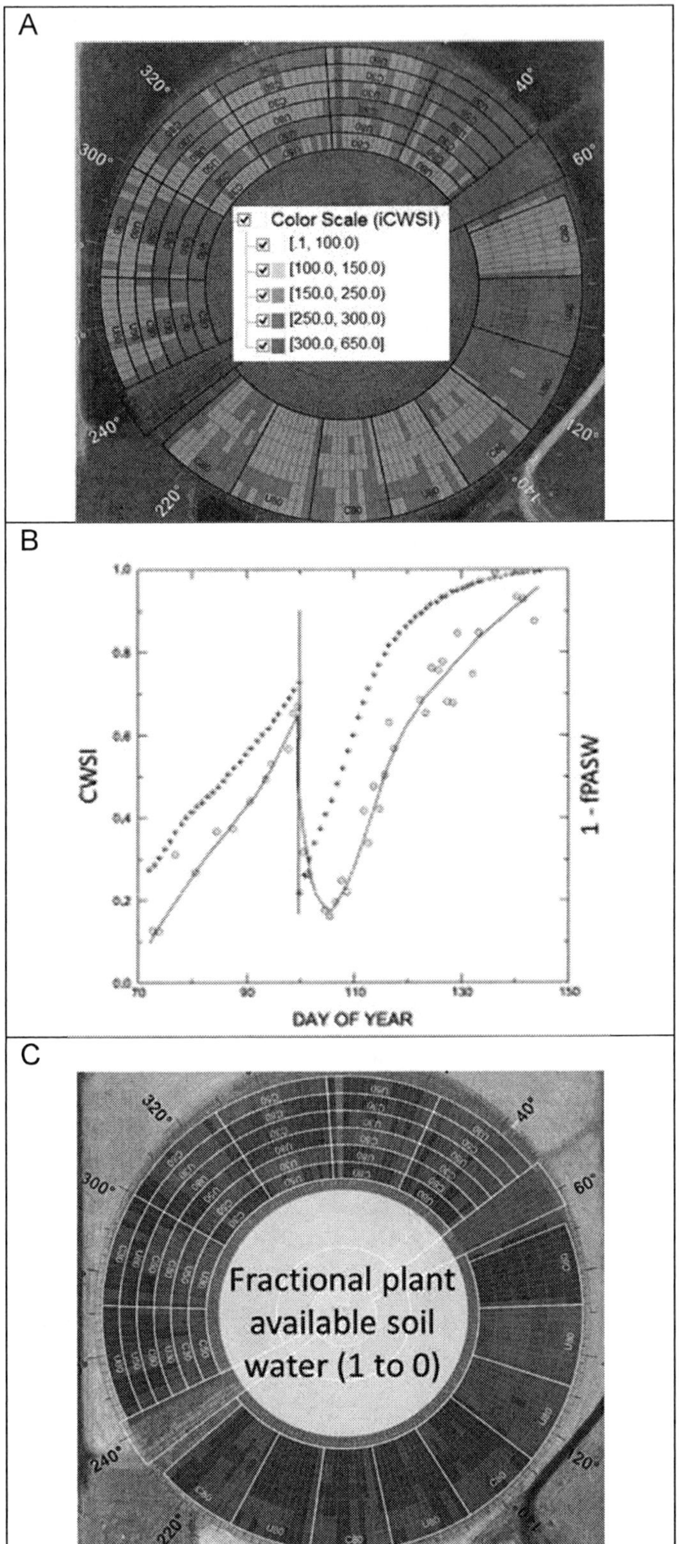

Figure 3. (A) Map of CWSI, (B) Relationship between CWSI and 1 – fPASW (Jackson et al., 1981), (C) Illustrative map of fPASW.

CWSI to produce a map of fractional PASW (Fig. 3, C). Studies have demonstrated how cokriging can be used to reduce the required intensity of expensive soil sampling or sensing methods by building a relationship between the sampled property and an easily measured one, and then repeating the easy sampling over space to achieve strong representation of the spatial variability of the less easily measured property (e.g., Goovaerts, 1998; Rab et al., 2009). Indeed, Yates (1986) and Yates and Warrick (1987) estimated the surface soil water content by cokriging using more numerous measurements of the soil surface temperature. A model that maps fraction of PASW is needed to overcome the infeasibility of installing numerous soil water sensors in a single field and to aid in mapping the fraction of PASW. A feasible model could incorporate cokriging of fractional PASW and CWSI combined with TSEB modeling of ET to determine the rate at which fractional PASW is changing, thus providing for determining and forecasting needed water application depths. Such a map could be uploaded as an instructive visual aid and used to develop a VRI prescription for whole-field irrigation management that provides water in precise amounts to precise locations, mitigating the limitations of the all-or-nothing prescription mapping described by Peters and Evett (2003) and others.

In previous research, cotton CWSI values were well correlated with soil profile water content within the root zone (Fig. 4), at least relatively later in the irrigation season (Evett et al., 2014b). Other research has shown that the crop temperature data can be used in energy and water balance models of crop water use (ET), which can be used to estimate changes in soil water content over time (Colaizzi et al., 2003), thus closing the circle between soil water sensing and crop water stress sensing.

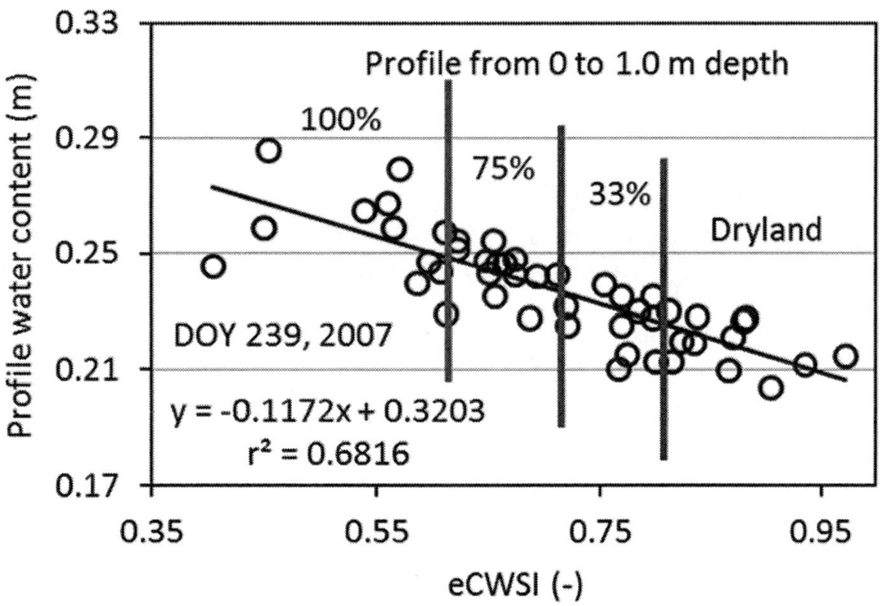

Figure 4. The soil profile water content within the crop root zone is correlated with the empirical crop water stress index (eCWSI). Different irrigation amount treatments produced groups of eCWSI values that did not overlap along the regression line, illustrating that eCWSI is a good surrogate for profile water content. Irrigation treatments were full (100%), 75% of full, 33% of full and dryland.

2016 Trials

The work of Jackson et al. (1981) was done using the neutron probe on a daily basis for a couple of irrigation cycles, but did not allow more thorough seasonal examination of the relationship between fPASW and CWSI. A variable rate irrigation study of corn response to varying irrigation levels provided a test bed for examining the relationship between fPASW and CWSI over a longer time scale at Bushland, Texas in 2016. Soil water sensors (Model TDR-315 true TDR sensors, Acclima, Inc., Meridian, ID) were installed at depths of 3.9, 7.9, 11.8, 17.7, 27.6, and 39.4 inches (10, 20, 30, 45, 70, and 100 cm) at each of three locations. The locations were irrigated weekly to replace either 30%, 50% or 80% of the weekly crop water use as determined using a field-calibrated neutron probe. At each location, infrared thermometers (model SAPIP-IRT, Dynamax, Inc., Houston, TX) were positioned to view the corn canopy above the soil water sensing profile from opposite oblique angles in order to reduce sun angle effects on the mean temperature. The daily integrated CWSI was calculated according to methods of O'Shaughnessy et al. (2016). From these values, a daily fractional iCWSI was computed as the ratio of iCWSI for a day to the value 370, which was slightly more than the maximal daily iCWSI determined for the season. Fractional soil water depletion (1 – fPASW) was calculated based on field capacity water content of 0.35 inch/inch and permanent wilting point water content of 0.19 inch/inch.

For the entire growing season, fractional soil water depletion was more significantly related to fractional iCWSI for the most heavily irrigated treatment (80%) and the importance of the relationship declined as deficit irrigation became more severe (Table 3). The implication of this is that the likelihood of developing a useful map of soil water depletion declines as irrigation deficit increases.

Table 3. Linear regression relationships between fractional soil water depletion (0-1, dimensionless) to fractional integrated crop water stress index (0-1, dimensionless) for corn over the entire growing season at Bushland, Texas in 2016. SE is the standard error

Irrigation Level	Slope (P value)	Intercept (P value)	r^2	SE	N
80%	0.276 (<0.01)	0.14 (<0.01)	0.45	0.06	55
50%	0.227 (<0.01)	0.43 (<0.01)	0.10	0.12	60
30%	0.181 (0.064)	0.34 (<0.01)	0.05	0.12	55

If data from only the latter part of the irrigation season were used (day of year 244 to 275, well after canopy closure), then the linear relationships between fractional soil water depletion and fractional iCWSI was much less strong for all three irrigation levels, with r^2 values of 0.21 for 80% irrigation, 0.05 for 50% irrigation, and 0.07 for 30% irrigation. However, polynomial relationships between fractional soil water depletion and iCWSI (not fractional) were strong for this latter part of the season (r^2 = 0.83, 0.69 and 0.55 for irrigation levels of 80%, 50% and 30%), and became more nonlinear for the more severely deficit irrigation levels, indicating a problem with a simple fractional soil water depletion metric for developing this relationship. As soil dries the relationship between soil water potential and water content becomes quite nonlinear, so it is likely that conversion of fractional PASW to a fractional soil water potential metric is necessary to develop a stronger relationship.

Summary

The first year of research to develop a relationship between fractional plant available soil water (or fractional soil water depletion) and a crop water stress index was exploratory but provided useful data for analysis. Further analysis will focus on using the relationship between soil water potential and soil water content to develop a fractional plant available soil water potential index that can reasonably be expected to relate to crop water stress. It appears that the effective soil water potential over the root zone will have to be computed in order to develop a stronger relationship with CWSI. This may require a computer modeling approach similar to that employed in the ENWATBAL model (Evett and Lascano, 1993) in which root length density changes with depth and over time are modeled and then used to scale effective soil water potential.

Acknowledgements

This research was supported in part by the Ogallala Aquifer Program, a consortium between USDA-Agricultural Research Service, Kansas State University, Texas AgriLife Research, Texas AgriLife Extension Service, Texas Tech University, and West Texas A&M University.

References

Colaizzi, Paul D.; Edward M. Barnes; Thomas R. Clarke; Christopher Y. Choi and Peter M. Waller. 2003. Estimating Soil Moisture under Low Frequency Surface Irrigation Using Crop Water Stress Index. J. Irrig. Drainage Eng., 129(1):27-35. Manuscript in Adobe format (157 kb)

Evett, S. R. and R.J. Lascano. 1993. ENWATBAL.BAS: A mechanistic evapotranspiration model written in compiled BASIC. Agron. J. 85(3):763-772.

Evett, S.R., R.T. Peters and T. A. Howell. 2006. Controlling water use efficiency with irrigation automation: Cases from drip and center pivot irrigation of corn and soybean. Proc. 28th Annual Southern Conservation Systems Conference, Amarillo TX, June 26-28, 2006. 57-66. Manuscript in Adobe format (980 KB)

Evett, S.R., L.K. Heng, P. Moutonnet and M.L. Nguyen (eds.). 2008. Field Estimation of Soil Water Content: A Practical Guide to Methods, Instrumentation, and Sensor Technology. IAEA-TCS-30. International Atomic Energy Agency, Vienna , Austria . ISSN 1018–5518. Manuscript in Adobe format (3 MB)

Evett, S.R., S.K. Anderson, J.J. Casanova and R.C. Schwartz. 2014a. Patent application serial No. 13/404,491, Entitled "Soil Water and Conductivity Sensing System", filed 24 February 2012 with claims priority to Provisional Serial No. 61/515,381 – Filed August 5, 2011. Patent allowed on 7 October 2014.

Evett, S.R., S.A. O'Shaughnessy, P.D. Colaizzi and R.C. Schwartz. 2014b. Soil water sensing: Implications of sensor capabilities for variable rate irrigation management. Technical Session Proc., Irrigation Show Education Conference, 19-20 Dec. 2014, Phoenix Convention Center, Arizona. Irrigation Association, Falls Church, Virginia.

Goovaerts, P. 1998. Geostatistical tools for characterizing the spatial variability of microbiological and physio-chemical soil properties. Biol. Fertil. Soils 27:315–334.

Hedley, C.B., and I.J. Yule. 2009. A method for spatial prediction of daily soil water status for precise irrigation scheduling. Agric. Water Manage. 96:1737-1745.

Jackson, R.D., S.B. Idso, R.J. Reginato and P.J. Pinter Jr. 1981. Canopy temperature as a crop water stress indicator. Water Resour. Res. 17(4):1133-1138.

Longchamps, L., R. Khosla, R. Reich, and D.W. Gui. 2015. Spatial and temporal variability of soil water content in leveled fields. Soil Sci. Soc. Am. J. 79:1446-1454.

O'Shaughnessy, S.A., S.R. Evett, P.D. Colaizzi and T.A. Howell. 2010. Automatic irrigation scheduling of grain sorghum using a CWSI and time threshold. Proc. 5th Decennial National Irrigation Proc., 5-8 December 2010, Phoenix, Arizona. Paper No. IRR10-9011. ASABE, St. Joseph, Mich. (CD-ROM). Manuscript in Adobe format (208 kb)

O'Shaughnessy, S.A., S.R. Evett, P.D. Colaizzi and T.A. Howell. 2012a. Grain sorghum response to irrigation scheduling with the time-temperature threshold method and deficit irrigation levels. Trans. ASABE 55(2):451-461. Manuscript in Adobe format (1 MB)

O'Shaughnessy, S.A., S.R. Evett, P.D. Colaizzi and T.A. Howell. 2012b. Performance of a wireless sensor network for crop water monitoring and irrigation control. Paper # 12-1338246. 2012 ASABE Annual International Meeting, Jul 29 - Aug 1 2012, Dallas, Texas. St. Joseph, Mich.: ASABE. Manuscript in Adobe format (703 kb)

O'Shaughnessy, S.A., S.R. Evett, A. Andrade, F. Workneh, J.A. Price and C.M. Rush. 2016. Site-specific variable rate irrigation as a means to enhance water use efficiency. Trans. ASABE 59(1):239-249. DOI 10.13031/trans.59.11165.

Padhi, J., R.K. Misra, and J.O. Payero. 2011. Estimation of soil water deficit in an irrigated cotton field with infrared thermography. Field Crops Res. 126:5-55.

Peters, R.T. and S.R. Evett. 2008. Automation of a center pivot using the temperature-time-threshold method of irrigation scheduling. J. Irrig. Drain. Engr. 134(3):286-291. Manuscript in Adobe format (242 kb)

Rab, M., P. Fisher, R. Armstrong, M. Abuzar, N. Robinson, and S. Chandra. 2009. Advances in precision agriculture in south-eastern Australia. IV. Spatial variability in plant-available water capacity of soil and its relationship with yield in site-specific management zones. Crop and Pasture Sci. 60:885-900.

Tolk, J.A., S.R. Evett and R.C. Schwartz. 2014. Field-measured, hourly soil water evaporation stages in relation to reference evapotranspiration rate and soil to air temperature ratio. Submitted to Vadose Zone J.

Yates, S.R. 1986. Disjunctive kriging 3. Cokriging. Water Resour. Res. 22:1371-1376.

Yates, S.R., and A.W. Warrick. 1987. Estimating soil water content using cokriging. Soil Sci. Soc. Am. J. 51:23-30.

This page intentionally left blank.

Advanced Controller Features Show Unexpected Results

[1]David W. Smith, Extension Program Specialist
Texas A&M AgriLife Extension Service, Biological & Agricultural Engineering, College Station, Texas
[2]Charles Swanson, Extension Program Specialist
Texas A&M AgriLife Extension Service, Biological & Agricultural Engineering, College Station, Texas
[3]Jennifer Nations, Water Resource Coordinator
City of College Station, Texas

Abstract. Modern residential irrigation controllers include several features that enable fine-tuning of irrigation schedules to accommodate different plant water needs and improve overall irrigation efficiency. These include the water budget or seasonal adjust option, rain sensor connection ports, and multiple programs and start times capability. What portion of residents actually use these features, and among those who do, are they applying less water compared to those who do not? Since 2010 College Station, Texas Water Utilities performed more than 500 residential landscape irrigation 'checkups'. Using data collected from the 'checkups' and monthly water records, irrigation was compared for customers who do and do not use these controller options. Surprisingly, irrigation applied from residents with rain sensors consistently exceeded those without rain sensors. Residents who used multiple cycles per irrigation day applied slightly more water than those who used only one cycle, and residents who used multiple programs irrigated slightly more compared to those who used a single program. Although the difference in irrigation applied between groups was not statistically significant (p<0.05), results suggests that, whether because of lack of knowledge or because programming is too complex, the water savings benefit of these controller features was not realized for these residential customers.

Keywords. Turf/Landscape (Residential), sprinkler, conservation, scheduling, controllers

INTRODUCTION

Landscape irrigation for residential and commercial properties, golf courses, athletic fields, and other types of recreation areas is estimated to be the third-largest user of water in Texas, behind only agriculture and municipal uses (Cabrera et al., 2013). With a growing population and competing interests for limited water supply, local communities employ various strategies to reduce potable water consumption including tiered water rates, prescriptive irrigation days, citations for water waste, and education campaigns. The greatest potential for municipal water conservation is in landscape irrigation. A study of monthly water use of 800 residences from 2000 through 2002 in College Station, Texas indicated that the average peak water consumption during the summer increased as much as 3.3 times the winter water use (White et al., 2004) as plant water requirements typically exceed precipitation. With in-ground, automated irrigation systems much of this water is wasted due to poorly-designed systems, improper scheduling (over-irrigation), and failure of system integrity. Furthermore, residential water customers find programming and operation of their controller difficult and confusing. This presents a real challenge for municipal and water utility driven water conservation efforts that encourage strategies such as potential evapotranspiration-based (ETo) scheduling, multi-cycling irrigation events to prevent runoff, and prescriptive, address-based weekly operation schedules. These strategies assume that customers are proficient in programming their controllers. In fact, failure to properly implement these recommendations can be counter-productive and actually increase overall irrigation use.

In 2010, College Station Water Utilities began providing free landscape irrigation 'checkups' to residential customers. To date, the City has performed more than 500 irrigation checkups, primarily for customers identified as having above average seasonal water use (Coleman, 2014). The checkup includes a general inspection of system components to identify damaged or broken hardware, documentation and evaluation of existing controller programming, and education on how to reduce runoff and install rain shut off sensors. Following the checkup a written report is delivered to the customer detailing significant findings along with recommendations for reducing irrigation use. Beginning in May 2012, College Station Water Utilities, in collaboration with the Texas A&M Department of Recreation, Parks, and Tourism provided additional resources to perform irrigation system checkups to meet increasing demand for this service. A licensed irrigator was hired to conduct irrigation inspections during the peak irrigation season.

The objective of this study was to evaluate water consumption records for residential customers who do and do not utilize advanced residential controller features to determine any differences, patterns, or trends in the amount of irrigation applied (when normalized by unit landscape area). Results of this study can inform future educational efforts by the City and others to encourage irrigation efficiency and water conservation. These results may also be helpful to determine whether residential customers properly employ such advanced features or whether, because of lack of understanding or training, application of these features actually increases irrigation use.

METHODOLOGY

Landscape Irrigation Checkup

College Station Water Utilities publicizes the free irrigation checkup service by distributing a letter to approximately 5,000 residential customers whose historical water consumption substantially exceeded their estimated water budget. The City also hosts a series of summer 'sprinkler system workshops' in the summer season for approximately 200 residents per year. Interested residents contact City staff either by email or telephone to schedule an appointment with the irrigation inspector. In 2012, 2013, 2014, and 2015 irrigation checkups were conducted for 211 residential customers (205 unique customers). Irrigation checkups were performed in 44 subdivisions within the city of College Station. Fifty-eight percent of all inspections were conducted in four subdivisions: Pebble Creek (58), Castlegate (34), Emerald Forest (20), and Edelweiss Estates (11).

Data collected during the checkup included the number and type of application devices, brand and model of controllers, irrigation start times, run times, presence of rain shut-off sensors, and inventory of hardware deficiencies and operational problems.

Data Collected

- Controller brand
- Current controller time/date
- Irrigation start times
- Irrigation programs being utilized ('A', 'B', 'C', etc.)

- Individual station run times
- Seasonal adjust or water budget setting
- Presence of controller backup battery
- Presence and functionality of rain shut-off sensor
- Type(s) of sprinkler heads (per station)
- Dominant plant type (turfgrass, shrubs, flowers, etc. per station)
- Description of area being irrigated per station
- Extent of sun exposure per station (full sun, part sun, shade)
- Integrity of system devices (backflow prevention device, solenoid valves, sprinkle heads)

Though not included in this report, data was also analyzed to determine any difference and/or trends in irrigation applied when comparing landscape size and age of property. This information too can be instrumental in prioritizing specific topics for future water conservation education, outreach, and training for residential customers.

Irrigation Use Analysis

All residences in this study were served by a single water meter that registered combined indoor and outdoor water consumption. College Station Water Utilities provided monthly water consumption data for the seven year period from 2008 through 2015. Irrigation use was calculated by subtracting average indoor water use from total metered water consumption on a monthly basis. For the purpose of this study irrigation use was investigated and compared for the typical growing season in College Station – April through September. Indoor water use was estimated to be the average monthly consumption for December, January, and February over the period from 2008 to 2015 or over the period of reliable record during this seven year period.

Landscape water use (irrigation) in 'gallons' was normalized for landscape size and converted to inches of water applied using the following equation.

Irrigation (inches) = irrigation (gallons) / [area of landscape (sqft) x 0.6234]

Estimate of residential landscape area was calculated as the total property size (in square feet) minus the residential footprint, space occupied by garages, out-buildings, patios, sidewalks, and driveways. Total property, garage, and patio area was acquired from the Brazos County Appraisal District, http://www.brazoscad.org/. Further deductions for sidewalks, driveways, and other non-pervious area were estimated using Google Earth satellite maps and area/distance calculator tools.

Area of landscape (sqft) = total property area (sqft) – non-pervious area (sqft) (including house, garage, patio, sidewalk, and driveway footprint)

Net Plant Water Requirement (Net-PWR) Estimation

Net plant water requirement was computed using a daily water balance approach utilizing measured evapotranspiration (ETo) and precipitation data, crop coefficients for warm season turfgrass, and soil water storage constraints assuming a 6-inch root zone depth and clay soil type. ETo data was acquired from two automated weather station locations – the Texas A&M University Golf Course and Texas A&M Turf Lab. Net plant water requirement (Net PWR) was calculated using the following relationship:

Net PWR (inches) = (ETo (inches) x Kc x Af) – Reff (inches)

Where:

Kc = monthly crop coefficient (dimensionless)
Af = allowable stress factor (dimensionless)
Reff = effective rainfall (inches)

Long term average monthly crop coefficients for College Station are referenced in the Texas Landscape Irrigation Auditing and Management Short Course Manual – Version 3 (Fipps et. al., 2009). For this analysis, a stress adjustment factor of 1 (no stress) was used.

Methodology for estimating effective rainfall followed that used for similar analyses performed by the Texas A&M School of Irrigation (Swanson, 2015).

IF R < 0.1, THEN Reff = '0'
IF 0.1 < R ≤ 1, THEN Reff = 'R'
IF 1 < R ≤ 2, THEN Reff = 'R x 0.67'
IF R > 2, THEN Reff = '2'

Where:

R = actual daily rainfall (inches)

Daily Net-PWR was further constrained by assuming that plant-available water could be stored within a 6-inch root zone and a clay soil. Total Net-PWR for irrigation season each year was calculated by summing daily Net-PWR from April through September.

Landscape Irrigation Ratio (LIR)

The LIR is one approach to quantifying landscape water use (or irrigation) efficiency (Glenn et. al., 2015). The LIR metric provides a means to evaluate and compare landscape water conservation potential for properties regardless of property size. It is calculated by dividing the volume or normalized equivalent depth of outdoor water use divided by the landscape water requirement over a certain time interval.

LIR = irrigation (inches) / Net-PWR (inches)

This study examined the LIR over the typical landscape irrigation season (April through September).

Glenn et al. (2015) used the LIR approach to assess landscape water use efficiency in single-family residences in Logan, Utah in 2004 and 2005. Category benchmarks, defined by LIR ranges, were specified as 'justifiable' and 'unjustifiable' water use and further classified as 'efficient', 'acceptable', 'inefficient', and 'excessive'. This classification system was used in this study to compare water use efficiency for residences over the typical irrigation season from 2008 to 2015.

Justifiable water use
    Efficient LIR ≤ 1
    Acceptable 1 < LIR ≤ 2

Unjustifiable water use

Inefficient $2 < LIR \leq 3$

Excessive $3 < LIR$

RESULTS

Characterization of Irrigation System Hardware and Controller Programming

Controllers and rain shut-off sensors – Of the 211 residential customers Toro®, Hunter®, and RainBird® model controllers were used by 78% (165) of residents. These controllers are similar in basic operation and feature multiple program options ('A', 'B', 'C', etc. programs) and multiple start times per program. Almost all controllers provided for a 9-volt backup battery intended to retain program settings in case of power loss. If a functional backup battery were not present, these controllers reverted to a default irrigation schedule of watering every day, 10 minutes per station, at 5:00 AM start time once power was restored after an outage. Only 18% (38) of irrigation systems inspected were equipped with a rain shut-off sensor. Of those, 63% (24) were wireless and 37% (14) were hard-wired.

Irrigation Stations – A total of 1,204 stations were inspected. The average number of stations per residence is 5.8. Seventy-three percent (154) of residents had 6 or fewer stations. Of these, pop-up fixed spray heads and rotor-type sprinkler heads were the most common representing 52% and 37% of all sprinkler head types. Other sprinkler devices noted include 'mixed' (a combination of multiple sprinkler head types), drip irrigation, and multi-stream application devices designed for slow-application rate.

Irrigation Schedules – A critical part of the irrigation checkup was to educate the resident on the capability and use of their irrigation controller in facilitating efficient irrigation practices such as adjusting individual station runtimes, utilizing multiple programs to compensate for different irrigation frequency needs, and setting multiple start times (multi-cycling) to prevent water runoff. Existing controller settings were documented and immediately brought to the attention of the resident. In most cases, residents were not familiar with their current controller settings and did not realize the implications for inefficient water use.

Station run times: An analysis of all residents suggests that, in general, stations with relatively high application rates were set with lower run times. For example, the average run time for pop-up spray sprinkler heads was 12 minutes while the average run time for rotor sprinkler heads was 17 minutes. Drip irrigation (characterized by relatively low water application rate) was usually set to run much longer. At a minimum, this illustrates that an attempt was being made to adjust individual station run times for different sprinkler types.

Irrigation days: Irrigation days were fairly well dispersed throughout the week with Mondays, Wednesdays, and Fridays being the most common. The calendar day option was obviously the most common selection for setting irrigation day, with less than 4% (23) of residents using the 'odd/ even day', or 'interval day' feature.

Program start times: Eighty-eight percent (347) of all program start times documented occurred between midnight and 8:00 AM. The most common start time was 5:00 AM. This was not surprising given that the default start time for most major controller brands was also 5:00 AM. Though irrigating in the early morning is strongly encouraged and essential for reducing evaporative losses, there may be a need in some locations or subdivisions to minimize peak morning demand to limit pressure drop.

Net-PWR (2008 – 2015)

Net plant water requirement was calculated using a daily water balance approach using measured evapotranspiration and precipitation, and constrained by an assumed root depth and soil type using the methodology previous defined. This approach was selected to limit the water contribution from heavy or intense rainfall events. During intense rainfall water is more likely to run off the landscape and/or water moves beyond the typical plant root zone thereby becoming unavailable to the plant. Figure 1 shows the normal pattern and overall trend of Net-PWR over the eight-year period. Residential water customers typically begin irrigating in April and continue through September or longer depending on weather trends and early fall tropical storm development in the Gulf of Mexico. Peak Net-PWR usually ranges between 4.0 and 4.5 inches per month in June, July, or August. Overall, Net-PWR trended upward over the study period, most likely a result of the extreme drought conditions in 2011 and 2013.

Figure 1 – Monthly Net-PWR from 2008 to 2015

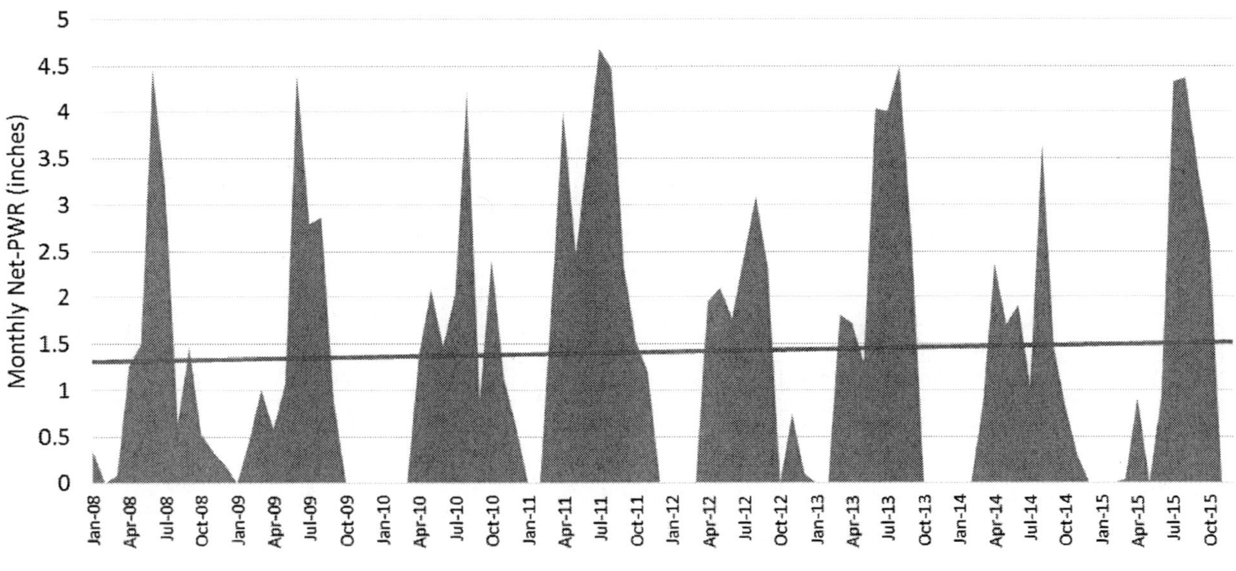

Water Use Analysis for Various Recommended Conservation Practices

There are several common *best management practices* used to encourage landscape water conservation and irrigation efficiency. Among these are: 1) installing rain shut-off devices that prohibit irrigation during and directly after rainfall events; 2) utilizing multiple programs ('A', 'B', 'C', etc.) that allow for adjusting irrigation frequency to account for different rooting depths and soil types; and 3) utilizing multiple start times or multi-cycling to reduce water runoff and promote infiltration. Water consumption data was analyzed to determine any differences between residents who did and did not adopt these practices.

Rain Sensors

Except for 13 arid counties located in the western part of the State, the Texas Commission on Environmental Quality requires rain or moisture sensor devices be installed on all new irrigation controllers. Many cities and municipalities also enforce ordinances that require rain sensors on irrigation systems as a condition of permitting and inspection. However, in this study many new and old properties lacked rain sensors, and of those who did have these devices, few were found on irrigation systems older than a few years. Of the residences included in this study, only 18% (38) of irrigation systems inspected were equipped with a rain shut-off sensor. Of those, 63% (24) were wireless and 37% (14) were hard-wired.

Table 1 lists the annual mean and range of irrigation applied for residents with and without rain shut off sensors. Surprisingly, irrigation applied from residents with rain sensors consistently exceeded those without rain sensors. However, statistical tests revealed no significant difference. This may be due to the large discrepancy between the number of residents in each group and the high variability within groups. In conversations with residents without rain sensors, many preferred to turn off their systems manually rather than rely upon a device which had to be maintained and periodically replaced. Some also commented that their landscape maintenance company or irrigation professionals had recommended against installing rain sensors due to their "limited life expectancy" and "unreliability".

Table 1. Comparison of irrigation applied for residents with and without rain sensors.

Year	[1]Net-PWR (in.)	Rain Sensor Installed			NO Rain Sensor		
		[2]Mean (in.)	[3]Range (in.)	[4]N	Mean (in.)	Range (in.)	N
2008	12.5	**30.7**	2.5 – 122.4	16	**21.0**	3.0 – 64.1	150
2009	12.6	**21.8**	1.9 – 60.1	18	**21.0**	3.5 – 80.4	153
2010	12.1	**22.2**	1.8 – 98.2	20	**17.7**	1.5 – 55.7	154
2011	21.5	**33.5**	2.1 – 187.1	25	**26.7**	4.2 – 75.6	153
2012	13.7	**19.5**	1.6 – 53.8	27	**18.7**	0.9 – 55.5	157
2013	18.1	**21.2**	2.5 – 49.4	27	**19.0**	1.9 – 73.5	154
2014	12.1	**16.6**	12.4 – 20.4	12	**15.1**	0.2 – 40.2	107
2015	14.0	**17.7**	0 – 37.6	35	**15.4**	0.7 – 45.8	157

[1]Net-PWR is the cumulative net plant water requirement (in inches) from April through September (typical irrigation season).
[2]Average irrigation applied (in inches) from April through September.
[3]Lowest to highest irrigation applied (in inches).
[4]Number of residents.

Multiple Programs

When properly used, multiple controller programs ('A', 'B', 'C', etc.) will allow a residential customer to fine tune their irrigation schedule to accommodate different irrigation frequency needs for various microclimate conditions and plant water requirements. A survey of residents in this study showed that only 16% (34) used more than one program (Figure 2), and 32% (67) practiced multi-cycling.

Figure 2. Number of residents using multiple programs and cycles per irrigation day

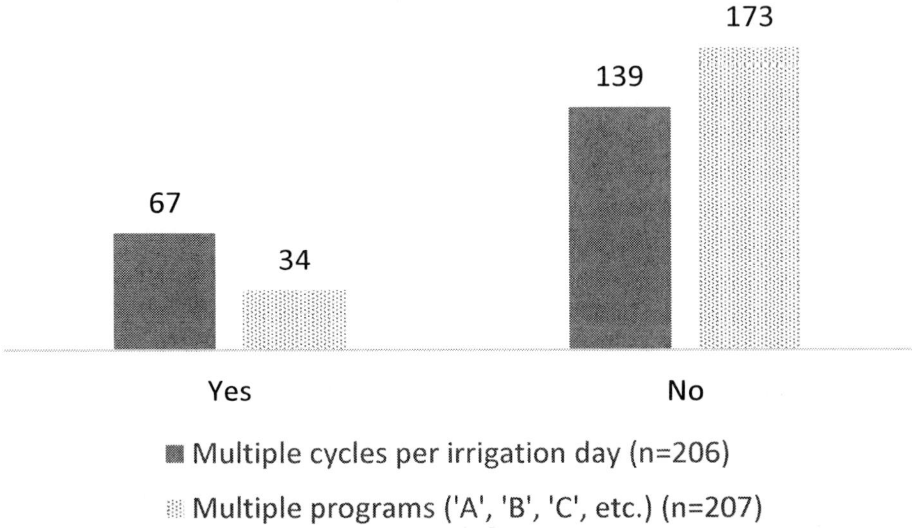

Table 2 compares average irrigation applied from residents who used multiple programs to those who used only one program. Residents who used multiple programs applied slightly more water during the growing season for each growing season, although the difference was not significant.

These results were unexpected, although there are a couple of possible explanations. First, multiple programs were often accidentally set by the resident. As electronic controllers age, their display can get difficult to read and function keys used to set the controller become less responsive. This often leads to unintentional programming errors. Second, despite the best intentions on part of some residents, programming the controller was confusing and absence programming instructions was not helpful.

Table 2. Comparison of irrigation applied for residents employing only one and multiple controller programs ('A', 'B', 'C', etc.).

Year	[1]Net-PWR (in.)	Use 1 Program Only			Use Multiple Programs		
		[2]Mean (in.)	[3]Range (in.)	[4]N	Mean (in.)	Range (in.)	N
2008	12.5	**21.4**	2.6 – 64.1	134	**25.4**	3.8 – 122.4	28
2009	12.6	**20.8**	1.9 – 80.4	138	**23.8**	7.3 – 60.1	28
2010	12.1	**17.6**	1.5 – 55.7	140	**22.6**	3.9 – 98.2	29
2011	21.5	**26.6**	2.1 – 75.6	146	**33.6**	10.0 – 187.1	29
2012	13.7	**18.2**	0.9 – 55.5	149	**22.3**	6.5 – 54.3	30
2013	18.1	**18.7**	3.0 – 73.5	145	**22.5**	1.9 – 47.3	31
2014	12.1	**14.8**	0.2 – 40.2	102	**17.0**	3.5 – 28.8	14
2015	14.0	**15.1**	0 – 36.8	156	**18.6**	1.8 – 37.6	32

[1]Net-PWR is the cumulative net plant water requirement (in inches) from April through September (typical irrigation season).
[2]Average irrigation applied (in inches) from April through September.
[3]Lowest to highest irrigation applied (in inches).
[4]Number of residents.

Multiple Cycles

Most residential controllers provide for multiple start times on a given irrigation day as a means to avoid excessive water runoff that may occur due to steep slopes or in soils with low infiltration rates. For example, instead of irrigating for 20 minutes for one cycle, this feature allows that 20 minutes to be distributed among two or more cycles (e.g., 10 minutes per cycle for two cycles). This allows water applied by sprinkler devices with high application rates a better chance to infiltrate into the soil between cycles. Table 3 compares average irrigation applied from residents who used multiple cycles per irrigation day to those who used only one cycle per irrigation day. Residents who used multiple cycles per irrigation day applied slightly more water than those irrigating only one cycle per irrigation day. Again, the difference between groups was not statistically significant.

Table 3. Comparison of irrigation applied by residents who use only <u>one</u> versus <u>multiple</u> cycles per irrigation day.

Year	[1]Net-PWR	Use 1 Cycle Only			Multi-cycle		
	(in.)	[2]Mean (in.)	[3]Range (in.)	[4]N	Mean (in.)	Range (in.)	N
2008	12.5	**21.7**	2.6 – 64.1	113	**22.8**	3.3 – 122.5	53
2009	12.6	**20.2**	1.9 – 80.4	116	**23.3**	5.6 – 60.1	54
2010	12.1	**17.5**	1.5 – 55.7	118	**20.1**	2.9 – 98.2	55
2011	21.5	**27.1**	2.1 – 75.6	122	**29.1**	4.2 – 187.1	56
2012	13.7	**18.8**	0.9 – 55.4	125	**19.1**	5.4 – 54.3	58
2013	18.1	**19.8**	3.0 – 73.5	122	**19.0**	1.9 – 43.4	57
2014	12.1	**15.3**	0.2 – 32.9	88	**14.9**	0.2 – 40.2	31
2015	14.0	**15.4**	0 – 45.8	132	**16.8**	0 – 37.4	60

[1]Net-PWR is the cumulative net plant water requirement (in inches) from April through September (typical irrigation season).
[2]Average irrigation applied (in inches) from April through September.
[3]Lowest to highest irrigation applied (in inches).
[4]Number of residents.

Landscape Irrigation Ratio (LIR)

Water use efficiency describes how closely irrigation applied matches plant water requirement. The LIR metric (used by Glenn et. al, 2015) (defined as the ratio of landscape water use divided by landscape water requirement) is one measure of water use efficiency. Although the choice of LIR classification is somewhat subjective, this methodology does provide a means to gauge the effectiveness of water conservation outreach, education, and awareness efforts among a large population.

LIR was computed for all properties and a comparison of LIRs for residences <u>with</u> and <u>without</u> rain sensors, and residences who <u>do</u> and <u>do not</u> use multiple programs and cycles per irrigation day when scheduling automated irrigation controllers (Figures 3, 4 and 5). Figure 3 illustrates that in all years. On average, water use efficiency was lowest (LIR highest) for properties equipped with rain sensors. Figure 4 demonstrates that water use efficiency was lower in all years for properties utilizing multiple controller programs as part of their controller irrigation schedule. Also, contrary to expectations, those residences utilizing multiple irrigation cycles per irrigation day used slightly more water than those using only one cycle (Figure 5). For all scenarios, there was overall increase in water use efficiency (lower LIR) over the 8-year study period.

Figure 3. LIR comparison and trend for residences with and without rain sensors.

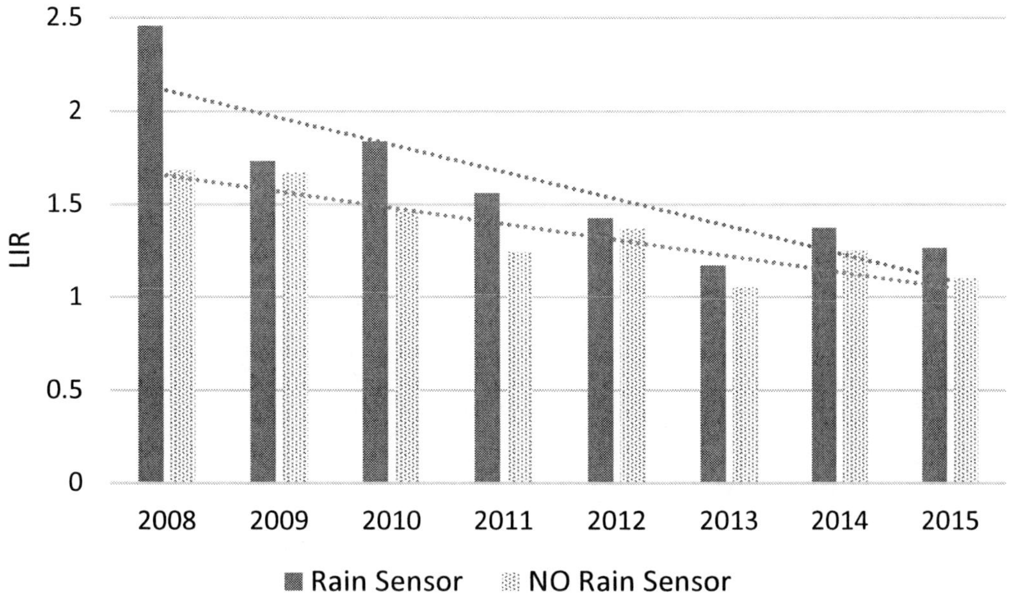

Figure 4. LIR comparison and trend for residences who do and do not utilize multiple programs.

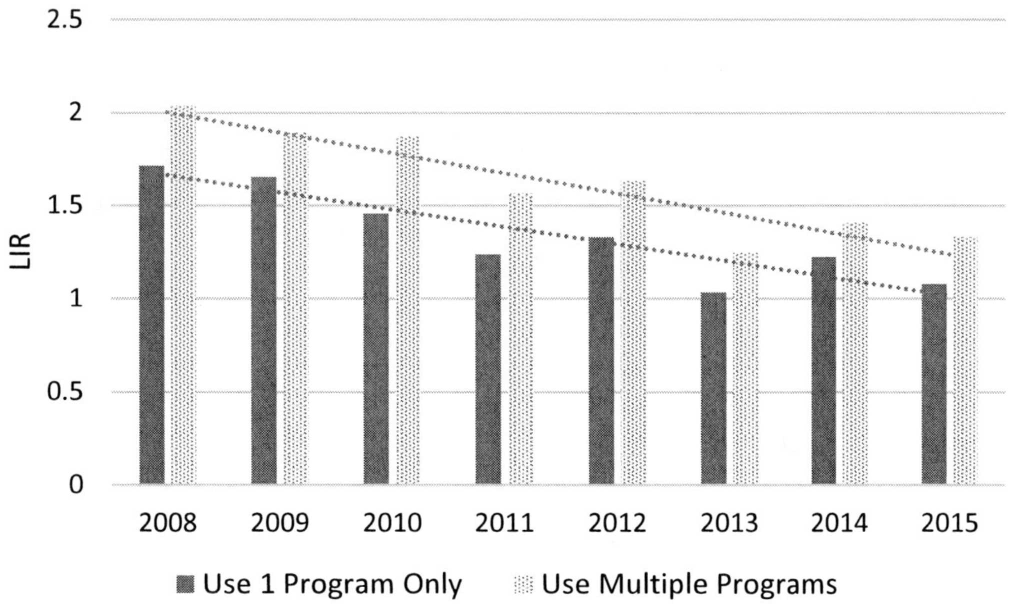

Figure 5. LIR comparison and trend for residences who do and do not utilize multiple cycles per irrigation day.

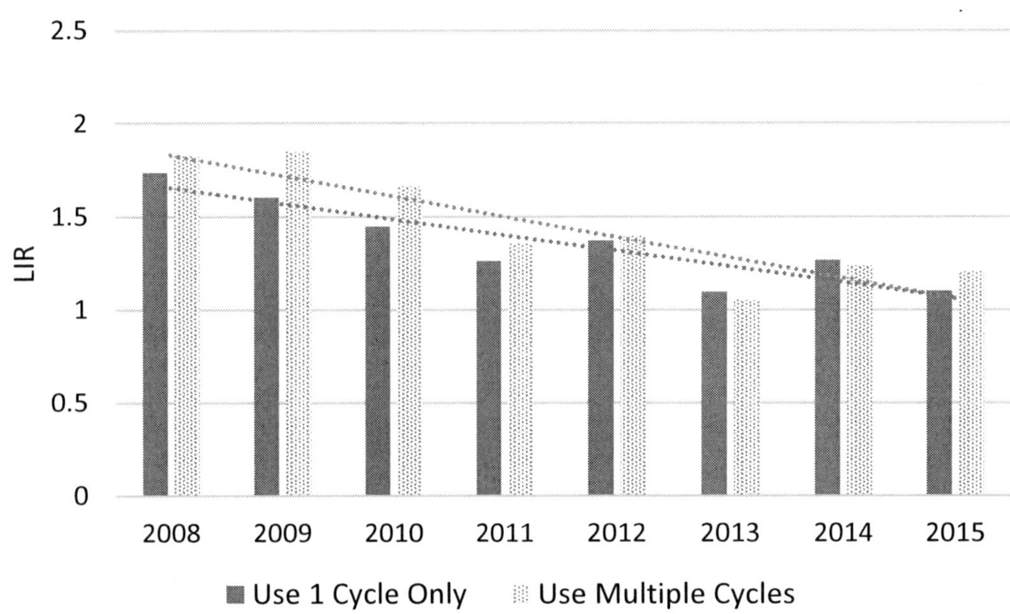

LIR for all residents included in this study were calculated and categorized in Table 4 using the classification system defined by Glenn et al. (2015). Overall, the percentage of residents classified as either 'efficient' or 'acceptable' increased from 70 percent to 91 percent from 2008 to 2015, with the highest percentages in these two categories occurring in 2011 (93 percent) and 2013 (95 percent), both extremely dry growing seasons. Furthermore, the number of properties classified as either 'inefficient' or 'excessive' dropped dramatically over the study period with less than 10 percent of residents falling into these categories.

Table 4. Distribution of residents by LIR category.

[1]LIR Category	Percentage of residents by LIR category							
	2008	**2009**	**2010**	**2011**	**2012**	**2013**	**2014**	**2015**
Justifiable water use								
Efficient LIR ≤ 1	19	19	26	31	26	51	35	41
Acceptable 1 < LIR ≤ 2	51	50	51	62	63	44	50	50
Unjustifiable water use								
Inefficient 2 < LIR ≤ 3	22	25	21	6	9	4	14	8
Excessive 3 < LIR	8	5	2	1	2	1	1	1
Total (%)	100	100	100	100	100	100	100	100
[2]N	167	170	173	178	183	179	119	193

[1]LIR is defined as the ratio of landscape water used divided by landscape water required (Net-PWR). Category designations defined by Glenn et al. (2015).
[2]Number of residents.

DISCUSSION

In 2012, 2013, 2014, and 2015 211 irrigation inspections for 205 unique customers were conducted as part of the College Station Water Utility's free residential irrigation checkup program. The objective of this study was to evaluate water consumption records for residential customers who do and do not utilize advanced residential controller features to identify any differences, patterns, or trends in the amount of irrigation applied (when normalized by unit landscape area). Unexpectedly, Irrigation applied (per unit landscape area) by residents with rain sensors consistently exceeded those without rain sensors. However, statistical tests revealed no significant difference. Irrigation applied by residents who used multiple programs and cycles per irrigation day was consistently higher than those who did not. Again, there was no significant difference among the groups. This data implies a lack of understanding or point of confusion among residents on how to properly utilize these advanced features intended to accommodate different irrigation scheduling needs and microclimates found in many residential properties. Water use efficiency, as measured by the Landscape Irrigation Ratio metric, showed an overall increase over the study period. Trends also show a decrease in the portion of residents classified as 'inefficient' or 'excessive' suggesting that the irrigation checkup service may have long term impact in reducing over-irrigation.

CONCLUSIONS

Results of this study profile existing irrigation system hardware and scheduling practices among residential customers in College Station, Texas. This study has shown that in order for 'best management practices' to be successful in reducing water use, customer education is imperative. Absent basic understanding of controller programming, attempts at implementing these practices may actually increase water use. Finally, although current education and landscape irrigation checkup service appear to be effective in increasing water use efficiency, there is an opportunity to further promote water conservation by delivering training focused on proper use of advanced controller features. There is also a need for education and training focused on proper installation, siting, maintenance, and operation of rain shut off sensors.

ACKNOWLEDGEMENTS

- College Station Water Utilities
- Texas A&M Department of Recreation, Park and Tourism Sciences
- Irrigation Technology Program, Texas ET Network

REFERENCES

Brazos County Appraisal District (2015). http://www.brazoscad.org/

Cabrera R.I., Wagner K.L., Wherley B. (2013): An evaluation of urban landscape water use in Texas. *Texas Water Journal*, 4, pp. 14-27.

Coleman, D. (2014). City of College Station Water Conservation Plan. May 2014. http://www.cstx.gov/Modules/ShowDocument.aspx?documentid=19343

Fipps, G., Welsh, D., Smith, D., and Swanson, C. (2009). Texas landscape irrigation auditing and management short course manual – version 3. Texas A&M AgriLife Extension Service.

Glenn, D.T., Endter-Wada, J., Kjelgren, R., Neale, C.M.U. (2015): Tools for evaluating and monitoring effectiveness of urban landscape water conservation interventions and programs, *Landscape and Urban Planning*, 139, pp. 82-93.

Texas ET Network (2015). http://texaset.tamu.edu/

White, R., Havlak, R., Nations, J., Pannkuk, T., Thomas, J., Chalmers, D., and Dewey, D. 2004. "How much water is enough? Using pet to develop water budgets for residential landscapes." Proc., Texas Water 2004, Texas Section American Water Works Association, Arlington, Tex., *Texas AWWA Paper No. TR-271*.

This page intentionally left blank.

Study of Moisture Sensors' Response to Drying Cycles of Soil

Pappu K. Yadav[1], Center for Irrigation Technology, California State University, Fresno

Charles Swanson[2], Texas A&M Agrilife Extension Service, Texas A&M at College Station

Tom Penning[3], President, Irrometer Company Inc.

Joe Wallace[4], CEO, Baseline Irrigation Solutions

E-mails: pyadav@csufresno.edu[1],clswanson@tamu.edu[2],tomp@irrometer.com[3],jwallace@baselinesystems.com[4]

Expeditious advancements in sensor technology to measure or estimate moisture in soil have taken place with time. There are numerous technologies to measure or estimate moisture in soil today. Sensors based on different technologies interact differently with the amount of moisture present in soil. The objective of this research was to study response of these sensors within normal (up to 50% of available moisture depletion) drying cycle of soil so as to effectively use them for landscape irrigation control. For this study, volumetric water content sensors based on Time Domain Transmission (WaterTec S100) and Soil Water Potential sensors based on electrical resistance (Watermark 200SS) were used at 0 dS/m (at 25 °C) in Sandy Loam textured soil. Three sensors of each type were used in containers packed with Sandy Loam soil in a temperature controlled environment. Each container was placed on a weighing scale to continuously monitor drying cycles over time. Total of four drying cycles were used, each cycle split into five levels of depletion (10% each) and at least one reading in each level from sensors was taken. Test results showed that these sensors' response to normal drying cycles were considerably repeatable, precise and less-variant.

Moisture, Sensor, Soil Moisture Sensor, Volumetric Water Content, Soil Water Potential, WaterTec, Watermark, Landscape, Irrigation, Controllers

Introduction

Total consumption of water for landscape irrigation in the United States equals to nearly nine (9) billion gallons per day (*EPA WaterSense, 2013*). However, due to recurring phenomena of drought and increasing demand for water with time, efficient irrigation of both agricultural fields and landscape with minimum use of water has become a necessity. It is estimated that each year, more than one-half of terrestrial earth is susceptible to drought (*Kogan,1997).*To meet this increasing demand of limited water resource, scheduling irrigation based on demand of crops has come into existence. The best way to meet this criterion is by measuring soil water content and scheduling irrigation based on the same. It is with regard to this, various manufactures have come up with different soil water content measuring devices which are popularly known as soil moisture sensors. Soil water content or soil moisture sensors have been in existence since as early as 1950s in the form of tensiometers by Irrometer and 1960s in the form of neutron probes (*Gardner and Klute, 1986*). These were mostly manual methods of measuring soil water content but with advancement of technology in the field of electronics over time, various such sensors have emerged that can measure soil water content automatically and in real-time. These automated moisture sensors not only can measure soil water content in real time but can also turn on and off irrigation with the help of interfacing devices that come along with them. With increment of many such moisture sensor based irrigation control technologies, and lack of any federal standards for such soil moisture based control technologies, United States Environmental Protection Agency's (U.S. EPA's) WaterSense program released a Notice of Intent(NOI) in May, 2013 to develop a draft specification for soil moisture based control technologies. As a result of this, *American Society of Agriculture and Biological Engineers (ASABE)* is working to develop two standards for such products: S633 (Testing of Soil Moisture Sensors for Landscape Irrigation) and S627 (Standardized Testing Protocol for Weather Based or Soil Moisture Based Landscape Irrigation Control Devices). This study was done as a precursor for developing S633 standard that is intended for bypass soil moisture based control technologies (*EPA-NOI, 2013*). For this, responses of moisture sensors based on two different technologies were studied during normal drying cycle (up to 50% available water depletion) of soil based on the method described. The whole objective of this study was intended to study variance, repeatability and hence precision of these sensors based on the method described and understand if it can be used as a base for developing S633 standard.

Soil Water Measurement

Direct Method

Soil water measurement refers to calculation of the amount of water (mg) present in a given mass of soil sample (mg). Hence, the unit of soil water content is mg mg^{-1}.However; this can also be expressed in terms of volume by calculating volume of water present (dividing the mass of water by density of water) in the given volume of soil sample. In this, case the unit for soil water content takes the form m^3m^{-3}. In this method, direct weight of soil sample is taken before and after drying in oven. Hence, the method is called gravimetric method of soil water measurement. For most of the applications, volumetric expression of water content is used. In short, it is called volumetric water content (θ_v). Mathematically, it is expressed as:

θ_v = Volume of water in given sample of soil(V_w) / Volume of soil sample(V_s) ....(i)

$= V_w / V_s$

$= (M_w/\rho_w)/ (M_s/\rho_s)$

Where M_w =Mass of water in gram

M_s = Mass of soil sample/bulk in gram

ρ_w = Density of water in gram/cubic centimeter = 1 at 25°C

ρ_s = Density of soil bulk in gram/cubic centimeter [also called bulk density(BD)]

$= (M_w/M_s) \times (\rho_s/ \rho_w)$

$\theta_v = (M_w/M_s) \times BD$ ....(ii)

To emphasize how volumetric water content (VWC) in a given sample of soil varies as a function of changing bulk density and amount of water, MATLAB simulation (Fig: 1) was done. For this, a standard amount of soil sample of 1000 grams was considered. Soil bulk density range of 1 to 1.4 gm/cc was considered as soil sample used for this study was sandy loam in texture and the ideal bulk density for plant growths in such soil is less than 1.4 gm/cc (*USDA-NRCS, 2008*).

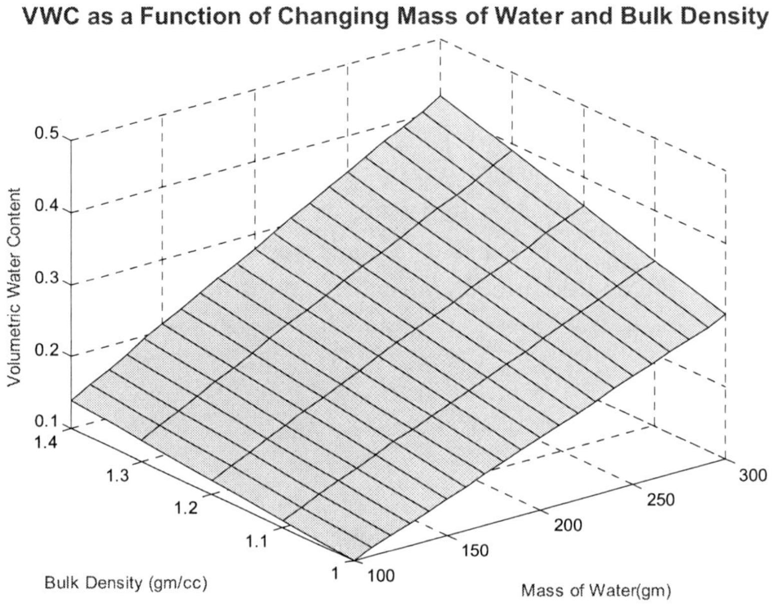

Fig 1: MATLAB Simulation of Changes in Volumetric Water Content (m^3m^{-3}) with Changing Bulk Density and Amount of Water in a Given Sample of Soil (1000 gram)

It can be seen (Fig: 1) that volumetric water content increases with increasing bulk density for the same amount of water in a given mass of soil. Similarly, for the same bulk density, for a given amount of soil, increase in amount of water increases volumetric water content of the soil. However, the rise in volumetric water content due to increase in bulk density has a higher slope value of 0.1 as compared to the slope value of 0.001 obtained as a result of rise in volumetric water content due to increase in amount of water in soil. This shows that bulk density of soil have much greater impact on volumetric water content of soil.

This method is considered to be reliable method of soil water measurement as it involves direct method of measuring water present in the soil. However, this method is limited to the position from where soil sample is taken as well as it is time consuming .This makes it practically impossible to be used for scheduling irrigation of agricultural farms or landscapes in real-time. As a result of this, soil moisture sensors are used which make estimate of volumetric water content in an indirect way.

Indirect Method

Indirect methods involve measuring of changes in certain physical quantities like electrical conductivity, resistance, soil-water tension, travel time of electromagnetic pulse, frequency of oscillating circuit, count of slow neutrons around a source of fast neutrons, etc. as results of changes in soil water content and then calibrating those measurements with respect to soil water content. Calibration equations may vary from being a simple linear equation to much more complex equations depending upon the type of technology used in moisture sensors. Although calibrated in factory, these sensors' variance , accuracy and repeatability(*please refer to section: variance ,repeatability and precision*) can vary widely depending upon soil and water properties like dielectric permittivity, electrical conductivity, soil bulk density, soil texture, etc. Nonetheless, gaps in such parameters can be minimized for in commercial products by allowing the users to set the irrigation ON and OFF points relative to the measurement for a particular installation based on manufacturer's recommendations. Out of the many commercially available soil moisture sensors for landscape irrigation, two brands were selected for this study: *Baseline's WaterTec S100 with biSensor* and *Irrometer's Watermark 200SS* (Table 1).

Table 1: Soil Moisture Sensors Used for Experiment

Manufacturer	Brand	Model	Sensor Technology	Digital Display
Baseline Irrigation	Baseline	WaterTec S100[a]	TDT[b]	Yes
Irrometer Inc.	Watermark	200SS	Resistance	Yes[c]

[a] Baseline's biSensor is connected to WaterTec S100

[b] Time Domain Transmission

[c] Digital Display is present when connected to Watermark Monitor (900M) data logger

Sensor Technologies Used

Time Domain Transmission (TDT)

Baseline's WaterTec S100 (with biSensor) soil moisture sensor is based on time domain transmission technology. In this, a high frequency electrical pulse signal is sent through wire path embedded in sensor's blade. This high frequency pulse causes sphere of influence of pulse move outside sensor's blade into the surrounding soil. As the pulse passes through moisture present in soil, it slows down. The sensor measures speed of this pulse by calculating transmission travel time and converts it into corresponding moisture content of the soil (volumetric). Unlike, transmission domain reflectometer (TDR) sensors which measure soil water content based on travel time of reflected electrical pulse signals, TDT sensors are relatively cheaper and easy to install with similar performance characteristics (*Robinson et al. , 2005*). Both TDR and TDT based sensors estimate dielectric permittivity of the medium in which they are buried from travel time measurement of electrical pulse signal and then use it in either Topp's equation (eqn. v) or Siddiqui and Drnevich's equation to calculate volumetric water content or gravimetric water content (*Xinbao Yu and Xiong Yu, 2006; Siddiqui et al., 1995; Topp et al., 1980*).

For TDR and TDT sensors, dielectric permittivity of medium is replaced by an equivalent term apparent permittivity (K_a). Mathematically, it is expressed as:

$K_a = (ct/2L)^2$ for TDR ....(iii)

$K_a = (ct/L)^2$ for TDT ....(iv)

where c= $3x10^8$ ms^{-1} is velocity of electrical pulse signal, t = pulse travel time in second and L= physical probe length of sensor in meter. This K_a is used in Topp's equation to calculate volumetric water content as given below:

$$\theta_v = 4.3x10^{-6}K_a^3 - 5.5x10^{-4}K_a^2 + 2.92x10^{-2}K_a - 5.3x10^{-2} \quad \text{.........(v)}$$

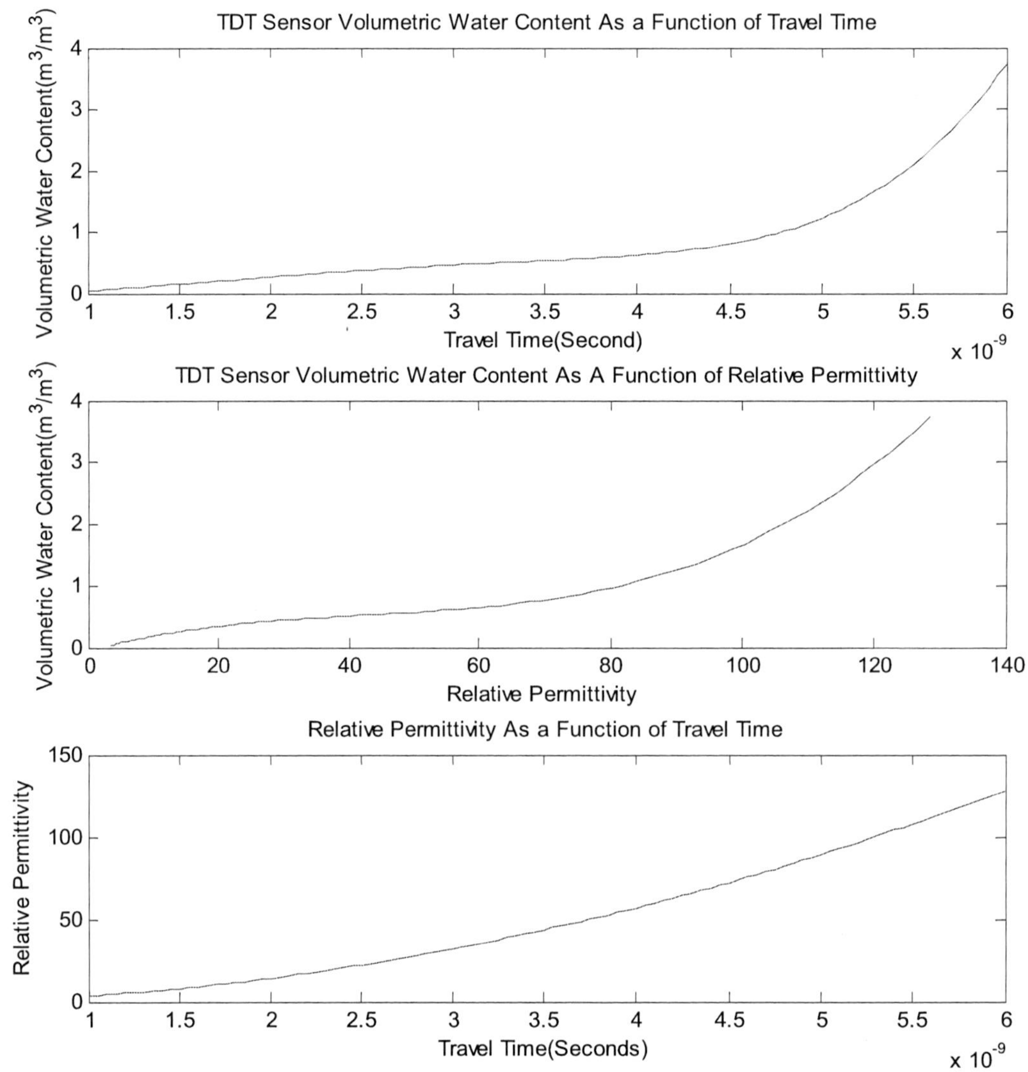

Fig 2: MATLAB Simulation of TDT Sensor for Volumetric Water Content as a Function of Travel Time (Top), Relative Permittivity (Middle) and Relative Permittivity as a Function of Travel Time (Bottom)

All the above charts (Fig: 2) were generated considering probe length of 6.25 inch (15.875 cm), which is the probe length of Baseline WaterTec S100 sensors (Fig 2). Top chart shows how volumetric water content varies with different measured travel time of electrical pulse signal. It clearly shows that more the travel time, higher is the relative permittivity (bottom chart, Fig: 2) and hence more is the water content (middle chart, Fig: 2) present in soil bulk where it is installed.

Similarly, MATLAB simulation (Fig: 3) was done to see how changes in probe length and travel time simultaneously cause change in volumetric water content.

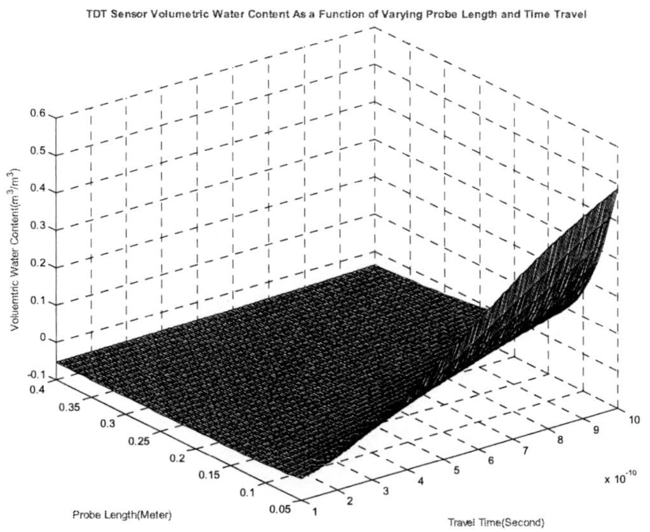

Fig 3: MATLAB Simulation of WaterTec S100 Sensor's Volumetric Water Content Readings with Respect to Changes in Probe Length and Travel Time Simultaneously

It can be seen (Fig:3) that for the same probe length, increasing travel time represents increasing volumetric water content while for the same travel time increasing probe length represents less volumetric water content. Highest volumetric water content is represented at smallest probe length with highest travel time.

Electrical Resistance

Irrometer's Watermark 200SS is an electrical resistance based sensor that measures soil moisture by estimation of soil water tension. Soil water tension (also sometimes referred to as soil water potential) is an indirect method of representing water content in soil. Soil water potential is defined as the total potential energy of unit amount of soil water with respect to potential energy of pure, free water at soil surface (i.e. reference potential energy of zero). Mathematically, total soil water potential (ψ_T) is expressed as:

$$\psi_T = \psi_M + \psi_P + \psi_O + \psi_Z \quad (N/m^2) \qquad \text{(vi)}$$

where ψ_M = Matric Potential related to capillary and absorptive forces(as a result of surface tension)

ψ_P = Pressure Potential related to changes in pressure

ψ_O = Osmotic Potential related to solute concentration

ψ_Z = Gravitational Potential related to gravitational field at earth's position

Among all of these, matric potential (ψ_M) is the one that best represents how readily soil water is available to a plant (*Evett, et al., 2008; Irrometer, 2016*). Unsaturated soil has water and air molecules present in soil pores. The air-water interface results in surface tension that is inversely proportional to surface area of soil pores. This surface tension results in matric potential. This matric potential in other words, is the energy invested in capillary force to push water upward towards air plus the energy of absorptive force. Hence, if there is more water present in soil, then there is less soil pore area void of water and hence more surface tension that ultimately results in less matric potential. At saturation, when there is almost no soil pores left unfilled, matric potential is zero ($\psi_M = 0$). Thus, as soil dries, ψ_M becomes more and more negative. Soil water potential is expressed in J/m^3 in SI unit which is equivalent to Pascal. However, Kilo Pascal (kPa) is the preferred SI unit for soil water potential. Bars and Centibars (CB) are other widely used units (1CB = 1 kPa).Irrometer's Watermark 200SS sensor is designed to give electrical resistance output relative to measured soil water matric potential (ψ_M) in centibars(CB) units. These sensors can measure soil matric potential from 0 to 239 centibars (CB). A typical calibration equation for converting measured electrical resistance (Kilo Ohm) into soil water matric potential (CB) with temperature compensation is given below (*Shock et. al, 1998*).

For Resistance >1 KΩ and Resistance <= 8 KΩ

$$\psi_M = (-3.213*R-4.093)/(1-((0.009733*R)-(0.01205*T))) \quad \text{(vii)}$$

For Resistance > 8 KΩ

$$\psi_M = -2.246 - 5.239 * R * (1 + 0.018 * (T - 24)) - 0.06756 * R * (1 + 0.018 * (T - 24)) \quad \text{(viii)}$$

where R = measured electrical resistance in Kilo ohm (KΩ)

T = Soil Temperature in °C

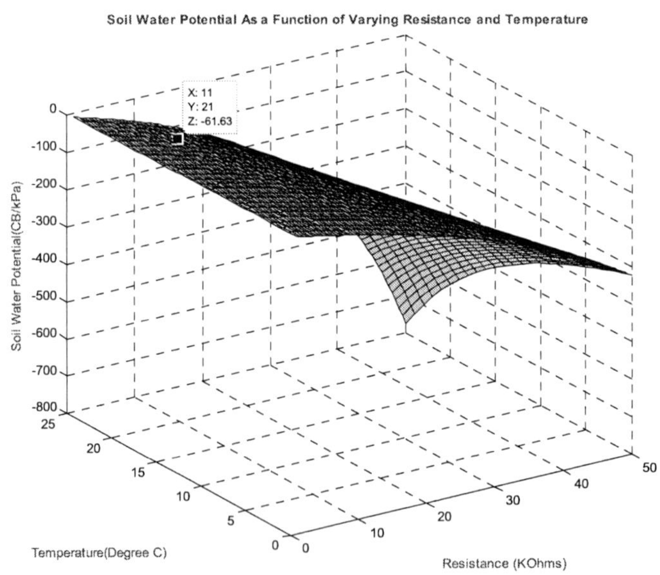

Fig 4: MATLAB Simulation of Soil Water Potential as a Function of Varying Measured Electrical Resistance and Temperature

It can be seen (Fig: 4) that at 21°C, measured electrical resistance is 11KΩ and hence; estimated soil water potential is calculated based on Shock's equation (viii) as resistance value is greater than 8 KΩ. Calculated soil water potential equal to -61.63 Centibars (CB). Similarly, for the same temperature, at resistance value of 6 KΩ, calculated soil water potential equals to -33.94 CB which is based on Shock's equation (vii).

Variance, Repeatability and Precision

Variance is a statistical measure that shows how far each value is from the mean from a set of data. In this study, there was one data point from each sensor in each set of readings (total of 3 data points in each set of readings). Hence, variance calculation for each set of readings reflects the amount by which each sensor's readings are away from the mean value. Higher values represent sensor's data are farther away from the mean value and vice versa. Similarly, repeatability of moisture sensors represents the ability of sensors to give repeated values over multiple measurements. In this study, a graph is plotted over each sensor's readings taken over three drying cycles (representing repeated readings) and corresponding percent depletion. Coefficient of determination (R^2) of this graph is taken as an indicator of degree of repeatability. Since, this value of R^2 is also an indicator of degree of scattering of data points; it may also be used as an indicator of precision of sensors (*Dukes et.al., 2014*).

Field Capacity, Wilting Point and Available Water

The maximum value of water content that can be held in soil without any rapid drainage is called field capacity of soil. In other words, it is the maximum water held by soil that is useful to plants. Although, soil can hold more water than the field capacity, that excess water usually drains within a day and hence is not available for plants. Field capacity differs depending on soil texture and is usually expressed in terms of volumetric water content ($m^3\ m^{-3}$).

As the soil dries out, soil water content decreases and water gets held by soil more tightly. At certain point, it is held so tightly that water is not available for plants. This minimum point of soil water content at which water is not available to plants is called wilting point of soil. This also varies depending on soil texture.

The difference between field capacity and wilting point is called available water. Table 2 shows most widely used field capacity, wilting point and available water values for different soil textures.

Table 2 Field Capacity, Wilting Point and Available Water for Different Soil Textures

Soil Texture	Field Capacity (m^3m^{-3})	Wilting Point (m^3m^{-3})	Available Water (m^3m^{-3})
Sandy Loam	0.20	0.08	0.12
Loam	0.27	0.12	0.15
Clay	0.40	0.25	0.15

Soil Water Depletion

Soil water depletion is defined as the amount of water loss with respect to field capacity of soil during drying process of soil. In other words, it is the difference of volumetric water content measured at particular moment of time and field capacity. Similar to volumetric water content[a], soil water depletion is also normally expressed in terms of percentage and hence called percent depletion.

Given below (Fig: 5) is a MATLAB simulation of changes in percent volumetric water content with increasing percent depletion (increasing drying process). For this, field capacity =23 % is taken as this was locally measured field capacity of sandy loam soil used for this study.

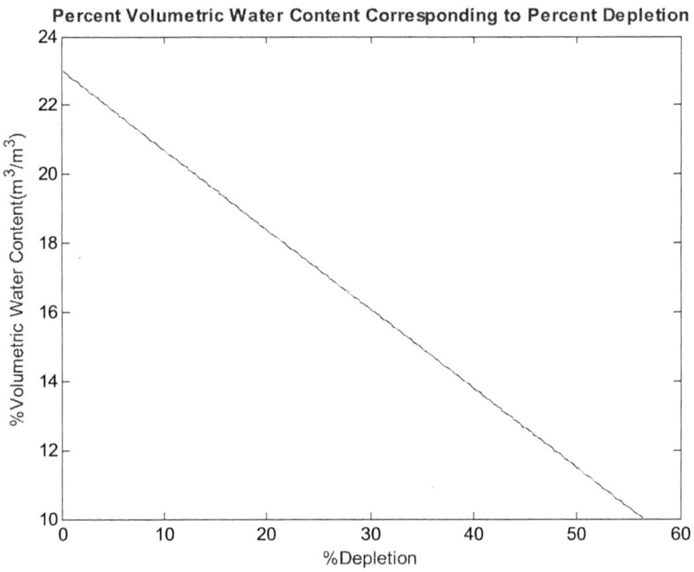

Fig 5: MATLAB Simulation of Changes in Percent Volumetric Water Content Corresponding To Percent Depletion for Sandy Loam Soil with Field Capacity of 0.23 (23%VWC)

Materials and Methods

As already mentioned above, for this experiment, two different types of moisture sensors were used; one based on TDT that measures volumetric water content(Baseline's WaterTec S100 with biSensor), while the other based on resistance that measures soil water potential (Irrometer's Watermark 200SS). In addition, Irrometer's tensiometers (206 RSU-C) were also used in conjunction with Watermark 200SS sensors to test sensors' accuracy as compared to tensiometers which give direct measure of actual soil water potential (*Irrometer Inc.)*. Sandy Loam texture soil with 67% sand, 21% silt and 12% clay and field capacity of 23 (%volumetric water content) was used. Deionized water at 0 dS/m was used for wetting of soil. Three sensors of each type were used. Two containers of different shape and size were used so as to contain entire sensor buried in soil with at least one inch of soil on top of them.

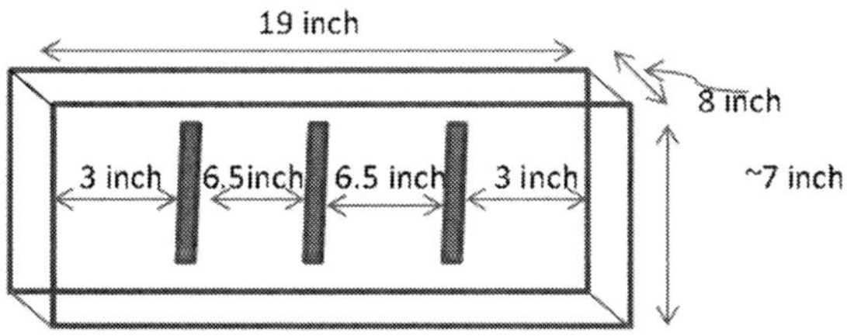

Fig: 6 Container Dimensions and Sensor Placement for Baseline Sensors

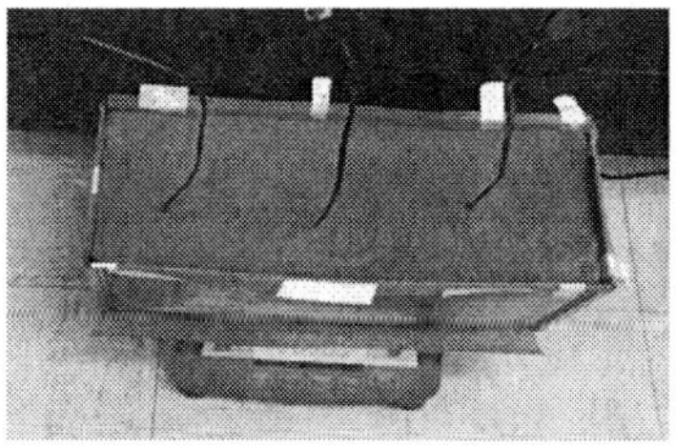

Fig: 7 Actual Picture of Baseline Sensors in Container

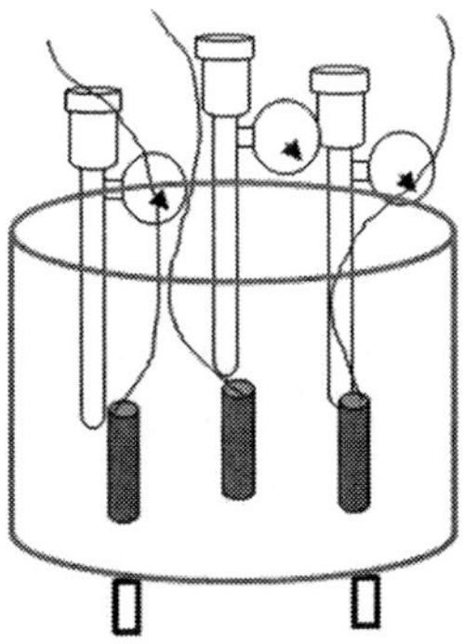

Fig: 8 Cylindrical Container[a] for Watermark Sensors and Tensiometers

Fig: 9 Actual Picture of Watermark Sensors and Tensiometers in Cylindrical Container

[a] *Radius = 3.8 Inch(9.652 cm), Height = 7.5 Inch(19.05 cm)*

Sensor placement was done such that their sphere of influence didn't interfere with each other based on manufacturer's recommendation(Fig: 6 and Fig: 8). A weighing scale was used to keep track of changes in weights during drying process of soil. WaterTec S100 gives digital reading of Baseline's biSensor connected to it. Similarly, Watermark Monitor 900M shows digital reading from Watermark 200SS sensors and tensiometers (206-RSU-C). Low from drip emitters were used for wetting soil in containers from top (Fig: 9). Weed control/filter paper was wrapped around inner walls and base of containers so that soil wasn't lost along with water during free drainage. The following steps describe test procedure:

- Sandy Loam soil was kept in oven for drying purpose at a temperature of 105°C for 24 hours. This was done to kill (if any) organic matters present in soil and to ensure soil totally dry.
- Soil was left to cool down for a bit (~30 minutes) and then grinded to obtain fine particles.
- A portion of this soil was taken to determine field capacity by packing an inch of soil in cylindrical container and wetting it from top using drip emitters until free drainage and then letting it dry for up to 48 hours. Mass of soil packed was such that a bulk density of 1.26 gm/cc was obtained. Difference in mass of dry soil and wet soil after 24 hours resulted (when free drainage seized) in field capacity of 23 (% vwc). Here, we prefer to call this field capacity as container capacity. Additionally, a portion of undisturbed soil sample was sent to a certified lab for texture analysis.
- Weed control papers / filter papers were wrapped around inner walls and base of container.

- Rectangular container was packed with prepared soil up to a height of 3.5 inch and cylindrical container was packed with soil up to a height of 5 inch. As a result, bulk densities of 1.46 gm/cc and 1.38 gm/cc were obtained respectively.
- Three Baseline's biSensors were installed in rectangular container placing them as per manufacturer's recommendation (as seen in figures 6 and 7). Similarly, three watermark sensors and three tensiometers (as seen in figures 8 and 9) were installed in cylindrical container.
- All the sensors were wired to their respective display units and logging system.
- Initial weights of both the containers(with soil+sensors+filter papers) were taken
- Deionized water at 0 dS/m (0 EC) was applied using drip emitters to both the containers for about an hour although free drainage was observed after 30 minutes.
- Water application was stopped after an hour and containers were left undisturbed for 24 hours.
- Soil water content (% volumetric) based on scale weight was determined and this was marked as container capacity (similar to field capacity).
- Scale readings as well as sensors' readings from their display units were taken at least once a day (if possible twice a day with a gap of four hours) for 10 days. With every scale readings taken, corresponding soil water content (called calculated soil water content) were determined (% vwc) and hence percent depletion. If 'x' is equal to calculated soil water content in %VWC and 'y' is equal to container capacity then %depletion= ((y-x)/y)*100 .
- Once, %depletion reached between (50-60) %, it marked the end of test cycle.
- Another test cycle was repeated with re-wetting of soil. A total number of four such test cycles were performed. For data analysis, all data from first test cycle were ignored.

All the tests were performed in a controlled temperature room at 25°C.

Results and Discussion

Baseline biSensors Data (WaterTec S100)

Data analysis was done to capture the variance among all three sensors for each cycle during the period of drying process of soil when compared against the calculated volumetric water content as obtained from the scale readings.

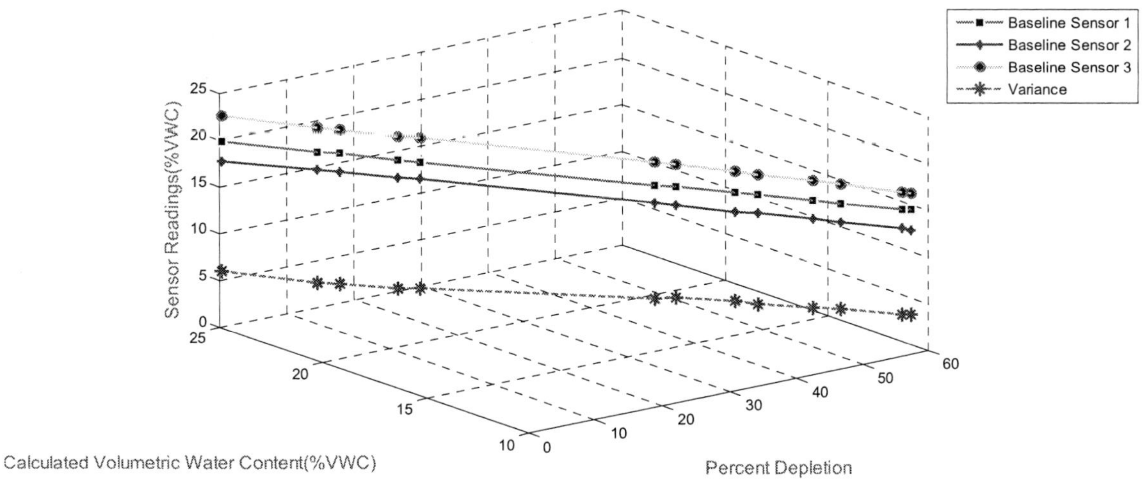

Fig 10: MATLAB Plot for Test of Variance among Three Baseline biSensors With Drying Cycle (Cycle#2) of Soil

For Cycle# 2 it was observed that variance among the sensors was maximum (6.043) when the soil was closer to saturation. Variance remained fairly high during (0-10) % depletion range. Variance kept decreasing with drying process of soil and was found to be lowest in depletion range of (50-60) % with minimum value of 3.823.In other words, there was more difference in reading among the sensors when soil was wet and the readings were found to be closer to each other as the soil dried (Fig: 10).

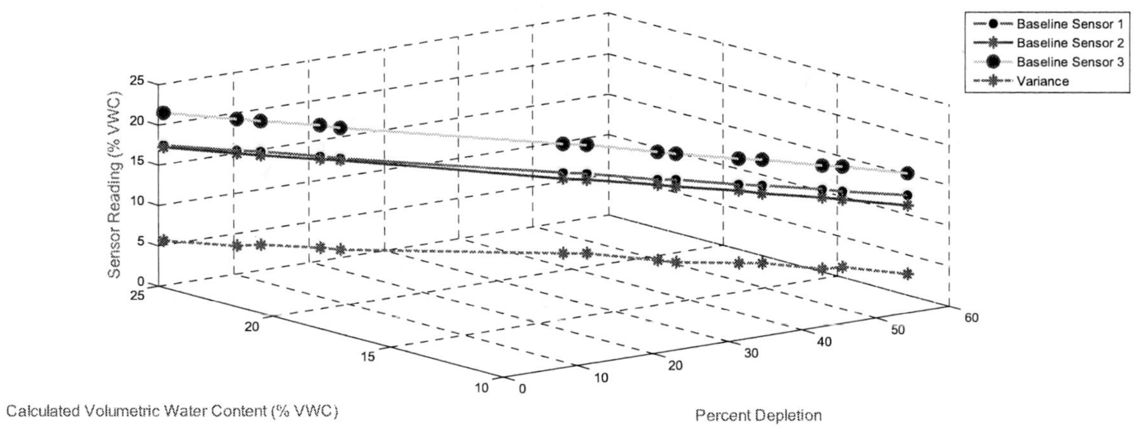

Fig 11: MATLAB Plot for Test of Variance among Three Baseline biSensors With Drying Cycle (Cycle#3) of Soil

In cycle# 3, again maximum variance between three sensors was found to be 5.763 when the soil was really wet i.e. close to its container capacity. With drying process, variance was found to decrease just as in cycle#2.The lowest variance was found to be equal to 3.99 in depletion range (50-60) % (Fig: 11).

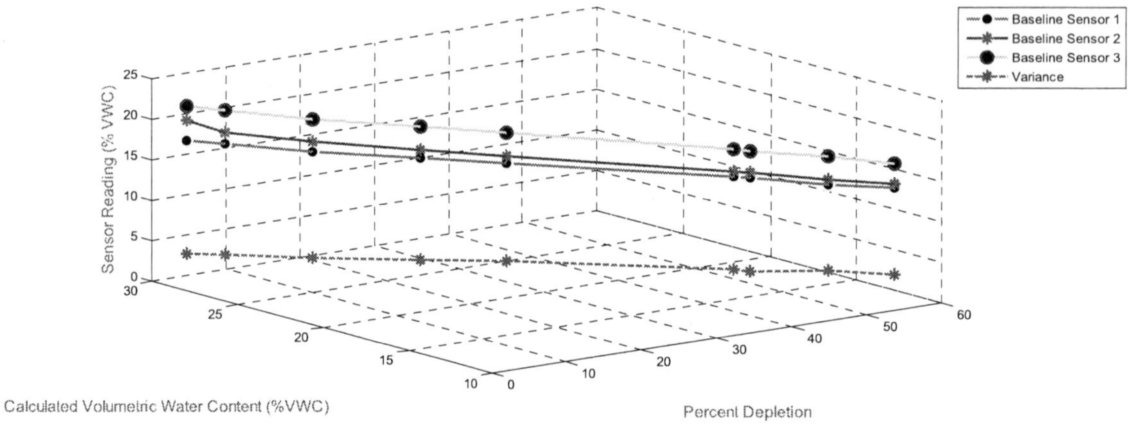

Fig 12: MATLAB Plot for Test of Variance among Three Baseline biSensors With Drying Cycle (Cycle#4) of Soil

Similarly, in cycle#4, maximum and minimum variance were found to be 4.47 and 2.77 respectively. Compared to previous two cycle, variance in this cycle was found to be less throughout the test cycle as a result of all three sensors reading values much closer to each other (Fig: 12).

Second part of data analysis was based on test of repeatability of sensors. Each sensor's repeatability test was done over three test cycles (cycle#2, cycle#3 and cycle#4). For this, each sensor's readings over drying period from all three test cycles were plotted (figures 13-15).

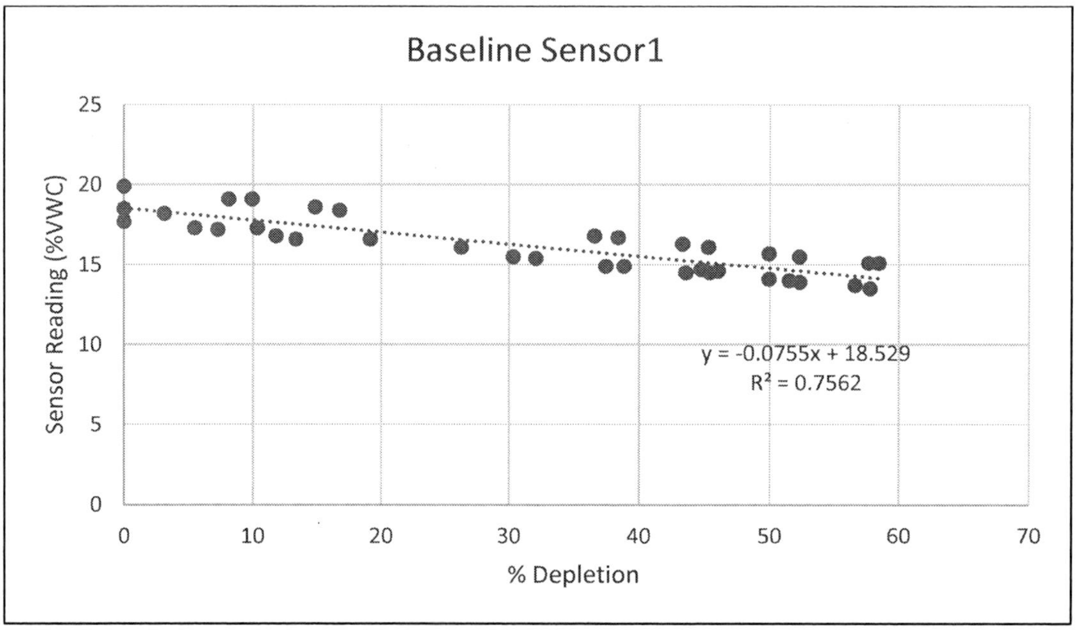

Fig 13: Repeatability Test for Baseline's Sensor#1 Over All three Test Cycles

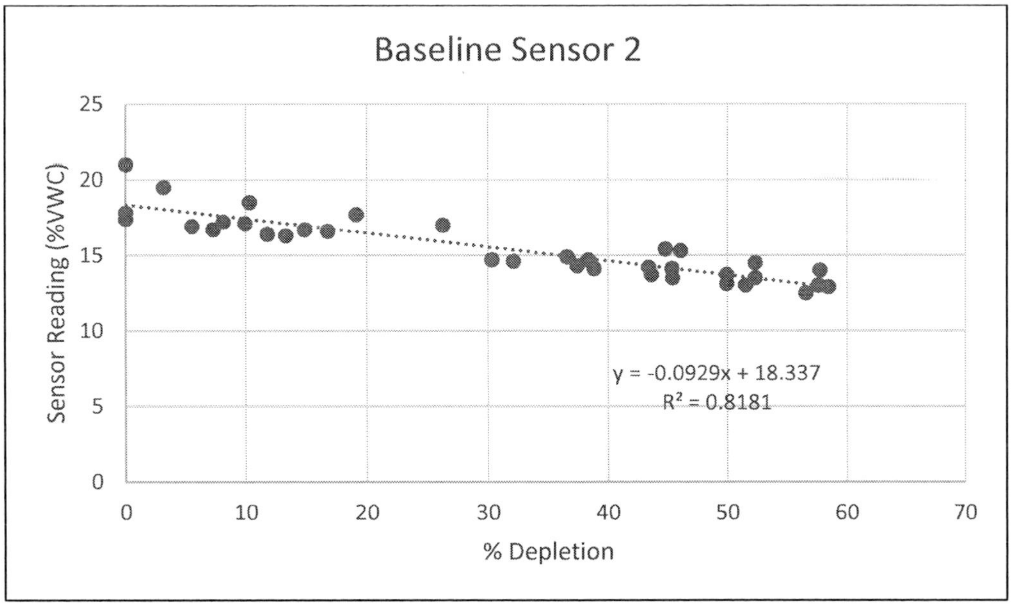

Fig 14: Repeatability Test for Baseline's Sensor#2 Over All three Test Cycles

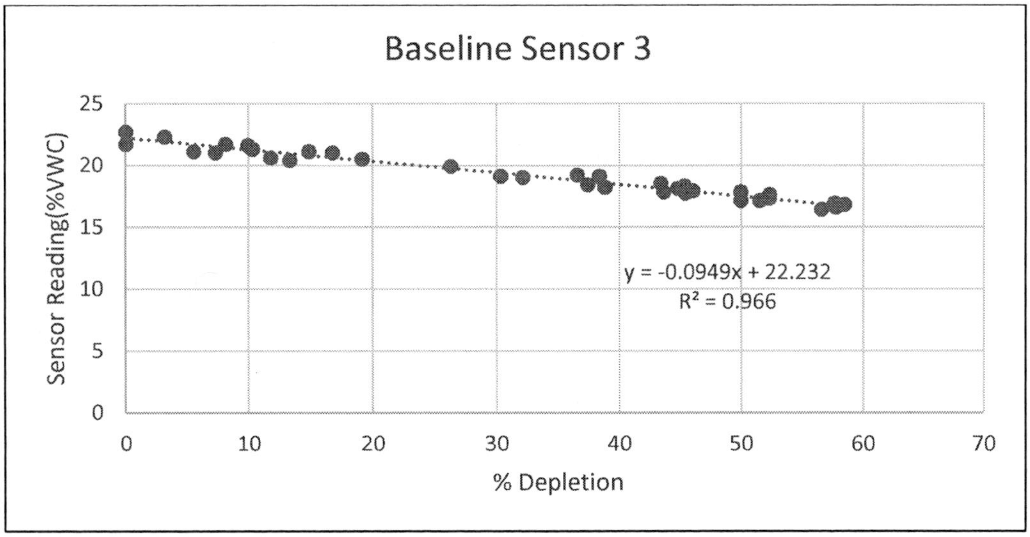

Fig 15: Repeatability Test for Baseline's Sensor#3 Over All three Test Cycles

Out of all three sensors, sensor#3 was found to be highly repeatable and hence more precise as compared to remaining two sensors with regression value (R^2) of 0.966. Sensor#2 was second most repeatable sensor with regression (R^2) value of 0.818. Sensor#1 was the least repeatable sensor and hence the least precise with regression (R^2) value of 0.756 (figures 13-15).

Irrometer's Watermark Data (With Tensiometers)

Similar to Baseline biSensors, first part of data analysis for Watermark 200SS sensors was test of variance among all three sensors for each cycle with drying process of soil as compared against the tensiometers.

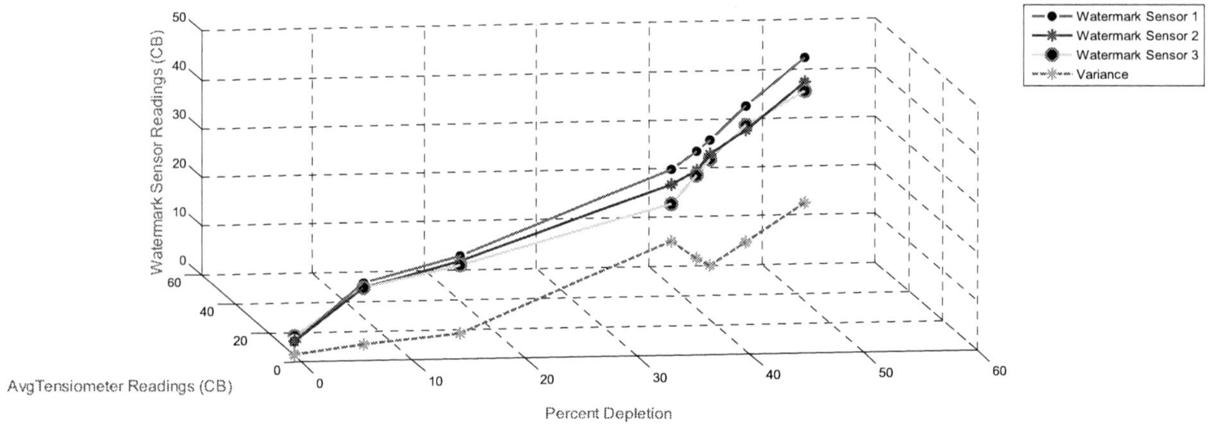

Fig 16: MATLAB Plot for Test of Variance among Three Watermark 200SS Sensors with Drying Cycle (Cycle#2) of Soil

For cycle #2, maximum variance of 13 was seen in depletion range (50-60) %. All three sensors read values very close to each other with minimal variance of values close to 1 up to depletion range (0-20) % (Fig: 16)

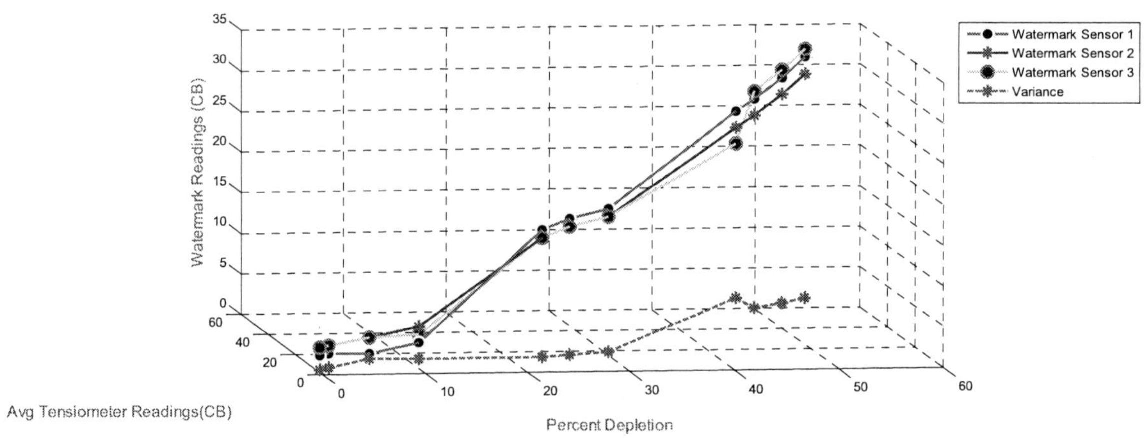

Fig 17: MATLAB Plot for Test of Variance among Three Watermark 200SS Sensors with Drying Cycle (Cycle#3) of Soil

In cycle#3, variance among sensors was around 0.33 up to 30% depletion. Maximum variance of 4 was found in (40-50) % depletion range (Fig: 21).

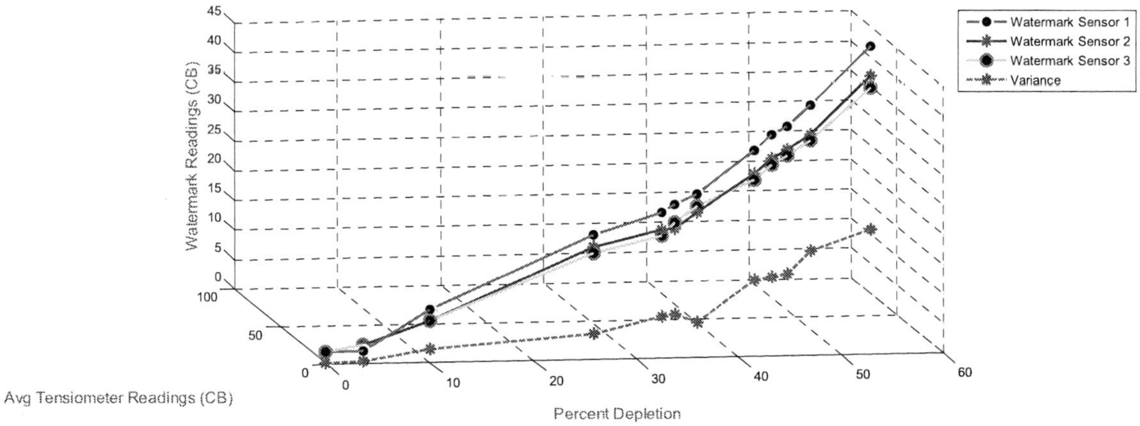

Fig 18: MATLAB Plot for Test of Variance among Three Watermark 200SS Sensors with Drying Cycle (Cycle#4) of Soil

Similar, in cycle #4 maximum variance of 4.33 was found up to depletion of 40% which means sensors read fairly closer values up to 40% depletion. However, variance was found to increase between (50-60) % depletion and reached a maximum value of 13 (Fig: 18).

Second part of data analysis was based on test of repeatability and hence precision of sensors like the biSensors above. Each sensor's repeatability test was done over three test cycles (cycle#2, cycle#3 and cycle#4). For this, each sensor's readings over drying period from all three test cycles were plotted (figures 19-21).

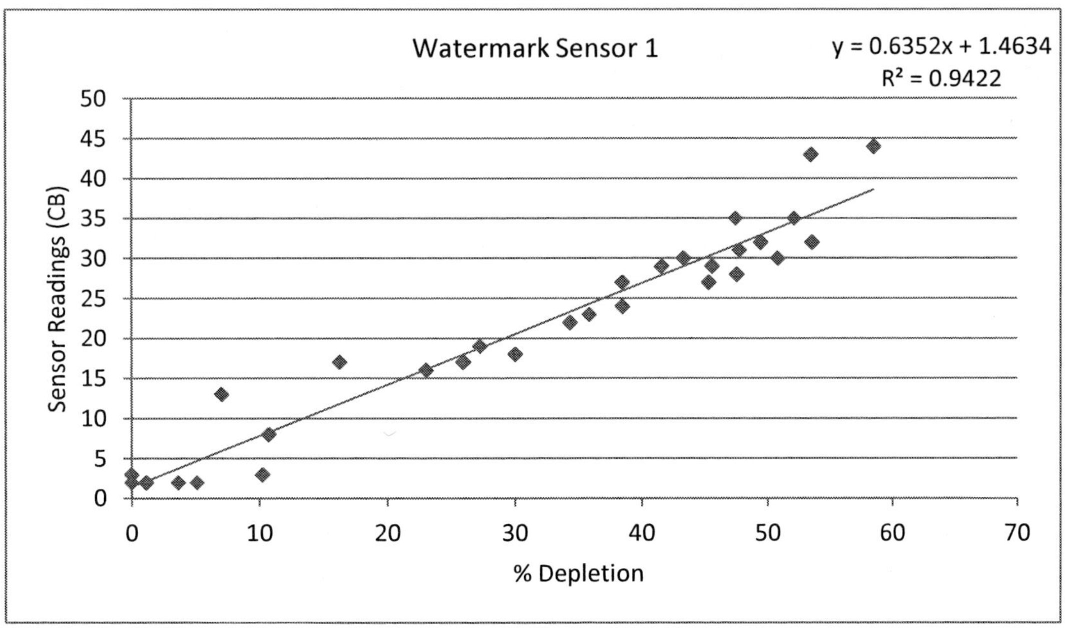

Fig 19: Repeatability Test for Watermark 200SS Sensor#1 Over All three Test Cycles

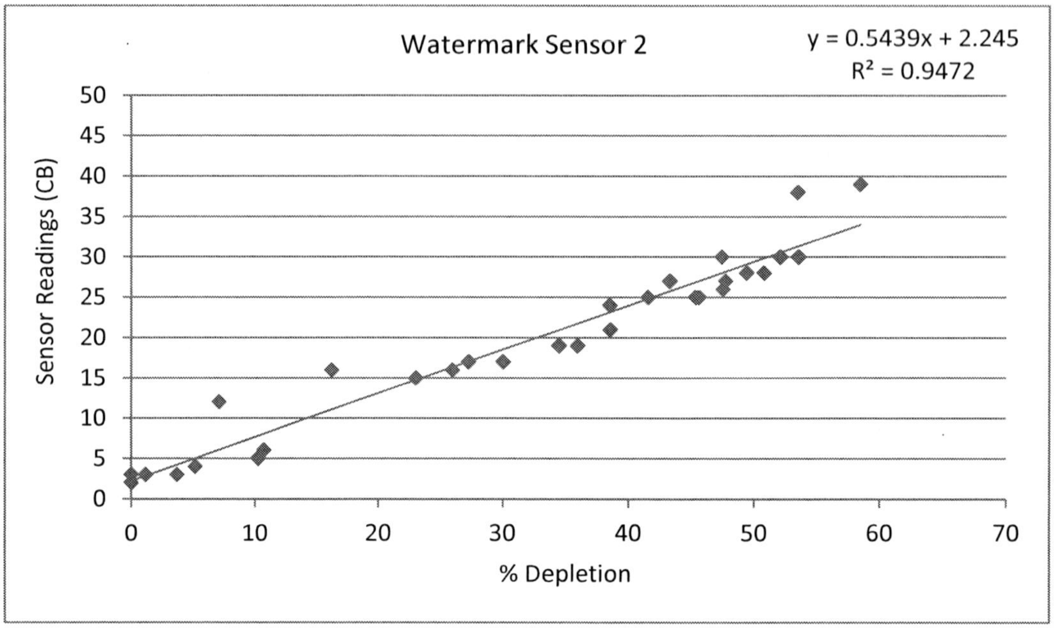

Fig 20: Repeatability Test for Watermark 200SS Sensor#2 Over All three Test Cycles

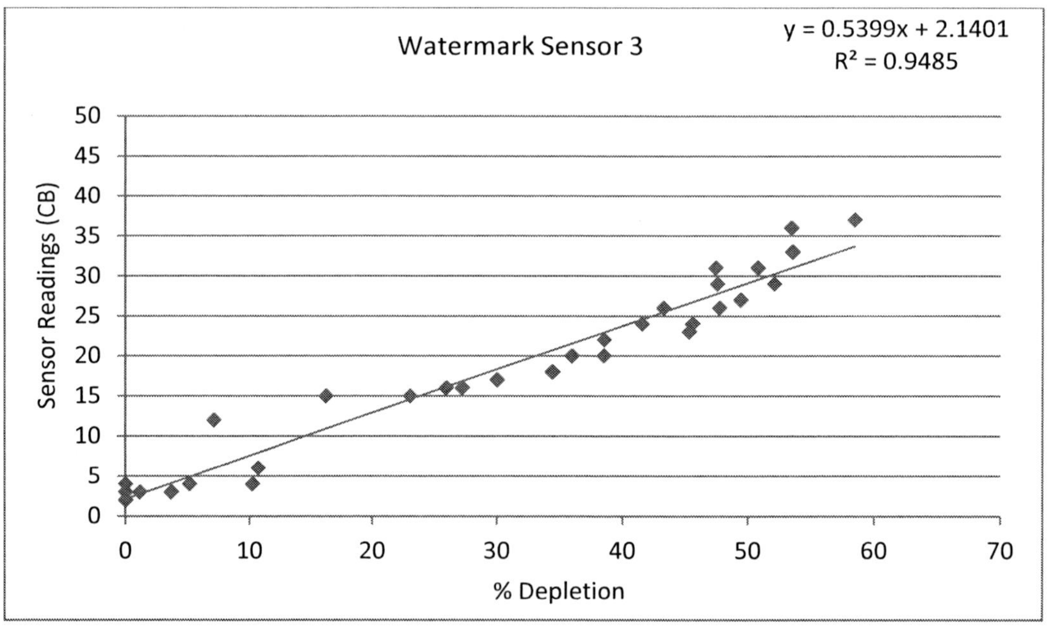

Fig 21: Repeatability Test for Watermark 200SS Sensor#3 Over All three Test Cycles

All three sensors were found to be highly repeatable and hence precise with regression values (R^2) of 0.9422, 0.9472 and 0.9485 for sensor#1; sensor#2 and sensor#3 respectively (figures 19-21).

From the data analysis performed on both types of sensors, variance seemed to vary with degree of wetness and with successive cycles. Initial findings from the data tend to suggest that as the soil settled and regained some of the structure back (which was lost due to the soil preparation process) from the successive wetting and drying cycles, the variance tends to improve. Additionally, we believe given the nature of the container and lack of any root structure, soil dries from outside-in, creating a profile or gradient with each sensor depending on their placements.

Conclusion

Based on the method described, behaviors of three moisture sensors of each technology (TDT and Resistance) were studied. Watermark sensors were found to be of higher degree in terms of repeatability and precision as compared to Baseline sensors. However, variances of Baseline sensors were comparatively consistent and linear throughout the test cycles. Based on the results obtained and method described, it was found that this method can be used to test proper working of moisture sensors' response to normal drying cycle of soil and consequently demonstrate their effectiveness of landscape irrigation control.

Future work

Based on the results obtained and method described, it was found that this method could be used to test proper working of moisture sensors' response to normal drying cycles of soil and consequently test their effectiveness for landscape irrigation control. However, additional work needs to done to address the proper packing of soil and sensors, which affects bulk density and hence sensors' readings (as explained in figure 1). Another area that needs some additional work is the wetting process, so that water could be applied and retained uniformly in the soil. Finally, further investigation about sensors' response variation with wetting cycle is required.

Acknowledgements

The authors would like to acknowledge the support of U.S.EPA , Eastern Research Group Inc., East Bay Municipal Utility District and the entire committee members of ASABE S633(including Dr. Diganta Adhikari, Dr. Michael Dukes and Brent Mecham) for their guidance in coming up with the testing methodology. The authors would also like to extend their sincere thanks to CIT staffs (Dr. David Zoldoske, Kate Norum and Dr. Kaomine Vang) for their persistent guidance and graduate laboratory students (Tou Y Thao, Janet Robles, Charles Cochran, and Rohan Pandit) for their help in collecting soil samples and preparing them for testing purpose.

Disclaimer

Findings and results mentioned in this paper are for educational purpose and hence, testing labs at Center for Irrigation Technology, Fresno State and Texas Agrilife and Extension Service don't guarantee or warranty any of the tested products. Usage of trade and brand names in this paper doesn't constitute endorsement by either of the labs and associated universities.

References

Blonquist Jr, J. M., Jones, S. B., & Robinson, D. A. (2005). A time domain transmission sensor with TDR performance characteristics. *Journal of hydrology*, *314*(1), 235-245.

Cardenas-Lailhacar, Bernard and Michael D. Dukes, 2014. "Effect of Temperature and Salinity on the Precision and Accuracy of Landscape Irrigation Soil Moisture Sensor Systems".

Deschamps, R. J., Siddiqui, S. I., & Drnevich, V. P. (2000). Time domain reflectometry development for use in geotechnical engineering.

Environmental Protection Agency (EPA) (2013). *Notice of Intent to Develop a Draft Specification for Soil Moisture-Based Control Technologies*. Retrieved October 15, 2016 from https://www3.epa.gov/watersense/docs/sms-notice-of-intent-final.pdf

Environmental Protection Agency (EPA) (2013). Reduce Your Outdoor Water Use. Retrieved October 8, 2016 from https://www3.epa.gov/watersense/docs/factsheet_outdoor_water_use_508.pdf

Evanylo, G. K., & McGuinn, R. (2009). Agricultural Management Practices and Soil Quality: measuring, assessing and comparing laboratory and field test kit indicators of soil quality attributes.

Evett, S and L.K. Heng, 2008. "Conventional Time Domain Reflectometry Systems", Chapter 4 in Field Estimation of Soil Water Content, 55-68

Evett, S, 2008. "Gravimetric and Volumetric Direct Measurement of Soil Water Content", Chapter 2 in Field Estimation of Soil Water Content, 23-35

Gardner, W. H., & Klute, A. (1986). Water content. *Methods of soil analysis. Part 1. Physical and mineralogical methods*, 493-544.

Heng, L.K. and S. Evett, 2008. "Tensiometers", Chapter 8 in Field Estimation of Soil Water Content, 113-120

Hignett , C and S. Evett, 2008. "Direct and Surrogate Measures of Soil Water Content", Chapter 1 in Field Estimation of Soil Water Content, 1-20.

Hignett ,Cand S. Evett, 2008. "Electrical Resistance Sensors for Soil Water Tension Estimates", Chapter 9 in Field Estimation of Soil Water Content, 123-129

Kogan, F. N. (1997). Global drought watch from space. *Bulletin of the American Meteorological Society*, *78*(4), 621-636.

Shock, C. C., Barnum, J. M., & Seddigh, M. (1998). Calibration of Watermark Soil Moisture Sensors for Irrigation Management.

Topp, G. C., Davis, J. L., & Annan, A. P. (1980). Electromagnetic determination of soil water content: Measurements in coaxial transmission lines. *Water resources research*, *16*(3), 574-582.

USDA Natural Resources Conservation Service (2008). Soil Quality Indicators. Retrieved October 12, 2016 from http://www.nrcs.usda.gov/Internet/FSE_DOCUMENTS/nrcs142p2_053256.pdf

Yu, X., & Yu, X. (2009). Time domain reflectometry automatic bridge scour measurement system: principles and potentials. *Structural Health Monitoring*, *8*(6), 463-476.

This page intentionally left blank.

Variation in Urban ET to Determine Weather Station Siting

Charles Swanson – Extension Program Specialist
Texas A&M AgriLife Extension Service
College Station, Texas

Seydou Traore – Extension Associate
Texas A&M AgriLife Extension Service
College Station, Texas

Dr. Guy Fipps, PE – Extension Agricultural Engineer
Texas A&M AgriLife Extension Service
College Station, Texas

Abstract. *The Water My Yard program was created to provide simple weekly evapotranspiration (ET) based irrigation recommendations for homeowners by providing irrigation runtimes (minutes) rather than irrigation amount (inches) needed. Since Texas has no state supported ET network, cities are required to purchase and install their own ET weather stations. However, during the planning stage prospective participants ask how many weather stations does the city need and where should they be placed? To answer these questions, a system was developed using landsat imaging to determine surface land temperatures and macro climates in an urban area.*

Keywords. Urban irrigation, evapotranspiration, weather stations, conservation programs.

Background

Many municipalities and water utilities have struggled to develop effective water conservation programs to address the excessive use of water in landscape irrigation. Most automatic irrigation systems are improperly programmed and "over-irrigate" (and waste) 20%-60% of the water applied. While research and demonstration projects have consistently shown that using ET (evapotranspiration) based irrigation schedules saves significant amounts of water, getting typical homeowners to understand and use ET-based irrigation schedules has proven challenging.

Conventional ET-based programs tell homeowners how much water (in inches) is needed to supplement rainfall for maintaining a healthy lawn. The problem is that each irrigation system has a different precipitation rate. Most homeowners do not know their precipitation rate, how to determine runtimes based on inches of water needed, or how to make adjustments to runtimes for soils and root zone depth.

The Water My Yard program was initiated in 2013 to help cities and utilities promote ET-based irrigation programs by providing homeowner weekly irrigation runtimes customized for their irrigation systems. However for a utility to participate in the Water My Yard program, ETo and rainfall data must be available for their service area. As Texas has no state funded ET Network, utilities are required to purchase and install their own weather stations for inclusion to the program. No methodology currently exists for determining the number and locations of weathers station in urban areas for irrigation.

Methodology

In the standardized Penman Monteith equation, the four climatic drivers of ETo are temperature, wind, relative humidity (dew point) and solar radiation. In this study, we focused on use of temperature. Space borne satellite thermal infrared imagery can be used to derive large aerial coverage of land surface temperatures (LST), making thermal image analysis a valuable tool for determining weather station siting. In this paper, thermal analysis was conducted for Dallas County.

Thermal image analysis and research consisted of the following:

- Acquiring satellite images on cloudless days during the turf irrigation season.
- Conducting radiometric calibration of images, atmospheric correction, bright/temperature processing, and surface temperature analysis.
- Collection of ground level temperatures at the time of the Landsat image flyover to validate the thermal images estimation.

Landsat 8 imagery was acquired during three satellite passes over Dallas as given in Table 1.

Table 1. Information on the LandSat-8 Satellite scene collected over Dallas County

County	City	Path/Row	Cloud Cover	LANDSAT8_SCENE_ID	DATE ACQUIRED
Dallas	Dallas	27/37	0.28	LC80270372015201LGN00	7/20/2015
Dallas	Dallas	27/37	0.53	LC80270372014294LGN00	10/21/2014
Dallas	Dallas	27/37	0.18	LC80270372013243LGN00	8/31/2013

Landsat-8 is the most recent Landsat satellite launched on February 2013. Its payload produces two imagers; the Operational Land Imager (OLI) that has nine shortwave bands (15-30m resolution) and the Thermal Infrared Sensor (TIRS) that has two longwave bands (Band 10 and Band 11 both at 100 m resolution). The TIRS have various applications in the field of agriculture, irrigation and water resources engineering. Landsat-8 has a temporal resolution of 16 days and collect more scenes than its predecessors: Landsat Multispectral Scanner (MSS) four spectral bands, Landsat Thematic Mapper (TM) seven spectral bands, and Landsat Enhanced Thematic Mapper Plus (ETM+) eight spectral bands. Land surface temperature (LST) is the temperature emitted by the surface, and its computation from satellite data requires thermal bands and surface emissivity.

Figure 1. Satellite Aerial Image of Dallas County

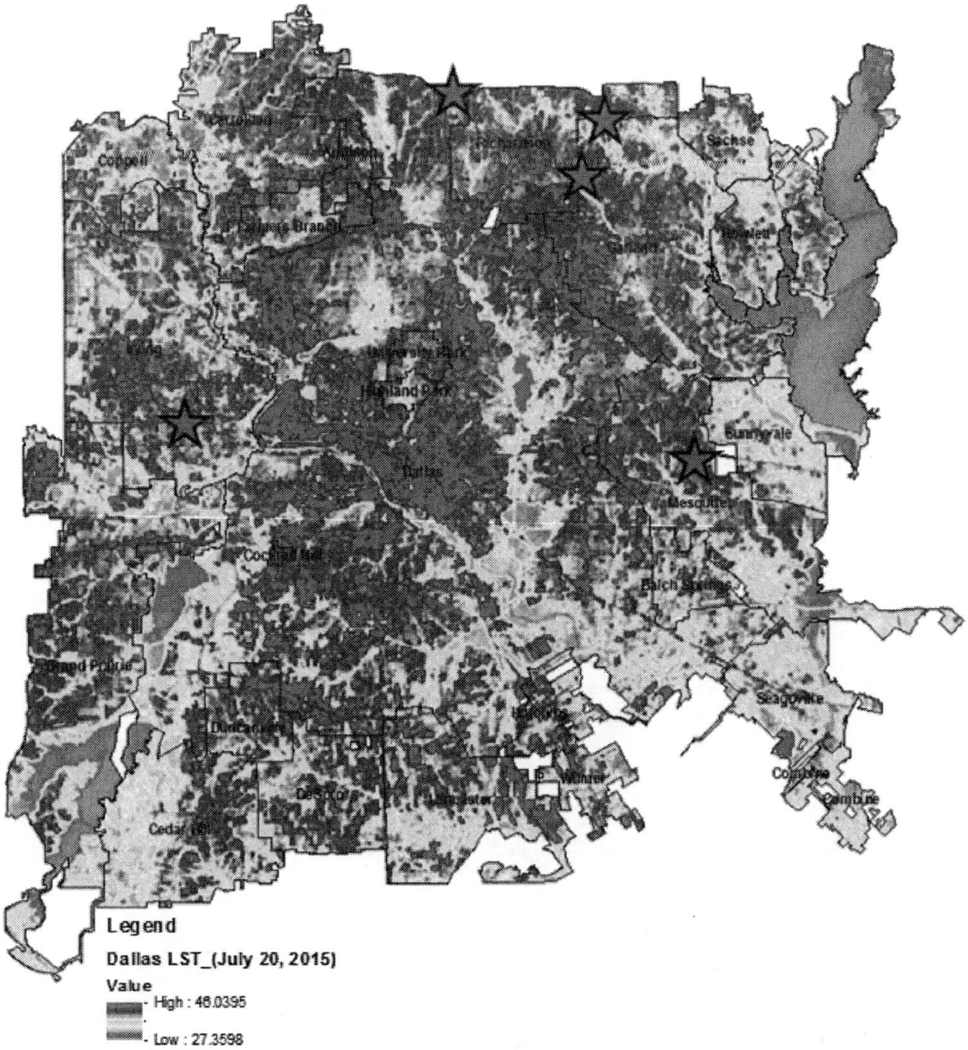

Figure 2. Land Surface Temperature Analysis of Dallas County (July 20, 2015)

In this present study, we used the TIRS band 10 to derive LST. Details of the image processing will be reported in a future paper. The ENVI5.3+IDL (Interactive data language) programming was used to compute the LST and land surface vegetation indexes. In addition, ArcGIS was used for the raster statistics to compute the minimum values, maximum, mean, standard deviation calculation and the correlation between two raster layers (LST and NDVI).

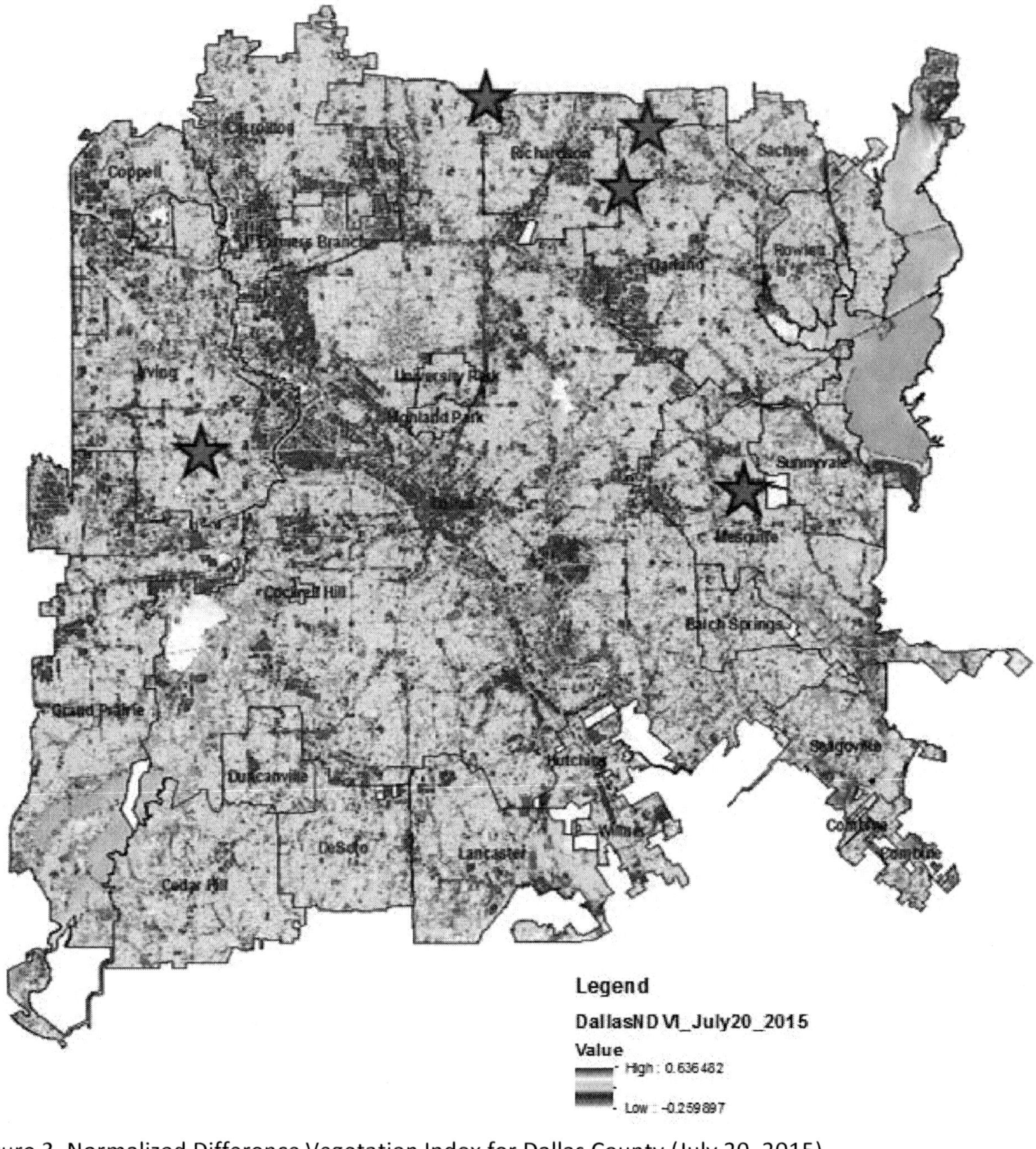

Figure 3. Normalized Difference Vegetation Index for Dallas County (July 20, 2015)

Results

A negative correlation was found between the LST and NDVI value in Dallas County (R^2=-0.19). The coefficients of correlations found showed clearly the degree of dependency between LST and NDVI layers. The negative correlations found in this study mean that the LST values change inversely to the NDVI over the urban landscape environment. Thus, vegetative cover as measured over the NDVI scale is directly related to surface temperatures. For example, the more vegetation measured, the lower the surface temperature, and the less vegetation (ie hardscapes & buildings) the higher the surface temperature. This relationship is extremely important when siting ET weather stations in urban area by providing a guide for areas to avoid and target areas for station siting.

Table 2. LST and NDVI spectral statistics for Dallas County (July 20, 2015).

City	Year	Layer	MIN	MAX	MEAN	STD	Correlation (LST vs. NDVI)
Dallas	2015	LST	27.3501	**53.8542**	38.1932	3.1403	**-0.198067**
Dallas	2015	NDVI	-0.2599	0.6365	0.2725	0.1365	
Dallas	2014	LST	28.2987	51.8112	38.8378	2.8682	**-0.10557**
Dallas	2014	NDVI	-0.2473	0.6159	0.2287	0.105	
Dallas	2013	LST	28.3100	51.7812	37.7218	2.9244	**-0.11652**
Dallas	2013	NDVI	0.2466	0.6159	0.3527	0.2036	

NB: LST (Land surface temperature) in degree Celsius; MIN, MAX and MEAN stand for Minimum, Maximum and Mean temperature, respectively. STD represent the standard deviation. NDVI stands for the normalized difference vegetation index that vary between -1 and+1 (Higher the NDVI index, greener is the land surface. A zero means no vegetation and close to +1 (0.8 - 0.9) indicates the highest possible density of green leaves.)

During the July 20, 2015 Landsat 8 path over Dallas County, multiple ground temperature measurements were collected at the projected path time. These measurements were compared to a thermal analysis to verify processing accuracy. Table 3 shows the ground measured versus landsat calculated surface temperature. Processing was shown to be very accurate, resulting in an average difference of 1.2 degree Celsius.

Table 3. Landsat8 temperature estimate versus and ground measured temperatures in Dallas (2015-07-29)

OBJECT ID	Latitude	Longitude	Ground measured temperature (°C)	Landsat Raster values (°C)	Difference (LandsatRaster – ground Measured)
1	33.00106	-96.81278	35.9	34.812469	-1.1
2	32.99382	-96.79143	43.0	39.306183	-3.7
3	32.97892	-96.76953	39.0	42.293518	3.3
5	32.98589	-96.76533	36.6	39.080597	2.5
6	32.98198	-96.72872	35.9	37.320099	1.4
7	33.00033	-96.73797	36.4	41.223785	4.8
Average			37.8	39.0	1.2

Conclusion

Image processing and ground measurement analysis show that satellite thermal imagery can be a valuable tool for measuring surface temperature. This process can be adopted by utilities and state agencies for identifying potential ET weather station sites when developing urban ET based water conservation programs. Further analysis is needed to develop processing indexes to define the characteristics of temperature based microclimates across urban areas.

This page intentionally left blank.

Full Season Comparison of Weather Based Smart Controllers

Mark A. Crookston, P.E., D.WRE, CAIS, CIC, CID, CIT, CLIA
Northern Water, 220 Water Ave, Berthoud, CO 80513, mcrookston@northernwater.org

Abstract

Several weather based smart controllers at Berthoud, Colorado were programmed to manage six virtual landscape zones over the 2015-2016 seasons. These virtual landscape zones covered a wide variety of soil types, plant materials, and irrigation methods but were the same six landscapes given each controller. The zone valve control outputs on each controller were connected to switches/relays monitored by a data logger which recorded the minutes of runtime for every irrigation cycle of all six zones on each controller. Controller performance was obtained by importing these irrigation events into a daily soil moisture depletion/balance spreadsheet using the dual Kc method for calculating plant water use.

The dual coefficient method partitions ET into evaporation from wetted surfaces and transpiration through vegetation. The increased evaporation following rain or irrigation events and the reduced transpiration resulting from soil moisture depletions were calculated day-by-day for the both the controller determined and the 'preferred' watering for each zone. Both were calculated using the same spreadsheet.

Controllers able to water deeply but less frequently to meet plant needs and/or that directly account for onsite rainfall compared closer to the 'preferred' watering needs than those which did not.

Keywords: Irrigation, Controller, Landscape, Smart, Scheduling, Water management

Weather Based Smart Controllers and Virtual Landscape Zones

The irrigation controllers utilized in this comparison are listed below in Table 1.

Table 1. Irrigation controllers

Manufacturer	Model	Weather Source
Rain Bird	ESP-SMTe	Onsite sensors for air temperature and tipping bucket rain
Irrisoft Rain Bird	Weather Reach Controller Link ESP-Me	Data from onsite weather station via Internet
Toro	Evolution	EVO-WS sensors for air temperature, sunlight, and rain delay
Weathermatic	Smart Line SL1600	SLW5 sensors for air temperature and rain delay
Irritrol	Rain Dial-R	Climate Logic sensors for air temperature, sunlight, and rain delay
HydroPoint	WeatherTRAK LC Central	ET Everywhere service - cellular

All six virtual landscapes used in this comparison are included below in Table 2. Their parameters vary significantly. However, to avoid becoming overly complex and redundant, only the results for Zone 1 are presented.

Table 2. Virtual landscape zones

	Zone 1	Zone 2	Zone 3	Zone 4	Zone 5	Zone 6
Plants	Cool season turf	Fescue	Ground cover	Trees	Woody shrubs / non-desert	Warm season turf
Veg factor, Kv	0.92	0.92	1.03	1.22	0.84	0.88
Density, Kd	1	1	0.85	0.98	0.9	1
Effective Root depth	10-inch	8-inch	9-inch	20-inch	15-inch	5-inch
Typical plant height	3.5-inch	3.5-inch	10-inch	177-inch	30-inch	3.5-inch
Managed stress	High	Ave	Low	Low	Low	Ave
Managed stress, Ksm	0.7	0.8	0.8	0.8	0.8	0.7
MAD	0.8	0.72	0.76	0.87	0.87	0.83
Fraction of organic mulch	n/a	n/a	0.25	0.75	0.75	n/a
Soil type	Silty clay	Loam	Loamy sand	Sandy loam	Silty clay loam	Clay
Slope	4%	8%	6%	2%	10%	12%
Exposure	Full sun	77% shaded	Full sun	50% shaded	Full sun	Full sun
Micro-climate factor, Kmc	1	0.65	1	0.77	1	1
Irrigation method	Spray sprinklers	Spray sprinklers	Spray sprinklers	Rotor sprinklers	Surface drip grid	Rotor sprinklers
Application rate	1.10-iph	1.35-iph	1.60-iph	0.60-iph	0.24-iph	0.42-iph
Irrigation application efficiency	70%	65%	60%	75%	80%	70%
Irrigation interval	MAD	MAD	MAD	MAD	MAD	MAD

Two Season Results

The soil moisture depletion/balance spreadsheet utilized for this comparison began on January 1st 2015 through October 31st 2016. However, the irrigation season was restricted to May 1st through October 31st of each year. Consequently, the following summary charts for Zone 1 cover the irrigation season only.

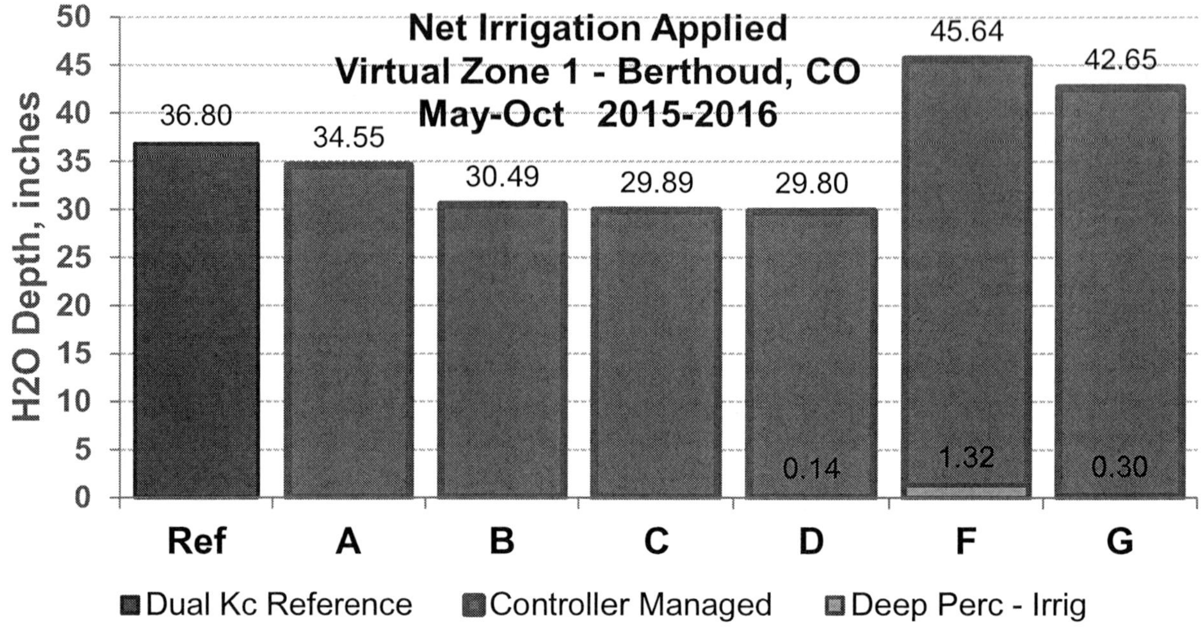

Figure 1. Net irrigation applied to Zone 1 during May-Oct of 2015 and 2016 seasons at Berthoud, Colorado. Depths are the sum of both seasons. Deep percolation losses from irrigation only occurred under controllers D, E, and F.

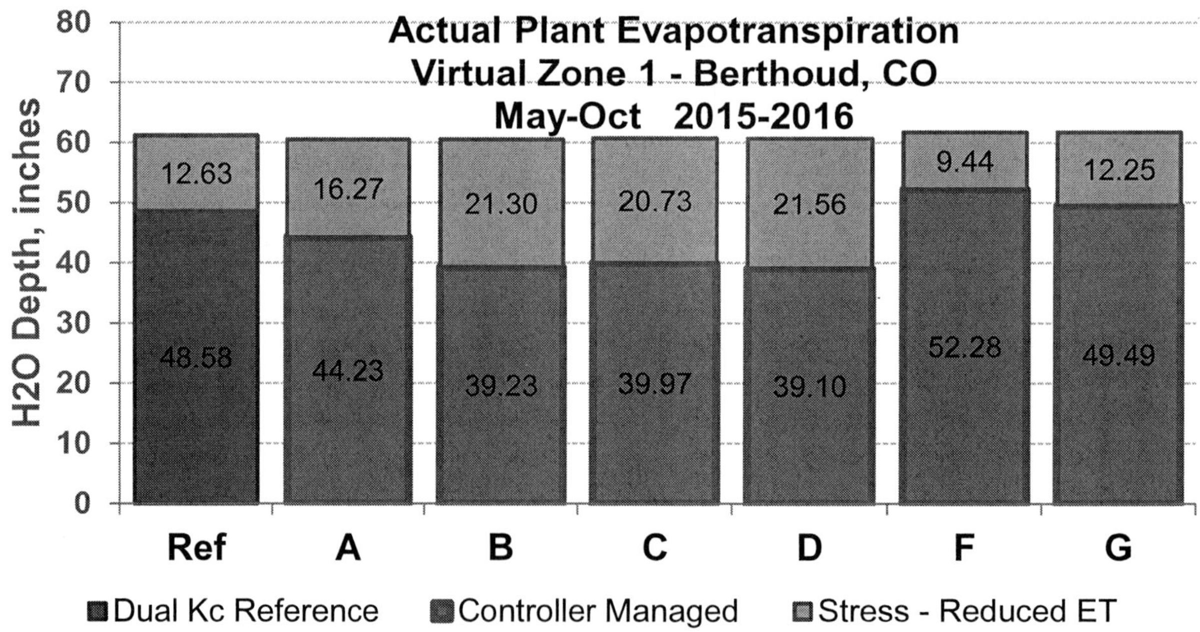

Figure 2. Plant evapotranspiration obtained from the dual Kc method for calculating plant water use Zone 1 during May-Oct of 2015 and 2016 seasons at Berthoud, Colorado. Depths are the sum of both seasons. Included in Figure 2 is the reduction in ET resulting from soil moisture depletions or water stress.

Daily Soil Moisture Depletions - 2015

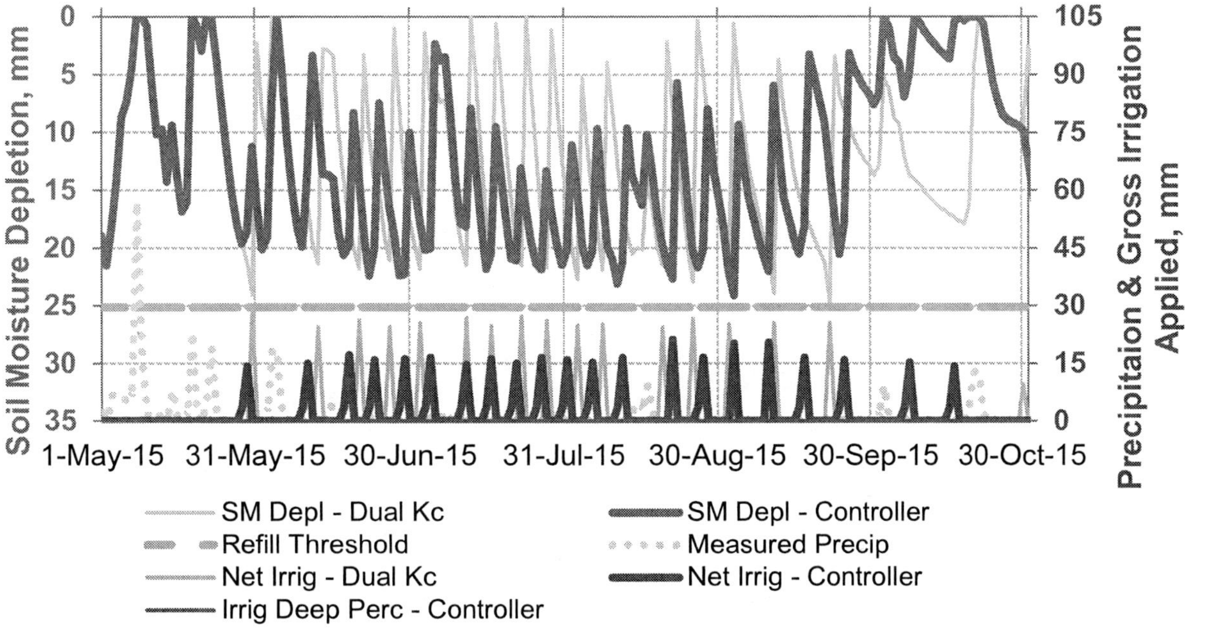

Figure 3. Daily soil moisture depletions for Controller A - Zone 1 during 2015 season at Berthoud, Colorado.

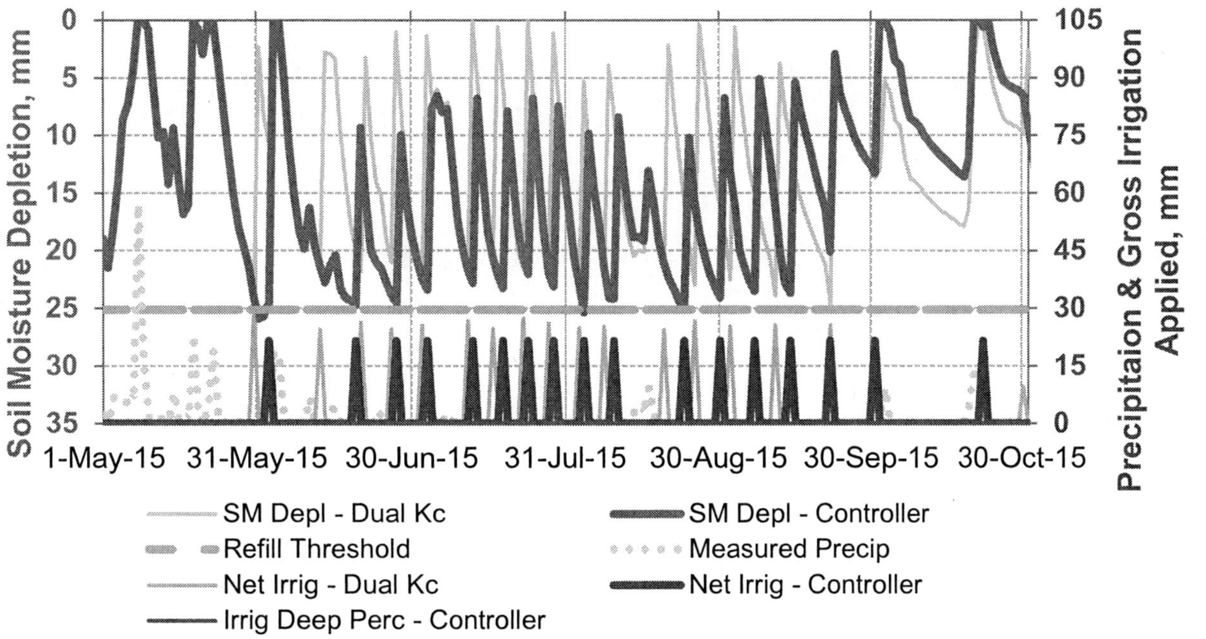

Figure 4. Daily soil moisture depletions for Controller B - Zone 1 during 2015 season at Berthoud, Colorado.

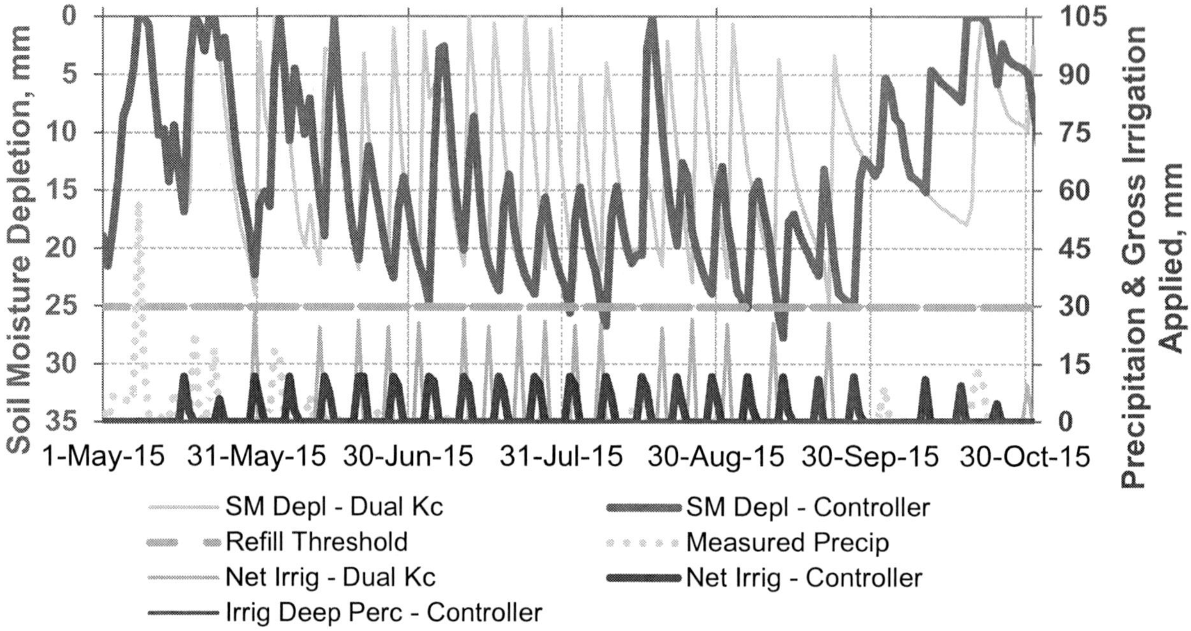

Figure 5. Daily soil moisture depletions for Controller C - Zone 1 during 2015 season at Berthoud, Colorado.

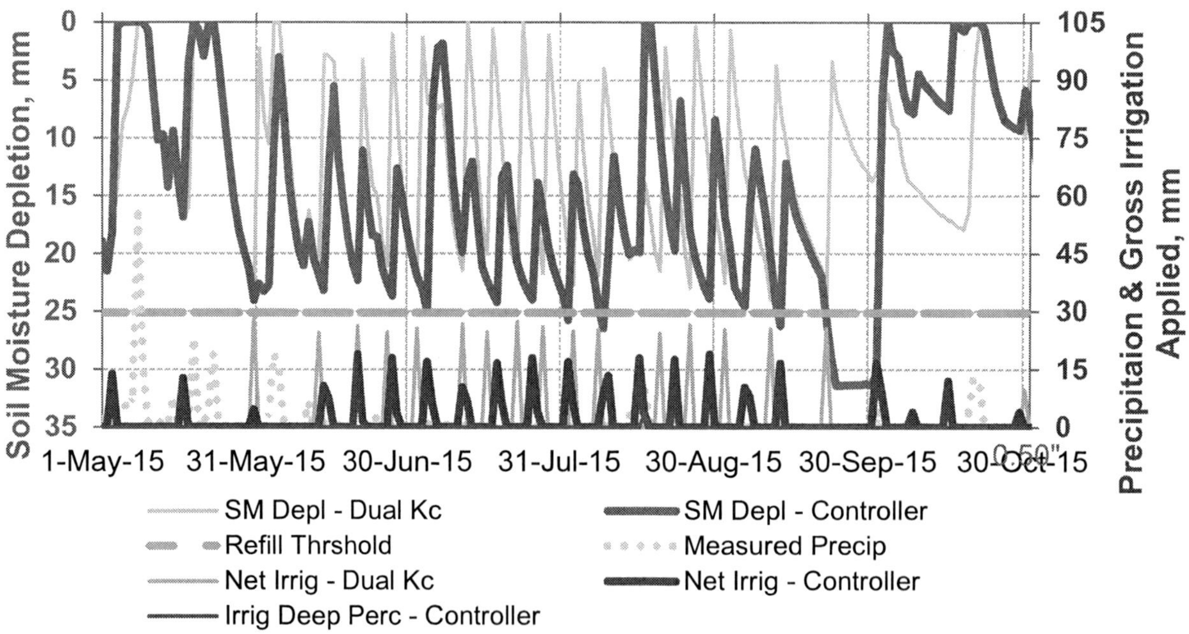

Figure 6. Daily soil moisture depletions for Controller D - Zone 1 during 2015 season at Berthoud, Colorado. The rain delay sensor precluded needed irrigation in late September.

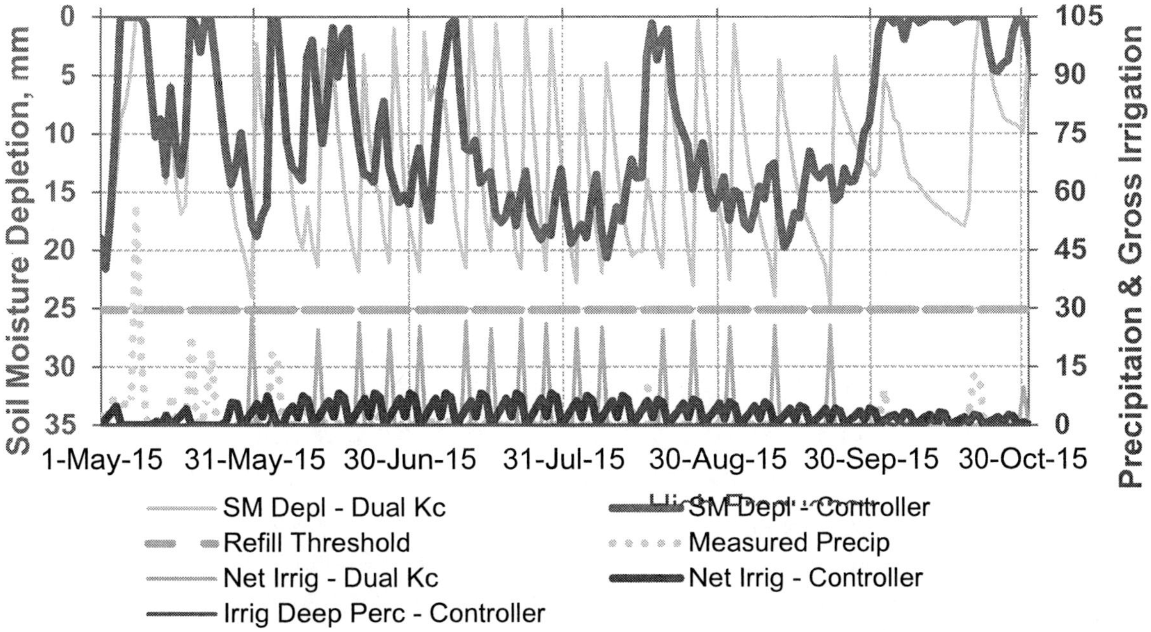

Figure 7. Daily soil moisture depletions for Controller F - Zone 1 during 2015 season at Berthoud, Colorado. This controller called for higher frequency of irrigation.

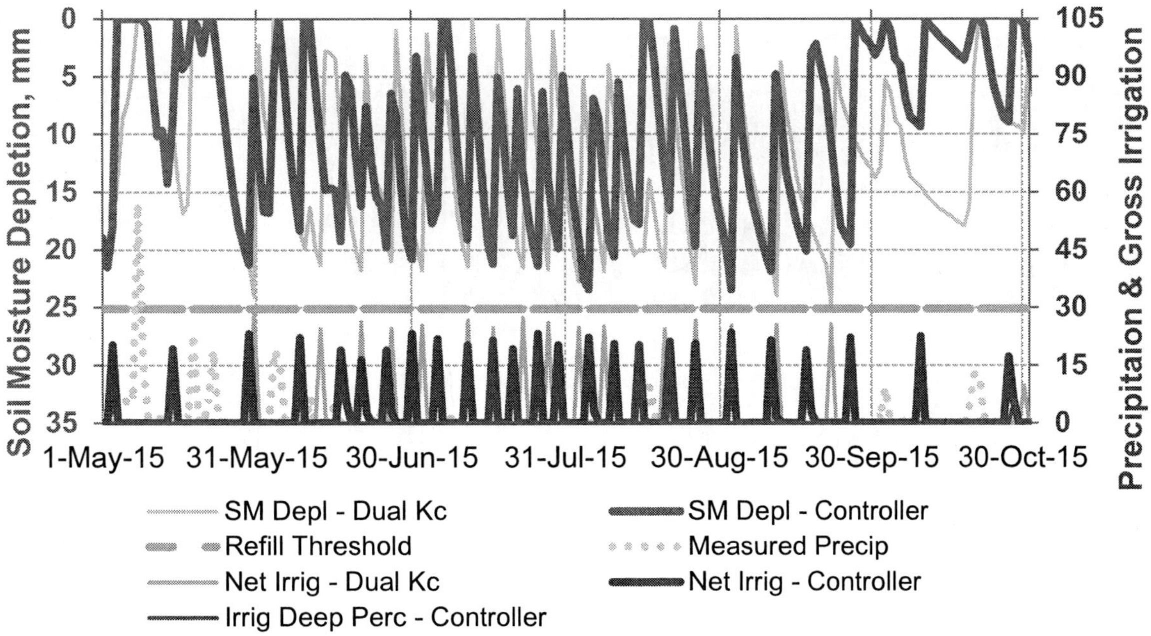

Figure 8. Daily soil moisture depletions for Controller G - Zone 1 during 2015 season at Berthoud, Colorado.

Daily Soil Moisture Depletions - 2016

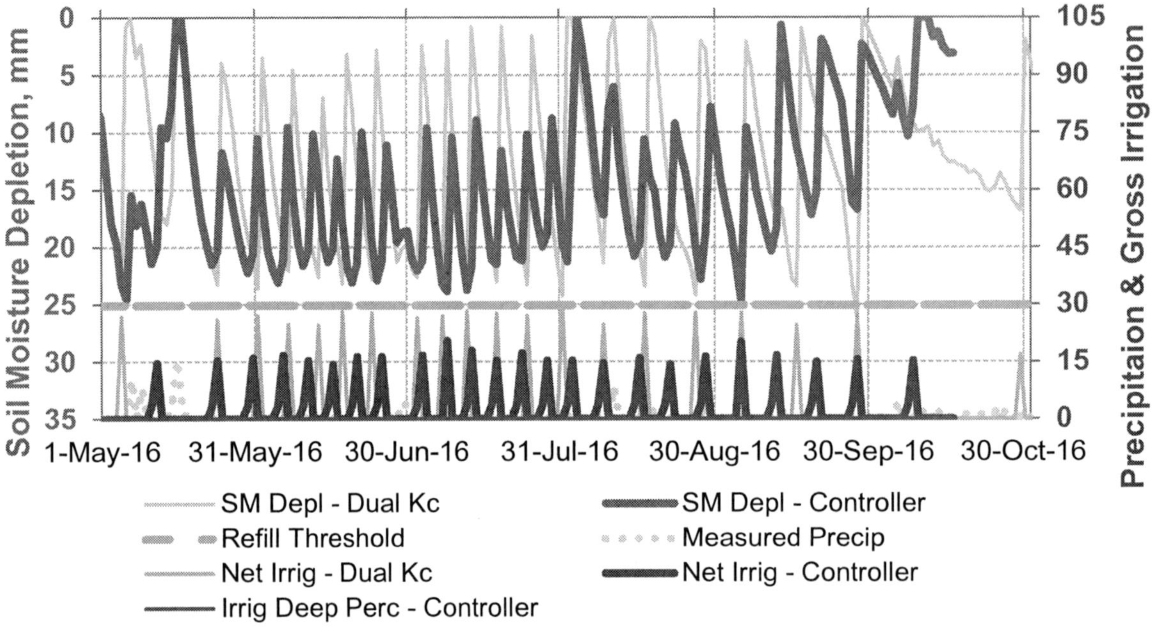

Figure 9. Daily soil moisture depletions for Controller A - Zone 1 during 2016 season at Berthoud, Colorado.

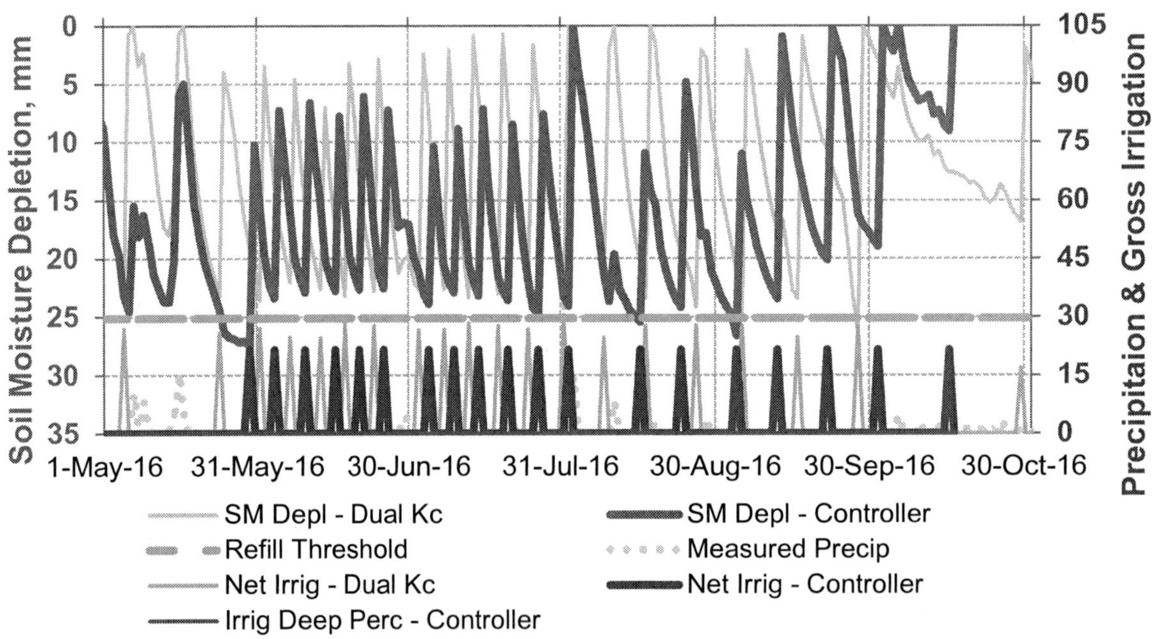

Figure 10. Daily soil moisture depletions for Controller B - Zone 1 during 2016 season at Berthoud, Colorado.

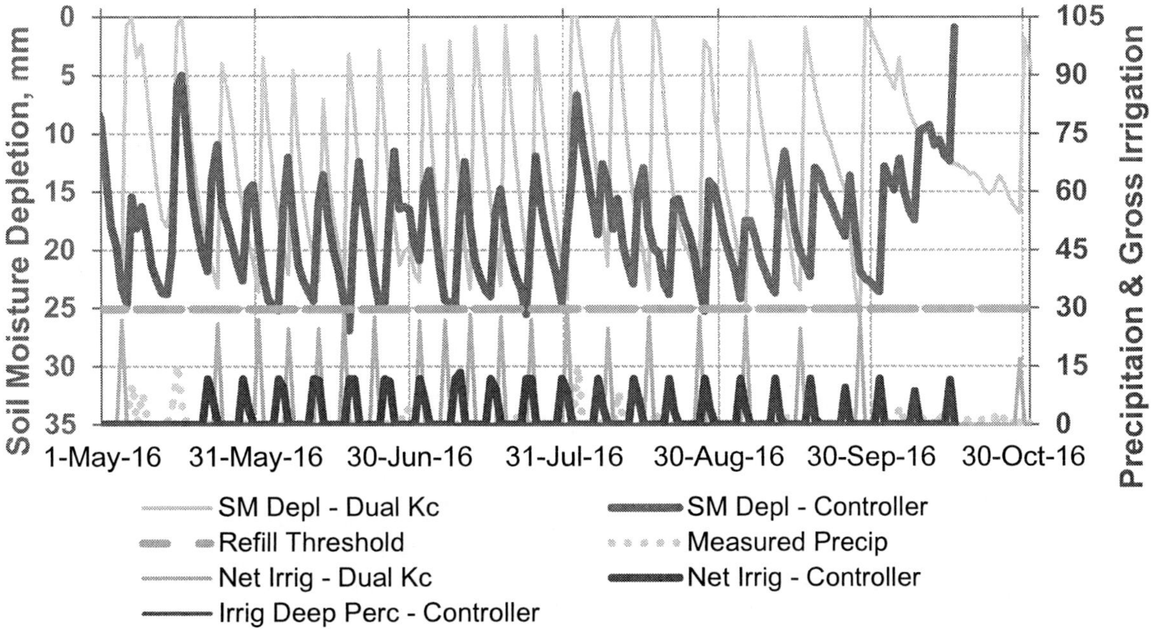

Figure 11. Daily soil moisture depletions for Controller C - Zone 1 during 2016 season at Berthoud, Colorado.

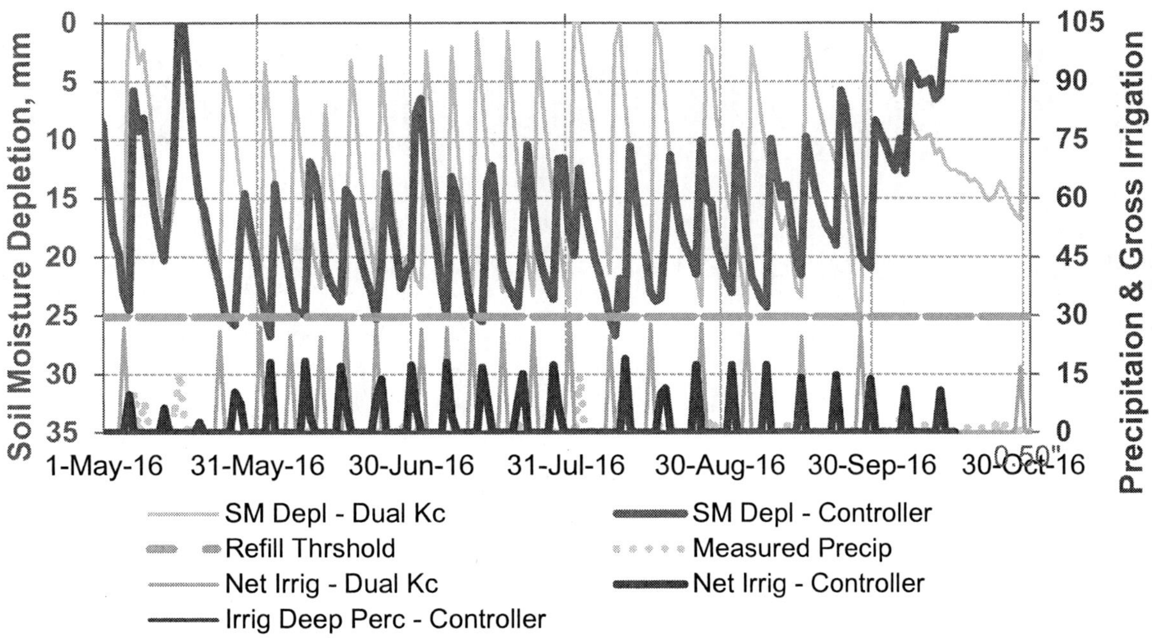

Figure 12. Daily soil moisture depletions for Controller D - Zone 1 during 2016 season at Berthoud, Colorado.

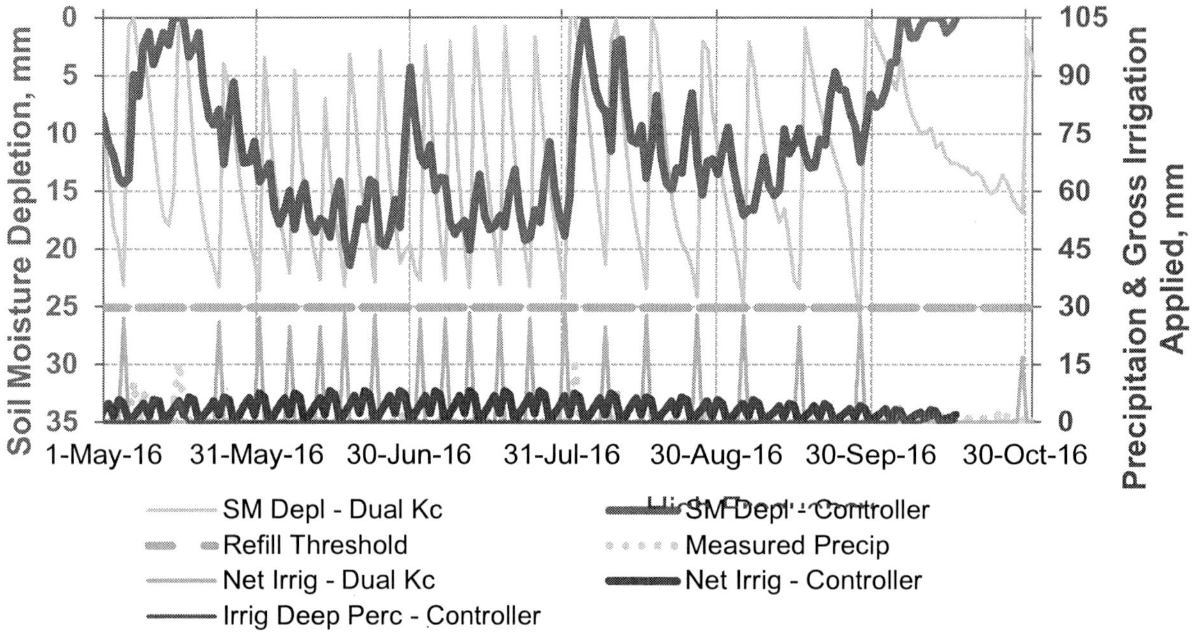

Figure 13. Daily soil moisture depletions for Controller F - Zone 1 during 2016 season at Berthoud, Colorado. This controller called for higher frequency of irrigation.

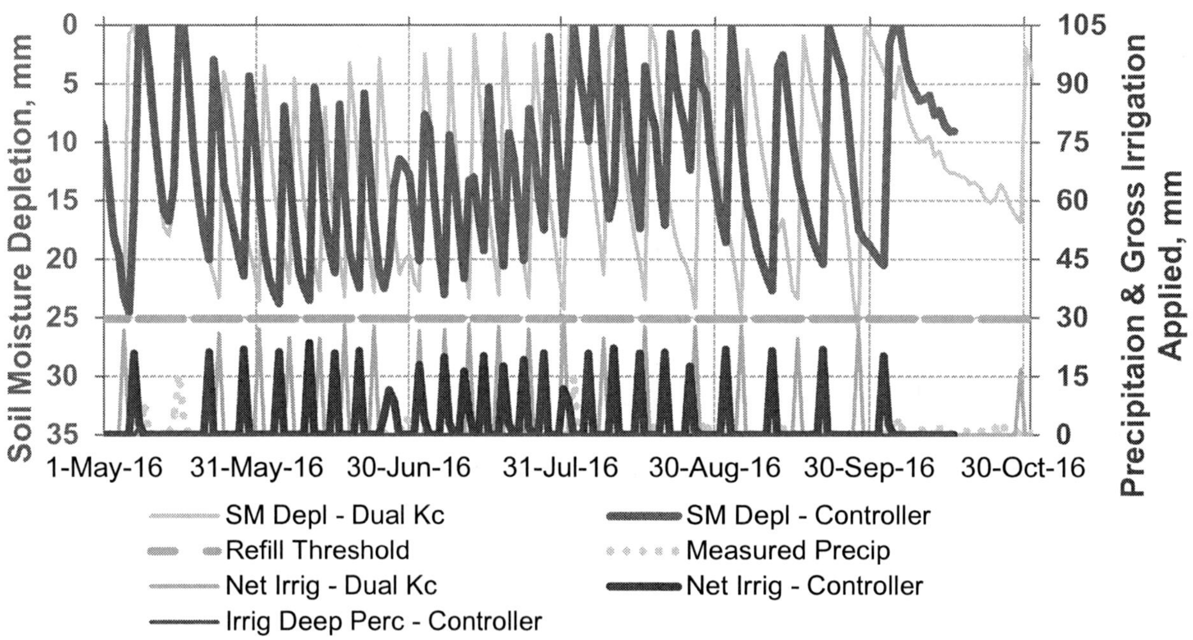

Figure 14. Daily soil moisture depletions for Controller G - Zone 1 during 2016 season at Berthoud, Colorado.

Reference ET

The ASCE standardized evapotranspiration equation (Allen, 2005) was utilized to provide reference evapotranspiration or ET_o. Weather data was obtained from an onsite station, owned and maintained by Northern Water staff. It is located over turf grass in a rural setting. All sensors are re-calibrated each year.

Figure 15 provides an example of the crop curve used for Zone 1 during 2015. Included are Kcb, Kcb x Ks, and final Kc or Kc actual curves. The beginning point for the vegetative growth period on the crop curve timeline was determined each year based on the average air temperature for the previous 30 days. Afterwards, growing degree days were summed to determine the beginning of the mid-season stage. The falling off of Kc values towards the end of season was determined by the lowering 30 day average air temperature and frost events.

Summary

Use of the dual Kc method (Allen, 2016) of calculating evapotranspiration enabled direct comparison of the irrigation management provided by each controller against the 'preferred' watering calculated by the soil moisture depletion/balance spreadsheet. Differences in actual evapotranspiration were realized from increased wet surface evaporation following irrigation events, particularly when more frequent irrigations increased the number of 'wet' days. Additionally, actual evapotranspiration decreased when soil moisture depletions increased and resulted in plant water stress.

Despite the long duration of this demonstration (two seasons) the controllers were generally able to maintain adequate soil moisture to preserve plant health. Soil moisture depletions exceeded the maximum depletion target by only minor amounts and occurred infrequently. Two exceptions were Controller B in late May 2016 (see Figure 10) and Controller D in late September 2015 (see Figure 6). No clear explanation for either has been determined. Controller D has a rain delay sensor but no measureable rain was recorded during either time period.

It would appear that seasonal crop curves were not utilized by the controllers. This is particularly evident for controller C in 2015 (see Figure 16). Soil moisture levels were generally high in the spring and fall periods when evapotranspiration demand was low. However, during mid-summer when both reference ET and the plant vegetative factor were highest, the controller struggled to keep up. Significant soil moisture depletions remained following every irrigation event but one until early October (timely precipitation helped). The use of seasonal crop curves would improve landscape plant health and appearance during the peak use months of July and August. Additionally the risk for plant loss (permanent wilting) could be avoided and desired landscape benefits preserved. Weather based smart controllers could then operate over longer time periods without needing intervention.

The proper utilization of smart irrigation controllers has great potential to achieve significant water conservation in urban landscapes.

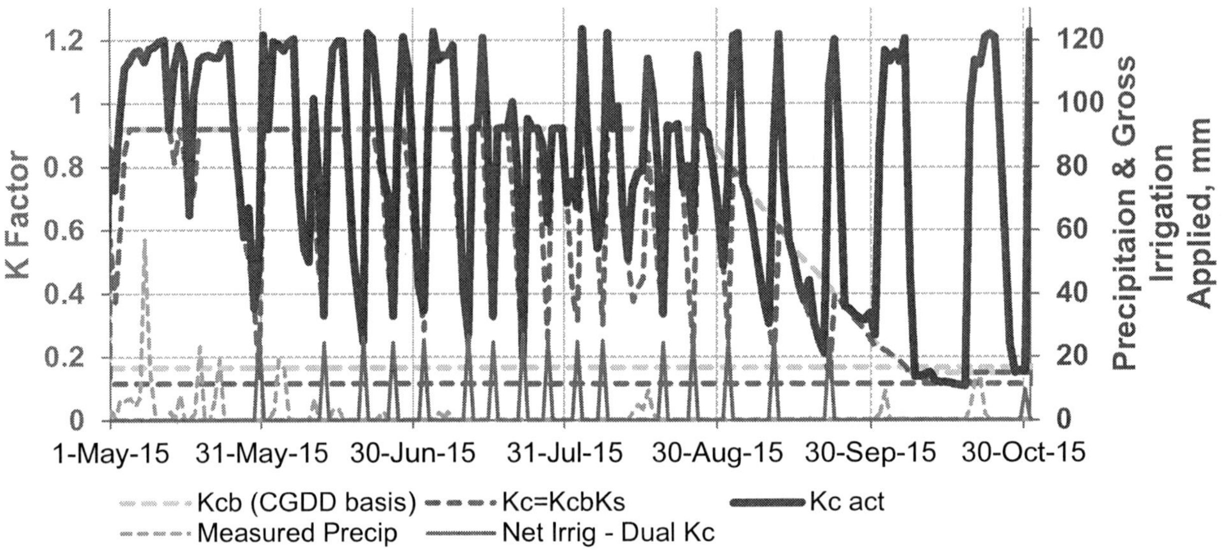

Figure 15. Crop curve coefficients for Zone 1 during 2015 season at Berthoud, Colorado.

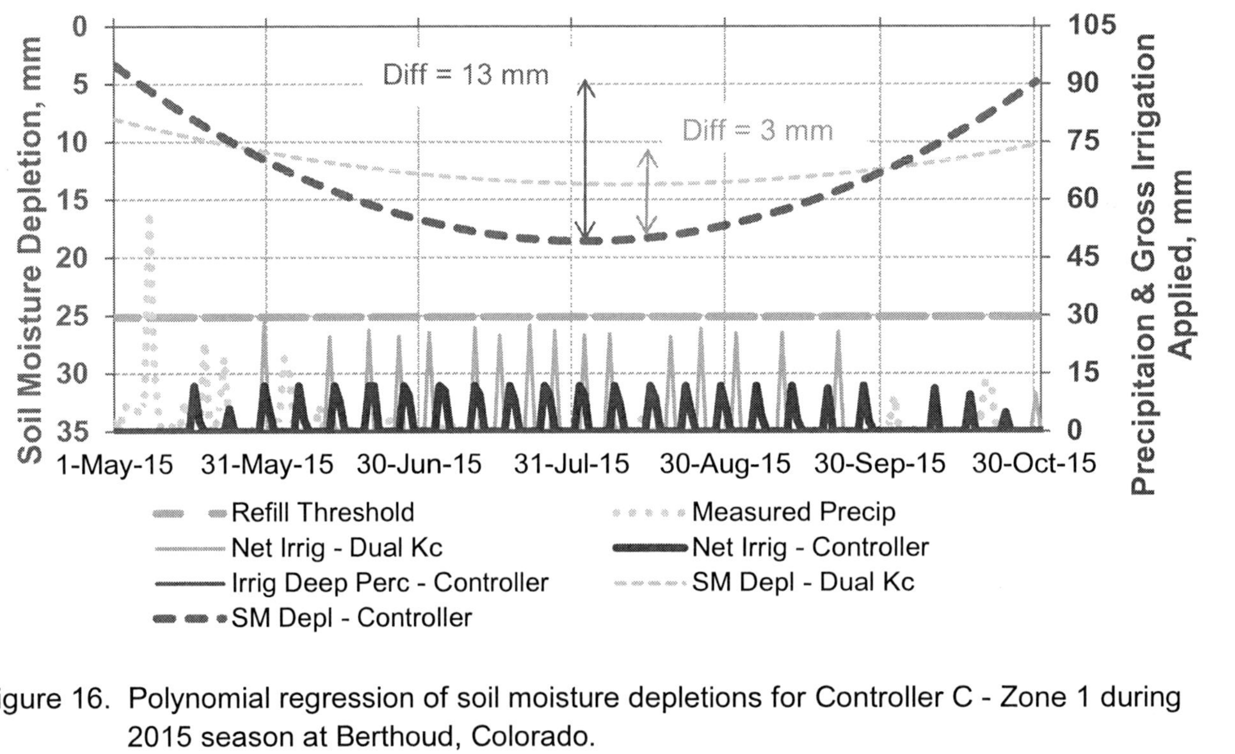

Figure 16. Polynomial regression of soil moisture depletions for Controller C - Zone 1 during 2015 season at Berthoud, Colorado.

References

Allen, R. G. et al. 2005. The ASCE Standardized Reference Evapotranspiration Equation, ASCE, Reston,VA.

Allen, R.G. and M.E. Jensen. 2016. Evaporation from Soil. Chapter 9 in Evaporation, Evapotranspiration, and Irrigation Water Requirements, 2nd ed., ASCE Manuals and Reports on Engineering Practice No. 70, pp.221-259, ASCE, Reston,VA.

Allen, R.G. and M.E. Jensen. 2016. Crop Coefficient Method. Chapter 10 in Evaporation, Evapotranspiration, and Irrigation Water Requirements, 2nd ed., ASCE Manuals and Reports on Engineering Practice No. 70, pp.221-259, ASCE, Reston,VA.

Creating a WaterSense Label for Efficient Landscape Sprinklers

Introduction

To help save water for future generations, the U.S. Environmental Protection Agency (EPA) created the WaterSense Program to help people save water by making it easier to identify water-efficient and high performing products. Products bearing the WaterSense label have been independently certified to use at least 20 percent less water and perform as well or better than standard models. Over the past 10 years, EPA has partnered with manufacturers, retailers and distributors, and utilities to bring millions of WaterSense labeled products to the marketplace. Since the program began in 2006, WaterSense has helped consumers save a cumulative 1.5 trillion gallons of water and more than $32.6 billion in water and energy bills.

Residential outdoor water use across the United States accounts for nearly 9 billion gallons of water each day, mainly for landscape irrigation. The average U.S. household uses more water outdoors than most American homes use for showering and washing clothes combined. Experts estimate that as much as 50 percent of this water is wasted due to overwatering caused by inefficiencies in irrigation methods and systems. To help consumers reduce outdoor water use, the WaterSense label can be found on weather-based irrigation controllers that use local climate and landscape data to determine when and how much to water. WaterSense is currently developing a labeling specification for soil moisture-based control technologies and landscape irrigation sprinklers.

Exploring Specification Development for Landscape Irrigation Sprinklers

In July 2014, WaterSense published a Notice of Intent (NOI) to develop a draft specification for landscape irrigation sprinklers. In the NOI, WaterSense defined a landscape irrigation sprinkler according to the American Society for Agricultural and Biological Engineers and International Code Council's *802-2014 Landscape Irrigation Sprinkler and Emitter Standard (ASABE/ICC 802-2014)*[1], "A sprinkler is a device consisting of a sprinkler body with one or more orifices (i.e., nozzles) to convert irrigation water pressure to high-velocity water discharge through the air, discharging a minimum of 0.5 gallons per minute (gpm) at the largest area of coverage available for the nozzle series when operated at 30 pounds per square inch (psi) or more with a full-circle pattern."

The NOI discussed two main components that influence the efficiency of a sprinkler: the nozzle and the body. The nozzle provides the pattern of water emitted from the sprinkler, either in a fan-like pattern (i.e., a spray nozzle) or by means of one or more moving streams [e.g., multi-stream, multi-trajectory (MSMT)]. The nozzle influences the uniformity of how water is applied. The body of the sprinkler, which houses the nozzle, can provide pressure regulation if applicable and can compensate for changes in inlet pressures. These two components are generally sold separately and are interchangeable between brands in some cases.

WaterSense initially recommended that its draft specification apply to both high-efficiency nozzles and pressure-regulating bodies of landscape irrigation sprinklers. It was EPA's intent to develop one specification that included separate criteria for each component (i.e., a set of nozzles criteria and a set

---

[1] Note that the standard was in draft form at the time of publication of the NOI, but the definitions and methodology regarding testing pressure regulation received only editorial changes from draft to final.

of bodies criteria). Each component would have been certified and labeled separately and could have either been purchased and used separately, or packaged and sold together as a WaterSense labeled landscape irrigation sprinkler.

Regarding high-efficiency nozzles, WaterSense proposed distribution uniformity (DU) as the appropriate performance measure. DU, as defined by *ASABE/ICC 802-2014*, is the measure of the uniformity of irrigation water applied to a defined area. Because field studies were lacking, the WaterSense NOI suggested incorporating DU into the irrigation schedule, thereby shortening irrigation run times and resulting in theoretical savings.

Regarding pressure-regulating bodies, the NOI proposed setting a performance threshold by developing an acceptable outlet pressure variance across a range of inlet pressures and using the test method for pressure regulation as outlined in *ASABE/ICC 802-2014*. WaterSense suggested calculating savings based on the reduction in flow when pressure regulation is in place, potentially capturing additional savings from devices that reduce flow when a nozzle is damaged or missing.

WaterSense listed several outstanding issues in the NOI regarding both nozzles and bodies and requested feedback during the public comment period on a variety of topics. More than two dozen public comments were received. In general, commenters supported moving forward with a specification for pressure-regulating bodies but expressed concern about high-efficiency nozzles and the use of DU as a performance measure. Specifically, commenters had concerns with WaterSense developing a specification for a product category based on theoretical savings based on improved DU. As discussed in the NOI, WaterSense identified two field studies, conducted by Southern Nevada Water Authority and San Antonio Water System, examining high-efficiency nozzles and savings in the field. While both studies measured an increase in DU with high-efficiency nozzle retrofits, neither resulted in the expected water savings.

Based on the lack of field studies demonstrating savings and the public comments received discouraging WaterSense from basing savings on theoretical calculations based on DU, WaterSense put specification development for high-efficiency nozzles on hold. WaterSense continues to collect data and would be interested in collaborating with the industry on field studies or other research that would assess tangible savings, develop consensus around a new performance measure, or demonstrate DU as a viable performance measure for high-efficiency nozzles.

Moving Forward With Pressure-Regulating Spray Sprinkler Bodies

WaterSense moved forward with specification development for pressure-regulating bodies (PRBs), based on the public comments received on the NOI and also potential savings that can be achieved by these products. Sprinklers are usually designed to operate within a range of pressures, and they have an optimum pressure under which the nozzle provides its best performance. Most sprinkler models available on the market have an operating pressure range between 15 and 75 psi, with an optimum pressure between 30 and 45 psi. In many cases, sprinklers are installed at sites where the system pressure is above this optimum operating range, resulting in wasted water.

High operating pressure can result in inefficiencies for a variety of reasons, including excessive flow rates, misting, fogging, and uneven distribution. By regulating system inlet pressure to an optimum level, a sprinkler with pressure regulation can increase efficiency in the irrigation system. The pressure-regulating feature, usually achieved by a device built in the stem, compensates for high inlet pressure and maintains the pressure at a relatively constant level. As a result, the flow through a sprinkler is also constant across a range of inlet pressures, resulting in more even performance and associated water savings. Additionally, by maintaining the pressure within a nozzle's operating range, the nozzle generates appropriate water droplet size and performs with high uniformity.

Although system pressure varies from site to site, high system pressure is not uncommon. Researchers from Utah State University have been conducting a landscape irrigation system evaluation program since 1999. In this program, researchers visit homes and commercial, industrial, and institutional sites to evaluate outdoor irrigation systems. During the visits, researchers collect system pressure at each site. The dataset currently holds 6,462 records[2], 29 percent of which have a pressure higher than 50 psi, including 10 percent that have pressures above 70 psi (see Figure 1).

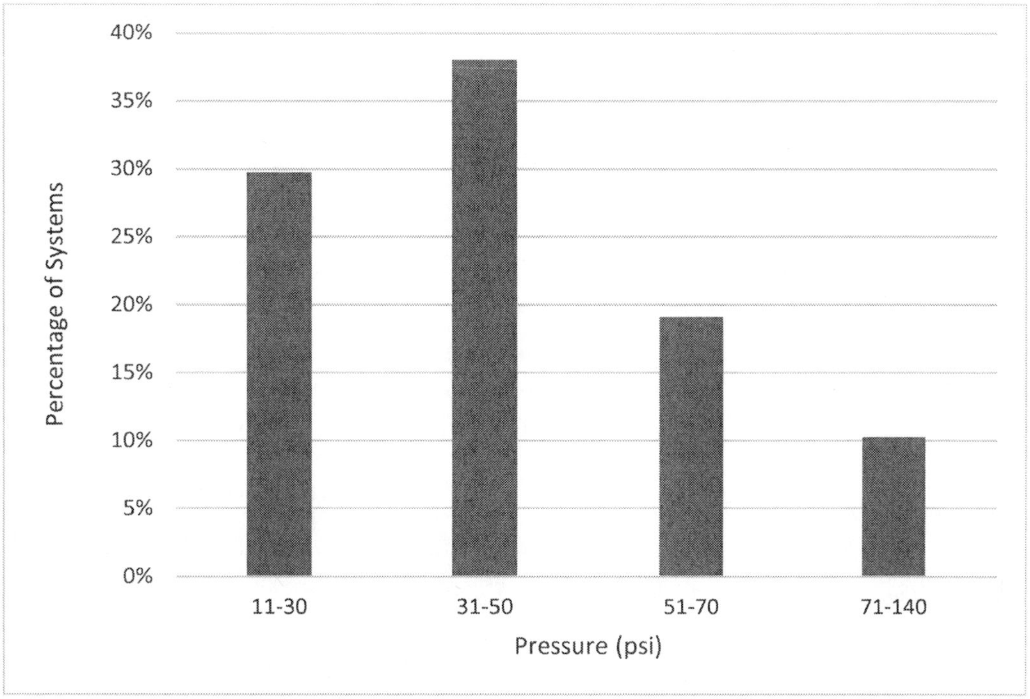

Figure 1. Irrigation System Pressure Data, Utah State University

Similarly, the Center for Resource Conservation in Boulder, Colorado, offers free onsite sprinkler consultations for residential properties. Trained irrigation auditors visit each property to conduct irrigation system inspections. During this process, sprinkler operating pressure is measured. According to the data gathered during these inspections (7,744 records in total)[3], 13 percent of them have a pressure higher than 50 psi, including 3 percent higher than 70 psi (see Figure 2).

[2] Updated data are currently under analysis and will be published at a later date.

[3] Updated data are currently under analysis and will be published at a later date.

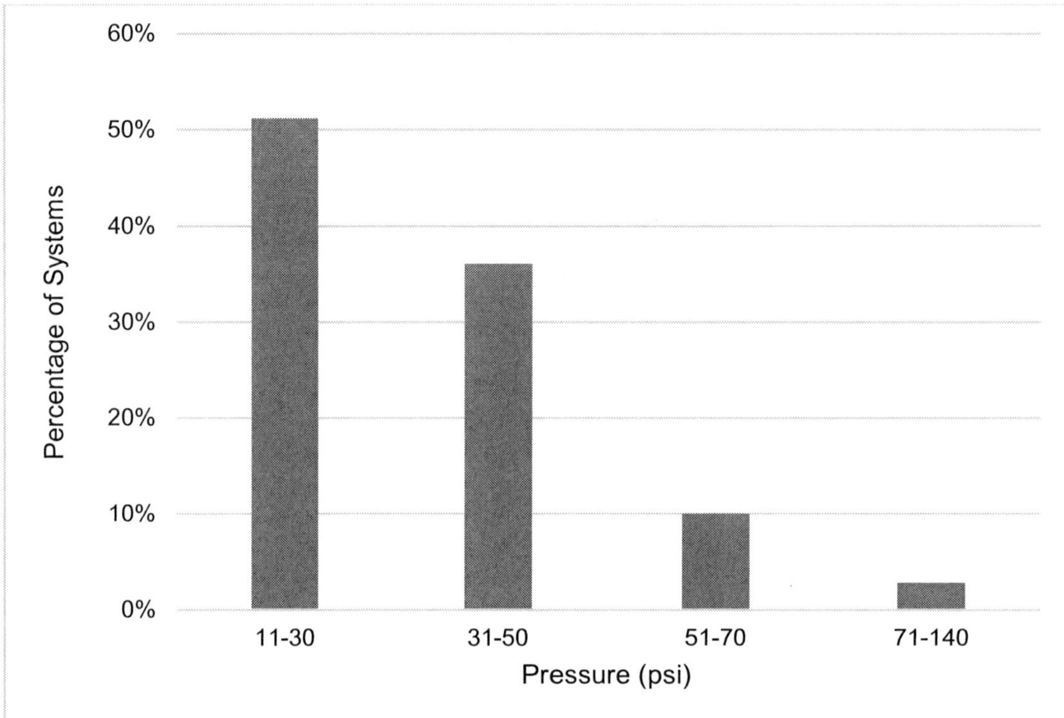

Figure 2. Irrigation System Pressure Data, Center for Resource Conservation

Additionally, the American Water Works Association Research Foundation published a table of water pressures in distribution systems for 15 cities across the United States and Canada in its *Residential End Uses of Water Study.*[4] Pressures ranged from 20 psi to 500 psi (see Table 1).

Table 1. Water Pressure Ranges in Distribution Systems

Utility/Provider	What are the range of pressures in your water distribution system?
Boulder, Colorado	80-160 psi
Cambridge, Ontario	20-100 psi
Waterloo, Ontario	20-100 psi
Denver, Colorado	40-110 psi
Eugene, Oregon	40-80 psi
Las Virgenes Municipal Water District (California)	30-500 psi
Lompoc, California	85-120 psi
Phoenix, Arizona	60-120 psi
Municipal Region of Waterloo (Ontario)	50-70 psi
San Diego, California	40-85 psi
Scottsdale, Arizona	40-120 psi
Seattle, Washington	40-80 psi
Tampa, Florida	20-65 psi (typical = 45 psi)

[4] Mayer, Peter W. and William B. DeOreo. American Water Works Association Research Foundation. 1999. *Residential End Uses of Water.*

Tempe, Arizona	50-90 psi
Walnut Valley Water District (California)	40-180 psi

With the prevalence of high system pressure, as demonstrated above, WaterSense anticipates that labeling and promoting PRBs can improve outdoor water efficiency in a wide range of service territories.

Development of a Test Method and Performance Data

In order for WaterSense to develop a specification for a product category, a repeatable test method must be available, or be developed. Additionally, a set of performance data resulting from the testing of several products according to the test method must be available to provide the basis for the performance and water efficiency criteria. Once WaterSense decided to move forward with specification development for PRBs, achieving these two goals was key to moving forward.

WaterSense began this process in early in 2015 by developing a method for performance testing that was heavily based on *ASABE/ICC 802-2014, Section 303.5.2 (pressure regulation)* with several modifications. First, stakeholders requested through public comment that a low and a high flow be tested. The standard only requires testing at one flow rate (1.5 gpm), so WaterSense incorporated testing at a high flow rate (3.5 gpm) as well. Second, stakeholders requested that outlet flow be measured in addition to outlet pressure, so WaterSense incorporated an outlet flow rate measurement. Additionally, WaterSense allowed the laboratories to use a variety of devices to control flow (e.g., needle valve, variable arc nozzle, or other device) instead of the standard orifice required by *ASABE/ICC 802-2014,* because the laboratories found the standard orifice to be onerous and unnecessary. WaterSense also reduced the number of pressure levels from 12, as specified in *ASABE/ICC 802-3014*, to five pressure levels. This change reduced the time required for each test, though it still allowed for each product to be tested at a range of pressures (i.e.,10 psi above the regulated pressure to the maximum operating pressure).

The three laboratories conducted performance testing using the revised methodology between April 2015 and April 2016. Each laboratory tested three models (three separate brands) of PRBs and three models of standard spray bodies of the same brands, with three samples of each model. Results from the performance testing demonstrated that the products perform as intended, though results were inconsistent among laboratories, indicating that the test method needed to be clarified in several sections. Therefore, WaterSense revised the test method to specify that a needle valve shall be used to control flow. Additionally, WaterSense revised the method to introduce a reduction to 0 psi between each pressure level to address hysteresis found in initial results. For additional information on the independent laboratory performance testing and subsequent test method revisions, please review *Landscape Irrigation Sprinklers: WaterSense Specification Update* on the WaterSense website, published in November 2015.

WaterSense then used the revised test method at the University of Florida to conduct a final round of performance testing on nine PRBs and three standard spray sprinkler bodies. This testing was conducted to determine a range of performance of PRBs using a consistent test method, as well as to determine the water savings of these products when compared to their standard counterparts (e.g., standard spray sprinkler bodies). The data from the University of Florida performance testing will form the basis of the water savings calculations included in WaterSense program materials, as well as the performance criteria included in the specification. The performance testing at the University of Florida was not complete at the time of the submission of this paper, but will be discussed during the technical presentation at the 2016 Irrigation Show & Education Conference (December 6, 2016).

Regarding the flow shut-off feature and associated missing nozzle test, WaterSense indicated in the NOI that products could be required to undergo a missing nozzle test included in *ASABE/ICC 802-2014, Section 303.5.6*, to determine how well the PRBs reduce flow in a situation where a nozzle is damaged or missing. This commonly occurs when a mower damages or completely severs the nozzle from the sprinkler body, among other causes. WaterSense included this test for two products in the initial performance testing conducted at the independent laboratories, as well as for four products at the University of Florida. Results indicate that products with flow shutoff can reduce flow 100 percent when the nozzle is damaged or missing. However, since PRBs without this feature can also significantly reduce the flow when compared to a standard spray sprinkler body, WaterSense has decided not to include this as an additional performance criterion in this version of the specification. Though this is an important water-saving feature, currently WaterSense is only aware of two products on the market that include flow shut-off technology and would like to see the market develop more in this arena before requiring this feature.

Draft Specification Publication

As of this writing, WaterSense is planning the release of a draft specification for PRBs in late 2016. This specification defines the scope of the product category, as well as the performance test method and water efficiency and performance criteria. General requirements regarding product marking and product certification are included as well. For details on the draft specification, visit the WaterSense website at www.epa.gov/watersense.

Next Steps

The public comment period associated with the draft specification is open through the end of January 2017. Two public meetings will be held during this time. The first will take place at the 2016 Irrigation Show & Education Conference in the Oasis (please see the IA Show guide for date and time), and the second will be a webinar (for a date and time, please visit the WaterSense website). Official public comments should be submitted in written form to watersense@epa.gov. Once the comment period closes, WaterSense will review all submissions and revise the draft specification as necessary. EPA is expecting to publish a final specification for PRBs in summer 2017.

This page intentionally left blank.

A New Way to Characterize Landscape Sprinklers Performance

Edward M Norum, Agricultural Engineer,
Center for Irrigation Technology,
California State University, Fresno

Brent Q. Mecham, CID, CLWM, CIC, CLIA, CAIS,
Industry Development Director, Irrigation Association

Over the years, there has been various ways used to characterize how landscape sprinklers perform and various ways to measure performance in the field. These methods have been adapted from agricultural irrigation to describe system performance. There have also been computerized programs that allows a designer to consider spacing and sprinkler configuration to determine the optimal spacing for best performance.

Christiansen (1942) developed a numerical index representing the system uniformity of overlapping sprinklers. This coefficient of uniformity (CU) is a percentage on a scale of 0 to 100 (absolute uniformity). It considers the average deviation and treats dry areas and wet areas equally.

$$CU = 100\left(1 - \frac{\sum x}{n \times m}\right)$$

CU = Equal distribution coefficient developed by Christiansen (%)
x = The total absolute value of deviations from average volume of water caught
m = Average amount of water (mm, mL)
n = The number of water accumulation containers

Distribution Uniformity lower quarter has been the metric most commonly used to measure sprinkler performance in landscape applications. It is focused on the areas receiving the least amount of water and compares the lowest 25 percent of catchments to the average of all of the catchments.

$$DU_{lq} = \frac{V_{lq}}{V_{avg}}$$

DU_{lq} = Distribution Uniformity lowest quarter expressed as a decimal fraction
V_{lq} = average volume of water of lowest 25% of catchments
V_{avg} = average volume of all catchments

In recent years there has been discussion that for landscape irrigation, using distribution uniformity lower half would be a better metric and especially when considering additional run time for irrigation stations. DU_{lh} provides a metric that is very similar to CU, especially in well-designed irrigation systems. The Irrigation Association introduced the concept of Scheduling

Multiplier first in the Golf Irrigation Auditor book and later in the in the Landscape Irrigation Auditor book to provide guidance on the amount of extra water or additional run time to compensate for the non-uniformity of water application and how it manifests itself in the appearance of the turfgrass. The SM is based on DU_{lq} and a simplified equation that would make it nearly equal to DU_{lh}, especially on good performing systems, whilc on poor performing systems the SM would reduce the extra amount of water or run time than just using DU_{lh}. The SM essentially made a "cap" on how much extra water or run time is added to the calculated depth of water or run time assuming nearly perfect conditions.

Scheduling coefficient is another metric that has been used to evaluate the effectiveness of a particular layout of sprinklers considering sprinkler spacing and sprinkler configuration such as square, rectangular or triangular. SC is calculated for landscape irrigation as the driest contiguous five percent of the area compared to the overall area. The ideal SC = 1.0. This particular metric is not measured in the field, but rather is a determined from computer programs that can use a sprinkler profile as shown in Figure 1 to create densograms as shown in Figures 2 and 3. The densograms provide a picture of the distribution of water with wet areas indicated by the darker shading and the drier areas indicated by lighter shading. Figures 1 and 2 show the same sprinkler but in different spacing configuration with the metrics of DU_{lq} and SC calculated.

Figure 1. Sprinkler profile

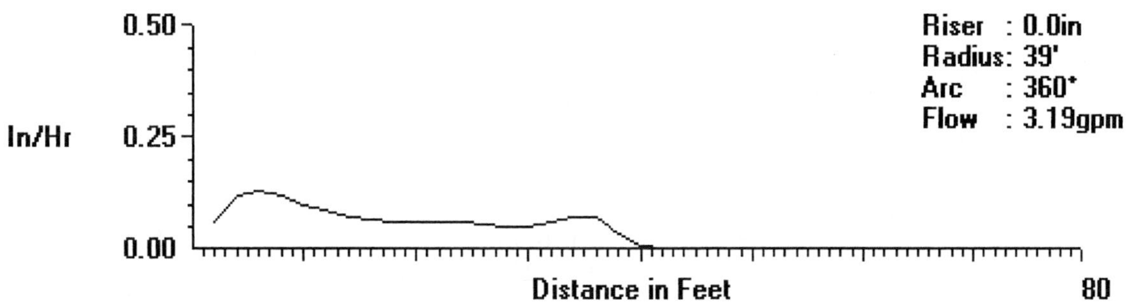

Figure 2. Triangle Spacing

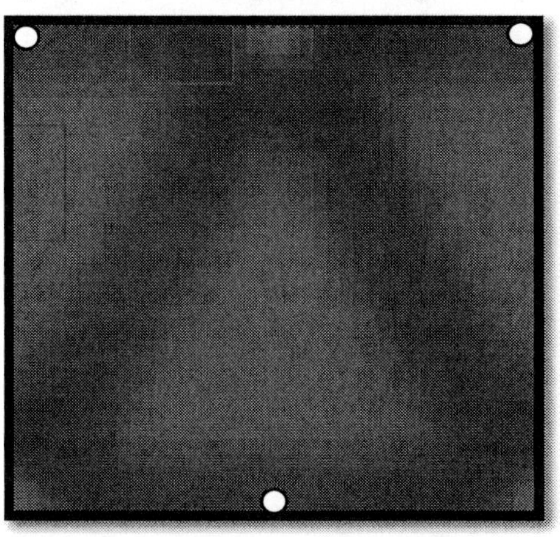

39' x 34' SC = 1.3 DU_{lq} = .82

Figure 3. Square Spacing

39' x 39' SC = 1.6 DU_{lq} = .73

The densograms show the distribution of water for the sprinkler indicated in the sprinkler profile with a maximum radius of throw of 39 feet. In this particular instance, equilateral triangular spacing provides a better distribution of water to the area rather than square spacing. This tool helps designers determine the optimal sprinkler spacing and configuration for each type of sprinkler and nozzle being considered for use in the field. This can change with each sprinkler and operating pressure, so it is difficult to provide a rule of thumb. A common design practice is to reduce sprinkler spacing by 10 percent to improve performance. However, the densograms don't show what happens to the water that is thrown off target and the computer program doesn't allow you to reduce the radius of throw as would be done in the field, so the results are often different than the design.

So while the various metrics for evaluating sprinkler performance have been used they have focused on the dry areas of coverage and then irrigation scheduling has been modified, usually with additional run times for the stations covering the area to reduce or eliminate any stressed areas for the best possible appearance. What is not measured is the amount of water that has been applied beyond the target area such as overspray or the amount of water that has percolated below the root zone. While overspray is visible, characterizing or accounting for deep percolation has not been evaluated in landscape irrigation.

A New Testing Methodology

Beginning in 2012, Smart Water Application Technologies began to develop a testing protocol for sprinkler nozzles. Originally, the intent was to test nozzles that were advertised or sold as being more efficient. A final testing protocol was published in April 2015. A few unique concepts with this testing protocol was to test sprinklers more as they are used in the field. Two defined areas based on the radius of throw of the sprinkler is a square that is twice the diameter of throw in dimensions and allows for four quarter-circle nozzles, four half-circle nozzles and one full- circle nozzle to create the test area, therefore a 15-foot radius nozzle would have a 30-foot by 30-foot square. The other shape is a circle, the diameter of the circle being twice the radius of throw and includes one full-circle nozzle and six part-circle nozzles with arcs adjusted to minimize overspray. The square shape being the one that should be optimal and the circle representing amoeba-shaped turf areas where keeping all of the water on target is a challenge.

The sprinkler nozzles would be evaluated for distribution uniformity and also sprinkler operational efficiency trying to characterize where all of the water is going.

In 2014, the Center for Irrigation Technology (CIT) was asked to develop a protocol that would be useful in administering sprinkler rebate programs. The objective of the program was to encourage the development of more efficient turf irrigation sprinklers. If successful in developing the test protocol, it could be administered by third-party testing agencies to pre-quality turf sprinklers for rebate programs. Threshold performance standards would be prepared by extensive testing of currently available sprinklers. This testing would establish the current state of the commercial art. Threshold performance values thus set should result in rebates being offered to encourage improved irrigation sprinkler operating efficiencies.

The challenge was no test protocol existed that provided a calculation of the sprinkler operating efficiency. Further, the current commonly used test protocol is scientifically suspect. This current protocol involves using a wetted radius lab study with computerized overlap simulations as a basis for system performance metric calculations. The protocol makes no allowance for jet mechanical interference and its effect on uniformity of application and other mechanics. This new protocol then uses a full grid testing layout with sprinklers located to duplicate actual field installations. Currently, used performance metrics such as distribution uniformity (DU) and coefficient uniformity (CU) were abandoned in this effort except for historic reference.

LABORATORY LAYOUT AND INSTRUMENTATION

Whenever feasible, products should be tested in a manner that duplicates their actual field use as closely as possible. The sprinklers in this study were all tested in a full-scale layout on the smooth concrete floor of the CIT sprinkler test building.

The sprinkler spacing was a square grid with a distance of 15 ft between sprinklers. The PVC piping network was sized to keep velocities below 3.0 fps. Test pressures were as registered to an accuracy of 0.5 percent in the plumbing network into which the sprinklers were attached. Rain gauges had a 4-in. diameter and recorded applications to the nearest 0.01 in. Flow measurement accuracy was to 1.0 percent. The building environment represents a zero wind environment. Sprinkler run times were set to provide an average catchment of 0.50 to 0.75 inches. Environmental measurements included temperature, humidity, and barometric pressure.

Grid rain gauge spacing was 3.0 ft by 3.0 ft. The target area was 30 ft by 30 ft representing a model yard and contained 100 evenly-spaced rain gauges (see Figure 1). The target area was surrounded by a single row of rain gauges. The gauges were spaced to represent the catchment within three feet of the target boundary. Virtually no water droplets were detected beyond the rain gauge grid geometry. A special valving arrangement allowed for nearly instantaneous system start up and shut down.

Figure 1. Sprinkler layout and catch device placement

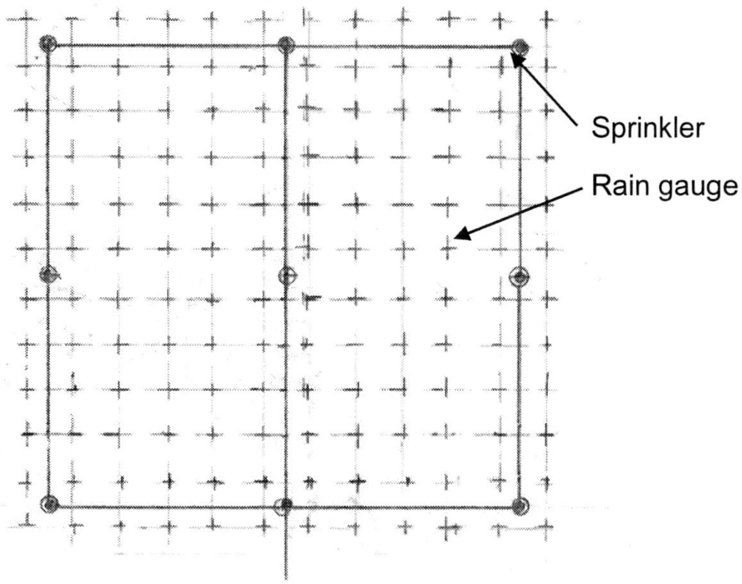

EVOLUTION OF PERFORMANCE PARAMETERS

Inefficiencies in turf sprinkler performance result from: losses to deep seepage caused by pattern non uniformities; losses due to over spraying of the target area; and losses to atmospheric evaporation. With the water distribution measured at the grass canopy, surface evaporation of drops that never reach the grass canopy is automatically accounted for. Strictly speaking, this evaporation loss should be accounted for because it could be caused by a variable in sprinkler design. Instrumentation to account for evaporation losses is prohibitively expensive.

Losses to deep seepage result from the repeated use of non-uniform patterns. Repeated use results in a tendency to index wet-on-wet and dry-on-dry spots between irrigation rounds. In practice, this is compensated for by over-irrigating the dry spot to maintain adequate dry spot quality. As a result of this over-irrigation, the wet spot will drive the surplus water through the wet spot into the subsoil. The formula for calculating this percolation loss (PL) is as follows:

$$PL = \frac{\sum_{1}^{75}(x......x_i)}{n(\overline{X})}$$

PL = Percolation losses
x = application rate of each individual catchment
x_i = application rate at 75% of area
n = number of catchments
$\overline{X}$ = average application rate

The calculation is shown graphically in Figure 2.

Figure 2 Graphic representation of percolation loss

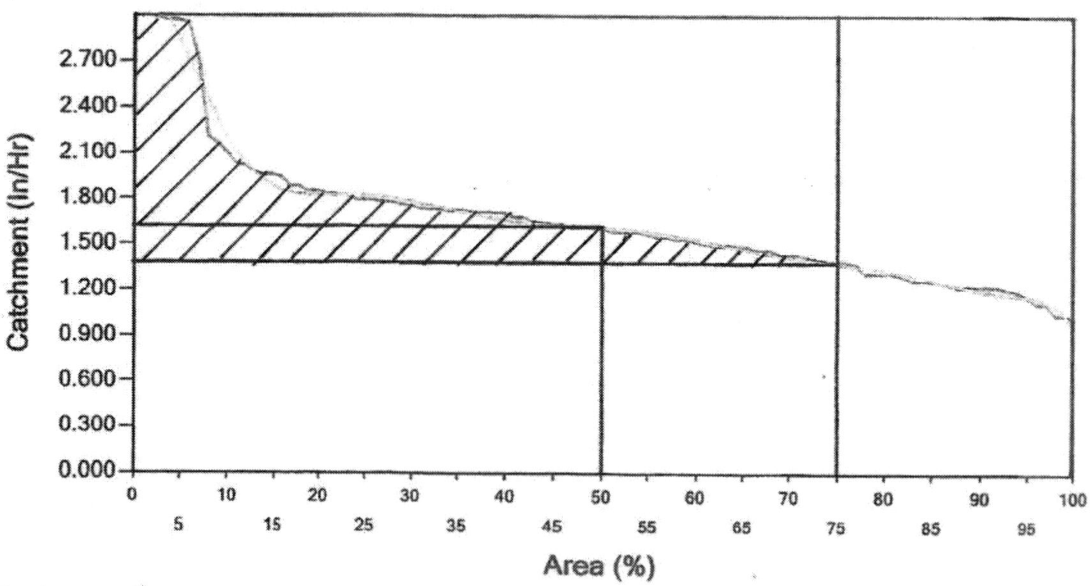

The 100 catchments are arrayed from wet (left side) to dry (right side). The percolation loss is represented by the shaded area in Figure 2. The concept makes the assumption that the commercial grass quality is adequate as long as 75 percent of the target area receives the scheduled amount of irrigation.

Overspray (OS) is directly related to the water caught in the rain gauges outside of the target area. The formula for the overspray losses is as follows:

$$OS = \frac{\sum os}{n(\bar{X}) + \sum os}$$

The sprinkler operating efficiency (S_{OE}) combines the percolation and overspray losses in the following formula:

$$S_{OE} = (1.0 - PL)(1.0 - OS)100$$

The sprinkler operating efficiency metric has physical significance and is useful in studies requiring a scientific characterization of the irrigation system water use efficiency.

GRAPHICAL CHARACTERIZATION OF SPRINKLER OPERATING EFFICIENCY

Figure 3 provides a graphical representation of the results suitable for determining the system required design parameters. Shown in Figure 3 is a 3D plot of a representative sprinkler pattern test.

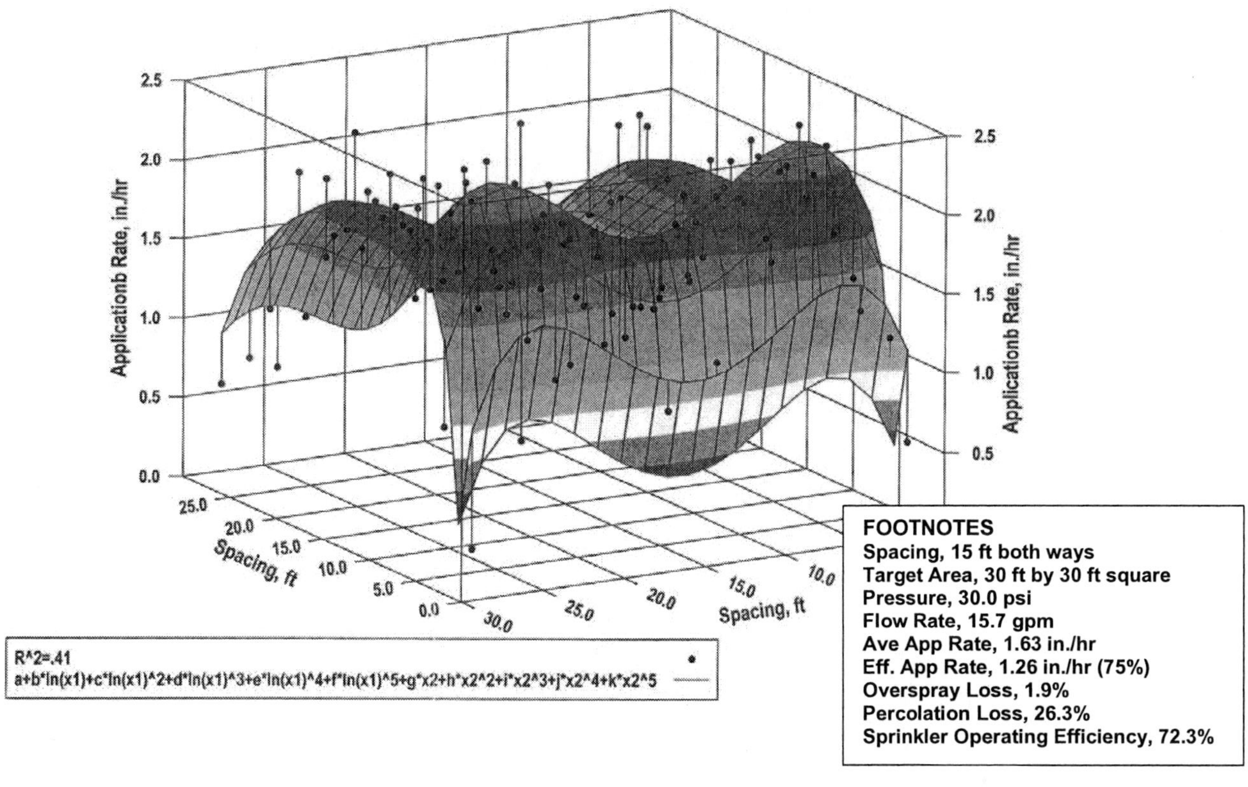

Figure 3. 3D plot of a representative nine sprinkler overlapped pattern

The sprinkler operating efficiency is 72.3 percent reflecting a percolation loss of 26.3 percent and an overspray loss of 1.9 percent. The plot is useful in experimenting with the overlapping of patterns to achieve better uniformity. This leads also to the best relationship between the flow rates of full and part circle sprinklers. It also graphically shows the jet interference phenomenon and the chronic problem of achieving satisfactory coverage next to the sprinklers. It may be possible to partially correct for this by a two-set system providing for different run times of full circle sprinklers complemented by a longer run time for the part circle boundary sprinklers.

REPRESENTATIVE METRICS

Table 1 shows the results of testing sprinklers in the manner proposed. The square target area is as shown in Figure 1. The round target area is as proposed in the SWAT testing protocol. The importance of combining the percolation loss and the overspray loss can be seen by comparing the results from square Test #2 with #3. Both tests have sprinkler operating efficiencies over 80 percent. In Test #2, the overspray loss was negligible at 0.1 percent. With Test #3 however, the overspray loss was 6.2 percent. This degree of overspray is apparently required to develop the designed-in uniformity of the target area. The difficulty of designing for coverage on the round area is shown with a relatively low overall average sprinkler operating efficiency of 68.4 percent (vs 78.6 percent for the square pattern).

Table 1. Selected Summary of Distribution Patterns – January 8, 2015

TEST #	SPRINKLER ID	TARGET SHAPE[1]	PRESSURE *psi*	FLOW RATE *gpm*	AVERAGE APPLICATION RATE[2] *in/hr*	EFFECTIVE APPLICATION RATE[3] *in/hr*	PERCOLATION LOSS[4] %	OVERSPRAY LOSS[5] %	SPRINKLER OPERATING EFFICIENCY[6] %	DU
1	Test #1SHV	□	30	15.50	1.616	1.384	20.1	1.0	79.1	74.0
2	Test #1SU	□	30	15.38	1.606	1.400	19.3	0.1	80.6	74.3
3	Test #3IR	□	40	6.20	0.611	0.556	12.7	6.2	81.9	78.7
4	Test #5 IPF	□	30	15.80	1.627	1.279	25.4	2.0	73.1	62.5
5	Test #4 USN	□	30	12.60	1.253	1.085	21.2	0.9	78.1	65.1
						AVERAGE	**19.7**	**2.04**	**78.6**	**70.9**
1	Test #1SHV (7)	O	30	14.50	1.749	1.468	24.2	7.0	70.5	63.0
2	Test #1SHV (8)	O	30	13.53	1.862	1.399	33.6	0.2	66.3	28.9
3	Test #5 IR	O	40	4.59	0.641	0.490	27.6	6.0	67.9	40.5
4	Test #1 UPS	O	30	7.00	0.902	0.732	30.8	1.1	68.4	54.7
5	Test #4 IPA	O	30	15.06	1.823	1.449	23.6	10.0	68.7	63.6
						AVERAGE	**28.0**	**4.9**	**68.4**	**50.2**

1 See SWAT protocol 2 Average application rate 3 Effective application rate (75%) 4 Percolation loss 5 Overspray loss 6 Sprinkler operating efficiency

The difficulty of indexing the jets to a round boundary is seen by the average overspray loss of 4.9 percent with the round area vs 2.0 percent with the square area.

JET INTERFERENCE PHENOMENA

The phenomenon of jet interference can be observed in Figure 4.

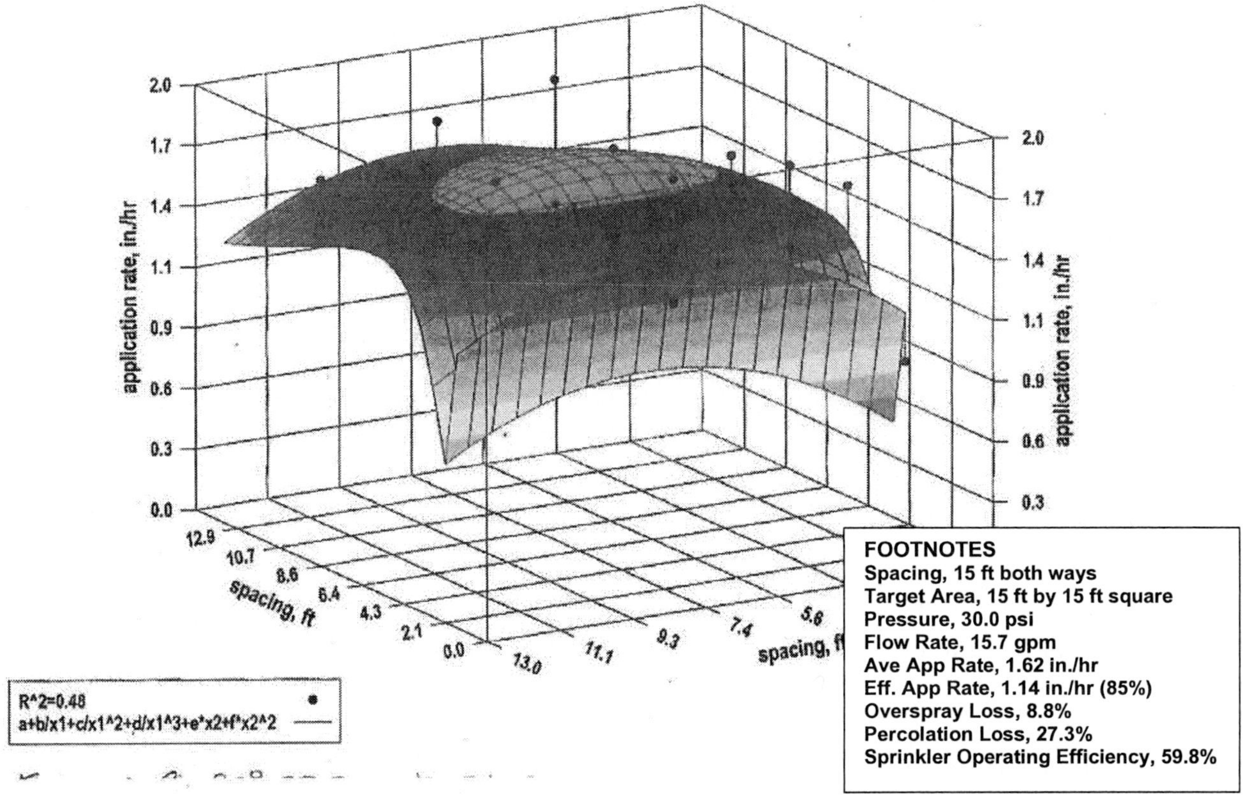

The 3D plot shows an accumulation of water deposited in a haystack fashion in the center of the pattern. This seems to be caused by the four opposing jets mechanically impacting each other. The haystacking effect is further demonstrated in Figure 5.

In the case shown in Figure 5, the overlapped pattern was developed by running catchment tests on the corners individually and overlapping them by hand. A sense of the improvement can be gotten by comparing the sprinkler operating efficiency of 59.8 percent to 76.3 percent that was achieved when the jet interference is avoided.

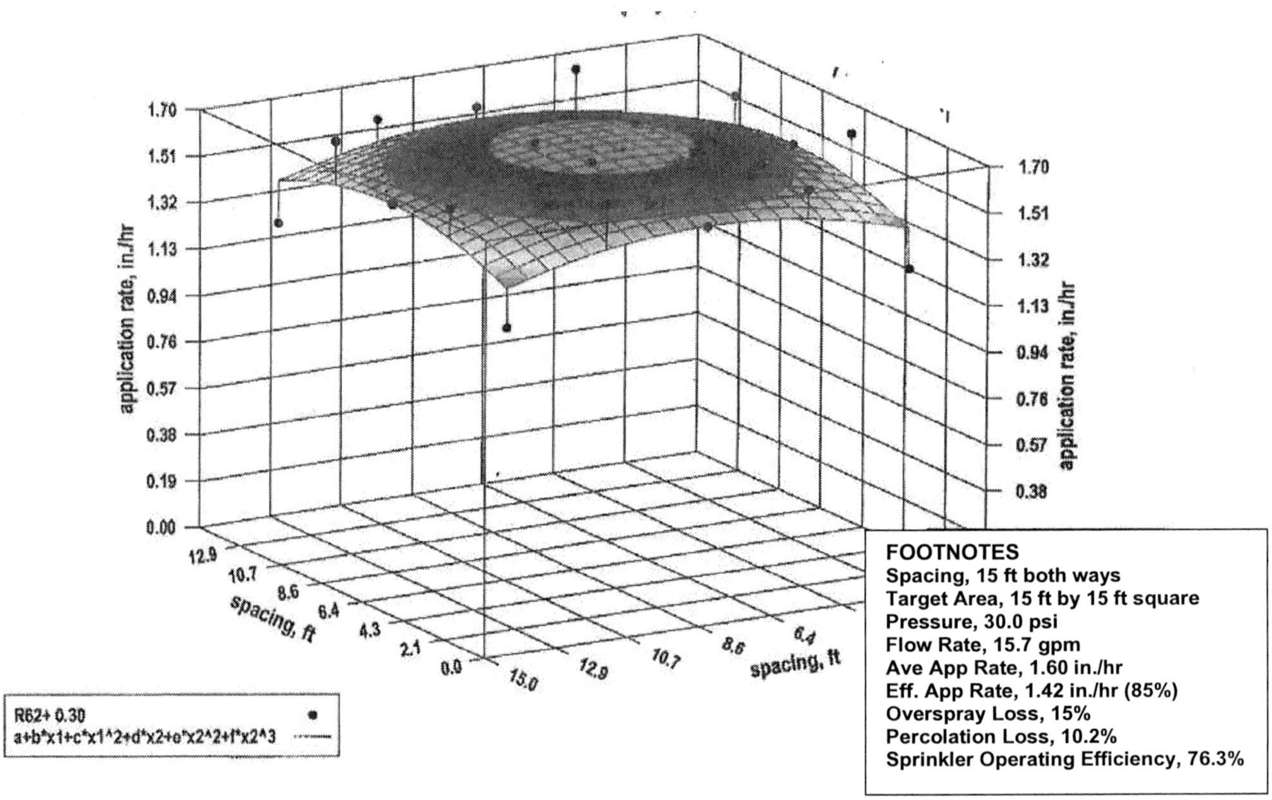

Figure 5. Overlapped pattern from Figure 4 developed by hand overlapping single catchment pattern

ENGINEERING IMPROVEMENTS

Figure 6 shows a 3D printout of a representative overlapped pattern. It provides a measurement of actual value in scheduling irrigations and characterizing the system's water application efficiency.

The actual value of sprinkler or programming changes can be quantified as relates to water management objectives. This evaluation concept provides a procedure for characterizing how efficiently sprinkler systems are applying water. This protocol, together with studies to determine the current state of the commercial art, will provide incentives for manufacturers to improve the efficiency of their products.

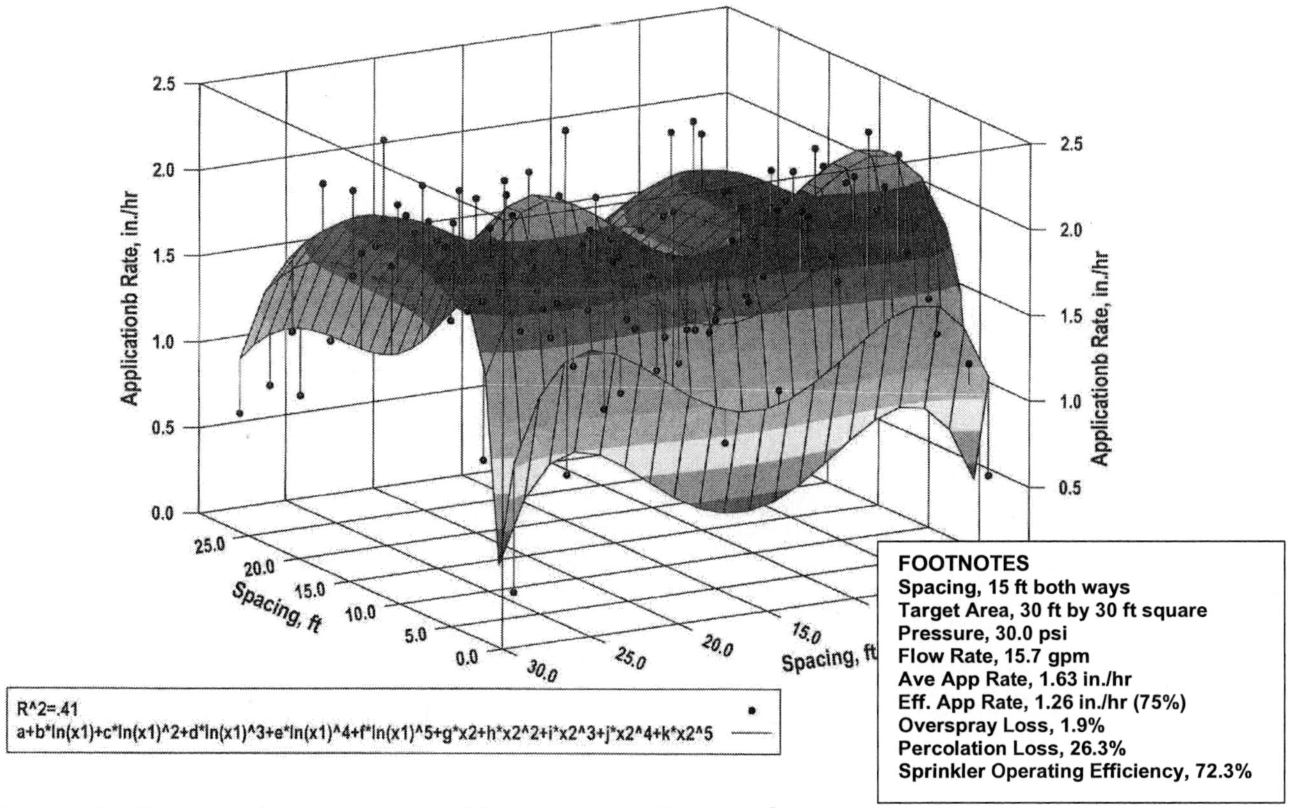

Figure 6. Representative nine-sprinkler square pattern performance

References:

Norum, E., 2015. A Project to Develop a Protocol to Determine the Sprinkler Operating Efficiency of Turf Sprinklers, The Center for Irrigation Technology, California State University, Fresno.

"Spray Head Sprinkler Nozzles, Performance Characteristics, Version 3.2", April 2015. Smart Water Application Technologies, Irrigation Association, Fairfax, Virginia

Landscape Irrigation Auditor, 3rd Edition, 2012. Irrigation Association, Fairfax, Virginia.

Oliphant, J.C. and D.F. Zoldoske, 1989. "Sprinkler Profile And Coverage Evaluation (SPACE)", Software and Documentation, CATI Publication No. 890403, Center For Irrigation Technology, California State University, Fresno, CA, 39 p.

Christiansen, J.E. 1942. Irrigation by Sprinkling. California Agricultural Experiment Station. Bulletin No. 670. Berkeley.

Drip System Design for Established Landscape Trees and Shrubs

Robert D. von Bernuth, PhD, PE, CIC, CID, CIT, CLIA, CLWM
Technical Advisor, Irrigation Association
Brent Q. Mecham, CAIS, CGIA, CIC, CID, CIT, CLIA, CLWM
Industry Development Director, Irrigation Association

Abstract

A procedure for designing drip systems for established trees and shrubs based on the ANSI/ASABE S623 OCT2015 Standard has been developed based on soil characteristics, emitter wetting patterns, and plant water needs. It is supported by an Excel spreadsheet and/or a set of design charts to ensure proper wetted pattern development and proper amount of water applied.

Good irrigation involves two fundamental concepts: putting water where it needs to be, and applying the right amount of water. At times keeping the water where you put it is also a factor. Drip irrigation amounts to a series or set of discrete points of application. Each point of application initially puts water on faster than the soil can accept it, so it spreads until the wetted area is equal to the application flow rate divided by the infiltration rate as given in equation 1.

$$r = 8.57 \times \sqrt{\frac{q_e}{IR}} \qquad \text{Equation 1}$$

where

r = radius of wetted soil {in.}
q_e = flow rate of the emitter {gph}
IR = infiltration rate of the soil {in./h}

This is the minimum that the area can be. Because the area of the wetted pattern is not immediately the size needed for the emitter flow and the soil, it will actually be larger. This suggests that the wetted pattern continues to grow for some time during and after the irrigation cycle. The amount of water needed is governed by the plant and ET_o and set by the flow rate and time of irrigation. The infiltration rate is governed by the soil. Hence several factors are in involved in the design of a drip system. Furthermore, usually one of the goals of a drip system on established trees to wet a portion of the soil surface. Rather, it is preferred to wet 70-75% of the surface as determined by the canopy drip line of the tree. This requires considering run time when designing where to place emitters.

Putting the water where it needs to be

Wetted Pattern (2-D: surface and 3-D: including depth)

Our approach has been to develop a model based on soil characteristics and emitter flow rate to determine the size of the wetted pattern as a function of time. Figure 1 shows the wetted pattern of a system run for a few minutes (left) and for a few hours (right).

Figure 1 showing wetted patterns after a few minutes (left) and a few hours (right).

The wetted pattern begins as a circle around the emission point and in time the circles overlap. If the emitter spacing is less than the row spacing as shown the figures, they will join to form first a scalloped edge and then a rectangle. If the row spacing is not too wide, the rows will eventually overlap. If the goal is to have the wetted area less than 100%, then the emitter spacing, row spacing, emitter flow rates, and run times must be chosen to achieve the desired results. As the wetted pattern spreads on the surface, the water is also moving downward in a hemispherical fashion. Eventually, as the wetted volume overlaps, the pattern becomes cylindrical. If the system is run long enough that the surface wetted pattern stabilizes, the cross-sectional depth profile is a 3-D rectangle with a hemispherical bottom as shown in Figure 2.

This model of water movement in the soil is dependent on the soil being very uniform in type, organic matter, and compaction. Often in landscape situations the surface area may have been tilled to improve soil structure, but deeper down the soil can be quite compacted. This compaction will affect water movement.

Figure 2. 3-D wetted pattern with stabilized surface wetted area.

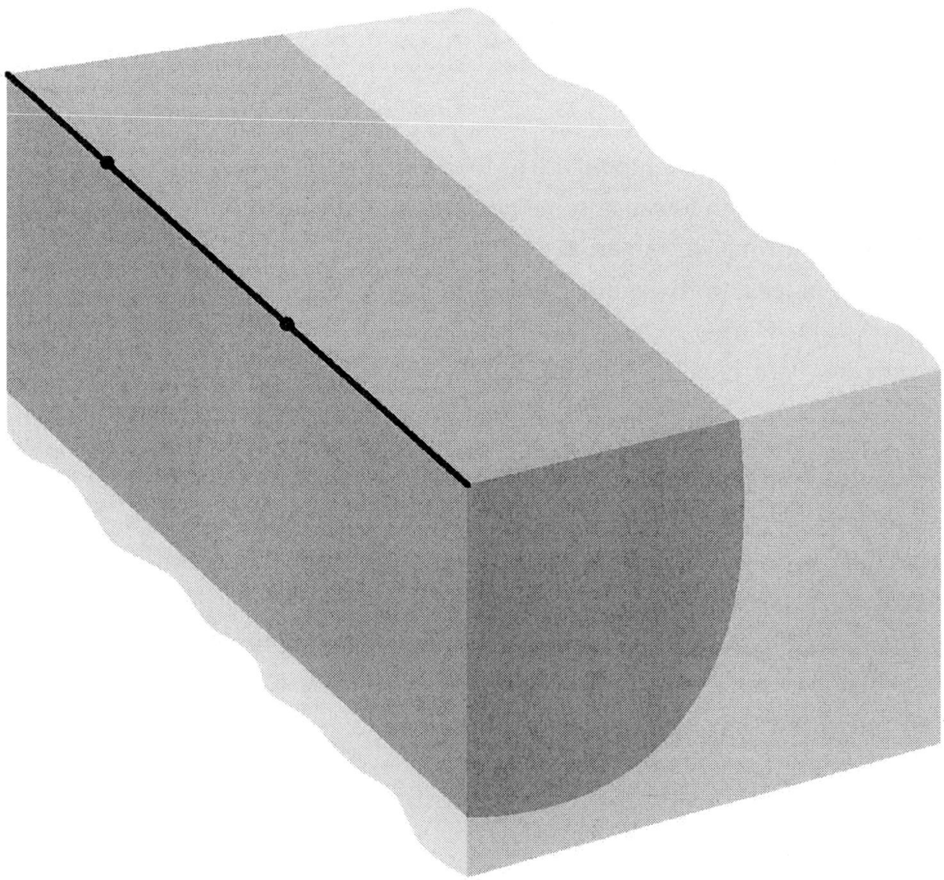

As the system is run, the wetted depth increases, so another challenge of managing the system is to avoid deep percolation. This is further complicated in that the grey zone is saturated, and the water will continued downward well after the irrigation event is stopped until the wetted area reaches field capacity. So...a management system must consider the eventual wetted depth. The amount of water the soil can accept depends not only on the soil characteristics but also on the wetness when the irrigation event was begun. This antecedent wetness is dependent on the management allowable depletion (MAD) employed.

Input Values to Model

Summarizing the discussion thus far, the following are important considerations in a drip wetted pattern model.

1. Soil characteristics
2. Row spacing
3. Desired width of the surface wetted pattern (sets percent of surface area wetted)
4. Root depth
5. ET_o
6. MAD
7. Emitter flow rate
8. Emitter spacing

Model Assumptions

1. The important assumptions made in the model are as follows:
2. The emitter flow rate is high enough to cause some ponding on the surface.
3. The ponded area spreads to form a wetted area.
4. The wetted area will grow at least until it equals the emitter flow rate divided by the infiltration rate, and normally grows much larger.
5. Vertical water movement is controlled by the saturated hydraulic conductivity.

Model features

The model was designed to develop the time relationship among wetted pattern radius, emitter flow rate and soil type. Figure 3 shows that relationship for three soils and two emitter flow rates. This figure shows that the wetted radius continues to grow in time until a maximum is reached. Higher flow rates lead to larger wetted patterns. Lower saturated hydraulic conductivities lead to larger wetted surface area. Given a soil, the wetted radius can be increased by running longer until a maximum is reached. That maximum can be increased by increasing the emitter flow rate. Note that the maximum wetted radius is larger than shown on the graph because radii continue to grow beyond the time shown in the graph.

Figure 3. Wetted radius for three soils and two emitter flow rates.

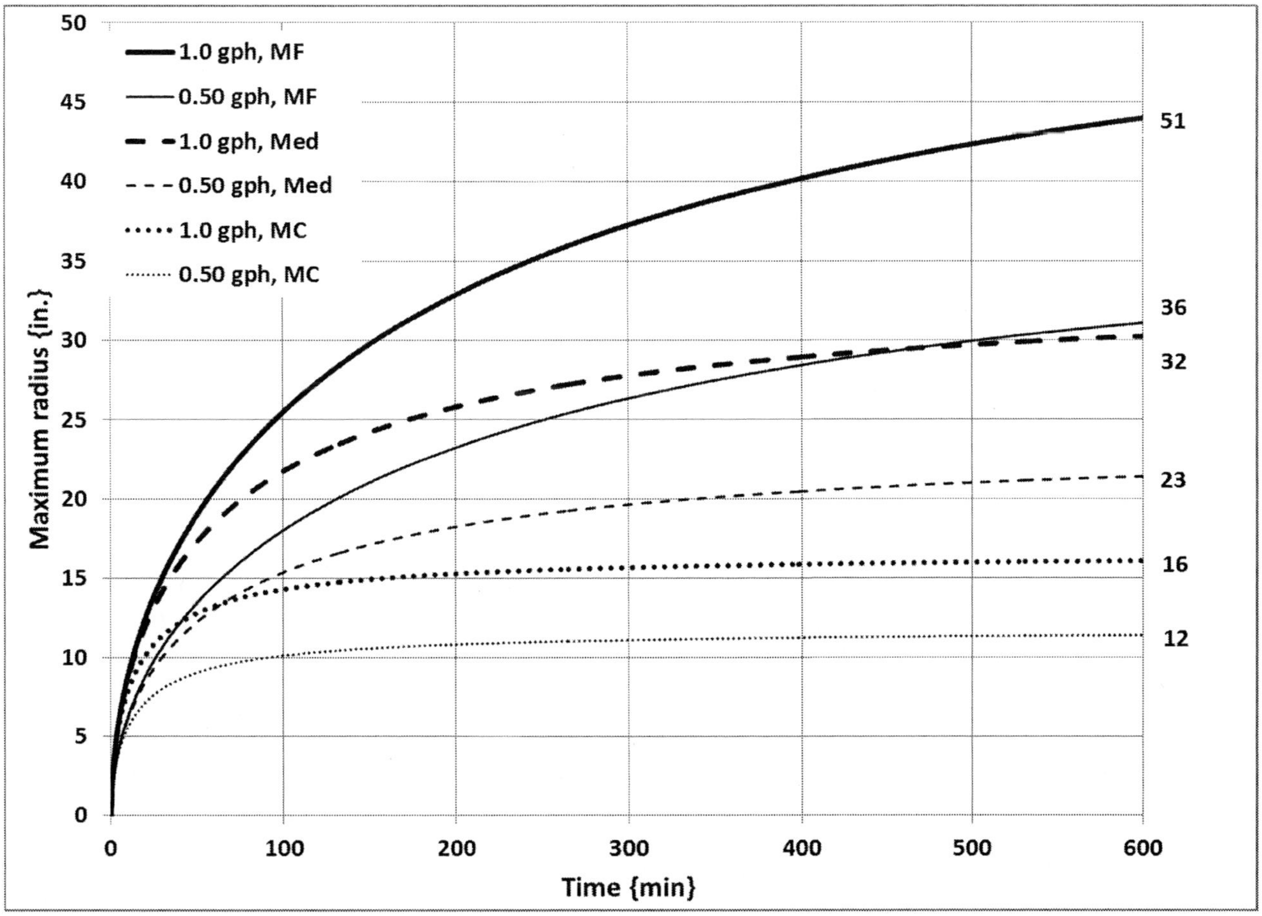

Where the water needs to be on established trees and shrubs

On established trees and shrubs, it is important to place the emitters where the roots are. Even in very arid climates, the roots tend to be located around the drip line of the tree. Most of the water absorbed by the roots will come from there, and not near the trunk. Figure 4 shows where an established tree will absorb water.

Figure 4. Emitter placement relative to tree drip line.

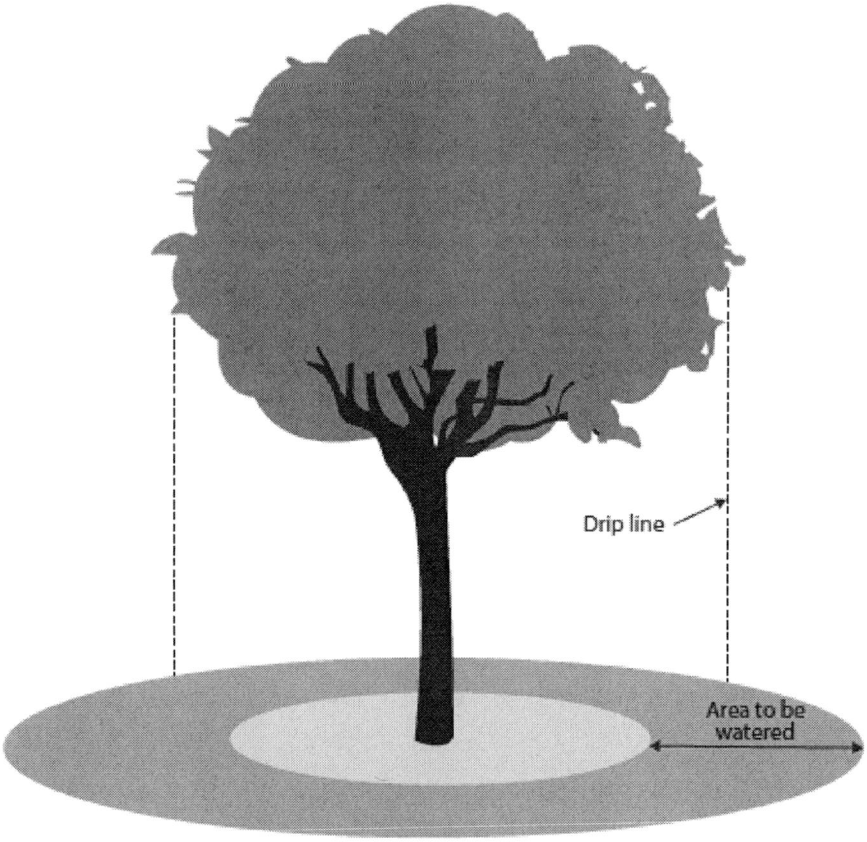

Tree water use

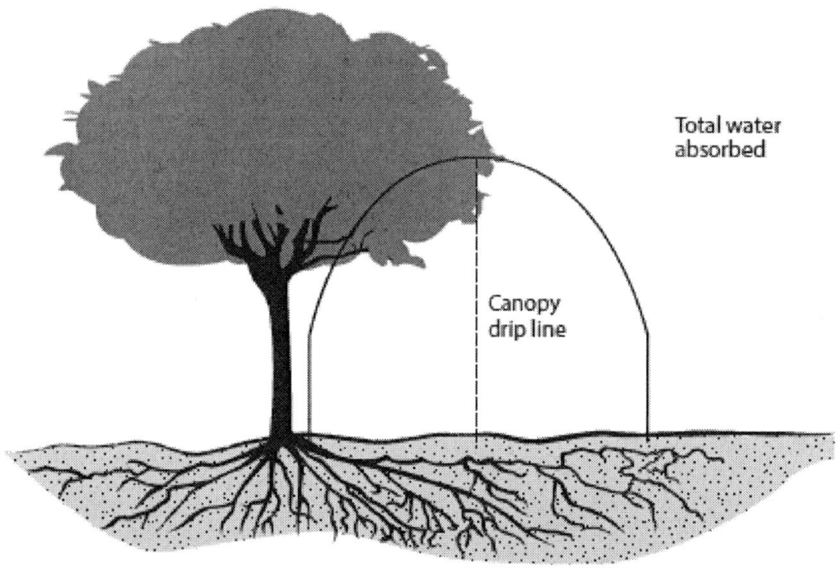

Tree roots can extend 1½–4 times beyond the canopy.

On an established tree, the area from which the tree draws water can be quite large. This probably involves several emitters on one, two or three concentric circles depending on the diameter of the tree canopy. This is another consideration for a design model, and should be added to the list.

Putting on the right amount of water

In many landscape systems, drip irrigation is designed per tree. Several key questions must be addressed in the design and operation. The general rule for trees that shade 80 percent or less of the area is to design the system so that about 75 percent of the canopy area is wetted. The number of emitters, flow rate of the emitters, and run time all affect the wetted area for any given soil.
A new standard has been developed by the American Society of Agricultural and Biological Engineers (ASABE) with active participation by the Irrigation Association (IA) that gives guidelines for the amount of water needed for turf, flowers, woody plants and desert plants. Table 1 gives the annual average fraction of ET_0 for acceptable appearance of established woody and desert landscape plants (ANSI/ASABE S623 OCT2015, Determining Landscape Plant Water Demands. *The ANSI designation indicates that the standard has been accepted by the American National Standards Institute*).

Plant type	Recommended plant factor
Woody plants and herbaceous perennials, wet*	0.7
Woody plants and herbaceous perennials, dry	0.5
Desert plants	0.3

* For tropical plants with precipitation the majority of months, a plant factor of 0.7 applies. Where monsoonal climates are present, 0.7 applies for the wet season and 0.5 during the dry season.

Source: ASABE S623

Areas with 20 inches or more of annual precipitation, especially if it occurs during the growing season, should be considered "wet." Areas with less than 20 inches but more than 10 inches of precipitation during the growing season should be considered "dry" and may require drought tolerant plants for survival without minimal irrigation. Areas with less than 10 inches of annual precipitation are considered desert and appropriate plants should be selected. Even they will need some irrigation at critical times to maintain an acceptable appearance.

Calculating Landscape Water Requirements

Before the number or size of emitters can be selected, the peak water demand for a period of time needs to be determined. The following equation can be used taking into account ET, landscape coefficient, and planting area and converting it to gallons per day or per week:

$$Q = \frac{A \times ET_o \times K_L \times 0.623}{E_a}$$

where
Q = gallons of water for a period of time such as a day or week
A = canopy area of plant × 0.75 for applying water to 75 percent of the area {ft2}
ET_o = peak ET for a specific period such as day or week
K_L = landscape coefficient
0.623 = conversion to gallons
Ea = irrigation application efficiency
0.85 for hot, arid climate
0.90 for moderate climate
0.95 for cool climate

Tree Drip Calculator Model

The tree drip calculator model incorporates the information provided by the designer as previously noted, considers ASABE Standard S623, and gives the designer alternatives for the design. A screen shot of the model is shown in Figure 5.

Figure 5. Tree Drip Calculator Model

A	soil	medium		IA Soil type
I-0	R_s	18.0	{in}	Row spacing
I-1	W_{Wdes}	19.8	{in}	Desired width of wetted pattern
I-2	canopy dia	10	{ft}	Tree canopy diameter
I-3	root depth	18	{in}	Root depth
I-4	ETo	0.23	{in/d}	Reference evapotransipration
I-5	Wet	0.70	{dec}	S623 climate description
I-6	q_{daily}	7.99	{gal/day}	Daily water requirement
B	E_s	18	{in}	Emitter spacing
B1	$E_{s\ Rec}$	11	{in}	Recommended emitter spacing
C	q_e	0.50	{gph}	Emitter flow rate
C1	$q_{e\ Rec}$	0.56	{gph}	Recommended emitter flow rate
R-1	AR	0.36	{in/hr}	Application rate
R-2	IR	0.40	{in/hr}	Infiltration rate
R-3	t_{max}	17.3h	1037 min	Max. time to avoid deep percolation
R-4	t_{daily}	0.46 h	27.4 min	Time daily to meet Eto
R-5	t_{rec}	0.50 h	30.2 min	Recommended run time
D	t_{sel}	28	{min}	Selected run time
R-6	W_{Wres}	19.2	{in}	Resulting width of wetted pattern
R-7	IN	1.02	{days}	Irrigation interval
R-8	# emitters	35	2 circles	Number of emitters used
R-9	circle dia's	9.0	7.5	{ ft}
R-10	flow per tree		0.29	{gpm}

IA paper for 2016 show, 10/18/16

The following are guidelines for the inputs and outputs. The "A" section is input. Yellow background cells are input values determined by the designer. Blue background cells are values that the designer can select from a list of values. The green values are calculated based on yellow inputs. The "B" and "C" sections have one blue selectable cell in which the designer should try to match the green recommended value. The results are shown in the "R" labeled values and one selection is required in "D" for the selected run time. In this example, the soil is medium, row spacing is 18 inches, the width of the desired pattern is 19.8 inches (the program defaults to 110% of row spacing to ensure overlap), canopy diameter is 10 feet, root depth is 18 inches, and ET_o is 0.23 in/day. The 75% wetted area requirement is met by the number of emitters in the given number of rows. Climate is "wet" by ASABE S623. The resulting values are $K_L = 0.70$, and q_{daily} is 7.99 gal/day. The recommended emitter spacing is 11 inches. The closest available value of 12 was selected. The recommended emitter flow rate is 0.56 gph, and the selected value of 0.50 is the closest available. The application rate is 0.36 in/hr which is less than the infiltration rate of 0.40 in/hr. The maximum time the system can be run without deep percolation is 17.3 hours, and the required daily time to meet ET is 27 minutes. The recommended time to achieve 75% area wetted is 30 minutes and 28 minutes was selected. This results in a width of the wetted pattern of 19.2 inches (which is slightly less than the desired width). The irrigation interval is 1.02 days, meaning that the system would be run daily. This requires 35 emitters in 2 rows, one at 7.5 feet diameter and one at 9.0 feet diameter.

An alternative would be to select an emitter spacing of 12 inches and an emitter flow rate of 0.40 gph and run the system 38 minutes three out of five days. Figure 6 shows this alternative. Note that this requires 53 emitters. This is probably not as good a design as the first alternative with 18 inch emitter spacing. A bit of experimentation with the model will soon lead the designer to understand that there are multiple viable designs.

Figure 6. Tree Drip Calculator, Alternative Choice

A	soil	medium		IA Soil type
I-0	R_s	18.0	{in}	Row spacing
I-1	W_{Wdes}	19.8	{in}	Desired width of wetted pattern
I-2	canopy dia	10	{ft}	Tree canopy diameter
I-3	root depth	18	{in}	Root depth
I-4	ETo	0.23	{in/d}	Reference evapotransipration
I-5	Wet	0.70	{dec}	S623 climate description
I-6	q_{daily}	7.99	{gal/day}	Daily water requirement
B	E_s	12	{in}	Emitter spacing
B1	$E_{s\ Rec}$	9	{in}	Recommended emitter spacing
C	q_e	0.40	{gph}	Emitter flow rate
C1	$q_{e\ Rec}$	0.37	{gph}	Recommended emitter flow rate
R-1	AR	0.43	{in/hr}	Application rate
R-2	IR	0.40	{in/hr}	Infiltration rate
R-3	t_{max}	17.3h	1037 min	Max. time to avoid deep percolation
R-4	t_{daily}	0.38 h	22.6 min	Time daily to meet Eto
R-5	t_{rec}	0.67 h	40.2 min	Recommended run time
D	t_{sel}	38	{min}	Selected run time
R-6	W_{Wres}	19.4	{in}	Resulting width of wetted pattern
R-7	IN	1.68	{days}	Irrigation interval
R-8	# emitters	53	2 circles	Number of emitters used
R-9	circle dia's	9.0	8.0	{ ft}
R-10	flow per tree		0.35	{gpm}
	Application Rate Exceeds Infiltration Rate			

Number of Rows of Emitters

As a general rule, if the tree canopy diameter is 16 feet or less, two rows are adequate. For canopy diameters 18 feet or more, three rows are needed. The row spacing is set equal to the emitter spacing to ensure overlap, and the center of the rows is set at the tree canopy diameter. The designer can override the row spacing and set it other than equal to the emitter spacing.

Design Charts

A series of design charts has been developed using the tree drip calculator. They are show in Figures 7, 8 , and 9 for three soils and three climate conditions per ANSI/ASABE S623.

Figure 7. Design/operating alternatives for "wet" climate. ETo = 7 in./month where dc {ft} is canopy diameter, di is the diameter of the inner circle, dm is the next circle, and do is the third circle (if necessary), all in ft.

IA soils – moderately coarse, wet										
dc	Es	qe	RT	IN	No.	Rows	di	dm	do	AR > IR
8	24	1.00	49	3	19	2	5.0	7.0		
10	24	1.00	57	3	25	2	7.0	9.0		
12	24	1.00	66	3	31	2	9.0	11.0		
14	24	1.00	50	2	38	2	11.0	13.0		
16	24	1.00	56	2	44	2	13.0	15.0		
18	24	1.00	66	3	71	3	13.0	15.0	17.0	
20	24	1.00	48	2	80	3	15.0	17.0	19.0	
22	24	1.00	52	2	90	3	17.0	19.0	21.0	
24	24	1.00	56	2	99	3	19.0	21.0	23.0	
IA soils – medium, wet										
dc	Es	qe	RT	IN	No.	Rows	di	dm	do	AR > IR
8	24	1.00	32	2	19	2	5.0	7.0		
10	24	1.00	38	2	25	2	7.0	9.0		
12	24	1.00	44	2	31	2	9.0	11.0		
14	24	1.00	50	2	38	2	11.0	13.0		
16	30	1.00	70	2	35	2	12.5	15.0		
18	30	1.00	57	2	55	3	12.0	14.5	17.0	
20	30	1.00	61	2	62	3	14.0	16.5	19.0	
22	36	1.00	82	2	57	3	15.0	18.0	21.0	
24	36	1.00	88	2	63	3	17.0	20.0	23.0	
IA soils – moderately fine, wet										
dc	Es	qe	RT	IN	No.	Rows	di	dm	do	AR > IR
8	36	1.00	52	2	12	2	4.0	7.0		X
10	36	1.00	60	2	16	2	6.0	9.0		X
12	36	1.00	70	2	20	2	8.0	11.0		X
14	36	1.00	78	2	24	2	10.0	13.0		X
16	36	1.00	87	2	28	2	12.0	15.0		X
18	36	1.00	70	2	44	3	11.0	14.0	17.0	X
20	36	1.00	76	2	50	3	13.0	16.0	19.0	X
22	36	1.00	82	2	57	3	15.0	18.0	21.0	X
24	36	1.00	88	2	63	3	17.0	20.0	23.0	X

Figure 8. Design/operating alternatives for "dry" climate. ETo = 7 in./month where dc {ft} is canopy diameter, di is the diameter of the inner circle, dm is the next circle, and do is the third circle (if necessary), all in ft.

IA soils – moderately coarse, dry										
dc	Es	qe	RT	IN	No.	Rows	di	dm	do	AR > IR
8	24	2.00	17	3	19	2	5.0	7.0		X
10	24	2.00	14	2	25	2	7.0	9.0		X
12	24	2.00	16	2	31	2	9.0	11.0		X
14	24	2.00	18	2	38	2	11.0	13.0		X
16	24	2.00	20	2	44	2	13.0	15.0		X
18	24	2.00	16	2	71	3	12.0	15.0	17.0	X
20	24	2.00	17	2	80	3	15.0	17.0	19.0	X
22	24	2.00	18	2	90	3	17.0	19.0	21.0	X
24	24	2.00	20	2	99	3	19.0	21.0	23.0	X
IA soils – medium, dry										
dc	Es	qe	RT	IN	No.	Rows	di	dm	do	AR > IR
8	30	3.00	23	3	14	2	4.5	7.0		X
10	30	2.00	27	3	19	2	6.5	9.0		X
12	30	2.00	20	2	25	2	8.5	11.0		X
14	30	2.00	22	2	30	2	10.5	13.0		X
16	30	2.00	25	2	35	2	12.5	15.0		X
18	30	2.00	20	2	55	3	12.0	14.5	17.0	X
20	30	2.00	22	2	62	3	14.0	16.5	19.0	X
22	30	2.00	24	2	70	3	16.0	18.5	21.0	X
24	30	2.00	26	2	77	3	18.0	20.5	23.0	X
IA soils – moderately fine, dry										
dc	Es	qe	RT	IN	No.	Rows	di	dm	do	AR > IR
8	36	1.00	55	3	12	2	4.0	7.0		X
10	36	1.00	64	3	16	2	6.0	9.0		X
12	36	1.00	50	2	20	2	8.0	11.0		X
14	36	1.00	56	2	24	2	10.0	13.0		X
16	36	1.00	62	2	28	2	12.0	15.0		X
18	36	1.00	56	2	44	3	11.0	14.0	17.0	X
20	36	1.00	56	2	50	3	13.0	16.0	19.0	X
22	36	1.00	58	2	57	3	15.0	18.0	21.0	X
24	36	1.00	62	2	63	3	17.0	20.0	23.0	X

Figure 9. Design/operating alternatives for "dry" climate. ETo = 7 in./month where dc {ft} is canopy diameter, di is the diameter of the inner circle, dm is the next circle, and do is the third circle (if necessary), all in ft.

IA soils – moderately coarse, desert										
dc	Es	qe	RT	IN	No.	Rows	di	dm	do	AR > IR
8	24	2.00	17	5	19	2	5.0	7.0		X
10	24	2.00	16	4	25	2	7.0	9.0		X
12	24	2.00	19	4	31	2	9.0	11.0		X
14	24	2.00	21	4	38	2	11.0	13.0		X
16	24	2.00	18	3	44	2	13.0	15.0		X
18	24	2.00	19	4	71	3	12.0	15.0	17.0	X
20	24	2.00	20	4	80	3	15.0	17.0	19.0	X
22	24	2.00	17	3	90	3	17.0	19.0	21.0	X
24	24	2.00	18	3	99	3	19.0	21.0	23.0	X
IA soils – medium, desert										
dc	Es	qe	RT	IN	No.	Rows	di	dm	do	AR > IR
8	30	2.00	19	4	14	2	4.5	7.0		X
10	30	2.00	22	4	19	2	6.5	9.0		X
12	30	2.00	24	4	25	2	8.5	11.0		X
14	30	2.00	20	3	30	2	10.5	13.0		X
16	30	2.00	23	3	35	2	12.5	15.0		X
18	30	2.00	24	4	55	3	12.0	14.5	17.0	X
20	30	2.00	20	3	62	3	14.0	16.5	19.0	X
22	30	2.00	21	3	70	3	16.0	18.5	21.0	X
24	30	2.00	23	3	77	3	18.0	20.5	23.0	X
IA soils – moderately fine, desert										
dc	Es	qe	RT	IN	No.	Rows	di	dm	do	AR > IR
8	36	1.00	55	5	12	2	4.0	7.0		X
10	36	1.00	51	4	16	2	6.0	9.0		X
12	36	1.00	59	4	20	2	8.0	11.0		X
14	36	1.00	51	3	24	2	10.0	13.0		X
16	36	1.00	56	3	28	2	12.0	15.0		X
18	36	1.00	60	4	44	3	11.0	14.0	17.0	X
20	36	1.00	49	3	50	3	13.0	16.0	19.0	X
22	36	1.00	52	3	57	3	15.0	18.0	21.0	X
24	36	1.00	56	3	63	3	17.0	20.0	23.0	X

Summary

In summary, a good drip irrigation design for established trees and shrubs involves choosing the emitter spacing, emitter flow rate and total number emitters so that the proper amount of water is placed where the tree needs it. The operation of the system must be such that the system is run long enough to achieve the desired wetted area and not exceed the root depth with a reasonable run time and irrigation interval.

An excel program, Tree Drip Calculator, was introduced which enables the designer and operator to evaluate choices in the design and operation of the system. A set of tables for assisting the design and operation without running the calculator was also presented.

This page intentionally left blank.

Turfgrass Selection to Maximize Water Use Efficiency

Bernd Leinauer, Matteo Serena, Elena Sevostianova, and Guillermo Alvarez
Extension Plant Sciences Department
New Mexico State University, Las Cruces, NM, 88003. USA

Abstract. *In arid and semi-arid climates, limited precipitation and uneven annual rainfall distribution can restrict adequate turfgrass growth and quality unless frequent irrigation is applied. Turfgrasses' water demands and irrigation requirements are measured as evapotranspiration (ET) and can vary greatly, depending on the local macro and micro-climate, turfgrass species and varieties used, quality expectations, intended use (traffic), and resulting maintenance level applied. Drought resistant turfgrasses that are adapted to the local climatic conditions and can sustain adequate quality on minimum irrigation can be used to maximize water use efficiency and to minimize irrigation requirements.*

Introduction

In arid and semi-arid regions annual precipitation amounts of 250 to 500 mm do not meet the estimated evaporative requirement of turf areas, which range from approximately 800 to 1200 mm. Consequently, 50% or more of urban domestic summer water use goes to landscape irrigation (Kjelgren et al., 2000, Devitt and Morris, 2008) to maintain turfgrass areas and lawns at desired aesthetic and functional levels.

Strategies aimed at conserving potable water use for turf irrigation include 1) replacing the potable water with recycled or other impaired water that usually does not meet standards for human consumption, 2) applying modern irrigation equipment, such as subsurface irrigation or new sprinkler and nozzle technology and/or by scheduling irrigation based on local ET or on soil sensor, and 3) the use of locally adapted, drought resistant turfgrasses that can sustain adequate quality on less irrigation than grasses currently used (Leinauer et al., 2012). However, many of the turfgrass areas have to survive and recover from significant traffic and very few plants besides turfgrasses can withstand the repeated pounding furnished by such activities as baseball, football, or soccer, and running kids and/or dogs. It is therefore no surprise that we routinely select bermudagrass, zoysiagrass, Kentucky bluegrass, perennial ryegrass, and tall fescue as grasses for our lawns, depending on the climatic conditions of the area. These grasses are the only ones that combine traffic tolerance with a dark green and uniform appearance that is aesthetically pleasing for many of us during most of the year.

Turfgrass Water Use and Drought Resistance

Several factors contribute to a turf stand's water requirement. First, water is taken up by roots and then lost to the atmosphere (transpiration) from the plants' green tissue. However, water is also lost from the soil surrounding the plants (called evaporation). The combined losses, referred to as evapotranspiration (ET), are commonly expressed in millimeters per day, and serve as a basis for a replacement requirement either from rainfall or from irrigation. Several authors have suggested that a turf plant's ET is genetically determined, and species and varieties can be ranked from high to low based on their ET rates. However, ET rates and subsequent water demands are not only influenced by the species or varieties present, but also by climate, soil type, maintenance intensity, quality expectation, and irrigation uniformity. Table 1 lists published ET values for cool and warm season grasses at different mowing heights, and in controlled or field environments at different geographical locations. Generally, water use rates are higher in dry, desert climates than in humid or temperate climates. Evapotranspiration rates are also higher when grasses are maintained at a higher rather than at lower mowing height. High fertility programs aimed at maintaining high quality and dark colored turf also influence ET rates of turfgrasses.

Evapotranspiration rates listed in Table 1 were predominately determined under well watered or non-limiting moisture conditions. The wide range of ET within each species indicates that water use rates are not only determined by genetic predisposition but also by the moisture availability in the rootzone (Leinauer et al., 2012). Consequently, in the context of water conservation, the question is not how much water do turfgrasses use, but what is the minimum amount of water they require to survive and meet desired quality expectations. All turfgrasses can maintain acceptable quality for a certain period of time when irrigation is less than 100% ET using physiological mechanisms that allow plants to adapt to drought. Irrigating below 100% ET replacement is called deficit irrigation and can be used as a practice to conserve irrigation water. Turfgrasses survive drought stress by means of drought resistance mechanisms or by successful recovery from longer term water deficits (Kneebone et al. 1992; Devitt and Morris, 2008). However, deficit irrigation is only effective if turf areas receive sufficient rainfall to occasionally recharge the soil profile (Shearman 2008). In desert areas where occasional natural precipitation is insufficient, drought periods need to be followed by periodic increased irrigation amounts for grasses to recover (Baird, et al., 2009; Devitt and Morris, 2008). The main drought resistance mechanism from a lack of sufficient irrigation may be dormancy which results in a loss in color and cover. This may not fulfill the aesthetic or the functional requirements of the area. Sevostianova et al. (2010) reported superior drought resistance in un-trafficked buffalograss compared to bermudagrass or zoysiagrass during a 3 year period of no supplemental irrigation in a desert climate. However, the decline of green cover from 100% to an average of 17% indicated that even buffalograss cannot maintain adequate turf quality on a long term basis in an arid climate without supplemental irrigation.

Conclusion

Water requirements of turfgrasses are influenced by several factors, including ET, quality expectations, traffic, water and soil quality, and irrigation efficiency. Great attention is given to low water use rates or ET when selecting turfgrasses for the purpose of conserving water. However, ET values of species and varieties can vary widely depending on climate conditions and maintenance intensities. Furthermore, some turfgrasses use dormancy as a mechanism to resist drought, but this may not be a viable option if green grass is needed throughout the year. Consequently, strategies aimed at conserving potable irrigation water cannot be based solely on selecting low water-use or drought resistant species, but need to include the use of efficient irrigation systems or switching to non-potable water sources (Leinauer et al., 2012, Leinauer and Devitt, 2013).

Table 1. Reported evapotranspiration rates for commonly used cool- and warm-season turfgrasses at different cutting heights and at different geographical locations.

Species	ET (mm day^{-1})	Cutting height (mm)	Varieties / ssp. included	Location	Reference
Festuca arundinacea	5.1 – 7.1	38	1	Texas	Kim and Beard, 1988
	5.8	n.l.	1	Colorado	Feldhake et al., 1983
	6.7 – 8	76	6	Nebraska	Kopec et al., 1988
	9.9 – 11.4	50	1	CE	Green at al., 1990
	10 – 13.5	50	20	CE	Bowman and Mcaulay, 1991
	10.6	40	1	Arizona	Kneebone and Pepper, 1982
	12.2	30	1	Israel	Biran et al., 1981
Lolium perenne	3.4 – 4.0	50	1	Rhode Island	Aronson et al., 1987
	4.9 – 10	50	12	Nebraska	Shearman, 1989
	9.1	50	1	CE	Green et al., 1990
	10.8	30	1	Israel	Biran et al., 1981
Poa pratensis	3.4 – 4.1	50	2	Rhode Island	Aronson et al., 1987
	3.9 – 6.3	50	20	Nebraska	Shearman, 1986
	4.1	63	1	Kansas	O'Neil and Carrow, 1982
	5.0 – 6.1	64	2	Colorado	Suplick-Ploense and Qian, 2005
	5.7	n.l.	1	Colorado	Feldhake et al., 1983
	5.4 – 6.8	45		CE	Ebdon et al., 1998
	11.0 – 12.4	50	3	CE	Green et al., 1990
Buchloe dactyloides	2.3	51	1	CE	Horst et al., 1997
	3.7 – 5.6	50	17	CE	Bowman et al., 1998
	4.4 – 5.3	38	1	Texas	Kim and Beard, 1988
	4.5	n.l.	1	Colorado	Feldhake et al., 1983
Cynodon dactylon	2.8 – 6.2	25	24	Texas	Beard et al., 1992
	3.0	64	1	Georgia	Carrow, 1995
	4.1 – 5.9	38	3	Texas	Kim and Beard, 1988
	4.5	n.l.	1	Colorado	Feldhake et al., 1983
	7.3 – 8.6	30	2	Israel	Biran et al., 1981
Cynodon dactylon x C. transvaalensis	3.1	64	1	Georgia	Carrow, 1995
	7.4	32	1	Arizona	Kopec et al., 2006
Paspalum vaginatum	4.7 – 6.2	38	1	Texas	Kim and Beard, 1988
	7.9	30	1	Israel	Biran et al., 1981
	8.2	32	1	Arizona	Kopec et al., 2006
Zoysia japonica	2.2	51	1	CE	Horst et al., 1997
	3.5	64	1	Georgia	Carrow, 1995
	4.7 – 6.5	38	2	Texas	Kim and Beard, 1988
	7.3	30	1	Israel	Biran et al., 1981

CE Controlled Environment
n.l. not listed

References

Aronson, L. J., A. J. Golf, and R. J. Hull. 1987. Cool-Season Turfgrass Responses to Drought Stress. Crop Sci. 27:1261-1266.

Baird, J. H., R. L. Green, S. Mitra, R. V. Plumb, G. J. Klein, and J. R. Frelich. 2009. Response of hybrid bluegrass, Kentucky bluegrass and tall fescue to short-term drought recovery in a Mediterranean climate. Int. Turfgrass Soc. Res. J. 11:1-7.

Beard, J. B., R. L. Green, and S. I. Sifers. 1992. Evapotranspiration and leaf extension rates of 24 well-watered, turf-type *Cynodon* genotypes. HortScience 27:986-988.

Biran, I., B. Bravdo, I. Bushkin-Harav, and E. Rawitz. 1981. Water consumption and growth rate of 11 turfgrasses as affected by mowing height, irrigation frequency, and soil moisture. Agron. J. 73: 85-90.

Bowman, D. C., D. A. Devitt, D. R. Huff, and W. W. Miller. 1998. Comparative evapotranspiration of seventeen buffalograss (*Buchloe dactyloides* (Nutt.) Engelm.) genotypes. J. Turfgrass Manage. 2(4): 1-10.

Bowman, D. C., and L. Macaulay. 1991. Comparative evapotranspiration rates of tall fescue cultivars. HortScience. 26:122-123.

Carow R.N. 1995. Drought resistance aspects of turfgrasses in the southeast: evapotranspiration and crop coefficients. Crop Sci. 35:1685-1690.

Devitt, D. A. and Morris, R. L. 2008. Urban Landscape Water Conservation and the Species Effect. In: Water Quality and Quantity Issues for Turfgrass in Urban Landscapes. J.B. Beard and M. P. Kenna (eds.). CAST Special Publication 27. 171-192.

Ebdon, J.S., A. M. Petrovic, and R. W. Zobel. 1998. Stability of evapotranspiration rates in Kentucky bluegrass cultivars across low and high evaporative environments. Crop Sci. 38:135-142.

Feldhake, C. M., R. E. Danielson, and J. D. Butler. 1983. Turfgrass evapotranspiration: I. Factors influencing rate in urban environments. Agron. J. 75: 824-830.

Green, R. L., J. B. Beard, and D. M. Casnoff. 1990. Leaf blade stomatal characterizations and evapotranspiration rates of 12 cool-season perennial grasses. HortScience 25:760-761.

Horst, G. L., D. H. Steinegger, and R. E. Gaussoin. 1997. Comparative evapotranspiration rates of some perennial herbaceous ground covers and turfgrass species under greenhouse conditions. Int. Turfgrass Soc. Res. J. 8:1274-1279.

Kim, K. S., and J. B. Beard. 1988. Comparative Turfgrass Evapotranspiration Rates and Associated Plant Morphological Characteristics. Crop Sci. 28:328-331.

Kjelgren, R., Rupp, L. and Kjelgren, D.. 2000. Water conservation in urban landscapes. HortSci 35: 1037-1040.

Kneebone, W. R., D. M. Kopec, and C. F. Mancino. 1992. Water requirements and irrigation. p. 441-472. *In* D.V Waddington, R.N. Carrow, and R.C. Shearman (eds.) Turfgrass. Agron. Monogr. 32. ASA, CSSA, & SSSA. Madison, WI.

Kneebone, W. R., and I. L. Pepper. 1982. Consumptive water use by sub-irrigated turfgrasses under desert conditions. Agron. J. 74:419-423.

Kopec, D. M., M. Pessarakli, J. Gilbert, and S. Nolan. 2006. Consumptive water use of inland saltgrass, seashore paspalum, and 'Tifway' bermudagrass in a semi-arid climate. 2006 USGA Turfgrass and Environmental Research Summary:60.

Kopec, D. M., R. C. Shearman, and T. P. Riordan. 1988. Evapotranspiration of Tall Fescue Turf. HortScience 23:300-301.

Leinauer, B. and D. Devitt. 2013. Irrigation science and technology. p. 1075 – 1133. *In* B. Horgan, J. Stier, S. Bonos (eds.) Agronomy Monograph 56. ASA, CSSA, and SSSA, Madison WI.

Leinauer, B., M. Serena, M. Schiavon, and E.Sevostianova. 2012. Water Demands and Water Conservation Strategies in Turfgrass Management. Acta Hort 938: 113-120.

O'Neil, K. J., and R. N. Carrow. 1982. Kentucky bluegrass growth and water use under different soil compaction and irrigation regimes. Agron. J. 74:933-936.

Sevostianova, E., B. Leinauer, and M. Serena. 2010. Performance of non–irrigated warm season turfgrasses in an arid climate. Proceedings 2nd European Turfgrass Society Conference:202-204.
Shearman, R. C. 2008. Turfgrass cultural practices for water conservation. p. 205-222. *In* J. B. Beard and M.P. Kenna (ed.) Water Quality and Quantity Issues for Turfgrasses in Urban Landscapes. Council for Agricultural Science and Technology, Ames, IA.
Shearman, R. C. 1989. Perennial Ryegrass Cultivar Evapotranspiration Rates. HortScience 24:767-769.
Shearman, R. C. 1986. Kentucky bluegrass cultivar evapotranspiration rates. HortScience. 21: 455-457.
Suplick-Ploense, M.R., and Y. Qian. 2005. Evapotranspiration, rooting characteristics, and dehydration avoidance: Comparisons between hybrid bluegrass and Kentucky bluegrass. Int. Turfgrass Soc. Res. J. 10:891-898.

This page intentionally left blank.

Water source effect on golf course soil quality in Oklahoma

Abstract

Reclaimed water use on golf courses is common in many areas of the Southwest United States, but is not common in Oklahoma. The objective of this study was to examine the effects of reclaimed water use on golf course soil quality in Oklahoma. This case study utilized soil and water quality samples collected from five golf courses in Oklahoma that used different water sources for turfgrass irrigation. This included Gaillardia Country Club, Quail Creek Country Club, Lincoln Park Golf Course, Lake Hefner Golf Course (all in Oklahoma City, OK) and the Jimmy Austin Golf Club (Norman, OK). The results from this case study indicate that irrigation water source (treated municipal water, untreated surface water, and groundwater) did affect soil test parameters, but there were not significant differences in nutrient and salt concentrations between certain parameters when irrigating with reclaimed water versus groundwater in Oklahoma. This case study data suggests reclaimed water can be beneficially used for golf course irrigation in Oklahoma in conjunction with supportive regulation and best management practices, such as aerification, leaching, choosing salt-tolerant turfgrass, applying proper applications of soil amendments, and consistently monitoring soil and irrigation water sources.

Introduction

Reclaimed water typically contains different levels of elements, such as nitrogen (N) and phosphorus (P), which can be beneficial to turfgrasses. Using the beneficial elements like N and P that already exist in reclaimed water can reduce the amount of fertilizers that golf courses use annually on their greens and fairways. In addition to these beneficial nutrients, reclaimed water can also contain high levels of total soluble salts, sodium (Na), and chloride, which can be damaging to plant and soil health. When using reclaimed water for irrigation, it is important to routinely monitor soil and water quality to properly manage the beneficial and harmful nutrients and elements.

Many of the studies that have been conducted on the use of reclaimed water for irrigation purposes have addressed use on golf courses in the southwest and arid regions of the United States (Hayes et al., 1990; Mancino and Pepper, 1992; Qian and Mecham, 2005; Lockett et al., 2008). Previous research has found that soil irrigated with reclaimed water contained elevated levels of soil electrical conductivity (EC), Na, macronutrients (N, P, K, Ca, P, Mg, and S), and micronutrients (Cl, Fe, Zn, B, and Mn) (Hayes et al., 1990; Mancino and Pepper, 1992; Qian and Mecham, 2005; Thomas et al., 2006; Lockett et al., 2008). Studies have also shown that proper irrigation management and soil and water monitoring can help balance out the excessive salts and nutrients. There is limited information regarding the effects of reclaimed water irrigation on soil chemical properties on golf courses in Oklahoma.

Currently in Oklahoma, reclaimed water is not used for golf course irrigation on a large scale. As drought conditions are frequent in Oklahoma, the use of reclaimed water for golf course irrigation is gaining interest from superintendents and municipalities. In this study, we examine the soil chemical properties of one golf course irrigated with reclaimed water in comparison to four other golf courses irrigated with different water sources (groundwater, untreated surface water, treated municipal water, and groundwater + reclaimed water mix).

Materials and Methods

This case study was conducted at five golf courses located in the Oklahoma City Metropolitan in central Oklahoma. Four of the five golf courses (Lincoln Park, Gaillardia, Quail Creek, and Lake Hefner) are located within Oklahoma City limits, and one golf course (Jimmie Austin OU) is located 25 miles south in Norman, Oklahoma. Each of the five golf courses uses a different water source to supply irrigation to their courses, including reclaimed water, treated municipal water, groundwater, and untreated surface water. The main soil series and texture classifications for each of the study sites was acquired through the assistance of the United States Department of Agriculture (USDA) and Natural Resources Conservation Service (NRCS) Web Soil Survey located in Table 1. The average annual precipitation for the central Oklahoma region is approximately 36 inches (Oklahoma Climatological Survey, 2012). The following is a brief list of attributes for each golf course. Gaillardia Country Club is an 18 hole private country club located in Oklahoma City, OK and has been irrigated with reclaimed municipal water since 1996. Lake Hefner Golf Club is a 36 hole public golf course and is irrigated with raw water from Lake Hefner in Oklahoma City, OK. Lincoln Park Golf Course is a 36 hole public golf course and is irrigated with municipal water purchased from the City of Oklahoma City, OK. Quail Creek Golf Course and Country Club is a 18 hole private golf course and is irrigated with groundwater wells. Jimmie Austin Golf Course at the University of Oklahoma is an 18 hole golf course irrigated with a mixture of 85% reclaimed water from the City of Norman and 15% groundwater from wells.

Soil samples and irrigation water samples were collected over two growing seasons at these golf courses and analyzed for the following parameters. The soil samples went through a soil fertility test, including the following parameters: pH, Soil Test Phosphorus (STP), Soil Test Potassium (STK), Surface Nitrate (NO_3), Surface Sulfur (S), Calcium (Ca), Magnesium (Mg), Iron (Fe), Zinc (Zn), Boron (B), and soil organic matter content (OM%). The soil samples also received a salinity management test (1:1 extraction), including the following parameters: Electrical Conductivity (EC), Sodium (Na), Calcium (Ca), Magnesium (Mg), Potassium (K), Boron (B), Total Soluble Salts (TSS), Sodium Adsorption Ratio (SAR), and Exchangeable Sodium Percentage (ESP). The irrigation water samples were tested for basic irrigation water quality and salinity management tests included pH, CO_3, HCO_3, EC, Na, Ca, Mg, K, B, NO_3-N, Cl, SO_4, Zn, Cu, Mn, Fe, NH_4-N, Hardness, Alkalinity, TSS, PAR, SAR, EPP, and ESP.

Statistical analysis was conducted to assess the interactions and effects of the independent variables (irrigation water sources and golf course greens, fairways, and non-irrigated roughs) on the dependent variables (soil chemical parameters) using Statistical Analysis Systems Software version 9.3 (SAS, Cary, NC, 27513) for the personal computer. An Analysis of Variance (ANOVA) procedure was performed using SAS 9.3 software, applying the General Linear Models Procedure, PROC GLM. The two-way factorial ANOVA procedure included a main effects analysis of the treatment (water source) and location (greens, fairways, and non-irrigated roughs) as well as an interaction of the main effects, treatment by location. The mean values of the soil properties from the interaction of the main effects that were statistically different at a p-value of 0.001 indicate that the data are consistent with the hypothesis that all soil chemical parameter means are significantly different for reclaimed water irrigation sources compared to the other irrigation water sources. Not all results of this work will be presented due to the time limit of the oral presentation at the Irrigation Show and Educational Conference Technical Sessions.

Results and Discussion

This case study evaluated the soil chemical properties and water quality properties of reclaimed water irrigation compared to untreated surface water, groundwater, and treated municipal irrigation on golf courses in Oklahoma. The results from the water quality tests showed that the highest concentrations of salts (TSS and EC) were found in the reclaimed water samples from Gaillardia, which was expected, but the reclaimed water source from Jimmie Austin OU contained half of the salt concentrations than that of Gaillardia's reclaimed water source. The nutrient concentrations varied amongst the water sources, with each of the water source results showing values above and below medium sufficiency ranges. Reclaimed water sources typically contain higher levels of P and NO_3-N, in which both of the reclaimed water sources contained the highest mean values for both of these nutrients The samples from the reclaimed water source from Jimmie Austin OU had the highest mean value for dissolved P, possibly resulting from only receiving secondary treatment before use. The samples from both of the reclaimed water sources had the highest mean values for NO_3-N, but both were within the normal range for irrigation water, 5-50 mg L^{-1}.

The results from soil quality tests suggest that the salts and nutrient concentrations from the interaction of water source and the location on each of the golf courses (greens, fairways, non-irrigated roughs) were not statistically different from each other for each soil chemical parameter for at least one of the golf course locations. The hypothesis for this case study that the chemical properties of soil irrigated with reclaimed water would be different from those chemical properties of soils irrigated from the other three water sources was proven false, as the soil chemical concentrations were different in value for all of the water sources, but not statistically different for the treatment by location interaction.

As the demand for potable water supplies increases among municipalities and industry, the use of reclaimed water for non-potable uses, such as landscape irrigation, will also increase. Golf courses are ideal candidates to use reclaimed water for irrigation purposes. Both opportunities and problems are evident when using reclaimed water for irrigation purposes. It is important to understand the constantly changing levels of soil chemical properties and water quality parameters when using reclaimed water for golf course irrigation. The results from this case study indicate that other water sources (treated municipal water, untreated surface water, and groundwater) are not different when discussing nutrient and salt concentrations, providing data that suggests reclaimed water can be beneficially used for golf course irrigation just as other water sources. Reclaimed water can be an effective source for golf course irrigation in Oklahoma in conjunction with supportive regulation and best management practices, such as aerification, leaching, choosing salt-tolerant turfgrass, applying proper applications of soil amendments, and consistently monitoring soil and irrigation water sources.

Literature Cited

Hayes, A.R., C.F. Mancino, and I.L. Pepper. 1990. Irrigation of turfgrass with secondary sewage effluent: I. soil and leachate water quality. Agron. J. 82:939-943. doi: 10.2134/agronj1990.00021962008200050020x.

Lockett, A.M., D.A. Devitt, and R.L. Morris. 2008. Impact of reuse water on golf course soil and turfgrass parameters monitored over a 4.5-year period. HortScience 43:2210-2218.

Mancino, C.F. and I.L. Pepper. 1992. Irrigation of turfgrass with secondary sewage effluent: soil quality. Agron. J. 84:650-654. doi: 10.2134/agronj1992.00021962008400040023x.

Oklahoma Climatological Survey. 2012. Normal annual precipitation 1981-2010. Oklahoma Climatological Survey. http://climate.ok.gov/index.php/site/page/climate_of_oklahoma (accessed 15 April 2016).

Qian, Y.L., and B. Mecham. 2005. Long-term effects of recycled wastewater irrigation on soil chemical properties on golf course fairways. Agron. J. 97:717-721. doi: 10.2134/agronj2004.0140.

Thomas, J.C., R.H. White, J.T. Vorheis, H.G. Harris, and K. Diehl. 2006. Environmental impact of irrigation turf with type I recycled water. Agron. J. 98:951-961. doi: 10.2134/agronj2005.0190.

Seasonal Curves for Turfgrass Using Dual Coefficient Method

Mark A. Crookston, P.E., D.WRE, CAIS, CIC, CID, CIT, CLIA
Northern Water, 220 Water Ave, Berthoud, CO 80513, mcrookston@northernwater.org

Abstract

Seasonal crop coefficient or Kc curves were developed for ten turfgrasses at Berthoud, Colorado utilizing the FAO-56 dual crop coefficient method. Nine grasses were cool-season and one was warm season. Actual turf evapotranspiration or ET was measured by small weighing lysimeters, with four replicates of each turfgrass (40 lysimeters total). All were seeded in 2010. Daily lysimeter measurements of ET during three seasons (2011-2013) were compared to daily ETo calculated using the ASCE standardized reference evapotranspiration equation.

The dual Kc method partitions ET into evaporation from wet soil/plant surfaces and transpiration from vegetation. This provides a substantially improved fit of simulated ET (using calculated ETo) to measured ET (from weighing lysimeters). The increased evaporation from wetted surfaces following a rain or irrigation event and the reduced transpiration resulting from soil moisture depletions are calculated on a day-to-day basis, not imbedded in averages as with the single Kc method. Consequently, developed Kc curves are not skewed by local rainfall frequency or irrigation management practices and are more readily transferable to other locations.

ET = Kc ETo = (Ks Kcb + Ke) ETo

Ks   *stress factor based on available soil moisture,*

Kcb   *basal Kc factor, visually dry surface soil, no stress, and*

Ke   *evaporation factor based on percent of soil surface wetted.*

The resultant seasonal Kc curves are expected to readily transfer to other similar locations and prove more accurate day-to-day than single Kc curves, particularly when rooting depths, frequency of rainfall, and/or irrigation management practices vary. The length of the four time periods within the seasonal curves can be readily adjusted to match localized plant growth and development.

Keywords: evapotranspiration, irrigation scheduling, lysimeters, soil moisture balance, turfgrass weighing lysimeters.

Procedures

Background

Use of weighing lysimeters provides a defensible basis for quantifying and comparing various inputs into the soil moisture balance calculations. Along with measurement of actual plant water use or $ET_{c\ act}$ (evapotranspiration), lysimeters may also aid in determination of net and effective rainfall, and net and effective irrigation. This improved input information may assist in the programming of weather-based smart controllers. It can also be utilized by municipalities to develop landscape water use standards and budgets in support of efficient water use and conservation.

A previous paper by Crookston, et al. (2010) included an overview of several previous studies regarding turfgrass water use. An additional paper by Crookston, et al. (2011) included preliminary results from work with the Berthoud lysimeter data. A presentation titled: Turfgrass ET from Small Weighing Lysimeters in Colorado (paper number 1862438) was made by Crookston on April 9, 2014 at the ASABE Symposium on Evapotranspiration: Challenges in Measurement and Modeling from Leaf to the Landscape Scale and Beyond, Raleigh, NC. A more recent paper was presented by Crookston (2016) at the 2016 ASABE Annual International Meeting, Orlando, FL, July 17-20, 2016, using the same information as is included herein.

Considerable information regarding utilization of the dual coefficient method of calculating evapotranspiration has been provided by Allen et al. (20016. 2011, 2005, and 1998). Interested professionals are referred to these reference documents for more in-depth explanations of appropriate methods and procedures.

Methods

In 2009, Northern Water staff commenced construction and installation of a 30-ft x 30-ft study plot for turfgrass lysimeters within its Conservation Gardens located at its headquarters in Berthoud, Colorado. A sandy clay loam soil was fully packed into all 44 lysimeters and also replaced the top 12 inches of soil in the entire lysimeter plot. The turfgrasses were seeded May 28 to June 2, 2010. Frequent sprinkler irrigation for establishment of the turfgrasses continued through most of July 2010. However, the turfgrasses had not yet filled the small gap surrounding each lysimeter and the top rims of most were still clearly visible. By the end of the 2010 growing season, all turfgrasses were well established in the lysimeters. The 2011 season became the first full season for evaluation of ETc act.

The lysimeter plot was divided into 4-ft x 4-ft sub-plots, separated by 1-inch x 6-inch PVC plastic composite decking/edging material. This edging clearly delineated the subplots and helped prevent the spread of one grass variety into another subplot. It also provided support for foot traffic by study technicians without compaction of the soil or inadvertently stepping onto a lysimeter. Turfgrasses were planted into 44 of the 49 sub-plots, with the four corners and center sub-plots excluded from the study, but planted to a bluegrass blend to maintain fetch. The lysimeter plot was divided into four blocks, with each block containing 11 randomized sub-plots, each with a lysimeter seeded to a different turfgrass. However, one of the turfgrasses (Ephraim Crested Wheatgrass) did not thrive and was subsequently eliminated from the study plot. Consequently, the study included four replicates of each of the following 10 turfgrasses:

- Blue gramma – buffalograss mix
- Drought hardy Kentucky bluegrass
- Fine fescue mix
- Kentucky bluegrass blend
- 'Low Grow' mix
- 'Natures Choice' - Arkansas Valley mix
- Perennial ryegrass
- Reubens Canada bluegrass
- Tall fescue
- Texas hybrid bluegrass blend

Equipment

The weighing platform for each lysimeter included a Revere PC6-100kg-C3 load cell transducer. Each load cell was connected to one of three AM 16/32 multiplexers, each connected to a Campbell Scientific CR10X data logger. Figure 1 is a diagram of the small turfgrass lysimeters and their arrangement within the lysimeter plot.

Every three seconds a weight measurement in inches of H2O was taken from each lysimeter load cell. These measurements were averaged every 60 seconds. This 1-minute average weight was time-stamped and stored in the data logger at the end of each 15-minute period. Stored data was automatically downloaded every 15 minutes to a desktop PC via an RF401 spread-spectrum radio. Hourly differences in lysimeter weights can be compared to calculated ETos utilizing data from the adjacent Campbell Scientific ET-106 weather station. The weather station instruments are calibrated annually. Hourly ETos can be obtained from the REF-ET software v.3.1.16
(http://extension.uidaho.edu/kimberly/2013/04/ref-et-reference-evapotranspiration-calculator/)

The weighing platforms for each lysimeter were calibrated in-place (without lysimeters loaded) in April 2009 over their full load range, using steel weights. The platforms were again re-calibrated in-place during March 2011 and March 2013, but only over their operational range (from dry soil to wet soil), again using steel weights. In-place full range re-calibration was again performed in March 2014. No problems were identified during the re-calibrations, and all weighing platforms were measuring the lysimeter weights properly. The calibration factors applied to each lysimeter were for the 11.875-inch inside diameter of the lysimeters. The outside diameter of the PVC reducer attached to the top of the outer shell containing the lysimeter was

13.188 inches, leaving a 0.656-inch space to be filled by turfgrass growing from inside and outside of the lysimeter. Consequently, the effective diameter for the ETc act from the turfgrass within each lysimeter was 12.531 inches. Changes in lysimeter weight were converted to ETc act by multiplying the weight difference by a factor of 0.898.

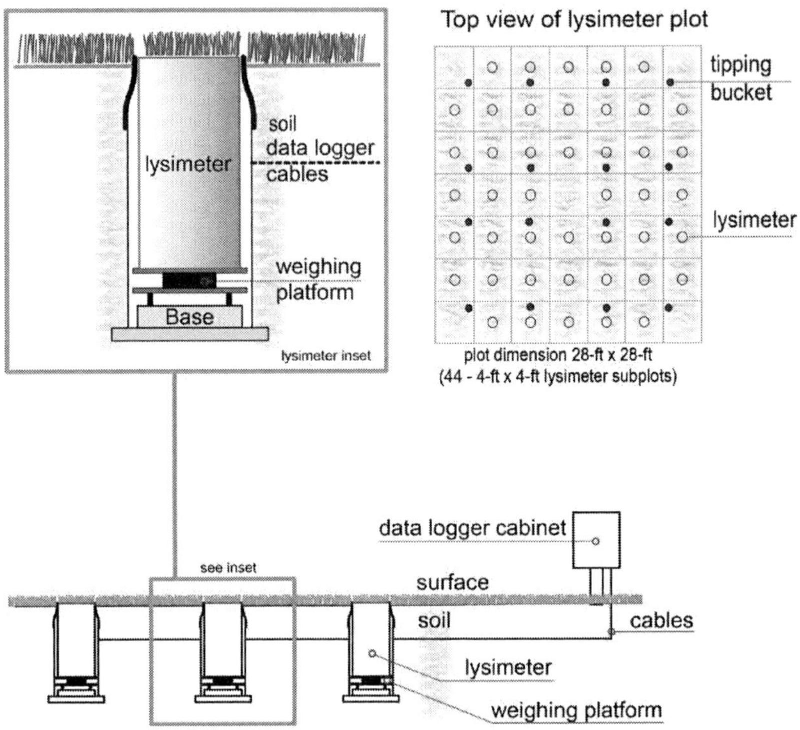

Figure 1. Diagram of small turfgrass lysimeters and arrangement of lysimeter study plot.

The entire lysimeter plot is on a single irrigation zone using Hunter MP Rotator 2000 sprinklers on 15-ft spacing. The gross application rate was 0.57 inch/hr. Daily gross irrigation applications were measured by a dedicated DLJ ¾-inch x ¾-inch brass multi-jet flow meter with pulse output connected to a Campbell Scientific data logger which measured all sprinkler irrigation applications to the lysimeter plot. In addition, fifteen Texas Electronics tipping bucket rain gauges were installed flush with the turf height throughout the lysimeter plot to measure net sprinkler irrigation application as well as rainfall. A second DLJ flow meter measured irrigation water applied by hand to bring each lysimeter up to field capacity following sprinkler irrigation.

Per field observation, the depth of the established turfgrass root zone was taken to be the depth of the lysimeter. Following removal of the lysimeters in November 2013, many lysimeters had significant root growth down to the filter fabric placed beneath the soil. During mid-season, all turfgrasses were mulched mowed to a height of 3 inches weekly. The observed steady infiltration rate was no more than 0.4 inch/hr.

Rainfall Measurements

Daily rainfall was measured every 15 minutes at an adjacent weather station having a weighing bucket precipitation gauge as well as a tipping bucket rain gauge. As noted above, combined daily rainfall and sprinkler irrigation catch was measured by fifteen tipping bucket rain gauges installed in-ground within the lysimeter grid with their collector rims at turf level.

Measurement agreement was typically very good between the weighing bucket precipitation gauge and the tipping bucket rain gauges. However, the weighing bucket precipitation gauge proved most consistent and reliable. The tipping bucket rain gauges under-measured catch during higher intensity rainfall events. The in-ground tipping buckets were also susceptible to partial plugging from windblown grass clipping debris. Catch readings would then be extended over longer time periods as the rain drained much more slowly through the collector funnel obstruction.

Deep Percolation

Deep percolation through the lysimeters was free draining and not measured directly. The sandy clay loam soil in each lysimeter was only 20-inches deep. Following field observations and inspection of the 15-minute lysimeter data, any deep percolation from irrigation was generally observed to be completed before sunrise. Turf water use during this nighttime drainage period was considered negligible. Beginning in late July 2010, all sprinkler irrigations were scheduled for after 9:00 P.M. and before midnight. Beginning in 2013, irrigations were cycle/soaked three times to insure infiltration without runoff. However, hand watering to bring each individual lysimeter up to field capacity did occur during daytime hours–usually the day following sprinkler irrigation. Any excessive percolate that ponded below a lysimeter following prolonged or heavy rainfall was removed as needed (rarely) through a manually-controlled vacuum extraction system.

Lysimeter Calculations

The following equation was utilized as the basis for the soil moisture balance in the root zone:

$$\text{LysWTend} = \text{LysWTstart} + \text{Rain} + \text{NetIrrig\_sprink} + \text{NetIrrig\_hand} - \text{ETc act} - \text{DeepPerc} \quad (1)$$

LysWT_{end} = Ending weight of lysimeter at midnight of current day, inches H_2O
LysWT_{start} = Beginning weight of lysimeter at midnight of previous day, inches H_2O
Rain = Measured weighing bucket precipitation, inches H_2O
NetIrrig_sprink= Net irrigation infiltrated from in-ground sprinkler system, inches H_2O (gross irrigation as measured by flow meter – wind drift and overspray losses)

NetIrrig_hand	= Net irrigation applied by hand to bring each lysimeter up to field capacity, inches H_2O
$ET_{c\ act}$	= ET_{os} x (K_s x K_{cb} + K_e), inches H_2O
DeepPerc	= Deep percolation losses to drainage when field capacity is exceeded, inches H_2O
ET_{os}	= Reference evapotranspiration for short reference crop (clipped grass) calculated hourly by the ASCE standardized reference evapotranspiration equation, then summed daily, inches H_2O
EffAreaFactor	= conversion factor of 0.898 utilized to convert lysimeter weight changes to $ET_{c\ act}$, ($LysWT_{end}$ – $LysWT_{begin}$) x EffAreaFactor = $ET_{c\ act}$, assuming no rain/irrigation or other losses

The net irrigation both by sprinklers and by hand was calculated for each lysimeter as the net positive change in lysimeter weight over each 15-min time period when irrigation was applied. The sprinkler applied irrigation correlated closely to the average irrigation derived from lysimeter gains, averaging nearly 92%.

Deep percolation was likewise estimated for each lysimeter as any unaccounted for weight loss in excess of [ETos x 1.25 x 1.1] following an irrigation or rain event. See Allen et al. (2011).

The following rearrangement of equation 1 was utilized to calculate daily Kc for 2011-2013:

$$K_c = [\ LysWT_{start} - LysWT_{end} + Rain + NetIrrig\_sprink + NetIrrig\_hand - DeepPerc\ ] / ET_{os} \quad (2)$$

Results

The nine cool season grasses were very similar in watering needs, with only minor differences. However the following ranking was applied:

Table 1. Ranking of cool season turfgrasses by water use.

High water use	Medium high water use	Medium water use	Medium low water use
'Natures Choice' Arkansas Valley mix	Fine fescue mix	Perennial ryegrass	Reubens Canada bluegrass
	Drought hardy Kentucky bluegrass	Kentucky bluegrass blend	Texas hybrid bluegrass blend
	'Low Grow' mix	Tall fescue	

The warm season blue gramma – buffalograss mix had the lowest water use. However it was only modestly lower than the Texas hybrid bluegrass blend.

Seasonal Kc curves were developed for both cool season and warm season turfgrasses. The timeline for thc Kc curves was determined seasonally based on the following parameters:

Table 2. Seasonal timeline for Kc curve.

	Cool season turfgrass	**Warm season turfgrass**
Initial	K_{cb} = 0.20	K_{cb} = 0.20
Start of development	Starts when average daily air temperature reaches 1 deg C. Ends when cumulative growing degree days (base 0) exceeds 530.	Starts when average daily air temperature reaches 8 deg C. Ends when cumulative growing degree days (base 0) exceeds 276.
Mid-season	K_{cb} = 0.921	K_{cb} = 0.881
Late season	Starts when minimum daily air temperature reaches -2.5 deg C. Ends when minimum daily air temperature reaches -15 deg C.	Starts when minimum daily air temperature reaches 0 deg C. Ends when minimum daily air temperature reaches -5 deg C.
Dormancy	K_{cb} = 0.20	K_{cb} = 0.20

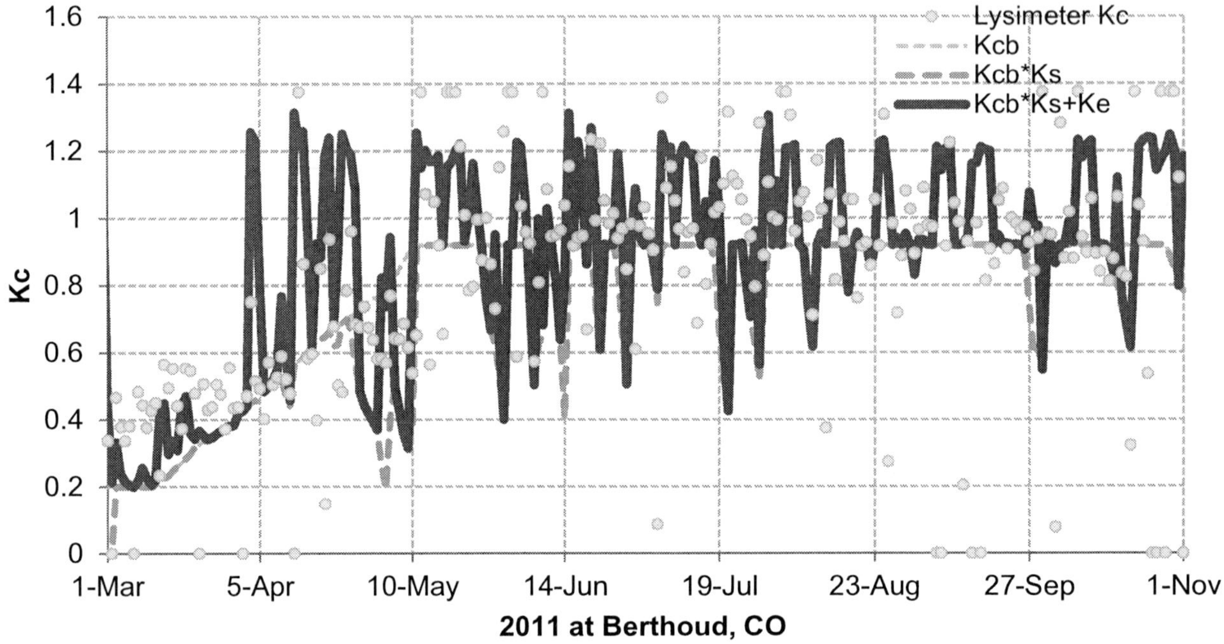

Figure 1. Seasonal Kc curve for cool season turfgrass at Berthoud, Colorado. Data from 40 small weighing lysimeters during the 2011 season.

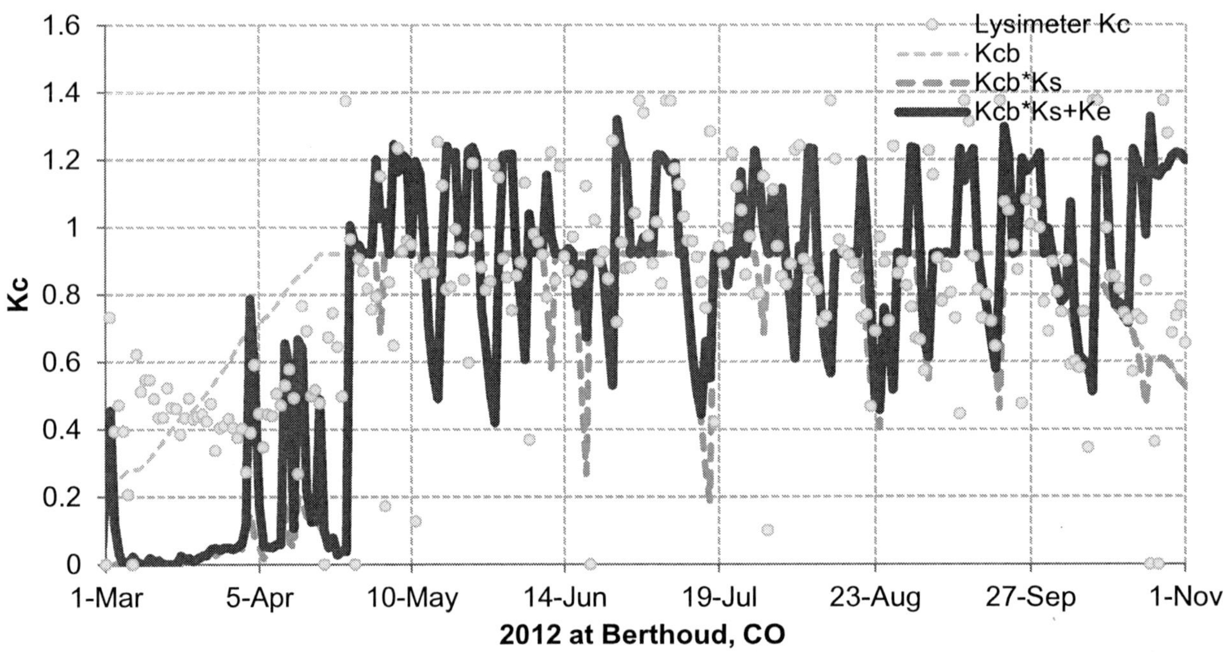

Figure 2. Seasonal Kc curve for cool season turfgrass at Berthoud, Colorado. Data from 40 small weighing lysimeters during the 2012 season.

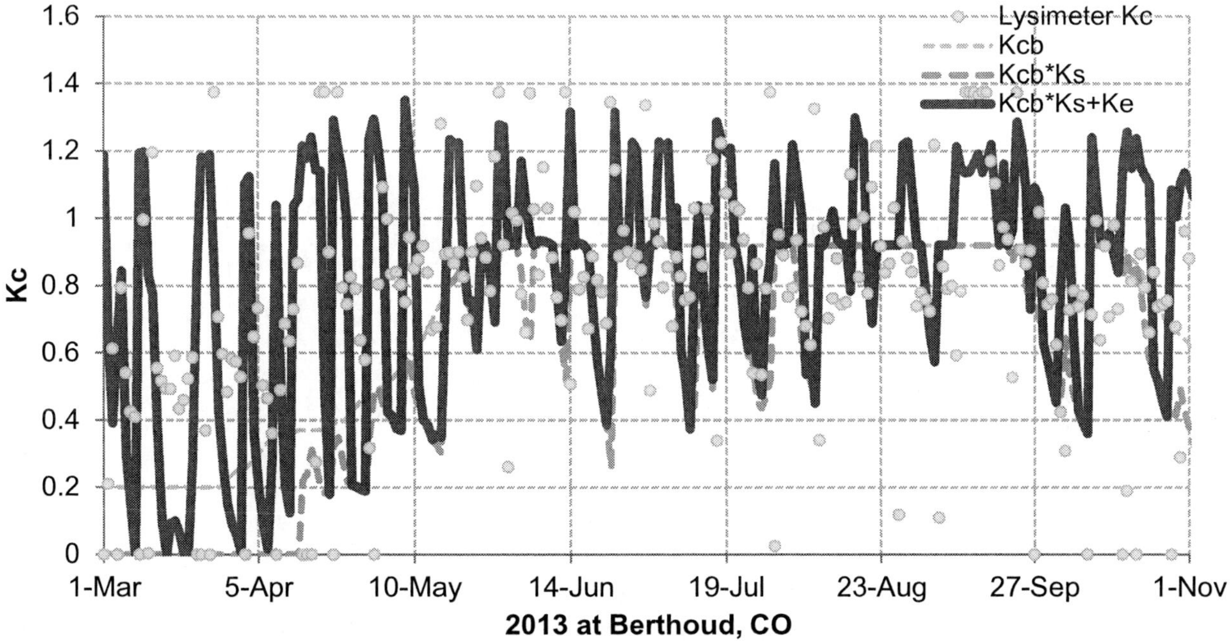

Figure 3. Seasonal Kc curve for cool season turfgrass at Berthoud, Colorado. Data from 40 small weighing lysimeters during the 2013 season.

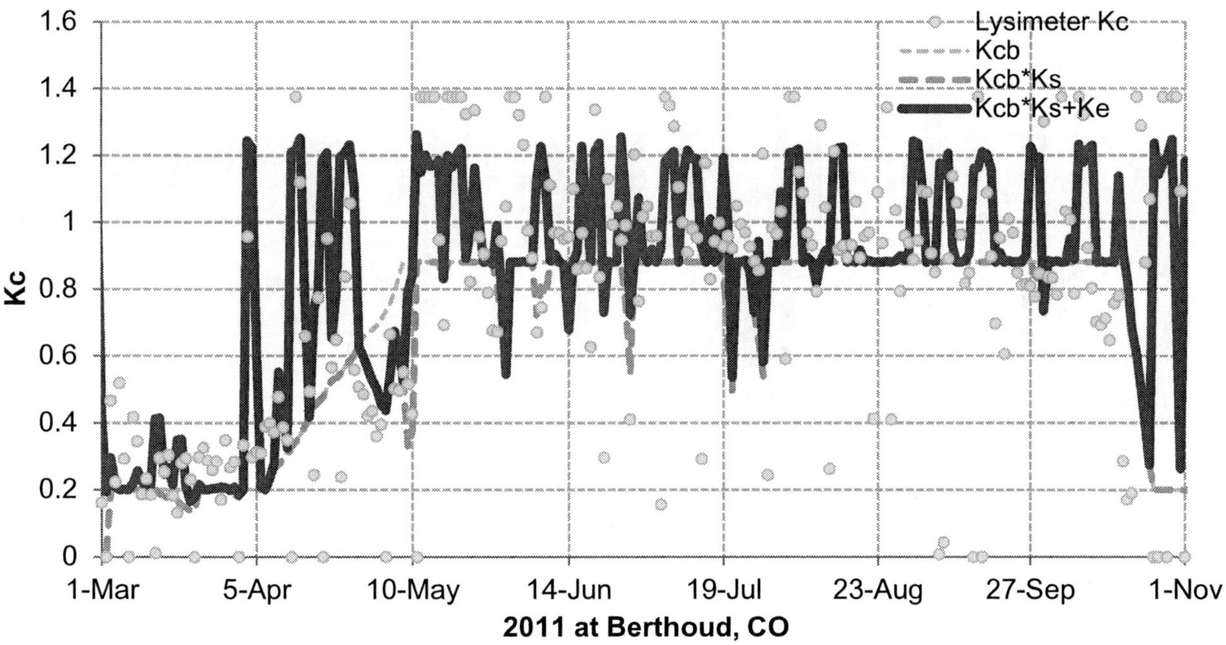

Figure 4. Seasonal Kc curve for warm season turfgrass at Berthoud, Colorado. Data from 40 small weighing lysimeters during the 2011 season.

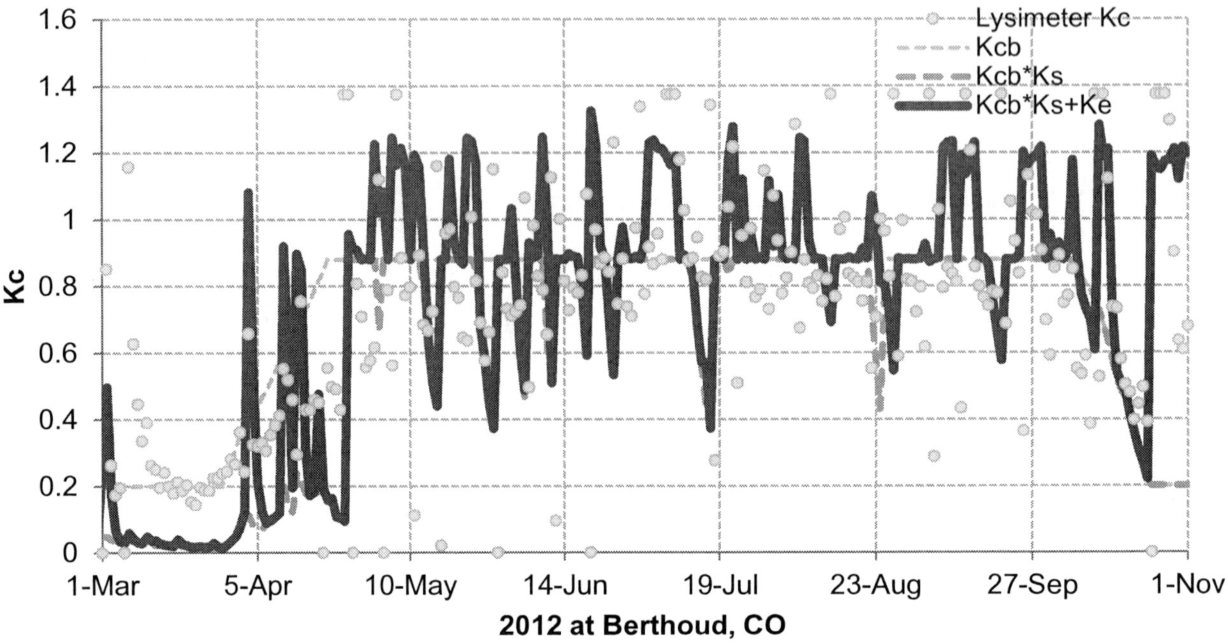

Figure 5. Seasonal Kc curve for warm season turfgrass at Berthoud, Colorado. Data from 40 small weighing lysimeters during the 2012 season.

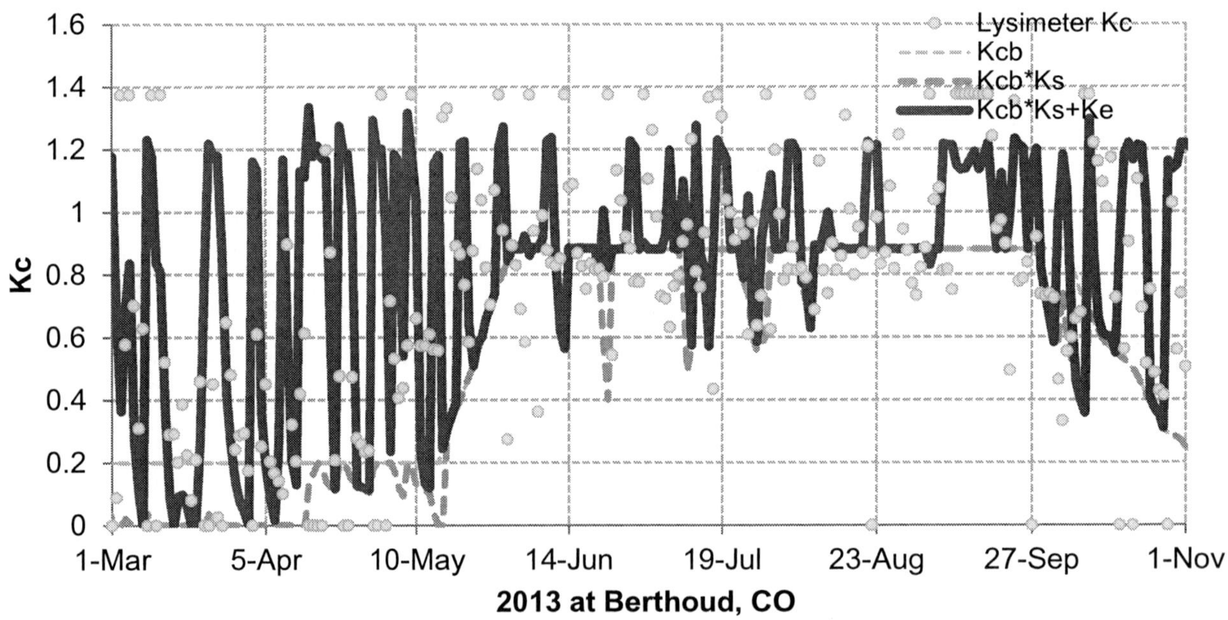

Figure 6. Seasonal Kc curve for warm season turfgrass at Berthoud, Colorado. Data from 40 small weighing lysimeters during the 2013 season.

Summary Conclusions

The application of the FAO-56 dual crop coefficient method for development of seasonal curves for both cool and warm season turf grasses proved workable and reasonably accurate. Incorporation of this method into the soil moisture balance computations of weather based smart controllers holds good potential for further improvement in the irrigation management of urban landscapes. The use of seasonally adjusted timelines for Kc curves should improve accuracy of plant water use calculations, particularly during the starting and ending of the irrigation season. Significant potential for increased water conservation may result.

Acknowledgements

Special recognition to the staff at Northern Water, particularly Ron Boyd, Chad Kuhnel, June Caves, and Lyndsey Lucia. These co-workers assisted in the installation of the lysimeter system, establishment and maintenance of the turfgrasses, routine collection of data, and/or service and calibration of each sensor annually. Without their cooperation and assistance, critical data would have been missing or compromised.

References

Allen, R.G. and M.E. Jensen. 2016. Evaporation from Soil. Chapter 9 in Evaporation, Evapotranspiration, and Irrigation Water Requirements, 2nd ed., ASCE Manuals and Reports on Engineering Practice No. 70, pp.221-259, ASCE, Reston, VA.

Allen, R.G. and M.E. Jensen. 2016. Crop Coefficient Method. Chapter 10 in Evaporation, Evapotranspiration, and Irrigation Water Requirements, 2nd ed., ASCE Manuals and Reports on Engineering Practice No. 70, pp.221-259, ASCE, Reston, VA.

Allen, R.G., T.A. Howell, and R.L. Snyder. 2011. Irrigation Water Requirements. Chapter 5 in Irrigation Sixth Edition, pp.91-172, Irrigation Association, Falls Church, VA 22042.

Allen, R.G., L.S. Pereira, M. Smith, D. Raes, and J.L. Wright. 2005. FAO-56 Dual Crop Coefficient Method for Estimating Evaporation from Soil and Application Extensions. J.Irrigation and Drainage Engineering, ASCE 131(1):2-13.

Allen, R.G., L.S. Pereira, D. Raes, and M. Smith. 1998. FAO-56, Crop Evapotranspiration, Guidelines for Computing Crop Water Requirements. Food and Agriculture Organization of the United Nations, Rome, Italy.

Allen, R.G., C.W. Robison, and J.L. Wright. 2007. Updated Procedures for Calculating State-Wide Consumptive Use in Idaho. USCID 4th International Conference on Irrigation and Drainage, proceedings pp. 189-212. USCID, Denver, CO 80202.

Allen, R.G. and Robison, C.W., 2006 rev. April 2007. Evapotranspiration and Consumptive Irrigation Requirements for Idaho. University of Idaho Research and Extension Center, Kimberly, ID. http://www.kimberly.uidaho.edu/ETIdaho/.

Crookston, M.A. 2015. Seasonal Kc Curves for Turfgrass Using FAO-56 Dual Crop Coefficient Method, 2016 ASABE Annual International Meeting, Orlando, FL, July 17 – 20, 2016.

Crookston, M.A. 2015. Net Infiltration from Rainfall as Measured with Small Weighing Lysimeters. Presented at the joint ASABE/IA Irrigation Symposium in Lon Beach, California, USA, November 10-12, 2015.

Crookston, M.A. and M. Hattendorf. 2011. Turfgrass ET from Small Weighing Lysimeters in Colorado: First Full Year Results. Presented at the Irrigation Show 2011 in San Diego, California, USA, November 7, 2011. Available from Irrigation Association, Falls Church, Virginia, USA 22042-6638, www.irrigation.org.

Crookston, M.A. and M. Hattendorf. 2010. Turfgrass ET from Small Lysimeters in Northeast Colorado. Presented at the USCID 2010 Fort Collins Conference, September 28 – October 1, 2010. Abstract on page 51 (full paper on accompanying CD). Available from USCID, 1616 Seventeenth Street #483, Denver, Colorado, USA 80202, www.uscid.org.

This page intentionally left blank.

A New Approach to Sustainable Irrigation Practices

Doug Pushard, Founder
Harvest H2o

Mike Collignon, Executive Director
Green Builder® Coalition

Outdoor irrigation has been a common practice amongst residential homes for decades, from the early days of watering cans and spray nozzles on hoses, to manually-placed rotary sprinklers, to manually operated spray systems, to automated drip and micro-drip systems powered by smart controllers. History shows us that refinements have been made, as we've learned more about average rainfall, evapotranspiration rates and the amount of water needed by the myriad of available plants at various stages of their maturity.

However, the irrigation industry faces a rising problem. Water scarcity is emerging as a critical challenge to the western United States. The limits of water could halt construction in the west, bringing economic development to a standstill. Mandates like MWELO in California threaten the irrigation industry, as they shine a spotlight on the massive amounts of water devoted to outdoor irrigation. Some homeowners and communities are being incentivized to remove turf altogether in order to comply.

A New Measurement Tool

It's been said that you can't manage what you don't measure. To that end, there's a new methodology to help everyone from an irrigation professional to a homeowner better understand what's going to happen onsite and in their community with regards to landscape water consumption, stormwater management and overall water use. This independent, predictive analysis can also preserve the freedom of the landscape architect/designer, which in turn can enable irrigation professionals to be seen as part of the solution instead of an endangered species.

The Water Efficiency Rating Score, or WERS, is a first-of-its-kind, predictive, performance-based approach to residential water efficiency and water resource management. The WERS is the culmination of calculations that consider the loading from principal plumbing fixtures, clothes washers, structural waste, and outdoor water management. Potential rainwater, greywater, stormwater and blackwater catchment are also calculated. The WERS Program is applicable for both new and existing single-family and multifamily residential properties. It can be used to model irrigation, landscape and source water changes. This can help determine the impact of these changes on water usage.

The WERS Tool also has a unique output that allows for comparison of properties. It uses a scoring scale of zero to 100, with zero being the most desirable and 100 representing the baseline home. In addition to the Score, the property owner also receives a daily, monthly and yearly projection of water usage. If water rates are entered into the program, financial savings (or expenditure) projections over the same time intervals are also projected. Finally, a dashboard contains a bar graph that breaks down the amount of rainwater, greywater, blackwater and stormwater (if applicable) generated and used, as well as a circle graph that displays the amount of potable water and alternative water used.

How WERS Can Be Used By Practitioners

Unlike a prescriptive program, a performance-based program gives all parties (architect/designer, builder, property owner) design and product flexibility. It doesn't require anything. Rather, it assesses the choices made. The same flexibility extends to the implementation of the WERS Program.

On its own, the WERS Program can help a property owner understand where and how water is being used. Without this knowledge, it's difficult to determine the most cost-effective conservation strategies. This lack of information can lead to poor decisions, from a property owner or an entire municipality.

While the WERS Program is a 3rd party certification executed by a WERS Verifier, the WERS Design Tool might be of equal or greater benefit to irrigation professionals. Project teams (architect/designer, builder, property owner, landscaper, etc.) will have the ability to use the WERS Design Tool to view initial estimates of the results of their proposed installed fixtures and appliances, as well as innovative water conservation strategies, without the involvement of a WERS Verifier.

How WERS Can Be Implemented

A voluntary modeling tool is just one way to use the WERS Program. Through this application, builders, landscape architects/designers and irrigation professionals can differentiate themselves in the market. By striving to keep their projects at or under a certain score, they would be able to demonstrate beautiful designs that are also sustainable. Building professionals can also leverage WERS as a first-step, market-driven, voluntary approach in communities or jurisdictions where water efficiency mandates or regulations are being considered.

It can be adopted as a regulation. The City of Santa Fe has become the first municipality in the nation to integrate a performance-based water efficiency requirement in its residential green building code. Their code now stipulates that all new single-family detached units must submit a preliminary WERS of 70 (or less) with their building permit application, and a verified 70 (or less) to obtain a certificate of occupancy. The County of Santa Fe is strongly considering the usage of WERS in a regulatory manner, if a project meets certain conditions.

WERS can assist in the pursuit of a financial incentive. The State of New Mexico allows a builder or new property owner to attach a WERS report to show compliance with the new water efficiency requirement of the State's very popular Sustainable Building Tax Credit. The WERS Program joins Build Green New Mexico and LEED for Homes as compliance paths for water. Other possible financial incentives on new construction projects could include a reduction in tap fees or stormwater impact fees, or a reduced permit review time.

The GreenHome Institute (GHI), after months of discussion, scrutiny and testing, has started the process of implementing the WERS Program as the water criteria for GreenStar, their residential green building program available nationwide. Their rationale was rooted in the movement towards modeling software in the energy world. GHI considered water to simply be the next frontier for such analysis. Projects that achieve certain WERS levels will obtain a majority of their water points for various certification levels.

Note about the author: Doug Pushard has been an EPA Outdoor Water Auditor for over a decade. He is well versed in both outdoor water usage and how to reduce onsite water use, and is an original member of the WERS Development Group.

Mike Collignon is the Chair of the WERS Development Group, as well as the Executive Director and Co-Founder of the Green Builder® Coalition. The Coalition has been involved with the development of the WERS Program since February 2014.

This page intentionally left blank.

Lawn Replacement Pilot Program

Final Report & Lessons Learned

October 23, 2015

LIVING WITH DROUGHT

GOODBYE, LAWN

Water Agency's free grass-replacement program kicks off in RP, Cotati

SONOMA COUNTY WATER AGENCY — SONOMA COUNTY YOUTH ECOLOGY CORPS — CONSERVATION CORPS NORTH BAY (A path to a better future.) — daily acts (because every choice matters)

Prepared by
Carl Shuller
Project Manager
Daily Acts

For
Carrie Pollard
Sonoma County Water Agency

Executive Summary

The Sonoma County Water Agency (SCWA) contracted Daily Acts and Conservation Corps North Bay (CCNB) to execute an eight (8) week pilot program to replace lawns in the cities of Rohnert Park and Cotati including the installation of up to five (5) model sites. The scope of the model sites went beyond lawn replacement and included design, planting, and drip irrigation installation. The program, and associated contracts, were approved by the Sonoma County Board of Supervisors on August 11, 2015. CCNB was contracted to provide labor for the project four days per week (Monday through Thursday) from 9am to 2pm. Daily Acts served as Project Manager and was responsible for training the crew, interfacing with property owners, performing site assessments, obtaining permission to perform work, ordering materials, managing landscape design professionals (in the case of the model sites), providing general oversight, quality control, photography, media coordination, and reporting.

High-Level Facts

- 30 (Business) Days
- 30 Residential Lawns Replaced
- 3 Amazing Model Sites
- 2 Educational Field Trips
- 43,050 Total Square Feet Replaced *(an American Football Field is 48,000 Square Feet)*
- 724,313 Gallons/Year of Water Saved *(an Olympic Swimming Pool is 660,430 Gallons)*
- 129 Yards of Compost *(or 1,548 Wheelbarrows)*
- 57 Rolls of Cardboard
- 392 Yards of Mulch *(or 4,704 Wheelbarrows)*
- 387 Native, Drought-Tolerant Plants and Trees
- [illegible] Fruit Trees

On Monday, August 17th, Daily Acts facilitated a full-day training/orientation for the CCNB crew members. This orientation took place at the Cavanagh Center Food Forest and Community Garden in Petaluma. A portion of the day was dedicated to the crew taking the QWEL exam in order to gauge pre-existing knowledge and serve as a baseline for learning. The team managed to remove 30 lawns and create three (3) new model sites in 30 business days (note that 20 days were dedicated to residential lawn removal and the team was able to convert 1.5 residential lawns per day with each lawn averaging 925 sq. feet).

The three (3) model sites that were installed are St. Elizabeth Seton Parish, Sonoma Mountain Village Office Park and Event Center, and Cotati City Hall. Each of these sites are highly visible to SCWA customers and will serve as a wonderful example for years to come.

The educational component of this project was an important one given the nature of the crew and CCNB's involvement. Daily Acts provided the crew members with invaluable insight into water conservation and sustainability in general. The crew participated in two field trips; one to tour Grab n' Grow's composting facilities in Santa Rosa, and one to the Permaculture Skills Center in Sebastopol where they learned more about basic Permaculture principles.

Overall the program was seen as a huge success for all stakeholders. Residential participants were thrilled with getting their lawns replaced for free and model site participants were equally thrilled to participate. CCNB gained a new core competency in sheet mulching and is actively pursuing funding to leverage this new-found skill. Daily Acts was honored to manage this important pilot program for SCWA and learned a lot about working with youth crews and ways to take these kind of water-saving efforts to scale.

Outreach and Participant Signup

- 134 signed up for the program via Google Form which worked great.
- Suggest adding field to capture email address in order to facilitate follow-up, especially to those on wait list.
- Went to "wait list" status within the first week of going live. SCWA staff followed up with those on waitlist when project.
- Consider making it more clear during signup what the specific criteria are for being selected.

Site Visits / Selection Criteria

- 37 sites were visited, 7 rejected due to lack of operational sprinklers, 30 were selected and subsequently completed.
- Both front and backyards were deemed eligible. Given desire for biggest impact consider limiting future lawn replacements to front yards only due to higher visibility.
- There was some concern regarding replacing "brown" lawns and the decision was made to move forward with these lawns if they had functioning irrigation. The number of "green" lawns is very limited in the Rohnert Park/Cotati and likely these residents are not (yet) ready to give up their lawn.
- Setting a boundary (Rohnert Park/Cotati only) was critical and made things manageable from a site scoping perspective.

Conservation Corps North Bay Crew

- The initial crew was comprised of the following: Daniel R. (Supervisor/Lead), Austin D., Isaiah M., Heaven H., David D., Hugo M., Michael S., and Calvin B.
- For the most part the crew remained intact for the duration of the pilot. There were a few days when there were only five (5) crew members which typically wouldn't have been an issue. Unfortunately, one of the days there was a small crew was the first day of the installation of the first model site which contributed to the project taking a bit longer than expected. Conservation Corps North Bay (CCNB) made some mid-course corrections and attendance improved.
- It isn't fair to compare productivity of this kind of crew against a professional, for-profit landscaping organization. When planning future projects it is important to assume a slightly lower level of productivity due to learning curves, substitute personnel, and the inevitable one or two individuals who from time to time are not interested in participating.
- Daily Acts was impressed with the crew's ability to quickly learn the new skills and the initiative that most crew members showed. Within the first two weeks crew members found their niche and hit the ground running. At times they moved a bit too fast and needed to be reminded of why they were doing what they were doing (i.e., trenching along a fence isn't necessary).
- Crew members had ample opportunity to interact with media, homeowners, and the general public. They all were able to articulate what they were doing and why.

- Daily Acts was very impressed with the respect the crew members demonstrated for each other (even on very hot days when even the best of us might be a bit agitated).
- The crew took great pride in their work and never complained when asked to make corrections (i.e., this trench needs to be deeper)
- Daniel (Danny) did a great job managing the crew. He provided regular progress updates and never hesitated to call if he had questions and/or concerns.

Educational Components

- The crew participated in a full day orientation on Monday, August 17th. In addition to learning the basics of sheet mulching and capping irrigation (major focus of program), the crew toured the Cavanagh Center food forest, community garden, water catchment systems (above and below ground) and toured Daily Acts Executive Director, Trathen Heckman's urban homestead.
- The QWEL exam was administered on the first day (see below for before/after results).
- A "Sustainability Assessment" was also completed by each crew member to determine the crew members sense of awareness, and a post-assessment was also completed on the second-to-last day of the pilot. Noticeable changes were seen in areas the crew received training and experience in, awareness around key sustainability issues/concepts increased marginally and confidence in various skills increased as well.
- Hands-on sheet mulching instruction was provided by Daily Acts staff, Kellen Watson and Carl Shuller, at the first residential site. The basics of laundry-to-landscape greywater systems was also demonstrated at one of the residential properties.
- Early on in the project the crew gained considerable experience in residential lawn irrigation repair. The subsequent decline in the number of repairs is proof that the crew learned from their "mistakes."
- The crew toured the composting facilities at Grab n' Grow and gained an appreciation for where the materials they were using on a daily basis came from.
- The field trip to Permaculture Skills Center was a huge success, with several crew members talking about it well after our visit. Crew members were given the opportunity to talk about the work they had been doing in the field to date and how it has personally impacted them. It was great for them to witness two young and enthusiastic leaders (Ryan Johnston and Sam Gerhard) working in a field they are passionate about. The tour opened their eyes to ways Permaculture can be applied with several crew members expressing an interest in taking what they are learning back to their homes and communities.
- Daily Acts staff, Brianna Schaefer and Carl Shuller, provided an introduction to the various plants that were specified for the various model site projects. The proper methods of planting and plant care were covered. Hands-on instruction in drip irrigation was also provided. An important by-product of this instruction was experience reading architectural drawings and learning about scale.

- One model site (St. Elizabeth Seton) provided the opportunity for the crew to lay out swales on contour using "A" frames. The concept of swales were discussed. Note that the client used heavy machinery to dig the 100+ feet of swales!
- Consider making expectations relating to education more tangible in terms of time budgeted. Finding a balance between getting work done and taking time to "educate" was difficult, and consequently, Daily Acts defaulted to getting work done. From a productivity standpoint arranging specific field trips probably worked better than many small training events.

Material Procurement / Vendor Assessment

- Grab n' Grow (A+)
  - Ordering materials at least a week or more in advance helped ensure materials were onsite when needed. There were zero instances of materials not being available. Brett and his team were excellent to work with and very accommodating in the few instances that materials needed to be delivered outside of their service level window.
- Wyatt Irrigation (A+)
  - It is critical to have knowledgeable vendors like Wyatt. Kris Loomis and her team were very helpful and responsive. Consider having an open PO that materials can be billed against moving forward in order to eliminate the need to "manually" pay each invoice.

Scheduling

- Scheduling went very well. Daily Acts left room in schedule to accommodate unknowns like slips in model site schedules. In the future be sure to leave some room to move things around as needed and set expectations with home owners that we reserve right to change dates. Daily Acts had to reschedule five (5) jobs for various reasons and each homeowner was more than amenable (as was vendor, Grab n' Grow).

Residential Sites

- Resources for residents:
  - Consider including signed disclaimer regarding Bermuda grass and likelihood that it will breach the confines of any/all cardboard eventually.
  - Consider creating a one-page document on drip irrigation planning/installation.
  - Consider creating a handful of design templates to help homeowners get a jump start on planting.
  - Consider including notice in waiver that a sign will be installed in yard upon completion. While there were no complaints, it would be good to have in writing so expectations are clear.
  - Include in 'thank you' letter a reminder for home owners to turn off their timers.

- It is very important to be mindful of property lines and anticipate potential issues, especially in cases where the client's lawn blends with a neighbor's lawn. It is important to initiate contact with the neighbor (and better yet ask the client to talk to their neighbor prior to work commencing) and inform them of what you will be doing. There was one case where the client's sprinkler heads where well over a foot inside their neighbor's yard. Digging a clean, straight trench along the property lines is critical in ensuring that mulch doesn't easily migrate into the neighbor's yard.
- Consider creating a dedicated tool box to irrigation repair parts, anticipating any number of possible scenarios. Daily Acts staff made ten (10) separate trips to local hardware stores over the course of the eight (8) week pilot.

Model Sites

- Three model sites were installed during the eight (8) week pilot: Cotati City Hall, Sonoma Mountain Village and St. Elizabeth Seton Church. Considerable time and effort was required to identify, vet, secure and design these sites. Special thanks to Patrick Picard at Equinox Design for his flexibility on such short notice and his ability to turn around two quality designs.
- Contingency planning is critical when working within such a short window of time. Make sure to leave free days in the schedule in case the crew needs to come back to complete tasks.
- Consider budgeting for plumbing/irrigation professionals to execute any/all work related to tapping into a client's existing system as there is a greater risk of something going wrong. Daily Acts did not experience any issues but in the case of the church, considerable time went into installing filters/pressure reducers in a new box.
- Remember to contact USA Dig whenever doing any major digging (better safe than sorry).
- In the future consider getting written agreements with all clients and subcontractors. While there were no issues, this pilot moved very fast (especially the development of the model sites) and it is good practice to get things in writing (in the form of an agreement/contract vs. email threads).
- When the scope of work includes planting and installing drip irrigation, make absolutely sure that there is ample time in the schedule to get water to the plants no later than 24 hours after planting (assuming that the plants get a good soak with the hose directly after planting). Daily Acts ran into delays at the Church that pushed planting to Thursday and due to the hot weather, staff had to visit the site over the weekend to ensure that the plants were getting the water they needed.
- Consider not including planting in scope during such warm months. On related note, Daily Acts has some concern regarding potential for model site stakeholders to not be totally vested.

Quality Control / Feedback

- All sites were inspected upon completion and any deficiencies addressed. It is important to note that there were very few issues with the quality of work performed. Main areas for improvement are consistency associated with "edging". Every site was left well swept and in general, better than it was found.
- Each homeowner was asked to fill out an online survey regarding their experience (see below for results of survey). Almost without exception expectations were exceeded in every aspect of the program.

Project Management

- It is critical to have an experienced project manager on the team who can anticipate potential issues before work starts and instruct the crew. Daily Acts was in the field almost every day to ensure things went smoothly.
- Communication was open and just frequent enough. It is a bit surprising to note that no face-to-face meetings were required beyond the initial kick-off meeting.

Media

- Press Democrat
- LA Times

Photography

- The pilot was extensively photographed (and video was captured as well). Pictures of the crew in action can be found online at https://www.flickr.com/photos/dailyacts/

This page intentionally left blank.

SNWA Rate of Play Study Using GPS: Finding Turf to Convert

Kent Sovocool, C.L.I.A., Master of Arts in Science
Southern Nevada Water Authority (SNWA), P.O. Box 99956, kent.sovocool@snwa.com

Michael Drinkwine, C.L.I.A., Master of Public Administration
Southern Nevada Water Authority (SNWA), P.O. Box 99956, michael.drinkwine@snwa.com

Abstract

In response to historic drought and subsequent regulatory action, golf courses in the Las Vegas area have converted over 38 million square feet of grass to water efficient landscaping in areas generally not impacting the playability of the course. The authors present how a method developed by Mr. Jim Moore of the USGA and adapted by the Southern Nevada Water Authority is helping courses to identify further areas that may be considered for conversion by tracking volunteer golfers on courses with inexpensive GPS recorders. The technique provides for far more sophisticated analyses and helps to evolve conversations from one of simple water-centric landscape conversion considerations to discussions of truly optimizing courses' design from a paradigm of functionalism with consideration of numerous inputs. The presentation will provide additional information about the method and annotated aerial imagery demonstrating its application.

Keywords

golf, GPS, turf removal, landscape conversion, xeriscape, conservation, sustainability, turf grass

Background

In 2003, in response to a drought of historic proportions on the Colorado River, the Las Vegas area's main water supply, the Southern Nevada Water Authority (SNWA) passed its Drought Plan. For this sector, conservation staff and community leaders determined that the day-of-week and time-of-day watering restrictions that worked for most of the community would not be practical for the golf course sector. They, in an occasionally contentious but ultimately collaborative effort with local course owners and superintendents proceeded to develop water budgeting policies that would result in significant water savings while maintaining needed flexibility in scheduling irrigation.

Upon activation of the plan by SNWA's Board and subsequent passage by its member stakeholder jurisdictions, golf courses faced first a budget threshold of 6.5 acre-feet per irrigated acre for Drought Watch then, as the drought worsened to Drought Alert it dropped to 6.3 acre-feet per irrigated acre. Penalties for going over the budget were dependent on the extent to which a course went over the threshold and the penalties were applied to this fraction. They ranged from as little as 2x the highest tier rate paid by the course up to 9x.

The water budgeting policy was designed to impact courses with the highest per unit area irrigation (with public supplies) while rewarding the majority of courses that were already earlier adopters of efficiency techniques. The drought dragged on and eventually in 2009 the community, having become accustomed to the restrictions in the Drought Plan, decided to place almost all of the restrictions into perpetuity, including golf course water budgets.

In addition to providing a regulatory framework designed to encourage courses to save water during drought, the founders of the Plan wanted to make those savings as assured and long-lasting as possible. To this end, the water budget was intentionally based on irrigated acreage as set at the time of drought level imposition (or construction for newer courses), not grass acreage. This permitted golf courses to be able to derive significant tangible benefit from conversion of low and no value grass to water efficient landscaping in generally non-play portions of the course. This is in contrast to approaches that, perhaps inadvertently, end up failing to reward and perhaps even penalize courses that implement what is often the strongest conservation measure by basing the budget on grass acreage.

The superintendents and owners responded resolutely with a number of strategies, turf reduction being one. Since the drought, Southern Nevada golf courses have become the most prolific converters of grass in the region. Collectively they have converted over 38 million square feet since program inception, the bulk of this coming after the Drought declarations and subsequent water budgeting policy implementation. In totality the landscape conversions alone are estimated to save over 2 billion gallons of water annually in this sector. Courses achieved this by finding convertible turf in low use or nonfunctional portions of their properties. In many cases this turf had served only aesthetic purposes and impacts on rate of play were very low to nonexistent. SNWA has been supportive of the golf courses in their turf reduction efforts and has provided over $39 million in landscape conversion rebates to courses over this time.

Eventually golf course conversion rates waned somewhat, with new conversions dropping below 1 million square feet per year in 2011. This was to be expected given economic distress in the community and the fact there is only so much grass available without getting into the common places golfers end up which could potentially impact the experience and rate of play, a crucial consideration for any course. In informal communication both SNWA staff and superintendents expressed their belief there was more turf conversion to be had in low use areas, but that determining where this was represented a challenge.

Method

As turf conversion was starting to taper off, SNWA staff became aware of research by the USGA's Dr. Jim Moore who was thinking about speed of play and helping courses to align with the changing landscape design ethic emphasizing the functionalism of various landscaping choices at courses. The USGA broadly recognized that a sustainable course in the future would be one where the landscape was coupled with the needs of both it and the community.

Dr. Moore developed the idea of using Geographic Position System (GPS) data as a means to solving the question of what parts of a course are used and what is not. He conceived of the idea of providing golfers that volunteered to participate with a small GPS recorder that they would carry in their pocket.

GPS dataloggers intended for fitness activity tracking, with their combination of durability (water resistance and shock), configurable data collection resolution and storage capabilities, +/- 2.5m accuracy, durability, and relatively small size, were found to be ideal. Upon return, the data was downloaded and the GPS records were used in conjunction with Geographic Information System (GIS) software and aerial imagery to plot the path the golfer(s) took on the course. With sufficient sample size he could develop a detailed picture of areas most and least used. The technique is elegant in its simplicity.

Once the paths are plotted and visualized with GIS software, the technique lends itself to further analyses to help reveal potential areas for alternative landscape treatments. With logical exceptions for the driving range, water hazards and such, turf areas where golfers never go are logical initial favorites for consideration. Next, areas rarely seeing golfers can be considered. For example, an area where only 2% of golfers go, might be better as rough than as maintained turfgrass. Now, with the areas of where golfers go revealed and the speed of play recorded, the Superintendent can start to consider the impacts of landscape conversion with an eye towards minimizing the impact of landscape conversion.

Combined with other information such as survey data, more advanced questions can be considered such as to what extent should a hole favor or disfavor golfers of a certain handicap level? Does the placement of greens align with the needs of golfers of a variety of stroke strengths? What holes do golfers unexpectedly slow down on and how might course modification improve this?

It should be noted that Dr. Moore's research in this area is not limited to just study of golfers. He has, with courses' permission, explored the maintenance aspects of areas too. The same technology can be used for tracking time devoted to an area for maintenance (where the GPS unit is carried by staff), time required for mowing (here, putting a unit on a lawn mower) and relative efficiencies of different mow patterns, and quantification of benefits of technology upgrades (for example, upgrading to a new mower). The list is only limited by the imagination and the reception of the tracking units.

Evolving the Discussion

When all this information is put together and combined with other data courses usually have, such as their water bills, per unit labor expenses, fertilizer costs and similar variables, the result is a powerful tool for building scenarios. These then can be analyzed and compared. Ideas can be developed and reimagined as the course management looks critically at the value of the various land areas.

Perhaps most importantly, this level of value scrutiny starts to move the conversation for superintendents and owners from the relatively modest goal of "Where can I remove grass to save water at my course?" to "How can I optimize the design and operation of the course to make the best use of the resources I invest in it?" They start to discover more benefit than the water savings alone suggests and they start to take realize innovative approaches distinct of the original discussions about water conservation.

While on balance the outcome of this work is supporting landscape designs for water conservation, it may be that in some places, the research shows that grass should remain or even should have remained from a course function perspective. That is fine. The point is that now there is a technique that provides superior abilities to consider suitability of the plantings. While most superintendents in the Las Vegas

Valley demonstrably had, and still have, an excellent feel about where to convert the bulk of their courses less functional turfgrass areas, cooperatively with SNWA, the opportunity exists to achieve further landscape conversion without negatively impacting the experience or the playability of the course.

Converting Kentucky Bluegrass to Native Grass in City Parks

Eric Becker, BA, CID, CLIA
Colorado Springs Parks
Water Conservation Specialist
ebecker@springsgov.com

ABSTRACT. *As a result of the severe budget constraints that have impacted the health of the City's parks, Colorado Springs Parks Department developed a strategy for reducing the amount of irrigated Kentucky bluegrass in order to help maintain a healthier and more sustainable parks system. The strategy includes converting low-use, high-maintenance Kentucky bluegrass to native grass to achieve significant water savings and reduce the overall water footprint of city parks and greenways. The results of converting of nearly 60 acres have been dramatic, proving very cost effective while providing more attractive landscapes. Learn which areas to convert, how much water you can save, and step-by-step details of how to make the conversion a success!*

Keywords. Conservation, turfgrass, removal, landscape, water, efficiency, sustainable

Sustainable landscape designs have greatly improved over the last two decades. This trend towards alternative landscape creation is driven in part by interest and need for resource conservation. More sustainable landscape designs can be seen in new construction along the Front Range corridor of Colorado and help address the dual needs to reduce water consumption and maintenance. The next challenge is to expand these benefits by exploring the potential for conversion of existing, conventional landscape treatments into more sustainable landscapes. Conversion of high water usage turfgrass areas to less water consumptive and lower maintenance areas is possible and can provide significant resource savings. While there are many viable alternatives for creating sustainable landscapes, the focus of this paper is to broaden the understanding and use of native grasses as a landscape option and a viable water and landscape management conservation strategy.

Situation

Colorado Springs is a growing city at the confluence of the Great Plains and Rocky Mountains. The city is almost completely dependent upon surface water resulting from snow melt for its water supply. It is the only large city in Colorado not built around a major river and for that reason the majority of its drinking water comes from approximately 200 miles away through a series of pipes, tunnels, reservoirs, and pumping stations (Colorado Springs Utilities, 2012). The city's unique geography and highly variable semi-arid climate combined with the complexity and scale of its water system means that the cost of water is high relative to other large providers on the Front Range of Colorado

As a result of the severe budget constraints that impacted the health of the City's parks, Colorado Springs Parks and Recreation Department (CSPRD) developed a comprehensive water management strategy to help maintain a healthier and more sustainable parks system, short and long term.

One significant part of the strategy consists in converting areas of low-use and/or poor performing Kentucky bluegrass to native grasses. By doing so we expected to reduce irrigation water usage and maintenance costs, and in many cases improve the overall health and appearance of the landscape area. We also anticipated that such conversions would provide a more sustainable parks system.

Identifying Areas for Conversion

From 1970-2000, Kentucky bluegrass was widely used as the landscape grass of choice in Colorado Springs parks and medians. Kentucky bluegrass has long been used as a durable turfgrass. It is still a good option for heavily used, active play areas and sports fields. However, many conventional bluegrass installations are low-use or not intended for active play. In

CSPRD's landscape conversions, we identified three major areas for conversions as follows: low-use conventional turf, restricted access turf, and problem maintenance areas.

Low-Use Conventional Turf

City parks have multiple functions for community citizens. Neighborhood parks provide relief from the built environment surrounding residents. They provide greenscape, recreation centers, sports fields and playgrounds. A greenscape is a landscaped park, often with large un-programmed turf areas, primarily for passive recreation. These large, low-use turf areas have little aesthetic diversity and offer an excellent opportunity for conversion. These areas also offer the greatest opportunity for water and maintenance savings.

Two, 7.5 acre low-use conventional turf sites converted in 2013.

Restricted Access Turf

Throughout Colorado Springs, we maintain a wide range of Kentucky bluegrass areas situated along roads and next to parking lots. These areas are not suitable for active play and mostly offer an aesthetic appeal to ongoing motorists. Many of these areas are a challenge to maintain and have received important consideration for conversion.

Two, restricted access turf sites converted from Kentucky bluegrass to native grasses.

Problem Maintenance Areas

Given our location on edge of the Great Plains, native grasses can offer a great alternative to difficult-to-maintain turfgrass and problem maintenance areas. Steep slopes, south-facing exposures, wet drainages and areas with heavy shade are some example problem maintenance areas that could be converted to achieve water and maintenance savings.

This heavily shaded area in front of the main Park & Rec office was converted to a shade-tolerant fine fescue blend in 2014.

Existing Site Condition Evaluation

Before selecting a conversion area, evaluate the physical characteristics and desired use. This site specific information will determine the design and native grass species. This section is intended to assist with identification of key considerations for a successful conversion.

- **Irrigation** – Renovation of an existing irrigation system will need to occur to support changes in landscape design. Conversion area installations generally require separate irrigation zones from non-converted areas. If only part of an irrigation turf area is to be converted, a dedicated zone could be an essential requirement. Other equipment changes could include retrofitting 4" pop-up heads to 6" pop-ups.
- **Soil**- When converting a landscape, the existing soil characteristics should be considered to assure a good match to the requirement of the native grass species. For example, buffalograss is not well adapted to sandy soils.

Irrigation modifications at Ford Frick Park

- **Topography and exposure** – Slopes tend to shed water quickly and thus tend to be drier. South and west facing slopes can become very dry, particularly during the winter when the sun angle is low. Grass selection is very important for these dry, exposed sites.
- **Existing vegetation** – When converting high water using turfgrass that contains established existing landscape vegetation (especially trees), special care should be given along with the use of a dedicated irrigation zone.

Grass Selection – Seed Mixes

The intended use and expectations should influence the selection of the grass species and desired mix. In order to select the best mix, potential uses should be identified and the previously mentioned site conditions evaluated. Once these have been identified, it will be possible to select an appropriate seed mix. The following two seed mixes are two of the primary conversion mixes we have used:

Midgrass Prairie Seed Mixture			
Common Name	**Grass Season**	**Improved Variety**	**PLS #/Acre**
Buffalograss	warm season	Texoka	4
Blue Grama	warm season	Hachita	4
Sand Dropseed	warm season	Native	2
Western Wheatgrass	cool season	Arriba	7
Sideoats Grama	warm season	Butte	6
Green Needlegrass	cool season	Lodorm	7
Seeding Rate: Total 30 PLS #/acre - Drill seeded at half rate; two different directions, perpendicular to one another.			

A seeded, native midgrass prairie at Keller and Wildflower Parks.

Wheatgrass Blend			
Common Name	**Grass Season**	**Improved Variety**	**% of mix by weight**
Pubescent wheatgrass	cool season	Luna	40
Western wheatgrass	cool season	Arriba	23
Streambank wheatgrass	cool season	Sodar	18
Slender wheatgrass	cool season	Pryor	19
Seeding Rate: Total 300 PLS # acre - Drill seeded at half rate; two different directions, perpendicular to one another.			

9.35 acres converted to wheatgrass at Memorial Park.

Wheatgrass area 8 weeks after seeding.

Conversion Method

We have experimented with a variety of conversion methods throughout the last four growing seasons. Below is the method we've found to be most effective and desirable. This method is more suitable for larger sites, but has been effective on smaller sites where equipment can be used. Conversion in areas with irrigation:

1. Initiate conversion when vegetation is actively growing. Generally speaking, May 1 – August 31st. To assure adequate time for establishment prior to the first hard frost, irrigated conversion areas should be seeded on or before August 31st. The majority of our projects were seeded in late July and August with great success. This period of time offers warms soil temperatures for quick and reliable germination, our most predictable moisture (monsoon) of the summer, with weed pressure tapering off in late August.
2. Irrigate conversion area well. If needed, allow more time and irrigation cycles to ensure conversion area is not drought stressed. You want vegetation that is in good growing condition at the time of herbicide treatment.
3. Thoroughly treat the conversion area with a glyphosate product at a 2-3 oz./1000 sq. ft. rate. Repeat with second application in 10-14 days to provide a complete kill. Any areas

that were missed or any remaining weeds should be treated with an additional application. This process takes about 4 to 5 weeks for a complete kill.

4. Mow conversion area at 2 to 3 inches or as short as possible.
5. Flag all irrigation heads, valve boxes, etc. to avoid damage.
6. Aerate using a hollow, tine core aerator that pulls a 2 to 3 inch plug. Make a minimum of three passes at different angles.
7. Broadcast seed using whatever means possible to assure seed is broadcasted over the conversion area. Native grass seed is quite large and can be difficult to apply using traditional rotary spreaders. Hand broadcasting seed can be an effective means.
8. Drill or slit seed conversion area with appropriate grassland drill or slit seeder. Seed the area twice in perpendicular directions using half the seed in each direction. The depth of the seed should be set at 0.25 to 0.5 inch depth.

9. Drag entire area thoroughly with drag mat.
10. Fertilize using a low fertility (8-2-0) slow release granular organic fertilizer at a rate of 0.5 to .75 #N/1000.
11. (Optional) The application of hydromulch may be necessary in some locations to prevent wind and water erosion. Straw netting has been effective on steep slopes for securing seed.
12. Fertilize using a low fertility slow release granular organic fertilizer at a rate of 0.5 to .75 #N/1000.
13. Irrigation programs for establishment should be carefully managed. It is critical to schedule irrigation based on the actual requirements of the seedbed soils. Irrigation should deliver consistent moisture to improve establishment results.

Water and Maintenance Savings – Midgrass Prairie

Annual maintenance activities for midgrass prairie conversion areas differ greatly from those required for conventional turf. Maintenance is adapted to the local, annual precipitation, requiring far less moisture than conventional turf. For highly visible locations, access to long-term consistent irrigation will ensure a healthy dense native stand. Expected water savings in comparison to maintaining conventional Kentucky bluegrass will range from 65% to 75%. Once established, no additional fertilizer, core aerating or overseeding is required. Mowing can be done a couple times during the growing season if desired. Unmowed midgrass prairie is visually attractive and will offer the greatest savings. Regular control of weeds is an essential annual activity. The following tables provide actual maintenance costs and savings numbers for the first couple of seasons, as well as anticipated savings for the next 10 years.

Briargate Roadway Maintenance Costs (.41 acres)			
2013-2016 Midgrass Prairie Annual Maintenance Costs	**Cost**	**2013-2016 Kentucky Bluegrass Annual Maintenance Costs**	**Cost**
Mowed once a season	$100	32 mowings (May – October)	$640
Herbicide (2 applications)	$300	Herbicide (2 applications)	$300
Fertilizer (not needed)	-	Fertilizer (3x/yr)	$90
Overseeding (not needed)	-	Overseeding (1x/yr)	$200
Aeration (not needed)	-	Aeration (2x/yr)	$100
10 Inches (anticipated usage)	$938	22 Inches of irrigation applied annually (historical avg. usage)	$2,064
Total	$1,338	Total	$3,394

Briargate Roadway 10-Yr Cost Benefit Analysis (.41 acres)	
10-Year Water Savings (CF):	**179,903**
10-Year Water Savings (GAL):	**1,345,673**
10-Year Annual Savings Average:	**$2,295**
Renovation Cost:	**$11,490**
Project Payback (Years):	**5.01**
Yearly Savings Per Acre	**$306.02**
Acre feet saved per year	**0.41**
Cost per acre foot saved	**$2,782**

Briargate Roadway Conversion Site

Keller Park Maintenance Costs (7.5 acres)					
2013-2016 Midgrass Prairie Annual Maintenance Costs	Cost per acre	Overall cost	2013-2016 Kentucky Bluegrass Annual Maintenance Costs	Cost per acre	Overall cost
Mowed 3x season	$100	$2,250	32 mowings (May – October)	$54	$12,960
Herbicide (2 applications)	$190	$2,850	Herbicide (2 applications)	$100	$1,500
Fertilizer (not needed)	-	-	Fertilizer (3x/yr)	$90	$2,025
Overseeding (not needed)	-	-	Overseeding (1x/yr)	$500	$3,750
Aeration (not needed)	-	-	Aeration (2x/yr)	$30	$225
10 Inches (anticipated usage)		$16,363	19 Inches of irrigation applied annually (historical avg. usage)		$31,090
	Total	$21,463		Total	$51,550

Keller Park 10-Yr Cost Benefit Analysis (7.5 acres)	
10-Year Water Savings (CF):	**2,885,850**
10-Year Water Savings (GAL):	**21,586,158**
10-Year Annual Savings Average:	**$33,820**
Renovation Cost:	**$16,052**
Project Payback (Years):	**0.47**
Yearly Savings Per Acre	**$4,509**
Acre feet saved per year	**6.63**
Cost per acre foot saved	**$173**

Keller Park Conversion Site

Water and Maintenance Savings – Wheatgrass

This cool season grass blend can be used to provide a range of cover from dense turfgrass to unmowed, naturalized grassland. After establishment, this blend of wheatgrasses will look strikingly similar to Kentucky bluegrass and will provide a very durable drought-tolerant turf which requires less maintenance. In comparison to maintaining conventional Kentucky bluegrass, expected water savings will range from 40 - 45%. These grasses germinate and establish very quickly. For wheatgrass conversion areas, annual maintenance activities differ from those required for conventional turf. Irrigated turf wheatgrass should be fertilized once a season using 1.0# N/1000 of a slow release nitrogen fertilizer, and core aerated once a year. These grasses tolerate frequent mowing, or they can be left unmowed, and generally perform better if not mowed shorter than 4 inches. The following tables provide actual maintenance costs and savings numbers for the first couple of seasons, as well as anticipated savings for the next 10 years.

Memorial Park Maintenance Costs (9.35 acres)					
2015-2016 Wheatgrass Annual Maintenance Costs	**Cost Per Acre**	**Overall Cost**	**2015-2016 Kentucky Bluegrass Annual Maintenance Costs**	**Cost Per Acre**	**Overall Cost**
Mowed bimonthly 16x season (May – October)	$54	$8,078	32 mowings (May – October)	$54	$16,156
Herbicide (2 applications)	$100	$1,500	Herbicide (2 applications)	$100	$1,500
Fertilizer (1x/yr)	$90	$842	Fertilizer (3x/yr)	$90	$2,525
Overseeding (not needed)	-	-	Overseeding (1x/yr)	$100	$935
Aeration (1x/yr)	$30	$281	Aeration (2x/yr)	$30	$562
14 Inches (anticipated usage)		$28,559	24 Inches of irrigation applied annually (historical avg. usage)		$48,595
	Total	$39,260		Total	$70,273

Memorial Park 10-Yr Cost Benefit Analysis (9.35 acres)	
10-Year Water Savings (CF):	**3,394,050**
10-Year Water Savings (GAL):	**25,387,494**
10-Year Annual Savings Average:	**$37,923**
Renovation Cost:	**$22,833**
Project Payback (Years):	**0.60**
Yearly Savings Per Acre:	**$5,056**
Acre Feet Saved Per Year:	**7.79**
Cost Per Acre Foot Saved:	**$147**

Summary

The City of Colorado Springs Parks Department developed a strategy for reducing the amount of irrigated Kentucky bluegrass to help maintain a healthier and more sustainable parks system. The successful strategy includes converting low-use, high-maintenance Kentucky bluegrass to native grass to achieve significant water savings and reduce the overall water footprint of city parks and greenways. The results of converting of nearly 60 acres have been dramatic, proving very cost effective while providing a more sustainable attractive landscape. This effort will continue as we look for additional opportunities for conversions and demonstrate thoughtful stewardship of our resources.

Acknowledgements

I want to acknowledge a couple of former colleagues at Colorado Springs Utilities that provided input for this paper but are not listed as authors; Julia Gallucci, Scott Winter and Catherine Moravec.

REFERENCES

Borland, D.F. and Keammerer, D.B. 2011. Sustainable Landscape Conversion Design and Irrigation

FROM AUDITOR TO CONSULTANT:

COMMUNICATING OUTDOOR WATER USE.

John Gebhart & Rick Alvarado – Conservation – Denver Water

DENVER WATER

Summer Watering Rules

- No watering between 10am – 6pm.
- Water no more than 3 days per week.
- Do not allow water to pool in gutters, streets or alleys.
- Do not waste water by letting it spray on concrete or asphalt.
- Repair leaking sprinkler systems within 10 days.
- Do not irrigate while it is raining or during high winds.
- Use a hose nozzle with a shut-off valve when washing your car.
- **21-day watering exemption for new seed and sod.**

Use Only What You Need

11/1/2016

DENVER WATER

Water Budgets

- Efficiency benchmark
  18 GPSF

- Target audience
  - Large irrigators (> 1 acre)
  - Schools
  - Government / Public Space
  - HOA's
  - Commercial

DENVER WATER

Thank you for participating in the Water Budget Program. The following is your customized water usage profile. This budget identifies the major water uses categories on your property, so that you can monitor water usage and maximize water use efficiency. All irrigation usage is based on the irrigated square footage of your property and weather data specifically collected for the Denver Water service area.

Customer Name:	Peoria Plaza
Address:	5476 PEORIA ST #2
Est. Irrigated Sq.Ft.	56,340

To-Date Gallons Per Sq.Ft.	20.48	GPSF
To Date Allowable Budget	3,199	kgals
To Date Consumption	3,080	kgals
To Date % of Budget	96%	

Water Use Categories	kgals	%
Indoor	2,136	67%
Irrigation	1,063	33%
Total	3,199	100%

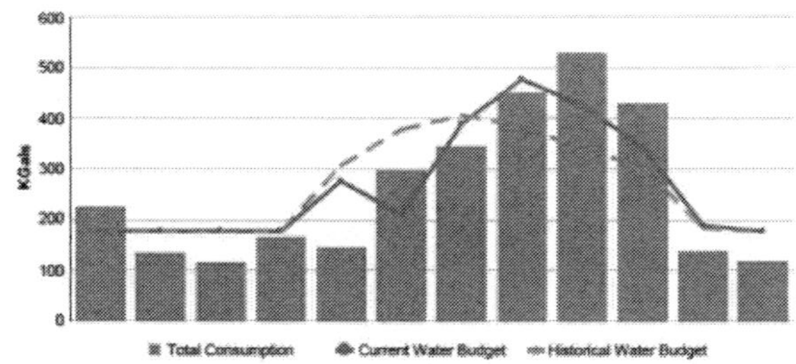

	1	2	3	4	5	6	7	8	9	10	11	12
Indoor Budget	178	178	178	178	178	178	178	178	178	178	178	178
Total Outdoor Budget	0	0	0	0	98	32	210	299	247	166	11	0
Current Water Budget	178	178	178	178	276	210	388	477	425	344	189	178
Total Consumption	225	134	114	164	145	296	343	451	527	427	136	118
% of Monthly Budget	126%	75%	64%	92%	53%	141%	88%	95%	124%	124%	72%	66%

The data in the report above reflects the most up to date data in our billing system at the time that the report was created (see bottom right). If you feel that your report is inaccurate or you have general questions regarding this program contact us at waterbudget@denverwater.org.

Thank you,

Denver Water Conservation Team

Dec 28, 2015

Water Budgets

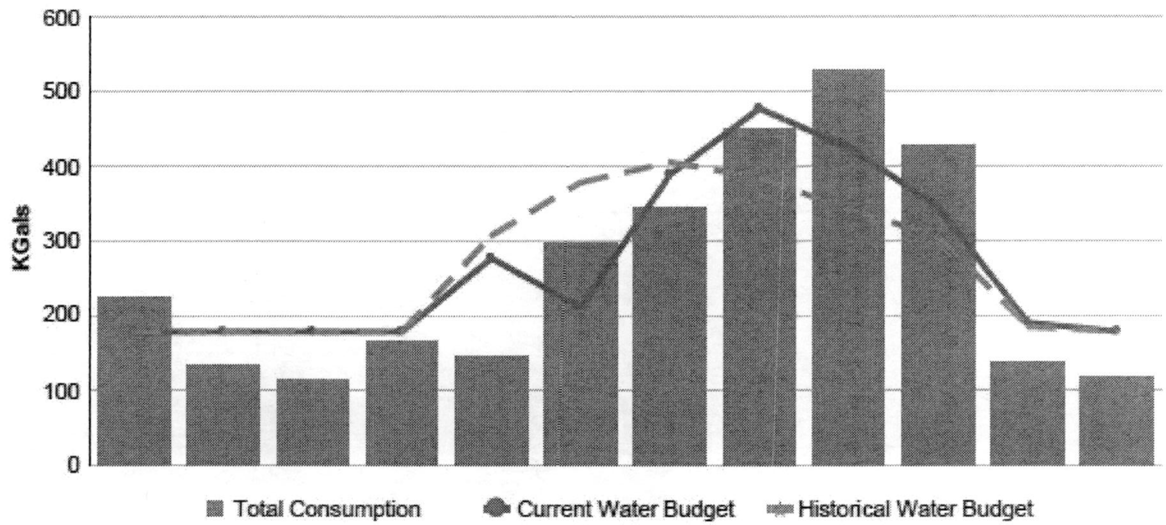

	1	2	3	4	5	6	7	8	9	10	11	12
Indoor Budget	178	178	178	178	178	178	178	178	178	178	178	178
Total Outdoor Budget	0	0	0	0	98	32	210	299	247	166	11	0
Current Water Budget	178	178	178	178	276	210	388	477	425	344	189	178
Total Consumption	225	134	114	164	145	296	343	451	527	427	136	118
% of Monthly Budget	126%	75%	64%	92%	53%	141%	88%	95%	124%	124%	72%	66%

11/1/2016 DENVER WATER

Single Family Residential (SFR) Letters

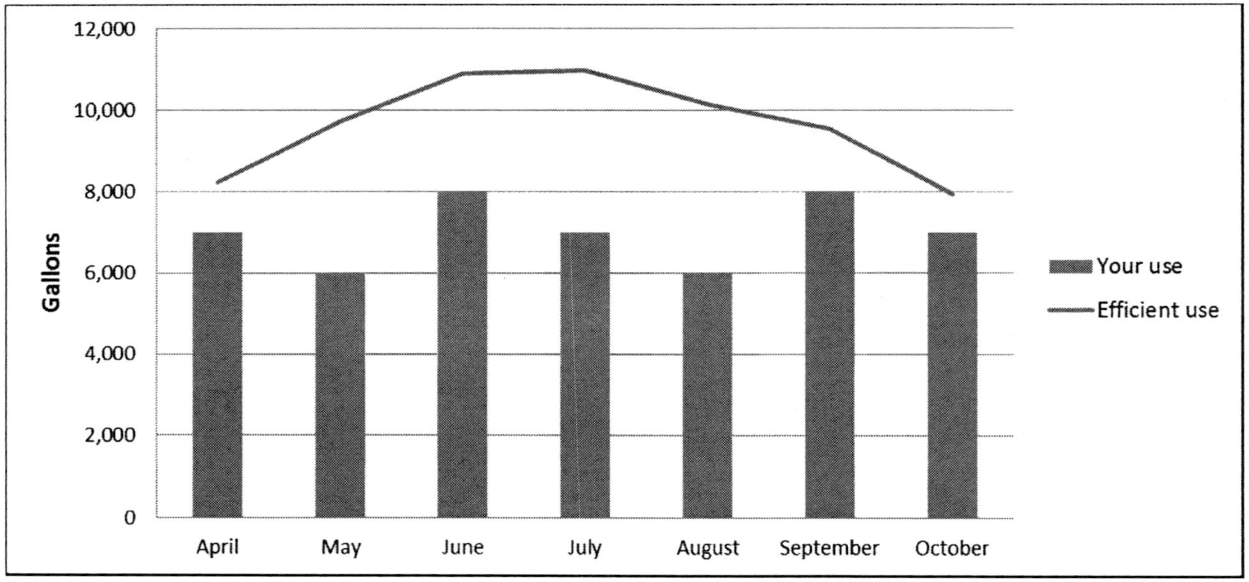

"**Your Use**" = billed consumption

"**Efficient Use**" is made up of two parts:

1. Indoor = Winter Use (capped at 7 kgals)
2. Outdoor = Pervious area *12 Gallons Per Square Foot

SFR Letters

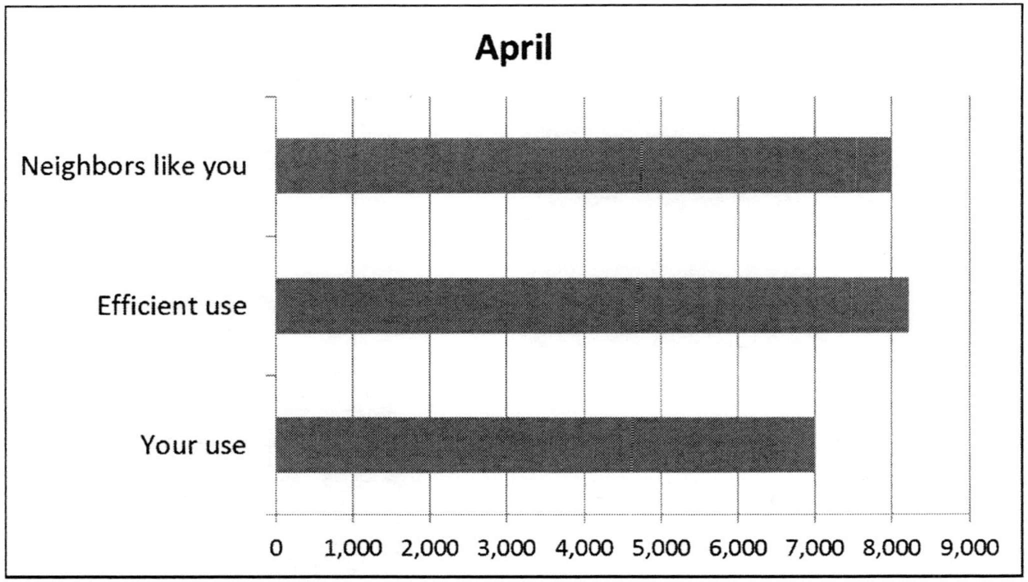

"**Your Use**" = billed consumption

"**Efficient Use**" = Same as first graph (red line)

"**Neighbors like you**" = Median consumption of group

Water Use Education

- Summer Water Savers
- Face-to-face interaction
- Promote watering rules
- Educate consumption

Consumption and Meter Read History

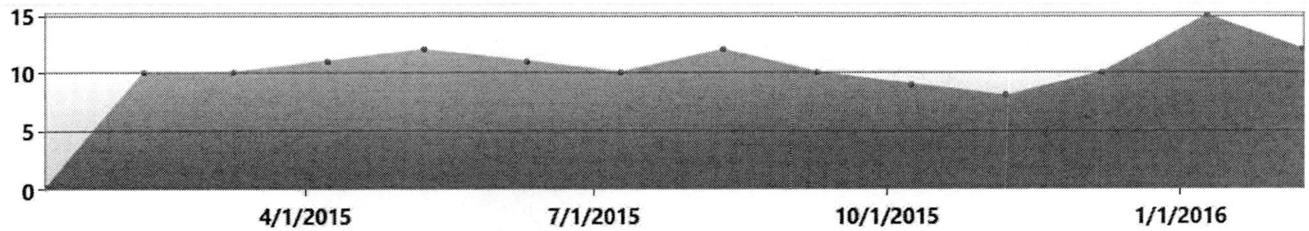

Badge #	Read Date & Time	Register Read	Read Type	UOM	Reading Diff.	Full Read
224663	2/8/2016 9:58:00 AM	641	Regular	SGAL	12	Full Read-641.5
224663	1/9/2016 6:38:18 AM	629	Regular	SGAL	15	Full Read-629.3
224663	12/7/2015 10:55:56 AM	614	Regular	SGAL	10	Full Read-614.55
224663	11/7/2015 8:08:32 AM	604	Regular	SGAL	8	Full Read-604.5
224663	10/8/2015 6:52:10 AM	596	Regular	SGAL	9	Full Read-596
224663	9/9/2015 7:37:04 AM	587	Regular	SGAL	10	Full Read-587.75
224663	8/10/2015 10:17:42 AM	577	Regular	SGAL	12	Full Read-577.1
224663	7/9/2015 10:04:10 AM	565	Regular	SGAL	10	Full Read-565.5
224663	6/9/2015 10:23:20 AM	555	Regular	SGAL	11	Full Read-555.75
224663	5/8/2015 8:45:30 AM	544	Regular	SGAL	12	Full Read-544.2
224663	4/8/2015 8:51:44 AM	532	Regular	SGAL	11	Full Read-532.1
224663	3/9/2015 6:47:36 AM	521	Regular	SGAL	10	Full Read-521.35
224663	2/9/2015 11:21:22 AM	511	Regular	SGAL	10	Full Read-511.4
224663	1/9/2015 10:20:10 AM	501	Regular	SGAL	0	Full Read-501.35

OK

11/1/2016 DENVER WATER

Water Efficiency Consultation

- Evaluating water-use efficiency and addressing customer's individual needs.

DENVER WATER

This page intentionally left blank.

A CASE FOR TARGETED WATER CONSERVATION OUTREACH

[1]David W. Smith, Extension Program Specialist
Texas A&M AgriLife Extension Service, Biological & Agricultural Engineering, College Station, Texas
[2]Charles Swanson, Extension Program Specialist
Texas A&M AgriLife Extension Service, Biological & Agricultural Engineering, College Station, Texas
[3]Jennifer Nations, Water Resource Coordinator
City of College Station, Texas

Abstract. Many cities and water utilities employ public awareness and education efforts for commercial and residential customers to promote water conservation and irrigation efficiency. These campaigns are usually intended for a wide audience and present brief, general messages to build awareness. This approach however may have limited impact on reducing water waste from landscape irrigation and at a high cost. An alternative strategy is to target education and provide assistance in a manner that achieves maximum water savings more economically by examining seasonal water use trends among different segments of the community. Since 2010 College Station, Texas Water Utilities performed more than 500 residential landscape irrigation 'checkups'. Using data collected from the 'checkups' and monthly water records, an analysis showed that small properties (<5,000 sqft) consistently applied more irrigation than large properties (>10,000 sqft) when normalized by landscape area. The difference was significant in four out of eight years examined. Newer properties (<10 years) consistently showed higher irrigation use compared to all other age groups. In fact, average irrigation applied from newer properties was significantly greater ($p<0.05$) than properties older than 30 years in four of eight years. These results can be useful for generating maximum impact when constrained by limited resources.

Keywords. Landscape, residential, conservation, education, irrigation efficiency, irrigation scheduling

INTRODUCTION

Landscape irrigation for residential and commercial properties, golf courses, athletic fields, and other types of recreation areas is estimated to be the third-largest user of water in Texas, behind only agriculture and municipal uses (Cabrera et al., 2013). With a growing population and competing interests for limited water supply, local communities employ various strategies to reduce potable water consumption including tiered water rates, prescriptive irrigation days, citations for water waste, and education campaigns. The greatest potential for municipal water conservation is in landscape irrigation. A study of monthly water use of 800 residences from 2000 through 2002 in College Station, Texas indicated that the average peak water consumption during the summer increased as much as 3.3 times the winter water use (White et al., 2004) as plant water requirements typically exceed precipitation. With in-ground, automated irrigation systems much of this water is wasted due to poorly-designed systems, improper scheduling (over-irrigation), and failure of system integrity. Furthermore, residential water customers find programming and operation of their controller difficult and confusing. This presents a real challenge for municipal and water utility driven water conservation efforts that encourage strategies such as potential evapotranspiration-based (ETo) scheduling, multi-cycling irrigation events to prevent runoff, and prescriptive, address-based weekly operation schedules. These strategies assume that customers are proficient in programming their controllers. In fact, failure to properly implement these recommendations can be counter-productive and actually increase overall irrigation use.

In 2010, College Station Water Utilities began providing free landscape irrigation 'checkups' to residential customers. To date, the City has performed more than 500 irrigation checkups, primarily for customers identified as having above average seasonal water use (Coleman, 2014). The checkup includes a general inspection of system components to identify damaged or broken hardware, documentation and evaluation of existing controller programming, and education on how to reduce runoff and install rain shut off sensors. Following the checkup a written report is delivered to the customer detailing significant findings along with recommendations for reducing irrigation use. Beginning in May 2012, College Station Water Utilities, in collaboration with the Texas A&M Department of Recreation, Parks, and Tourism provided additional resources to perform irrigation system checkups to meet increasing demand for this service. A licensed irrigator was hired to conduct irrigation inspections during the peak irrigation season.

The objective of this study was to identify any discernable difference or trend in irrigation applied (normalized by landscape area) among residential properties due to the size of the landscape and age of property. If such differences are found, this data can be used to strategically plan outdoor water conservation efforts and direct limited funding and personnel to that segment of the customer base where water savings is more promising.

METHODOLOGY

Landscape Irrigation Checkup

College Station Water Utilities publicizes the free irrigation checkup service by distributing a letter to approximately 5,000 residential customers whose historical water consumption substantially exceeded their estimated water budget. The City also hosts a series of summer 'sprinkler system workshops' in the summer season for approximately 200 residents per year. Interested residents contact City staff either by email or telephone to schedule an appointment with the irrigation inspector. In 2012, 2013, 2014, and 2015 irrigation checkups were conducted for 211 residential customers (205 unique customers). Irrigation checkups were performed in 44 subdivisions within the city of College Station. Fifty-eight percent of all inspections were conducted in four subdivisions: Pebble Creek (58), Castlegate (34), Emerald Forest (20), and Edelweiss Estates (11).

Data collected during the checkup included the number and type of application devices, brand and model of controllers, irrigation start times, run times, presence of rain shut-off sensors, and inventory of hardware deficiencies and operational problems.

Data Collected

- Controller brand
- Current controller time/date
- Irrigation start times
- Irrigation programs being utilized ('A', 'B', 'C', etc.)
- Individual station run times
- Seasonal adjust or water budget setting

- Presence of controller backup battery
- Presence and functionality of rain shut-off sensor
- Type(s) of sprinkler heads (per station)
- Dominant plant type (turfgrass, shrubs, flowers, etc. per station)
- Description of area being irrigated per station
- Extent of sun exposure per station (full sun, part sun, shade)
- Integrity of system devices (backflow prevention device, solenoid valves, sprinkle heads)

Though not included in this report, data was compiled and analyzed to determine commonalities in system equipment and design, general tendencies in controller programming, and occurrence of hardware and system performance problems. This information too can be instrumental in prioritizing specific topics for future water conservation education, outreach, and training for residential customers.

Irrigation Use Analysis

All residences in this study were served by a single water meter that registered combined indoor and outdoor water consumption. College Station Water Utilities provided monthly water consumption data for the seven year period from 2008 through 2015. Irrigation use was calculated by subtracting average indoor water use from total metered water consumption on a monthly basis. For the purpose of this study irrigation use was investigated and compared for the typical growing season in College Station – April through September. Indoor water use was estimated to be the average monthly consumption for December, January, and February over the period from 2008 to 2015 or over the period of reliable record during this seven year period.

Landscape water use (irrigation) in 'gallons' was normalized for landscape size and converted to inches of water applied using the following equation.

Irrigation (inches) = irrigation (gallons) / [area of landscape (sqft) x 0.6234]

Estimate of residential landscape area was calculated as the total property size (in square feet) minus the residential footprint, space occupied by garages, out-buildings, patios, sidewalks, and driveways. Total property, garage, and patio area was acquired from the Brazos County Appraisal District, http://www.brazoscad.org/. Further deductions for sidewalks, driveways, and other non-pervious area were estimated using Google Earth satellite maps and area/distance calculator tools.

Area of landscape (sqft) = total property area (sqft) – non-pervious area (sqft) (including house, garage, patio, sidewalk, and driveway footprint)

Net Plant Water Requirement (Net-PWR) Estimation

Net plant water requirement was computed using a daily water balance approach utilizing measured evapotranspiration (ETo) and precipitation data, crop coefficients for warm season turfgrass, and soil water storage constraints assuming a 6-inch root zone depth and clay soil type. ETo data was acquired from two automated weather station locations – the Texas A&M University Golf Course and Texas A&M Turf Lab. Net plant water requirement (Net PWR) was calculated using the following relationship:

Net PWR (inches) = (ETo (inches) x Kc x Af) – Reff (inches)

Where:

Kc = monthly crop coefficient (dimensionless)
Af = allowable stress factor (dimensionless)
Reff = effective rainfall (inches)

Long term average monthly crop coefficients for College Station are referenced in the Texas Landscape Irrigation Auditing and Management Short Course Manual – Version 3 (Fipps et. al., 2009). For this analysis, a stress adjustment factor of 1 (no stress) was used.

Methodology for estimating effective rainfall followed that used for similar analyses performed by the Texas A&M School of Irrigation (Swanson, 2015).

IF $R < 0.1$, THEN Reff = '0'
IF $0.1 < R \leq 1$, THEN Reff = 'R'
IF $1 < R \leq 2$, THEN Reff = 'R x 0.67'
IF $R > 2$, THEN Reff = '2'

Where:

R = actual daily rainfall (inches)

Daily Net-PWR was further constrained by assuming that plant-available water could be stored within a 6-inch root zone and a clay soil. Total Net-PWR for irrigation season each year was calculated by summing daily Net-PWR from April through September.

Landscape Irrigation Ratio (LIR)

The LIR is one approach to quantifying landscape water use (or irrigation) efficiency (Glenn et. al., 2015). The LIR metric provides a means to evaluate and compare landscape water conservation potential for properties regardless of property size. It is calculated by dividing the volume or normalized equivalent depth of outdoor water use divided by the landscape water requirement over a certain time interval.

LIR = irrigation (inches) / Net-PWR (inches)

This study examined the LIR over the typical landscape irrigation season (April through September).

Glenn et al. (2015) used the LIR approach to assess landscape water use efficiency in single-family residences in Logan, Utah in 2004 and 2005. Category benchmarks, defined by LIR ranges, were specified as 'justifiable' and 'unjustifiable' water use and further classified as 'efficient', 'acceptable', 'inefficient', and 'excessive'. This classification system was used in this study to compare water use efficiency for residences over the typical irrigation season from 2008 to 2015.

Justifiable water use
- Efficient $LIR \leq 1$
- Acceptable $1 < LIR \leq 2$

Unjustifiable water use

Inefficient $2 < LIR \leq 3$

Excessive $3 < LIR$

RESULTS

Characterization of Irrigation System Hardware and Controller Programming

Controllers and rain shut-off sensors – Of the 211 residential customers Toro®, Hunter®, and RainBird® model controllers were used by 78% (165) of residents. These controllers are similar in basic operation and feature multiple program options ('A', 'B', 'C', etc. programs) and multiple start times per program. Almost all controllers provided for a 9-volt backup battery intended to retain program settings in case of power loss. If a functional backup battery were not present, these controllers reverted to a default irrigation schedule of watering every day, 10 minutes per station, at 5:00 AM start time once power was restored after an outage. Only 18% (38) of irrigation systems inspected were equipped with a rain shut-off sensor. Of those, 63% (24) were wireless and 37% (14) were hard-wired.

Irrigation Stations – A total of 1,204 stations were inspected. The average number of stations per residence is 5.8. Seventy-three percent (154) of residents had 6 or fewer stations. Of these, pop-up fixed spray heads and rotor-type sprinkler heads were the most common representing 52% and 37% of all sprinkler head types. Other sprinkler devices noted include 'mixed' (a combination of multiple sprinkler head types), drip irrigation, and multi-stream application devices designed for slow-application rate.

Irrigation Schedules – A critical part of the irrigation checkup was to educate the resident on the capability and use of their irrigation controller in facilitating efficient irrigation practices such as adjusting individual station runtimes, utilizing multiple programs to compensate for different irrigation frequency needs, and setting multiple start times (multi-cycling) to prevent water runoff. Existing controller settings were documented and immediately brought to the attention of the resident. In most cases, residents were not familiar with their current controller settings and did not realize the implications for inefficient water use.

<u>Station run times:</u> An analysis of all residents suggests that, in general, stations with relatively high application rates were set with lower run times. For example, the average run time for pop-up spray sprinkler heads was 12 minutes while the average run time for rotor sprinkler heads was 17 minutes. Drip irrigation (characterized by relatively low water application rate) was usually set to run much longer. At a minimum, this illustrates that an attempt was being made to adjust individual station run times for different sprinkler types.

<u>Irrigation days:</u> Irrigation days were fairly well dispersed throughout the week with Mondays, Wednesdays, and Fridays being the most common. The calendar day option was obviously the most common selection for setting irrigation day, with less than 4% (23) of residents using the 'odd/ even day', or 'interval day' feature.

<u>Program start times:</u> Eighty-eight percent (347) of all program start times documented occurred between midnight and 8:00 AM. The most common start time was 5:00 AM. This was not surprising given that the default start time for most major controller brands was also 5:00 AM. Though irrigating in the early morning is strongly encouraged and essential for reducing

evaporative losses, there may be a need in some locations or subdivisions to minimize peak morning demand to limit pressure drop.

Multiple programs: When properly used, multiple controller programs ('A', 'B', 'C', etc.) and multi-cycling on irrigation days can minimize over-irrigation and help prevent water runoff. A survey of residents showed that only 16% (34) used more than one program, and 32% (67) practiced multi-cycling.

Net-PWR (2008 – 2015)

Net plant water requirement was calculated using a daily water balance approach using measured evapotranspiration and precipitation, and constrained by an assumed root depth and soil type using the methodology previous defined. This approach was selected to limit the water contribution from heavy or intense rainfall events. During intense rainfall water is more likely to run off the landscape and/or water moves beyond the typical plant root zone thereby becoming unavailable to the plant. Figure 1 shows the normal pattern and overall trend of Net-PWR over the eight-year period. Residential water customers typically begin irrigating in April and continue through September or longer depending on weather trends and early fall tropical storm development in the Gulf of Mexico. Peak Net-PWR usually ranges between 4.0 and 4.5 inches per month in June, July, or August. Overall, Net-PWR trended upward over the study period, most likely a result of the extreme drought conditions in 2011 and 2013.

Figure 1 – Monthly Net-PWR from 2008 to 2015

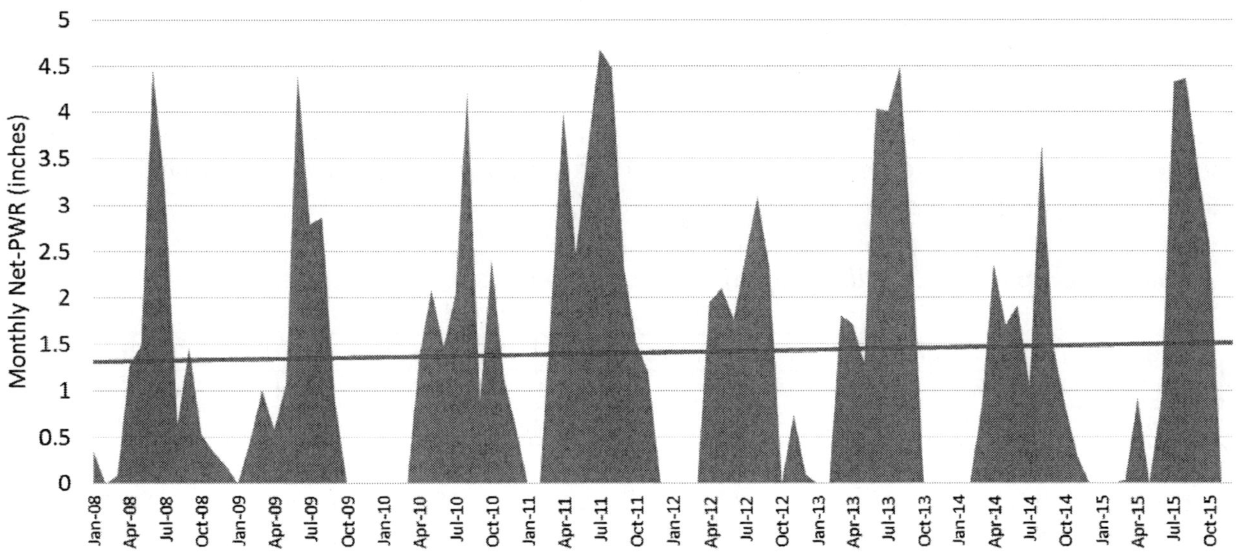

Irrigation Comparison of Landscape Size and Age of Property

Landscape size for each property was calculated and placed into categories (<5,000 sqft, 5,001 to 10,000 sqft, and >10,001 sqft). The mean and range of average irrigation applied from April through September in each year (normalized on a sqft basis) are listed in Table 1 according to size category. Interestingly, for all years irrigation use (expressed in inches applied) was greatest for small landscapes (less than 5,000 square feet). There was also a significant difference ($p<0.05$) between the smallest and largest landscapes for half of the years studied (2008, 2009, 2010, and 2012). General observations were that the smaller landscapes (and properties) were also those located in relatively new subdivisions where plants were less mature and subject to more frequent watering to establish plants.

Table 1. Comparison of irrigation applied (in inches) by landscape size.

Year	<5,0000 sqft			5,001 – 10,000 sqft			> 10,001 sqft		
	[1]M	[2]R	[3]N	M	R	N	M	R	N
2008	**26.4**	3.3 – 122.5	35	**22.6**	3.0 – 50.9	93	**16.4**	2.6 – 29.2	39
2009	**23.9**	1.1 – 80.4	36	**21.9**	4.0 – 55.0	97	**16.2**	1.9 – 30.1	38
2010	**22.3**	1.8 – 98.2	37	**18.4**	1.5 – 35.1	97	**14.3**	2.1 – 33.3	39
2011	**31.0**	2.1 – 187.1	40	**28.0**	4.2 – 75.6	100	**23.0**	8.3 – 41.0	39
2012	**20.6**	1.6 – 55.5	42	**19.3**	3.8 – 54.3	102	**16.0**	0.9 – 25.9	39
2013	**21.2**	3.3 – 73.5	41	**20.0**	3.0 – 46.7	99	**16.5**	1.9 – 38.6	39
2014	**16.1**	0.9 – 40.2	24	**15.9**	0.2 – 32.9	71	**12.3**	0.2 – 25.0	24
2015	**15.4**	0.4 – 36.8	48	**16.1**	0 – 37.4	103	**15.6**	0 – 45.8	41

[1]Average irrigation applied (in inches) from April through September.
[2]Lowest to highest irrigation applied (in inches).
[3]Number of residents.

A similar comparison is shown in Table 2 with delineation by age of property. Younger properties (10 years and younger) showed consistently higher irrigation use (per square foot landscape area) when compared to all other age groups. In fact, average irrigation applied was significantly greater ($p<0.05$) than properties older than 30 years in 2008, 2011, 2013, (relatively dry years), as well as 2014 and 2015. This may be due to the higher irrigation frequency required (and thus a greater chance for water loss) during the establishment phase of new landscape plants. This difference may also be a function of more shade (provided by mature trees and shrubs), and relatively deeper root systems in established turfgrass that characterize older properties.

Table 2. Comparison of irrigation applied on properties by age delineation.

Year	1 – 10 years		11 – 20 years		21 – 30 years		>30 years	
	[1]Mean (in.)	[2]N	Mean (in.)	N	Mean (in.)	N	Mean (in.)	N
2008	**29.7**	14	**22.2**	93	**22.2**	36	**16.7**	23
2009	**26.0**	19	**20.9**	93	**21.0**	36	**18.8**	33
2010	**20.2**	21	**18.5**	93	**18.5**	36	**15.5**	23
2011	**28.8**	26	**29.1**	93	**26.6**	36	**22.5**	23
2012	**19.5**	29	**19.5**	95	**18.9**	36	**15.6**	23
2013	**21.5**	30	**19.6**	91	**20.2**	36	**15.5**	22
2014	**18.5**	16	**14.5**	67	**17.1**	21	**12.2**	15
2015	**18.0**	39	**17.4**	95	**15.0**	36	**13.0**	23

[1]Average irrigation applied (in inches).
[2]Number of residents counted in this category.

Landscape Irrigation Ratio (LIR)

Water use efficiency describes how closely irrigation applied matches plant water requirement. The LIR metric (used by Glenn et. al, 2015) (defined as the ratio of landscape water use divided by landscape water requirement) is one measure of water use efficiency. Although the choice of LIR classification is somewhat subjective, this methodology does provide a means to gauge the effectiveness of water conservation outreach, education, and awareness efforts among a large population.

LIR was computed for all properties, landscape size classifications, and age categories. Figure 2 illustrates that water use efficiency for these properties is typically a function of landscape size, with the smaller properties having lower water use efficiency compared to larger properties. Figure 2 also shows and overall increasing trend in water use efficiency (lower LIR) over the 8-year study period, with water use efficiency increasing more sharply for landscapes less than 5,000 square feet.

Figure 2. LIR comparison and trend for various landscape sizes.

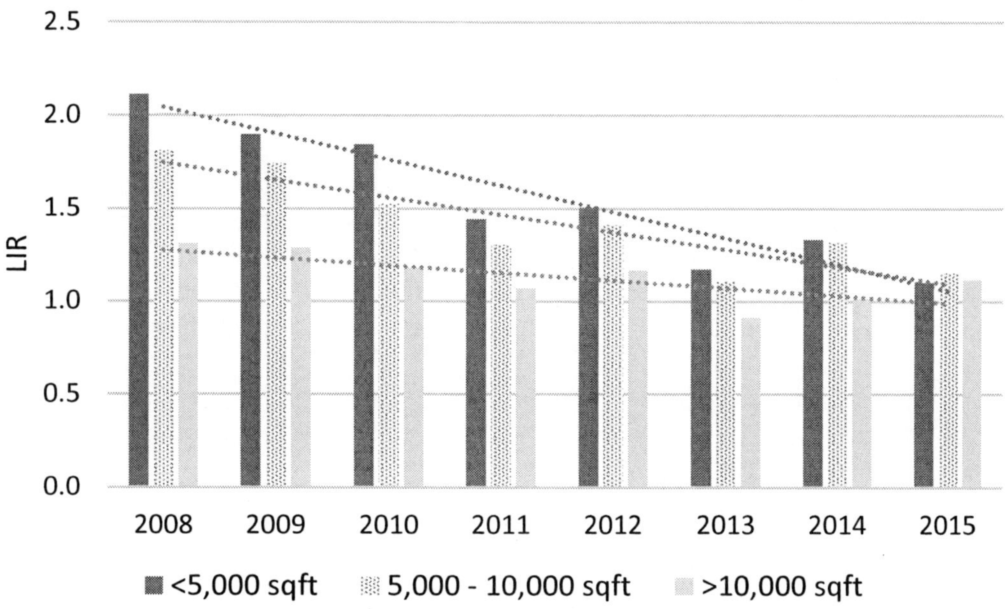

Figure 3 shows a similar trend with increasing water use efficiency over time for all properties independent of age. However, properties less than or equal to 10 years of age are consistently less water efficient compared to older properties, particularly compared to the oldest age group (30 years and older). Water use efficiency appears to be increasing most dramatically (decreasing LIR) over the 8-year study period for those properties of the youngest age group.

Figure 3. LIR comparison and trend with age of property.

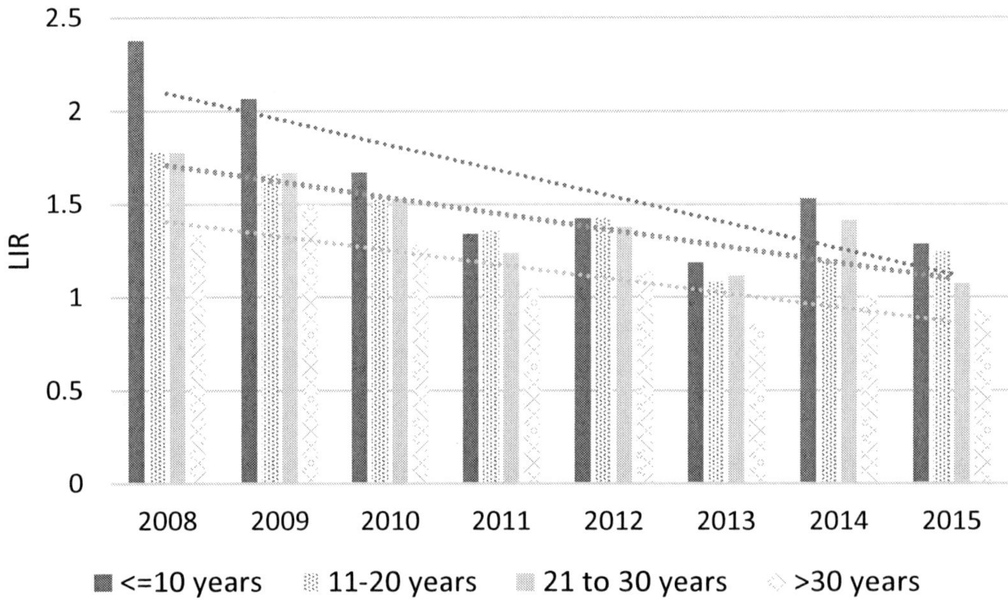

LIR for all residents included in this study were calculated and categorized using the classification system defined by Glenn et al. (2015). Overall, the percentage of residents classified as either 'efficient' or 'acceptable' increased from 70 percent to 91 percent from 2008 to 2015, with the highest percentages in these two categories occurring in 2011 (93 percent) and 2013 (95 percent), both extremely dry growing seasons. Furthermore, the number of properties classified as either 'inefficient' or 'excessive' dropped dramatically over the study period with less than 10 percent of residents falling into these categories.

Table 3. Distribution of residents by LIR category.

	Percentage of residents by LIR category							
[1]LIR Category	2008	2009	2010	2011	2012	2013	2014	2015
Justifiable water use								
Efficient LIR ≤ 1	19	19	26	31	26	51	35	41
Acceptable 1 < LIR ≤ 2	51	50	51	62	63	44	50	50
Unjustifiable water use								
Inefficient 2 < LIR ≤ 3	22	25	21	6	9	4	14	8
Excessive 3 < LIR	8	5	2	1	2	1	1	1
Total (%)	100	100	100	100	100	100	100	100
[2]N	167	170	173	178	183	179	119	193

[1]LIR is defined as the ratio of landscape water used divided by landscape water required (Net-PWR). Category designations defined by Glenn et al. (2015).
[2]Number of residents.

DISCUSSION

In 2012, 2013, 2014, and 2015 211 irrigation inspections for 205 unique customers were conducted as part of the College Station Water Utility's free residential irrigation checkup program. The objective of this study was to identify any discernable difference or trend in irrigation applied (normalized by landscape area) among residential properties due to the size of the landscape and age of property. Landscape size appears to play a significant role in the amount of irrigation applied. Average irrigation applied during the growing season was significantly higher for the smallest landscapes compared to the largest landscapes in 2008, 2009, 2010, and 2012 (relatively wet years). Younger properties (10 years and younger) showed consistently higher irrigation use when compared to all other age groups. In fact, average irrigation applied was significantly greater ($p<0.05$) than properties older than 30 years in 2008, 2011, 2013, (relatively dry years), as well as 2014 and 2015. Water use efficiency, as measured by the Landscape Irrigation Ratio metric, showed an overall increase over the study period. Trends also show a decrease in the portion of residents classified as 'inefficient' or 'excessive' suggesting that the irrigation checkup service may have long term impact in reducing over-irrigation.

CONCLUSIONS

Results of this study reveal that although current education and the landscape irrigation checkup service appear to be effective in increasing water use efficiency, future efforts should focus on younger and smaller properties to achieve greatest savings on a per unit area basis and to maximize limited financial and personnel resources. Younger properties are often characterized by new plants with limited root zones and usable water storage capacity, as well as limited shade which increases overall evaporative losses. Use of automated irrigation systems designed and managed for established landscapes on establishing landscapes can lead to significant water waste (water that is not utilized by new plants). Irrigation applied to younger landscapes is necessarily increased to accommodate these special needs and consequently increases the potential for water loss due to evaporation, runoff, and wind drift. The increasing presence of smaller properties being built in new subdivisions in College Station, Texas helps to explain why the smaller properties in this study also had the highest water use when normalized on a square foot basis. This study supports the need to target newer properties for outdoor water conservation, and to educate homeowners on methods to minimize water loss while their landscapes are becoming established and increasing soil water storage capacity over time.

ACKNOWLEDGEMENTS

- College Station Water Utilities
- Texas A&M Department of Recreation, Park and Tourism Sciences
- Irrigation Technology Program, Texas ET Network

REFERENCES

Brazos County Appraisal District (2015). http://www.brazoscad.org/

Cabrera R.I., Wagner K.L., Wherley B. (2013): An evaluation of urban landscape water use in Texas. *Texas Water Journal*, 4, pp. 14-27.

Coleman, D. (2014). City of College Station Water Conservation Plan. May 2014. http://www.cstx.gov/Modules/ShowDocument.aspx?documentid=19343

Fipps, G., Welsh, D., Smith, D., and Swanson, C. (2009). Texas landscape irrigation auditing and management short course manual – version 3. Texas A&M AgriLife Extension Service.

Glenn, D.T., Endter-Wada, J., Kjelgren, R., Neale, C.M.U. (2015): Tools for evaluating and monitoring effectiveness of urban landscape water conservation interventions and programs, *Landscape and Urban Planning*, 139, pp. 82-93.

Texas ET Network (2015). http://texaset.tamu.edu/

White, R., Havlak, R., Nations, J., Pannkuk, T., Thomas, J., Chalmers, D., and Dewey, D. 2004. "How much water is enough? Using pet to develop water budgets for residential landscapes." Proc., Texas Water 2004, Texas Section American Water Works Association, Arlington, Tex., *Texas AWWA Paper No. TR-271*.

Irrigation Association (IA)
8280 Willow Oaks Corporate Drive
Suite 400
Fairfax, Virginia 22031
USA

ISBN 978-1-5108-4188-8

9 781510 841888

CURRAN ASSOCIATES INC.
proceedings.com